Risk Management Technologies for Deep Excavations in Water-Rich Areas

Risk Management Technologies for Deep Excavations in Water-Rich Areas

Editors

Yixian Wang
Panpan Guo
Hang Lin
Yanlin Zhao

MDPI

Basel • Beijing • Wuhan • Barcelona • Belgrade • Novi Sad • Cluj • Manchester

Editors

Yixian Wang
School of Civil Engineering
Hefei University of Technology
Hefei
China

Panpan Guo
School of Civil Engineering
Hefei University of Technology
Hefei
China

Hang Lin
School of Resources and Safety Engineering
Central South University
Changsha
China

Yanlin Zhao
School of Resource & Environment and Safety Engineering
Hunan University of Science and Technology
Xiangtan
China

Editorial Office
MDPI
St. Alban-Anlage 66
4052 Basel, Switzerland

This is a reprint of articles from the Special Issue published online in the open access journal *Water* (ISSN 2073-4441) (available at: www.mdpi.com/journal/water/special_issues/V8W2XJ3ZP4).

For citation purposes, cite each article independently as indicated on the article page online and as indicated below:

Lastname, A.A.; Lastname, B.B. Article Title. *Journal Name* **Year**, *Volume Number*, Page Range.

ISBN 978-3-7258-0460-3 (Hbk)
ISBN 978-3-7258-0459-7 (PDF)
doi.org/10.3390/books978-3-7258-0459-7

Contents

About the Editors

Yixian Wang

Yixian Wang is a full professor of geotechnical engineering at the School of Civil Engineering, Hefei University of Technology. He is engaged in scientific research relating to geotechnical engineering, bridge and tunnel engineering, etc., focusing on scientific research in transportation infrastructure construction technology (e.g., subway construction, road, bridge and tunnel construction impact and disaster control), geotechnical engineering technology application, and solid waste resource utilization. He has hosted one National Defense Science and Technology Foundation Strengthening Program, four National Natural Science Foundation of China projects, more than 10 other programs or platform-related projects, and more than 40 joint scientific and technological research projects between universities and enterprises. He has published over 130 SCI papers (over 40 in the past 5 years), with a total of over 3000 citations, a h-index of 34, 9 highly cited ESI papers worldwide (top 1% globally), and 3 ESI hot topic papers (top 1 ‰ globally). He has received more than 10 national-level industry association and provincial association/association-level science and technology awards, including the second prize of the Hunan Province Natural Science Award, the third prize of Anhui Province Science and Technology Progress Award, the Science and Technology Award of China Construction Enterprise Management Association, and the Science and Technology Progress Award of China Rock Mechanics and Engineering Society (including 1 special provincial natural award). He has been accredited 34 national invention patents. He has compiled 4 textbooks and formulated 5 provincial-level standards. He is serving as an Editorial Board Member of Applied Rheology, as well as a young and middle-aged Editorial Board Member of the International Journal of Mining Science and Technology, Deep Underground Science and Engineering, and Journal of Southwest Jiaotong University.

Panpan Guo

Panpan Guo is a lecturer of geotechnical engineering at the School of Civil Engineering, Hefei University of Technology. He obtained a PhD in Geotechnical Engineering from Zhejiang University in 2022. His research interests include intelligent construction of urban underground engineering, research and development of new building materials, low-carbon technology, and multi-scale simulation of particulate matter. He has hosted six scientific research projects, including the National Natural Science Foundation of China and Anhui Provincial Natural Science Foundation, and participated in five scientific research projects. He has published over 40 SCI/EI papers with a total of over 1000 citations. He is serving as a member of the American Society of Civil Engineers, the Chinese Society of Civil Engineers, the Chinese Society of Rock Mechanics and Engineering, and the Chinese Highway Society. He is serving as a young Editorial Board Member of Deep Underground Science and Engineering.

Hang Lin

Hang Lin is a full professor at the School of Resources and Safety Engineering, Central South University. He engages in teaching and research in mining and geotechnical engineering numerical simulation, slope stability and reinforcement, tunnel and underground engineering. He has studied abroad at Colorado University of Mining and Technology in the United States. He has been selected as a Young Scholar in the "Lotus Scholar" Award Program in Hunan Province. He serves as the deputy leader of the technical support team for the first national comprehensive natural disaster risk survey in Hunan Province, executive director of the Hunan Society of Mechanics, director of the Hunan

Society of Rock Mechanics, and expert of the Hunan Provincial Safety Production Committee. He has published over 100 papers, among which 8 papers were selected as the top 1% highly cited papers in ESI, 3 papers were selected as the top 1 ‰ hot topic papers in ESI, and 4 papers were selected as the leader F5000 top academic papers in China's high-quality scientific and technological journals. He has been selected as the 2022 China Highly Cited Scholars List (Elsevier). As an Editor in Chief, he has edited 2 academic monographs and 2 professional textbooks. He has received the National Xu Zhilun Mechanics Excellent Teacher Award, the Outstanding Young Science and Technology Talent Award in the fields of mining, petroleum, and safety engineering, the First Prize for Scientific and Technological Progress of the Chinese Society of Rock Mechanics, the Second Prize for Natural Science of Hunan Province, the Second Prize for Natural Science of the Chinese Society of Rock Mechanics and Engineering, and the Excellent Master's Thesis Guidance Teacher of Hunan Province. He has hosted 4 projects funded by the National Natural Science Foundation of China, 2 projects funded by the Hunan Provincial Natural Science Foundation, and 1 project funded by the Ministry of Education for doctoral programs.

Yanlin Zhao

Yanlin Zhao holds a doctoral degree and is a professor and doctoral supervisor at Hunan University of Science and Technology. He is also the Vice Dean of the School of Resources, Environment, and Safety Engineering at Hunan University of Science and Technology. He is a young backbone teacher at ordinary universities in Hunan Province, an outstanding expert in Xiangtan City, and a recipient of excellent doctoral dissertations in Hunan Province. He is also a visiting Editor in Chief of international SCI journals Frontiers in Earth Science and Geofluids, a discipline leader (Xiangjiang Scholar) at Hunan University of Science and Technology, and an excellent master's thesis supervisor in Hunan Province. He was selected for the 2020 Elsevier High Citation List of Chinese Scholars and Top 2% Global Scientists in the field of Mining Engineering. He graduated from Taiyuan University of Technology with a Master's degree in Mining Engineering in 2006 (under the guidance of Academician Zhao Yangsheng). In 2009, he graduated from Central South University with a Ph.D. in Geotechnical Engineering. His doctoral thesis was awarded the Outstanding Doctoral Dissertation in Hunan Province in 2011. From March 2015 to March 2016, he worked as a visiting scholar at the University of Arizona in the United States. He is currently a reviewer for several authoritative international and domestic academic journals, and has been dedicated to the study of multi-field coupling theory and engineering response of mining rock masses for many years. He has hosted 2 general projects funded by the National Natural Science Foundation of China, as well as more than 10 provincial- and ministerial-level research projects and multiple horizontal projects funded by the Hunan Provincial Natural Science Foundation.

water

Editorial

Risk Management Technologies for Deep Excavations in Water-Rich Areas

Yixian Wang [1], Panpan Guo [1,*], Hang Lin [2] and Yanlin Zhao [3]

1 School of Civil Engineering, Hefei University of Technology, Hefei 230009, China; wangyixian2012@hfut.edu.cn
2 School of Resources and Safety Engineering, Central South University, Changsha 410083, China; hanglin@csu.edu.cn
3 Hunan Provincial Key Laboratory of Safe Mining Techniques of Coal Mines, Hunan University of Science and Technology, Xiangtan 411201, China; yanlin_8@163.com
* Correspondence: guopanpan@hfut.edu.cn

check for updates

Citation: Wang, Y.; Guo, P.; Lin, H.; Zhao, Y. Risk Management Technologies for Deep Excavations in Water-Rich Areas. *Water* **2024**, *16*, 323. https://doi.org/10.3390/w16020323

Received: 25 December 2023
Revised: 29 December 2023
Accepted: 12 January 2024
Published: 18 January 2024

1. Introduction

Deep excavations play an important role in the construction of urban infrastructures such as metro stations and high-rise buildings [1–3]. In recent decades, the depth and plan view size of most of the deep excavations in megalopolises are becoming unprecedented, due to growing usage requirements [4–7]. For example, the depth of the deepest excavation for a metro station in China, the Binjiang Station of Nanjing Metro Line 2, reached 51 m. Such a large excavation depth will inevitably bring challenges in controlling the safety and stability of the excavating work and its surroundings [8–10]. Moreover, the geological and hydrological conditions faced by deep excavations are increasingly complicated. Soft soils and high groundwater levels are not uncommon in deep excavation [11–13]. Therefore, the safety risk of deep excavations remains high, and various types of deep-excavation-related accidents have occurred in many parts of the world, causing enormous economic losses and casualties [14–18]. Figure 1 shows photos of typical cases of deep-excavation-related accidents [19].

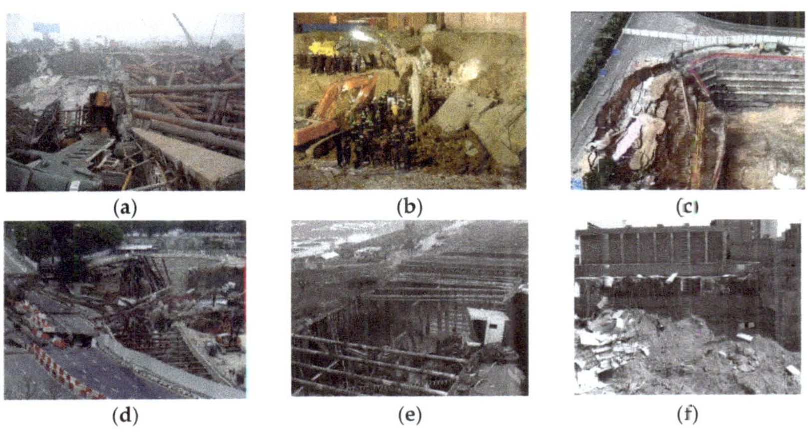

Figure 1. Photos of typical cases of deep-excavation-related accidents in many regions: (**a**) Hangzhou; (**b**) Harbin; (**c**) Nanning; (**d**) Singapore; (**e**) Wuhan; (**f**) Xining (after Guo [19]).

In water-rich areas, most deep excavations are affected by groundwater. During the construction of deep excavations in water-rich areas, for the sake of preventing seepage failure, effective dewatering measures have to be taken [20,21]. These measures provide a dry construction environment, which is beneficial for the stabilization of the excavation slope. However, dewatering can be detrimental in that it induces considerable and differential ground settlements, complicating the stress and seepage fields surrounding the site. On the one hand, dewatering can increase the effective stress in the soil skeleton surrounding the deep excavation, which induces ground settlement [22,23]. On the other hand, the dynamic water pressure generated by dewatering-induced waterhead difference will further exacerbate ground settlement [24,25]. Excessive and differential ground settlements can induce cracking, structure tilts, pipeline fractures, and other disasters in buildings. Additionally, the coupling of the seepage field and stress field during dewatering is rather complicated. Thus far, its mechanism and effect on the stability and deformation behavior of the deep excavation have not been well understood [26–28]. Thus, the safety risk of deep excavations in water-rich areas is extremely severe compared to that in arid areas.

The safety risk of deep excavations in water-rich areas has aroused extensive research interest among geotechnical engineers and researchers worldwide. Since the pioneering work of Einstein [29], who introduced the risk management concept to the field of underground engineering, significant advances have been made in risk management technologies for deep excavations. In detail, risk management technologies contain three critical aspects: risk identification, risk assessment, and risk control. Obviously, this is the primary step in identifying the various risk factors for ensuring the safety of deep excavations in water-rich areas. On this basis, risk assessment needs to be performed to ascertain the risk grade and the high-risk factors. This step involves the establishment of a risk assessment model capable of considering the causality between different risk factors. For example, Li [30] has proposed an improved risk assessment model, as depicted in Figure 2, based on the Multi-Attribute Border Approximation Area Comparison (MABAC) method. Finally, a risk control procedure is conducted to obtain the optimal countermeasure scheme for maintaining a good balance between restraining the high-risk factors and reducing the project's cost. However, due to the systematic construction of deep excavations, and the complexity of the dewatering-induced coupling of the seepage and stress fields, severe accidents related to deep excavations still happen in water-rich areas. Consequently, in-depth investigations into risk management technologies for deep excavations in water-rich areas are urgently required.

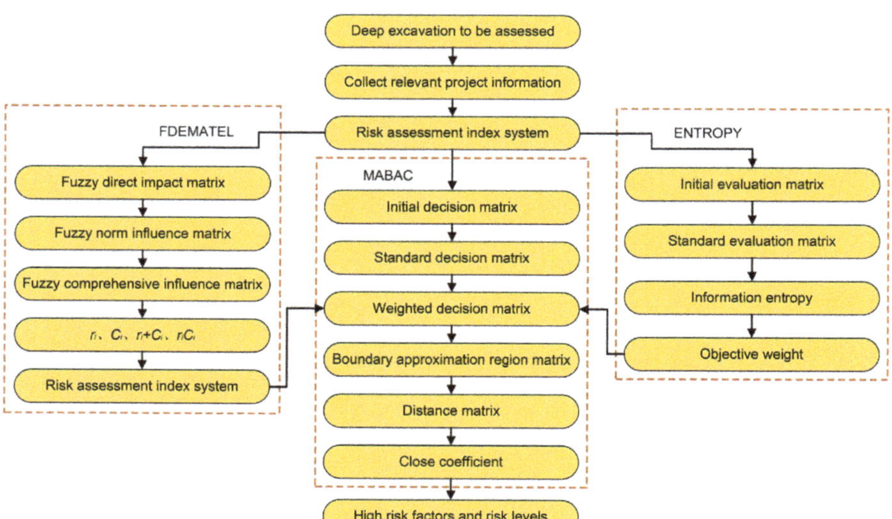

Figure 2. Flow chart of the improved risk assessment model proposed by Li [30].

This Special Issue has attracted many submissions and publications focusing on new and important contributions to the field of risk management technologies for deep excavation in water-rich areas. It contains eighteen articles and one correction. In particular, we are honored and grateful that four articles among them have been awarded the status of 'Feature Paper', a title that indicates advanced research with significant potential for high impact in the field. In the next section, we will briefly describe the contributions of the published articles. More importantly, the purpose of this Editorial is to encourage the readers to actively explore these articles.

2. An Overview of the Published Articles

Tang et al.'s article (Contribution 1) discusses the safety risk assessment method for pedestrian escape in subway stations subjected to flood disasters. Based on an actual subway station engineering project, the authors conducted a series of numerical simulations considering different conditions. A method for calculating the maximum pedestrian capacity, critical flood height, and minimum pedestrian escape speed is proposed. This research is of practical significance for achieving a safe escape from flooded subway stations.

The article by Wang et al. (contribution 2) is focused on analytically predicting the ground surface settlements caused by deep excavations in flood detention zones. The proposed method takes the effect of embankment surcharge load near deep excavations into account. As excessive ground surface settlements are detrimental to the safety of adjacent infrastructures, this method can be used for rapidly estimating the degree of risk in deep excavations in water-rich soft ground.

The third article, from Li et al. (contribution 3), aims to analyze the stability of a deep braced excavation spanning two shallowly buried large-diameter pressurized pipelines. By performing three-dimensional finite element analysis, the deformation and internal force behavior of the excavation support structures were captured. A suspension structure is also proposed for protecting the in situ pipelines and is demonstrated to be effective for reducing the risk of pipeline damage during the construction of deep excavations.

The fourth text published in this Special Issue is a case study by Sun et al. (contribution 4) on the performance of a soft soil deep excavation supported by a composite structural system. The system, comprising SMW piles, concrete struts, and rotary spray anchor cables, is a relatively new excavation support system. By analyzing the performance data from on-site monitoring, the behavior of this excavation support system can be understood, which provides a significant basis for risk control and design optimization.

The article by Yuan et al. (contribution 5) proposes a new machine-learning-based method for identifying the engineering risk of goaf. By performing principal component analysis, both the data-processing speed and prediction accuracy are improved. The risks of goaf are clarified with the help of the support vector machine method. The proposed method is demonstrated to be effective for goaf risk identification.

In the sixth article, Lu et al. (contribution 6) focus on the application of the frequency selection method to shallow groundwater exploration. They established a simplified geophysical model for a low-resistivity conductive sphere in homogeneous half space. The authors proved that the frequency selection method is effective for shallow groundwater exploration.

The research from Shakya et al. (contribution 7) delves into the environmental influences of boreholes installed for the demolition of existing pile foundations. Finite element analysis was performed to determine the ground settlements and distribution ranges for three different soil saturation and loading conditions. This research can provide a reference for risk control during pile removal.

Liu et al.'s study (contribution 8) focuses on water-resisting coal pillars, aiming to comprehend their failure behavior under stress–seepage coupling. A series of stress–seepage coupling tests were performed to investigate the evolution law of the stress–seepage coupling characteristics of coal rock. Based on the numerical analysis results, a method for determining reasonable coal pillar widths is proposed. This study is helpful for risk control in coal seam water inrush.

Liu et al.'s work (contribution 9) is devoted to understanding the effects of mining leachates on the hydraulic performance of geosynthetic clay liners under different temperatures. The research reveals that an increase in the temperature can restrain the ion exchange between bentonite and the mixture. It was also found that the swelling index of bentonite increases with increasing temperature. This research promotes the application of geosynthetic clay liners in mining.

The study by Wu et al. (contribution 10) focuses on a case study of a deep excavation in Lanzhou, China. This deep excavation faced complex environmental conditions comprising a composite stratum, dense buildings, and a high groundwater level. To reduce the safety risks during construction, an automated monitoring system was used. This research is beneficial for risk control practice in similar cases.

The eleventh text published in this Special Issue is the research by Lu et al. (contribution 11) that focuses on detecting the landfill leakage points. The effectiveness and resolution in detecting the landfill leakage points are compared between the opposing-coils transient electromagnetic method and the electrical resistivity tomography method. This article provides a new basis for the detection of leakage points in landfills.

The article by Liu et al. (contribution 12) is a correction to a figure in an article by the same authors (contribution 8). The text "Mongolia" in the initial figure has been revised to be "Inner Mongolia". This correction does not affect the academic content.

Shu et al.'s article (contribution 13) discusses the stability of river embankment seepage affected by underwater shield tunnelling. They performed finite element analysis considering various magnitudes and thicknesses of underwater tunnel overburden. The findings of this study can serve as a guide for reducing the risks of river embankment erosion and seepage instabilities during underwater shield tunnelling.

In the fourteenth article, Wang et al. (contribution 14) report a case history of a deep excavation for the construction of Hefei metro station. The risks associated with this deep excavation, in terms of self-stability, groundwater, and surrounding structures, are assessed. Field monitoring data were analyzed to capture the behavior of the excavation support system and the adjacent buildings. This research is helpful for implementing deep excavation design optimization in similar cases.

The research from Tu et al. (contribution 15) is devoted to investigating the performance of a deep excavation in a water-rich area in Fuzhou, China. This deep excavation is supported by cast-in-place piles combined with internal struts. The dewatering scheme and field monitoring results on excavation performance are elaborately described. This research concludes that tube-well dewatering is effective for deep excavations in water-rich areas.

The article by Wang et al. (contribution 16) is focused on the performance of a deep braced excavation in a thick silty clay stratum. The monitored behavior of the deep excavation, in terms of ground movements, structure deformation, and internal forces in struts, is revealed. Three-dimensional finite element analysis is also performed. The findings from this study provide a reliable scientific basis for risk control.

Chen et al.'s article (contribution 17) proposes a simplified method for estimating the reliability of a three-dimensional slope. This method is within the framework of limit equilibrium analysis. Through two typical slope examples, the effectiveness of the proposed method has been validated. This method can be conveniently used in real engineering practices.

The article by Lu et al. (contribution 18) delves into the applicability of the frequency selection method in the exploration of underground hot water in the Maoyanhe scenic spot. This research is based on a geological investigation in the field and on existing geological data. Using two verification boreholes, the effectiveness of the frequency selection method is demonstrated.

The final article published in this Special Issue is the study by Tu et al. (contribution 19), which focuses on a case history of deep braced excavation in soft soils in Fuzhou, China.

Based on a thorough analysis of the observed performance, the influence of deep excavation on the displacements and internal force characteristics of supporting structures, as well as surrounding buildings, roads, and buried pipelines, is captured. This contribution provides a reference for design optimization and risk control in deep excavations in soft soil areas.

3. Future Research Directions

The nineteen articles published in this Special Issue make a significant contribution to advancing the field of risk management technologies for deep excavation in water-rich areas. Meanwhile, several research gaps can be also detected in exploring the content of these articles, which will direct future research work in this field. For the benefit of readers interested in this field, the guest editors of this Special Issue are glad to present their opinions on several potential directions for further research in the future. The potential avenues that could be explored in the future, in order to achieve a high level of safety in deep excavations in water-rich areas, may include the following:

- An exploration of how seepage and stress fields are coupled during the entire construction period of deep excavation in water-rich areas, and this effects the stability and deformation behavior of the deep excavation;
- The establishment of risk evaluation index systems that are comprehensive, representative, and scientific;
- The development of risk assessment models and methods able to take into account the correlations between parameters and the dynamic evolution of risk indicators during the construction of deep excavations in water-rich areas;
- An exploration of how to shorten the period of risk assessment so as to reserve sufficient time for taking risk prevention measures;
- The optimization of risk control technologies, especially for the methods controlling deep excavation deformation, with the aim of promoting risk control technologies that are more efficient, intelligent, green, and low-carbon;
- The development of a real-time monitoring system and risk alertness forecasting system that can be conveniently used on site in deep excavations in water-rich areas;
- The application of artificial intelligence to specifically enhance the efficiency, precision, and effectiveness, and reduce the cost of risk management technologies, including risk identification, risk assessment, and risk control.

As a final note, we would like to express our sincere gratitude to all the authors of the forty articles submitted for this Special Issue, as well as all the Assistant Editors and Academic Editors of *Water* for their contributions in ensuring its success.

Author Contributions: Conceptualization, P.G. and Y.W.; investigation, H.L. and Y.Z.; writing—original draft preparation, P.G.; writing—review and editing, Y.W., H.L. and Y.Z.; funding acquisition, P.G. and Y.W. All authors have read and agreed to the published version of the manuscript.

Funding: This research was funded by the National Natural Science Foundation of China (52308344, 52374085); the Anhui Provincial Natural Science Foundation (2308085QE190); the Opening Project of State Key Laboratory of Explosion Science and Technology (Beijing Institute of Technology) (KFJJ23-04M, KFJJ23-05M); and the Fundamental Research Funds for the Central Universities (JZ2023HGTA0193, JZ2023HGQA0094).

Acknowledgments: As Guest Editors of "Risk Management Technologies for Deep Excavations in Water-Rich Areas", we would like to express our deep gratitude to all the authors whose valuable work was published under this Special Issue and who thus contributed to the success of the edition. In addition, we thank Haotian Yu for drawing the figures.

Conflicts of Interest: The authors declare no conflict of interest.

List of Contributions

1. Tang, Y.; Zhou, T.; Zhong, Y.; Hu, S.; Lin, J.; Lin, Z.; Liu, H.; Liu, B.; Zhao, Y.; Wang, Y.; et al. Risk Assessment for Critical Flood Height of Pedestrian Escape in Subway Station. *Water* **2022**, *14*, 3409. https://doi.org/10.3390/w14213409.
2. Wang, Y.; Chen, S.; Ouyang, J.; Li, J.; Zhao, Y.; Lin, H.; Guo, P. Predicting Ground Surface Settlements Induced by Deep Excavation under Embankment Surcharge Load in Flood Detention Zone. *Water* **2022**, *14*, 3868. https://doi.org/10.3390/w14233868.
3. Li, F.; Guo, P.; Geng, N.; Mao, L.; Lin, F.; Zhao, Y.; Lin, H.; Wang, Y. Stability of Braced Excavation Underneath Crossing Underground Large Pressurized Pipelines. *Water* **2022**, *14*, 3867. https://doi.org/10.3390/w14233867.
4. Sun, L.; Mao, K.; Wang, Z.; Ye, S.; Su, T.; Dai, G.; Xu, G.; Sun, J. Design and Field Monitoring of a Pile–Anchor–Brace Supporting System in a Soft Soil Area. *Water* **2022**, *14*, 3949. https://doi.org/10.3390/w14233949.
5. Yuan, H.; Cao, Z.; Xiong, L.; Li, H.; Wang, Y. A Machine Learning Method for Engineering Risk Identification of Goaf. *Water* **2022**, *14*, 4075. https://doi.org/10.3390/w14244075.
6. Lu, Y.; Yang, T.; Tizro, A.T.; Liu, Y. Fast Recognition on Shallow Groundwater and Anomaly Analysis Using Frequency Selection Sounding Method. *Water* **2023**, *15*, 96. https://doi.org/10.3390/w15010096.
7. Shakya, S.; Nakao, K.; Kuwahara, S.; Inazumi, S. Impact of the Boreholes on the Surrounding Ground. *Water* **2023**, *15*, 188. https://doi.org/10.3390/w15010188.
8. Liu, Q.; Xue, Y.; Ma, D.; Li, Q. Failure Characteristics of the Water-Resisting Coal Pillar under Stress-Seepage Coupling and Determination of Reasonable Coal Pillar Width. *Water* **2023**, *15*, 1002. https://doi.org/10.3390/w15051002.
9. Liu, Y.; Li, X.; Tu, Y.; Lu, Y. Mining Leachates Effect on the Hydraulic Performance of Geosynthetic Clay Liners under Different Temperatures. *Water* **2023**, *15*, 1132. https://doi.org/10.3390/w15061132.
10. Wu, J.; Ye, S.; Wang, Z.; Yang, D. Application and Automatic Monitoring and Analysis of Hybrid Support Structure in Ultra-DEEP Foundation Pit Engineering in the Lanzhou Area under Complex Environmental Conditions. *Water* **2023**, *15*, 1335. https://doi.org/10.3390/w15071335.
11. Lu, Y.; Tao, J.; Cao, C.; Liu, H.; Liu, Y.; Ge, Z. Detection of Landfill Leachate Leakage Based on ERT and OCTEM. *Water* **2023**, *15*, 1778. https://doi.org/10.3390/w15091778.
12. Liu, Q.; Xue, Y.; Ma, D.; Li, Q. Correction: Liu et al. Failure Characteristics of the Water-Resisting Coal Pillar under Stress-Seepage Coupling and Determination of Reasonable Coal Pillar Width. *Water* 2023, *15*, 1002. *Water* **2023**, *15*, 1804. https://doi.org/10.3390/w15101804.
13. Shu, W.; Ma, J.; Geng, N.; Xiang, Y.; Ma, S.; Li, X.; Tong, F.; Fang, S. Influences of Underwater Shield Tunnelling on River Embankment Seepage Stability Considering Various Overburden Thickness. *Water* **2023**, *15*, 2346. https://doi.org/10.3390/w15132346.
14. Wang, D.; Ye, S.; Zhang, J. Risk Reduction Measures and Monitoring Analysis of Deep Foundation Pit with Water in a Metro Station in Hefei. *Water* **2023**, *15*, 3007. https://doi.org/10.3390/w15163007.
15. Tu, B.; Zheng, J.; Shen, M.; Ni, W. Monitoring Analysis of a Deep Foundation Pit with Water Supported by Cast-in-Place Pile and Internal Bracing in a Soft Soil Area of Fuzhou. *Water* **2023**, *15*, 3008. https://doi.org/10.3390/w15163008.
16. Wang, D.; Ye, S.; Xin, L. Study on the Analysis of Pile Foundation Deformation and Control Methods during the Excavation of Deep and Thick Sludge Pits. *Water* **2023**, *15*, 3121. https://doi.org/10.3390/w15173121.
17. Chen, J.; Zhu, D.; Zhu, Y. A Simplified Method for Effective Calculation of 3D Slope Reliability. *Water* **2023**, *15*, 3139. https://doi.org/10.3390/w15173139.

18. Lu, Y.; Ding, H.; Yang, T.; Liu, Y. Geothermal Water Exploration of the Maoyanhe Hot Spring Scenic Spot in Zhangjiajie Using the Natural Electric Field Frequency Selection Method. *Water* **2023**, *15*, 3418. https://doi.org/10.3390/w15193418.
19. Tu, B.; Zheng, J.; Ye, S.; Shen, M. Study on Excavation Response of Deep Foundation Pit Supported by SMW Piles Combined with Internal Support in Soft Soil Area. *Water* **2023**, *15*, 3430. https://doi.org/10.3390/w15193430.

References

1. Wang, J.; Ma, Z.; Zeng, J.; Chen, Z.; Li, G. Numerical study on the influence of aquitard layer distribution and permeability parameters on foundation pit dewatering. *Water* **2023**, *15*, 3722. [CrossRef]
2. Zhang, Y.; Yi, L.; Zhang, L.; Yang, Y.; Hao, X.; Li, H.; Ma, H. Causation identification and control measures of deformation by integrated dewatering-excavation process simulation of a T-shaped deep foundation pit. *Water* **2022**, *14*, 535.
3. Shi, C.; Sun, X.; Liu, S.; Cao, C.; Liu, L.; Lei, M. Analysis of seepage characteristics of a foundation pit with horizontal waterproof curtain in highly permeable strata. *Water* **2021**, *13*, 1303.
4. Guo, P.; Gong, X.; Wang, Y.; Lin, H.; Zhao, Y. Analysis of observed performance of a deep excavation straddled by shallowly buried pressurized pipelines and underneath traversed by planned tunnels. *Tunn. Undergr. Space Technol.* **2023**, *132*, 104946. [CrossRef]
5. Guo, P.; Gong, X.; Wang, Y.; Lin, H.; Zhao, Y. Minimum cover depth estimation for underwater shield tunnels. *Tunn. Undergr. Space Technol.* **2021**, *115*, 104027. [CrossRef]
6. Guo, P.; Gong, X.; Wang, Y. Displacement and force analyses of braced structure of deep excavation considering unsymmetrical surcharge effect. *Comput. Geotech.* **2019**, *113*, 103102. [CrossRef]
7. Yang, Y.; Chen, C.; Liu, C.; Huang, L.; Chen, W.; Lin, N.; Cui, J.; Xie, W. Performance of a deep excavation and the influence on adjacent piles: A case history in karst region covered by clay and sand. *Undergr. Space* **2023**, *8*, 45–60. [CrossRef]
8. Xu, W.; Zhang, D.; Zhang, Q. Deformation behaviors and control indexes of metro-station deep excavations based on case histories. *Tunn. Undergr. Space Technol.* **2022**, *122*, 104400.
9. Zhang, W.; Han, L.; Gu, X.; Wang, L.; Chen, F.; Liu, H. Tunneling and deep excavations in spatially variable soil and rock masses: A short review. *Undergr. Space* **2022**, *7*, 380–407. [CrossRef]
10. Mu, L.; Zhang, P.; Shi, Z.; Zhu, M.; Gu, Z. Predicting longitudinal tunnel deformation due to deep excavation-induced ground movement. *Tunn. Undergr. Space Technol.* **2023**, *131*, 104793.
11. Liu, J.-C.; Tan, Y. Review of through-wall leaking incidents during excavation of the subway stations of Nantong metro line 1 in thick water-rich sandy strata. *Tunn. Undergr. Space Technol.* **2023**, *135*, 105056. [CrossRef]
12. Bozkurt, S.; Abed, A.; Karstunen, M. Finite element analysis for a deep excavation in soft clay supported by lime-cement columns. *Comput. Geotech.* **2023**, *162*, 105687. [CrossRef]
13. Tanoli, A.Y.; Yan, B.; Xiong, Y.L.; Ye, G.L.; Khalid, U.; Xu, Z.H. Numerical analysis on zone-divided deep excavation in soft clays using a new small strain elasto–plastic constitutive model. *Undergr. Space* **2022**, *7*, 19–36. [CrossRef]
14. Yi, F.; Su, J.; Zheng, G.; Cheng, X.S.; Pei, H.T.; Liu, X.M.; Jia, J.W. Progressive collapse analysis and robustness evaluation of a propped excavation failed due to inappropriate strut removal. *Acta Geotech.* **2023**, *18*, 6775–6801.
15. Yi, F.; Su, J.; Zheng, G.; Cheng, X.S.; Zhang, J.T.; Lei, Y.W. Overturning progressive collapse mechanism and control methods of excavations retained by cantilever piles. *Eng. Fail. Anal.* **2022**, *140*, 106591. [CrossRef]
16. Chen, R.P.; Li, Z.C.; Chen, Y.M.; Ou, C.Y.; Hu, Q.; Rao, M. Failure investigation at a collapsed deep excavation in very sensitive organic soft clay. *J. Perform. Constr. Facil.* **2015**, *29*, 04014078. [CrossRef]
17. Do, T.N.; Ou, C.Y.; Chen, R.P. A study of failure mechanisms of deep excavations in soft clay using the finite element method. *Comput. Geotech.* **2016**, *73*, 153–163.
18. Choosrithong, K.; Schweiger, H.F. Numerical investigation of sequential strut failure on performance of deep excavations in soft soil. *Int. J. Geomech.* **2020**, *20*, 04020063.
19. Guo, P. Characteristics and Control Method for Deformation of Shallowly Buried Pipeline Induced by Braced Excavation. Ph.D. Thesis, Zhejiang University, Hangzhou, China, 2 June 2022.
20. Pujades, E.; Vàzquez-Suñé, E.; Carrera, J.; Jurado, A. Dewatering of a deep excavation undertaken in a layered soil. *Eng. Geol.* **2014**, *178*, 15–27. [CrossRef]
21. Wu, Y.X.; Lyu, H.M.; Han, J.; Shen, S.L. Dewatering–induced building settlement around a deep excavation in soft deposit in Tianjin, China. *J. Geotech. Geoenviron. Eng.* **2019**, *145*, 05019003. [CrossRef]
22. Liu, N.W.; Peng, C.X.; Li, M.G.; Chen, J.J. Hydro-mechanical behavior of a deep excavation with dewatering and recharge in soft deposits. *Eng. Geol.* **2022**, *307*, 106780. [CrossRef]
23. Zhang, Y.Q.; Wang, J.H.; Li, M.G. Effect of dewatering in a confined aquifer on ground settlement in deep excavations. *Int. J. Geomech.* **2018**, *18*, 04018120. [CrossRef]
24. Tan, Y.; Lu, Y.; Wang, D. Deep excavation of the Gate of the Orient in Suzhou stiff clay: Composite earth-retaining systems and dewatering plans. *J. Geotech. Geoenviron. Eng.* **2018**, *144*, 05017009.
25. Wang, J.; Feng, B.; Yu, H.; Guo, T.; Yang, G.; Tang, J. Numerical study of dewatering in a large deep foundation pit. *Environ. Earth Sci.* **2013**, *69*, 863–872.

26. Peng, C.X.; Liu, N.W.; Li, M.G.; Zhen, L.; Chen, J.J. Hydro-mechanical coupled analyses on wall deformations caused by deep excavations and dewatering in a confined aquifer. *Acta Geotech.* **2022**, *17*, 2465–2479.
27. Liu, F.; Xing, H.; Wu, B.; Liu, G.; Ma, S.; Huang, H.; Meng, G. Study on the effect of dewatering-recharge-excavation of deep foundation pits on soil-structure-groundwater under a deep confined water environment. *Arab. J. Geosci.* **2022**, *15*, 710. [CrossRef]
28. Zheng, G.; Zeng, C.F.; Diao, Y.; Xue, X.L. Test and numerical research on wall deflections induced by pre-excavation dewatering. *Comput. Geotech.* **2014**, *62*, 244–256.
29. Einstein, H.H. Risk and risk analysis in rock engineering. *Tunn. Undergr. Technol.* **1996**, *11*, 141–155.
30. Li, X. Study on Safety Risk Assessment and Control of Foundation Pit Construction in Subway Stations with Water-Rich Stratum. Master's Thesis, Guangxi University, Nanning, China, 1 June 2022.

water

MDPI

Article

Study on Excavation Response of Deep Foundation Pit Supported by SMW Piles Combined with Internal Support in Soft Soil Area

Bingxiong Tu [1,*], Jinhuo Zheng [2,*], Shuaihua Ye [3,4] and Minglong Shen [2]

1 College of Civil Engineering, Huaqiao University, Xiamen 361021, China
2 Fujian Provincial Institute of Architectural Design and Research Co., Ltd., Fuzhou 350108, China; m15392008809@163.com
3 Key Laboratory of Disaster Mitigation in Civil Engineering of Gansu Province, Lanzhou University of Technology, Lanzhou 730050, China; yeshuaihua@163.com
4 Western Center of Disaster Mitigation in Civil Engineering, Ministry of Education, Lanzhou University of Technology, Lanzhou 730050, China
* Correspondence: tubingxiong@hqu.edu.cn (B.T.); ak9206@126.com (J.Z.)

Abstract: Based on a foundation pit project in Fuzhou, China, the influence of foundation pit excavation on the supporting structure and surrounding environment in a soft-soil area is studied. This study was based on actual monitoring data and investigated the variations in the supporting structure, surrounding constructions, and groundwater levels during excavation. The analysis of the monitoring data demonstrates the presence of pronounced 'spatial effects' and 'temporal effects' on the deformation of the support structure and surrounding structures. The deformation of the support structure and surrounding structures exhibits distinct spatial distribution characteristics at different locations along the excavation pit wall. Typically, more significant deformations are observed in the middle section of the pit wall, while deformations decrease as the distance from the pit corner decreases. The support structure's and surrounding structures' deformation characteristics vary during different construction stages. During the excavation phase, the rate of deformation increase in the support structure and surrounding structures is notably higher. In contrast, during the construction of the underground basement floor and the backfilling phase of the excavation pit, the rate of deformation increase in the support structure and surrounding structures is relatively lower. Throughout the entire construction period of the excavation pit, the groundwater level in the vicinity of the pit exhibits a fluctuating trend. Apart from the influence of rainfall, the overall variation in groundwater level is minimal, indicating the effective water-sealing performance of the combined Soil-Mixing Wall (SMW) support structure within the circular enclosure.

Keywords: deformation; settlement; excavation; field monitoring; groundwater

check for updates

Citation: Tu, B.; Zheng, J.; Ye, S.; Shen, M. Study on Excavation Response of Deep Foundation Pit Supported by SMW Piles Combined with Internal Support in Soft Soil Area. *Water* **2023**, *15*, 3430. https://doi.org/10.3390/w15193430

Academic Editor: Glen R. Walker

Received: 31 August 2023
Revised: 25 September 2023
Accepted: 28 September 2023
Published: 29 September 2023

1. Introduction

The continuous development of high-rise buildings and underground space construction has led to many deep foundation pit projects. These pits exhibit notable features, such as a wide excavation area, great excavation depth, and a complex surrounding environment. Due to these characteristics, such projects present unique and significant challenges for foundation pit construction and design. These challenges make foundation pit support structures and surrounding structures susceptible to significant deformation or settlement. To ensure the safety of foundation pits and surrounding structures during excavation in soft-soil areas and reduce unnecessary economic losses, it is essential to conduct research on the excavation response and analyze the stress and deformation of the pit and surrounding structures during excavation.

Many scholars have carried out research on the stress and deformation of excavation engineering. Ou [1] conducted a study on the characteristics of ground settlement during the excavation process of foundation pits by analyzing monitoring data from multiple engineering examples of excavation projects. Wong et al. [2] derived empirical formulas for the deformation of enclosure structures and ground settlement by analyzing monitoring data from multiple excavation engineering projects in Singapore. Through analysis of the support issues in deep foundation pit engineering in soft-soil areas in China, Huang et al. [3] identified vital problems in foundation pit support, including analysis methods, construction techniques, deformation monitoring, and the construction environment. Ye et al. [4] monitored the horizontal and vertical displacement of pile tops, surface settlement, axial force of steel supports, and groundwater level during the construction process of a subway station excavation project. They systematically analyzed the stress and deformation characteristics of the support structure during the excavation process and comprehensively summarized the stress and deformation characteristics of the internal support structure. Fook et al. [5] studied the excavation response of multi-level support structures through numerical simulation and field measurement data and analyzed the spatial effect of the internal support structure of the excavation support. In studying internal support structures for concrete rings, Cheng et al. [6] investigated ultra-deep foundation pits supported by circular internal supports using an engineering case study. They conducted a systematic analysis of the stress conditions of the circular internal support structure and identified critical areas of attention for the engineering application of circular internal supports from both construction and design perspectives. Li et al. [7] analyzed an asymmetric circular internal support structure and proposed design concepts and analytical methods for asymmetric circular lining-supported excavations based on engineering monitoring data. Liu et al. [8] studied the excavation response of a super-large irregular foundation pit in soft-soil areas. By analyzing monitoring data, they discovered that the construction period and pit size have varying impacts on the internal support structure of the concrete and the surrounding environment. In general, more extended construction periods and larger pit sizes result in a more significant deformation of the pit and surrounding structures. Several researchers [9–17] have investigated the stress and deformation of support structures during the excavation of foundation pits. However, these studies mainly focused on small- to medium-sized foundation pits and were conducted for foundation pits with better soil conditions. There is a dearth of research on the excavation response of foundation pits in soft-soil areas. Due to their remarkable ability to provide effective waterproofing and structural support, SMW support structures have found extensive application in areas with high water levels. Chen and Cui [18] monitored a deep excavation supported by a combination of SMW and anchor cables as the support structure.

Based on field monitoring data, Feng et al. [19] used the numerical simulation method to study the deformation law of the retaining structure and surrounding soil during the excavation of a soft-soil foundation pit. Chen et al. [20] observed the deformation characteristics of deep foundation pits and the interaction between adjacent foundation pits via a monitoring system. Ki et al. [21] measured the parameters of a soil hardening model (HSM) of soft clay through an experimental method, which provided a method for the study of soil constitutive models in soft-soil areas. Based on the stress and deformation characteristics of deep foundation pit excavations in soft-soil areas, Panchal et al. [22] studied the construction technology and equipment for deep foundation pit excavation in soft-soil areas. An analysis of the monitoring data on the anchor cables' internal forces, the pile tops' horizontal displacement, and the groundwater levels revealed a close correlation between the maximum horizontal displacement of the support structure and the excavation depth and time. Furthermore, fluctuations in groundwater levels significantly influenced the operational behavior of the SMW and anchor cable combined support structure. Ding et al. [23] employed a novel monitoring technology to achieve real-time monitoring of the stress and deformation of SMW support structures under complex conditions. Real-time data analysis enabled the identification of stress and deformation characteristics of SMWs. Xian et al. [24] conducted a study on the application of SMW support structures in sandy areas and investigated the mechanical behavior of SMW support structures during excavation and backfilling processes based on actual engineering cases, providing a reference for the design and construction of SMW support structures in sandy areas. Chen et al. [25] conducted model experiments to measure the horizontal displacement of pile tops in SMW support structures during excavation. Through the analysis of experimental observations, they revealed the deformation and failure characteristics of excavation pits supported by SMW in soft-soil areas. Research has also been carried out on the influence of foundation pits on adjacent structures. Finno et al. [26] and Son et al. [27] put forward the damage standard and judgment method of buildings affected by foundation pit excavation through theoretical analysis. Asker et al. [28] studied the influence of foundation pit excavation on the surrounding environment using numerical simulation. Sharma et al. [29] used two-dimensional finite element simulation software, taking a project in Singapore as the background, combined with measured data to analyze the deformation and bending moment of a side tunnel affected by a foundation pit. Huang et al. [30] also carried out a similar study through a centrifuge model test, and considered that the deformation shape of the tunnel uplift conformed to the Gaussian distribution curve, and preliminarily analyzed the affected area of the tunnel.

At present, many scholars have studied internal ring bracings and SMW-pile-retaining structures, but most of them analyze the two retaining structures separately, and most of them study foundation pits with good soil conditions. The geological conditions of soft-soil areas are usually complex, the soil layer may be uneven, the water content is high, and it is easy to cause soil settlement and flow problems. SMW-pile-retaining structures with inner ring bracing are complex composite structures, and the interaction between structures has an important influence on the deformation of foundation pits and the surrounding environment. Previous studies have shown that changes in groundwater level have a significant impact on the support performance of SMW, but most monitoring data analysis studies have not analyzed the changes in groundwater level. In order to ensure the safety of foundation pit excavation and improve the economy of foundation pit construction, this paper takes the deep excavation project using the combination of inner-ring support and SMW support in soft-soil area as the background; analyzes the monitoring data of the support structure, surrounding buildings, and groundwater level during excavation; studies the stress and deformation characteristics of inner ring support combined with an SMW support structure during excavation; and provides guidance and reference for the design and construction of an internal ring support combined with an SMW support structure in soft-soil areas.

2. Project Overview

2.1. Site Overview

This paper describes a proposed construction project in Taijiang District, Fuzhou, China, along Jiangbin Middle Road. The chosen location borders the Zhonggeng Real Estate development to the north, Jiangbin Avenue to the south, the completed Huarun Binjiang Garden residential complex to the west, and a Guoyou building to the east. The project is expected to have a total built area of 85,898 square meters, with 59,898 square meters designated for above-ground structures and 26,000 square meters allocated for underground facilities where the ±0.000 benchmark corresponds to an elevation of 9.000 m above sea level. This project consists of two underground stations, with the second basement's bottom slab at a height of −9.000 m. The excavation depth for the bottom layer of the foundation pit is approximately 8.700 m below the benchmark, while the footing layer's bottom depth is about 10.100 m below the standard.

2.2. Hydrogeological Conditions

The maximum excavation depth of the foundation pit project was approximately 10 m. The site's stratigraphy comprises 10 distinct layers within the exploration depth range. The primary soil layers within the pit's influence range are layers 1 to 6, as shown in Table 1. The heaviness γ, thickness, moisture content, cohesion c, and internal friction angle φ of each soil layer are shown in Table 1. Groundwater is mainly pore-type and submersed in 3 medium sand, the depth of the stabilized water level is 3.50–6.30 m, and the elevation of the stabilized water level is 0.43–3.64 m. The permeability coefficient of the project is 10.63 m/d. The depth of excavation of the foundation pit in this project is about 10 m, and the excavation of the foundation pit was carried out with artificial descending measures, including pipe wells with descending pressure and water descending in an open ditch with water-collecting wells.

Table 1. Basic information of each soil layer.

Soil Layer	Layer Name	Thickness/m	Heaviness γ/kN/m³	Moisture Content/%	Cohesion c/kPa	Internal Friction Angle φ/°	Compression Modulus E/Mpa
1	Miscellaneous fill (Q_4^{ml})	1.40~7.40	17.5	/	9	10.0	3
2	Muddy soil (Q_4^{m})	0.60~2.20	17.2	29.9	12.3	7.07	3.02
3	Medium sand (Q_4^{al})	7.30~19.40	19.0	43.7	5	26	17
3A	Silty soil (sandy) (Q_4^{m})	0.80~5.00	16.4	32.6	14.66	9.91	3.61
3B	Silt (Q_4^{al})	2.10~3.50	19.0	28.1	5	23	11
4	Silty soil (sandy) (Q_4^{m})	1.20~13.00	16.5	33.3	13.71	10.61	3.69
5	Silty clay (Q_4^{al})	1.20~4.80	19.2	32.9	16.1	17.06	15
6	Medium sand (Q_4^{al})	11.40~20.20	19.3	28.9	5	30	8.47

3. Support Structure Scheme and Construction Conditions

3.1. Support Structure Form

As shown in Figure 1, the foundation pit excavation is supported by SMW combined with an internal concrete support. Within a specific range of the central part of the east side of the foundation pit is sectional view diagram 2-2, while the rest of the locations are in sectional view diagram 1-1. (1) The retaining piles are all SMWs, and the mixing piles are three-axis cement-mixing piles with a length of 21 m (locally 24 m). (2) The horizontal internal support structures are all cast-in-place reinforced-concrete structures, with one support structure installed at a depth of 1.5 m. (3) The columns are composed of steel lattice columns for the upper portion, and the bottom portions are supported by driven (drilled) cast-in-place piles. The support cross-sections are illustrated in Figure 2a,b. The quality of cast-in-place piles can be significantly affected by underground water, as it has a diluting effect on the drilling mud, making it difficult to remove debris and maintain the stability of the borehole walls. This can lead to collapsed sand and gravel layers along the pile wall. Therefore, it was necessary to carry out effective dewatering and waterproofing measures during the excavation of the foundation pit. It was also essential to monitor changes in the groundwater level to ensure the stability of the pile foundation. The excavation of the foundation pit was carried out in sections using mechanical excavation methods. The excavation work began in early January 2020 and continued until the basement slab was poured in early June 2020. The basement construction then proceeded gradually up to the elevation of ±0.00 in late June 2020. The space between the outer wall and the foundation pit was gradually backfilled with soil during this time. In early September 2020, the steel beams began to be removed, and this process continued until late November 2020, marking the completion of the foundation pit construction project.

Figure 1. Birds-eye view of the layout of the support structure.

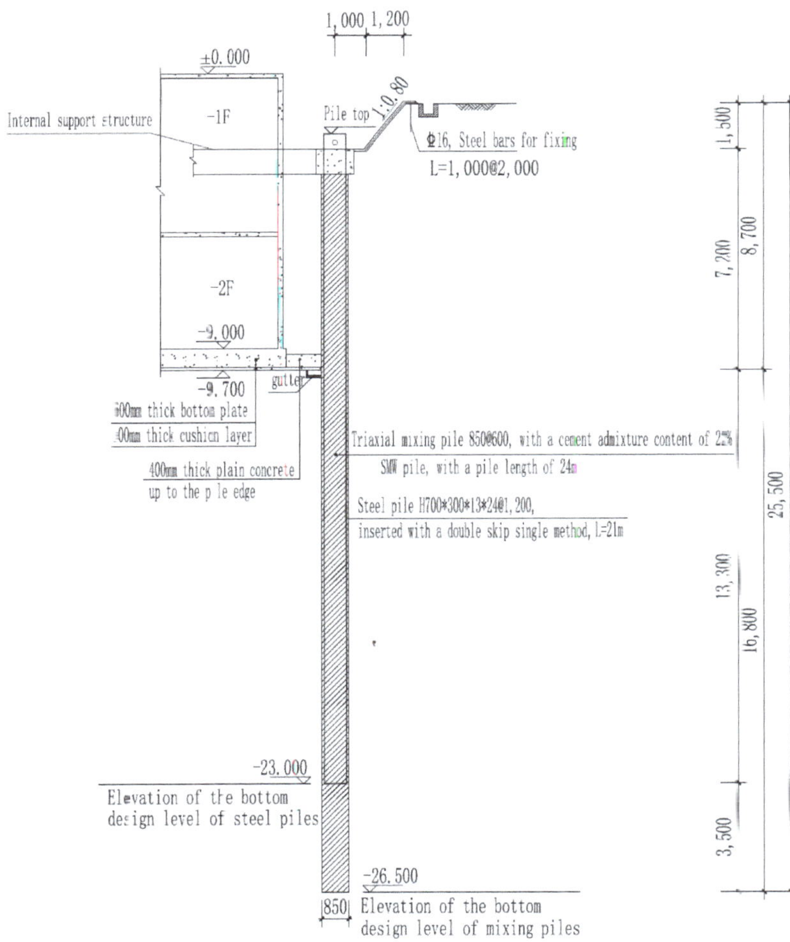

(a)

Figure 2. *Cont.*

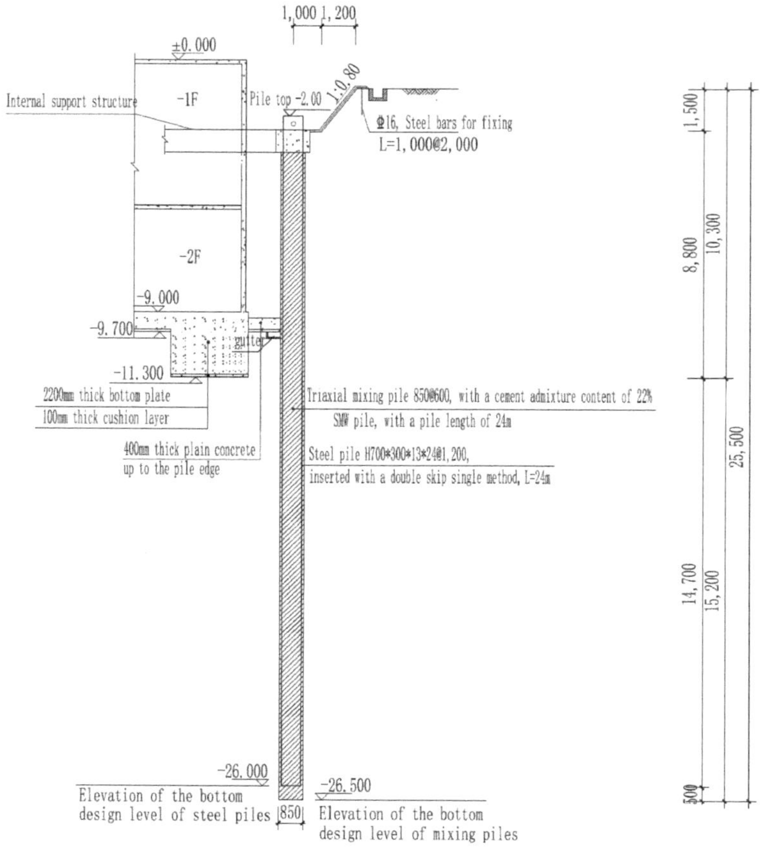

(b)

Figure 2. Sectional view diagrams of the supporting structure: (**a**) Sectional view diagram 1-1. (**b**) Sectional view diagram 2-2.

3.2. SMW

The SMW method is the abbreviation of Soil-Mixing Wall, which was firstly developed successfully by Narayuki Industry Co., Fuzhou, China. The SMW method uses a special multi-shaft mixer to drill into the cutting soil in situ, and at the same time inject cement slurry into the soil at the end of the drill bit, and then insert H-beam steel or other profiles into the mixing piles after sufficient mixing to form a continuous underground wall that can be used as a direct soil-retaining and water-stopping structure. Its main features are a simple structure, good water-stopping performance, short construction period, low cost, and low environmental pollution, and it is especially suitable for foundation pit projects in cities. Based on the actual conditions of this project, the mixing piles used in the foundation pit were Φ850@600 three-axis cement mixing piles, and H70030013*24 steel beams were inserted into the mixing piles with 1200 mm/600 mm spacing.

3.3. Reinforced-Concrete Inner Support

In order to effectively control the deformation of the excavation and ensure the stability and safety of the foundation pit while considering the engineering cost and various factors, a reinforced-concrete internal support was installed at a distance of 1.5 m from the ground level of the site in this project. The internal support system consisted of a combination of ring braces, angle braces, and diagonal braces, and columns were set beneath each intersection of the braces. The design parameters of the supporting structure are shown in Table 2.

Table 2. Design parameters.

Type	Size ($H'B$)	Concrete Strength	Reinforcement
ring braces	900 mm′2000 mm	C30	32C25
angle braces	800 mm′900 mm	C30	22C25
diagonal braces	700 mm′800 mm	C30	20C25
crown beam	900 mm′1300 mm	C30	14C25 + 4C20 + 8C22

4. Excavation Monitoring and Data Analysis

4.1. Monitoring Scheme and Layout of Monitoring Points

Due to the extensive excavation depth and complex surrounding environment of this foundation pit project, it was necessary to monitor the supporting structure and the surrounding environment during the excavation period to ensure the safety of the excavation support structure as well as the surrounding roads, underground pipelines, and other protected structures. Figure 3a shows the plan layout of monitoring points for the excavation project, including 12 monitoring points for deep horizontal displacement (C1 to C12), 26 monitoring points for enclosure settlement and horizontal displacement (P1 to P26), 36 monitoring points for column settlement (L1 to L36), and 4 monitoring points for groundwater level (S1 to S4). Figure 3b shows the plan layout of environmental monitoring points around the excavation project, including 13 monitoring points for adjacent building inclination (QX1 to QX13), 26 monitoring points for adjacent building displacement (J1 to J26), and 14 monitoring points for road and surrounding pipeline settlement and horizontal displacement (D1 to D14). Z1 to Z6 are internal-support-stress-monitoring points.

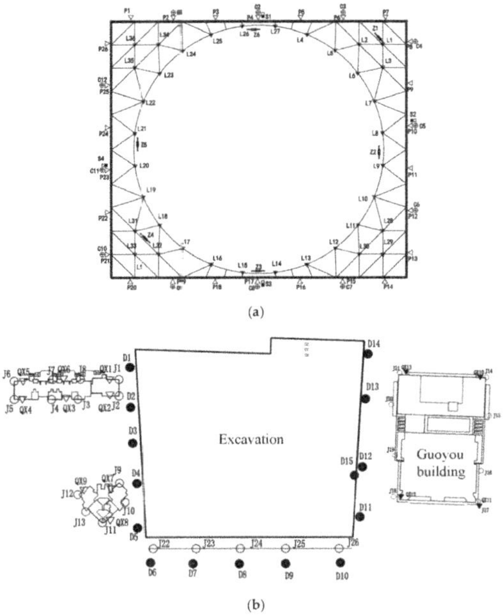

Figure 3. Layout plan of monitoring points: (**a**) Layout plan for excavation monitoring points. (**b**) Layout plan for monitoring points of surrounding structures.

The monitoring method is as follows. An inclinometer was used for the horizontal displacement monitoring of deep soil mass, 12 inclined points were arranged for monitoring, and a borehole-embedding method was adopted. The embedding depth of the inclinometer tube was 20~21 m, and the embedding was completed before the foundation pit excavation. For monitoring the slope top horizontal displacement, measuring points were set along the crown beam of the foundation pit, observed with the small-angle method, and monitored 108 times in total. The horizontal displacement of the slope top of the supporting structure was monitored by a total station. The observations of column settlement were tested using a precision level, and measuring points were embedded at the position of each support beam column according to design requirements. According to the actual situation around the site, two buildings on the west side and one building on the east side of the foundation pit were monitored. According to the scale of the buildings, 5~8 settlement observation points and 3~6 inclination observation points were arranged for each building. The settlement observation of surrounding roads and underground pipelines was tested using a precision level. For underground water level monitoring, four underground water level observation points were arranged around the foundation pit, and the water level observation was conducted using a steel ruler water level gauge.

4.2. Data Analysis

In order to study the response characteristics of foundation pit excavations in soft-soil areas, this section analyzes the deep horizontal displacement of the support structure, pile top displacement, settlement of surrounding structures, column settlement, and groundwater level changes. Existing studies have shown that there are apparent 'time-space effects' during the excavation of foundation pits, and there are some differences in the deformation characteristics at different periods and monitoring points, so the monitoring data of different points and periods were selected for comparative analysis. Foundation pits are often considered a plane-strain problem in calculation and design. However, the deformation of the retaining structure is different in the same direction of pit wall, and has obvious spatial distribution characteristics. It has been shown that the negative corner of a foundation pit can restrain deformation, and the influence of the corner effect will decrease with the increase in distance from the negative corner. According to this principle, some scholars adopt the concept of PSR (maximum displacement of retaining wall at a certain position and plane strain state of retaining wall) to quantitatively describe the spatial effect of foundation pits. Therefore, this paper describes the degree of spatial effects of the foundation pit by comparing the displacement of the supporting structure at different positions.

(1) Deep Horizontal Displacement

The excavation site had 12 monitoring points for measuring deep horizontal displacement. Taking monitoring points C4 to C6 as an example, the deep horizontal displacement variation curves over different periods are shown in Figure 4. Figure 4 shows that the deep horizontal displacement curves measured at various time points exhibit a bow-shaped distribution, with the displacement values initially increasing and then decreasing as the depth below the excavation site surface increases. With increasing excavation depth, the location of the maximum deep horizontal displacement at each monitoring point gradually shifted downward. Following completion of the excavation, the maximum deep horizontal displacement tended to remain stable and was typically located at a depth of around 6 m. In addition, prior to 16 March 2020, the deep horizontal displacement values at each monitoring point were noticeably smaller due to the relatively shallow excavation depth. Between 26 March and 20 April, the deep horizontal displacement increased rapidly as the excavation progressed. After 20 April, as the excavation was essentially completed, the deep horizontal displacement at each monitoring point gradually approached a state of stability. The deep horizontal displacement of the piles is closely related to the construction conditions of the excavation pit. As the excavation depth gradually increased, the deep horizontal displacement exhibited a progressive change. Therefore, it was necessary to pay special attention to the changes in deep horizontal displacement during the excavation process and strengthen the monitoring frequency during the later stages of excavation. By comparing the deep horizontal displacement monitoring values of monitoring points C4, C5, and C6, it can be observed that the deep horizontal displacement monitoring value of C5 is significantly larger than that of C4 and C6. The maximum deep horizontal displacement value at the C6 monitoring point reached 34.52 mm, while the maximum deep horizontal displacement value at the C4 and C6 monitoring points was only about 17 mm. This is because, generally, the corner of the excavation pit limits the deformation of the excavation pit, and the closer the corner, the smaller the deformation. The C6 monitoring point was located in the middle of the pit wall, far from the corner of the excavation pit, and was less affected by the corner effect. In contrast, the C4 and C6 monitoring points were located at the corner of the excavation pit and were more affected by the corner effect. Hence, the deformation at the C6 monitoring point was more significant than that at the C4 and C6 monitoring points.

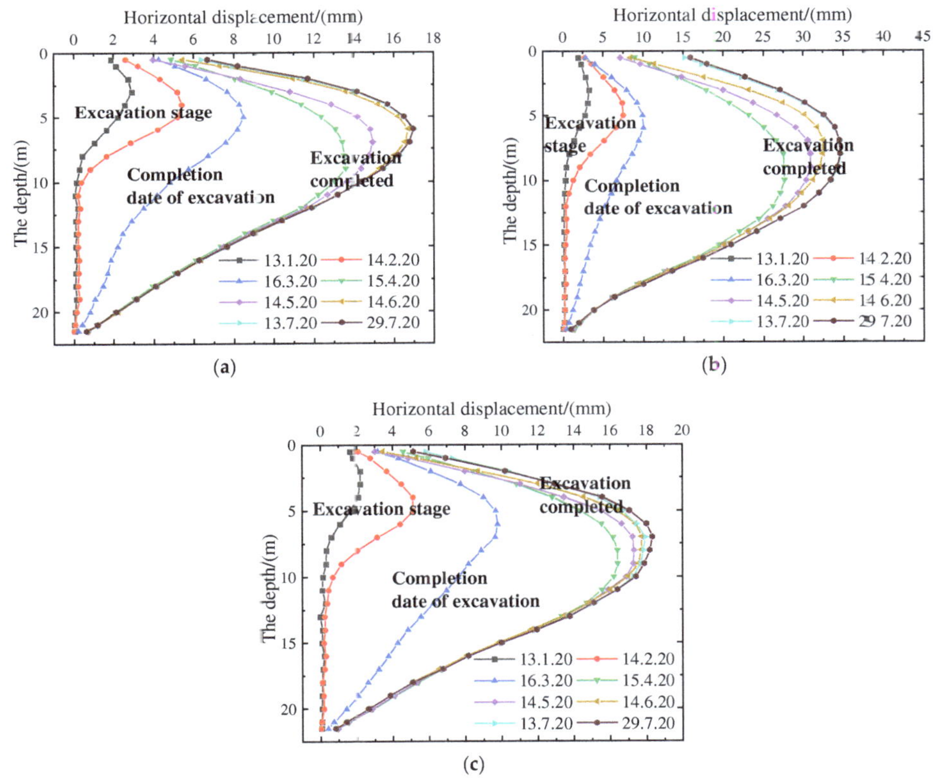

Figure 4. The deep horizontal displacement curve: (**a**) C4. (**b**) C5. (**c**) C6.

Based on the analysis of the horizontal displacement of the top of the pile, it is evident that the bottomless foundation pit supported by the SMW pile combination exhibited a significant 'spatial effect'. Therefore, in actual engineering, the support structure design in the middle of the foundation pit can be strengthened, and the strength at the corners of the foundation pit can be reduced, thereby reducing the cost of foundation pit construction.

(2) Analysis of pile top horizontal displacement and settlement

There were a total of 26 monitoring points for pile top displacement (P1~P26), with displacement in the direction inside the excavation pit considered positive. Figure 5 shows the time-dependent curve of pile top horizontal displacement during the excavation process. From Figure 5, it is evident that the horizontal displacement of the pile tops exhibited an apparent 'time effect' characteristic during the excavation process. That is, the horizontal displacement of pile tops showed variations over time at different excavation stages of the foundation pit. Around 20 days before the excavation of the foundation pit, there was a significant sharp increase in the horizontal displacement of the pile top. However, during the period from the 20th to the 50th day of excavation, the increase in the horizontal displacement of the pile top was not significant. The reason for this phenomenon is that the stress in the site was released during the initial excavation of the foundation pit, resulting in a sharp increase in the horizontal displacement of the pile top. However, with time, the internal stress redistribution of the soil was gradually completed. In addition, during the period of about 20 to 50 days, the concrete support was cast inside the pit, and the excavation of the foundation pit was not continued, which resulted in the relatively smooth variation in pile top horizontal displacement during this period. After 50 days, as the foundation pit excavation resumed, the pile top's horizontal displacement again showed a rapid increasing trend. After the completion of the excavation of the foundation pit, with the construction of the foundation slab, the increase in the horizontal displacement of the pile top gradually tended to stabilize. The corresponding construction stages at different times are shown in Figure 5a.

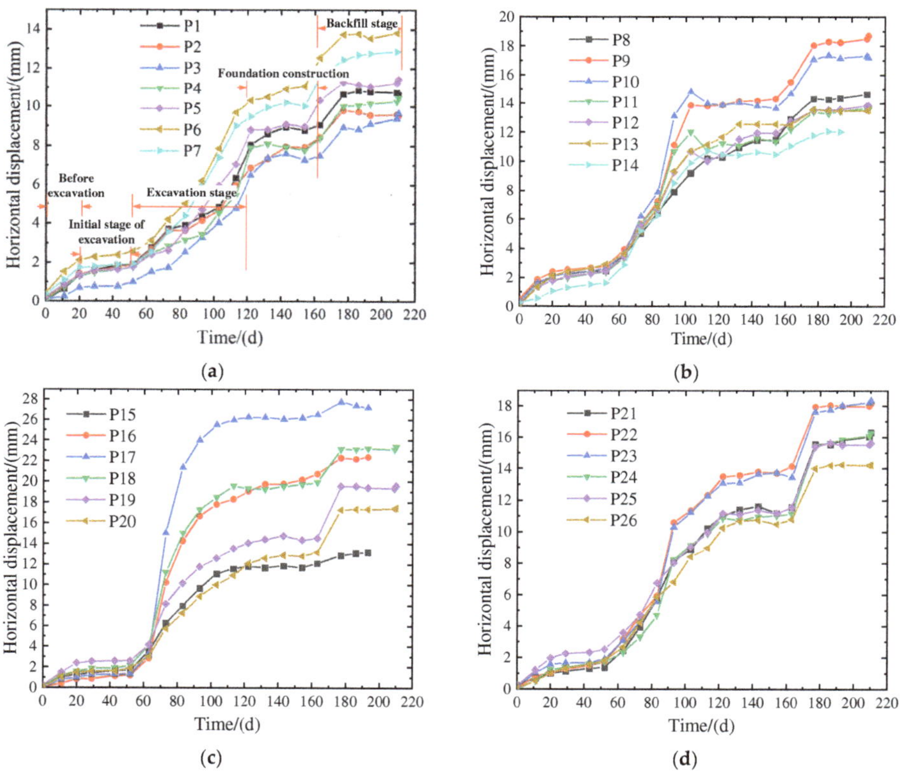

Figure 5. The curve of horizontal displacement variation at pile top: (**a**) P1 to P7. (**b**) P8 to P14. (**c**) P15 to P20. (**d**) P21 to P26.

Furthermore, the 'spatial effects' were also evident in this excavation. Apart from the northeast corner of the excavation, the horizontal displacement at the mid-position of the excavation wall was significantly greater than that at the corner. This is due to the deformation restriction effect at the excavation's corner. Usually, the closer the position is to the corner of the excavation, the smaller the deformation. However, due to the presence of existing structures and the loading on the top of the excavation at the northeast corner, the horizontal displacement at P6–P9 in this excavation was relatively large.

The settlement variation curve of the pile top is presented in Figure 6, which also shows prominent distributions characteristic of the 'time effect' and 'spatial effect' Initially, the pile top settlement rapidly increased due to the excavation, but the curve gradually became smoother due to the construction of concrete support. However, with the continued excavation of the foundation pit, the pile top settlement rapidly increased again. After the completion of the excavation, the settlement variation curve gradually became smoother, and the rate of increase decreased during the construction process of the foundation slab. The spatial distribution pattern of the pile top settlement is consistent with that of the horizontal displacement. Settlement decreases as the location becomes closer to the excavation corner, while it increases as it becomes closer to the middle position of the excavation wall. It also can be seen from Figure 6a that the settlement of point P1 was significantly greater than that of P2, P3, and P4. The reason for this phenomenon is that there were already buildings on the west side of point P1, and these buildings acted on the top of the foundation pit in the form of load, thus causing the settlement of point P1 to increase significantly. This is why the settlement of point P1 was still greater than that of P2, P3, and P4 although the corner was displaced.

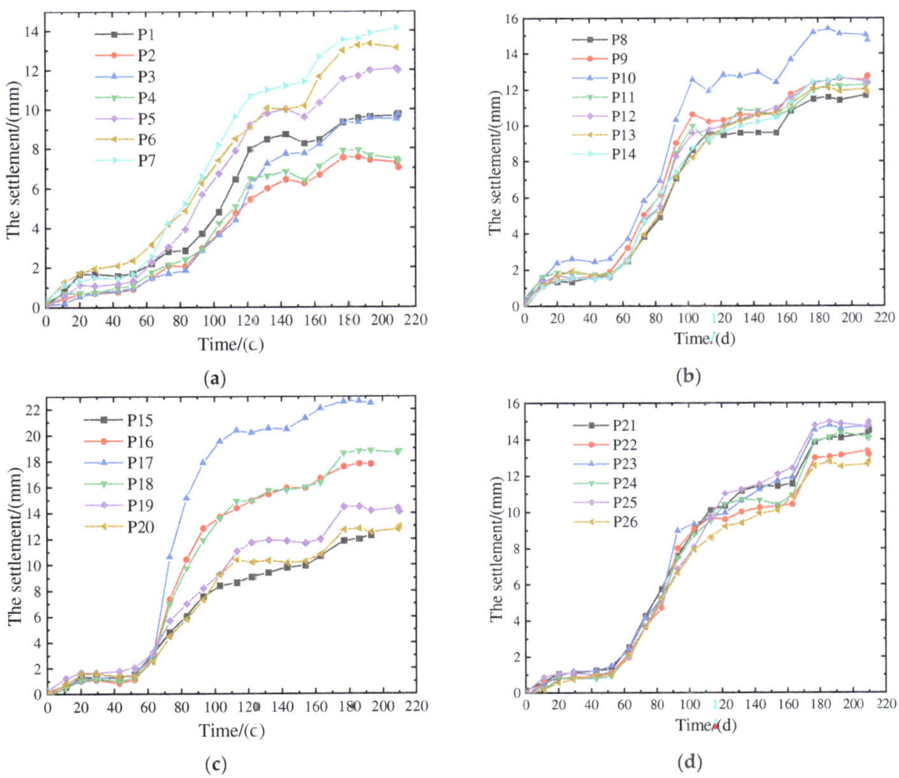

Figure 6. The curve of settlement variation at pile top: (**a**) P1 to P7. (**b**) P8 to P14. (**c**) P15 to P20. (**d**) P21 to P26.

(3) Settlement Analysis of Adjacent Buildings

North of the foundation pit excavation is an area of real estate under construction, south is Jiangbin Avenue, west is the built Hairun Riverside Garden Community, and east is a state-owned building. Both the state-owned assets building and the Garden Community are high-rise buildings, which adopt pile foundations. Figure 7 shows the settlement curves of adjacent buildings over time, recorded 50 days before the foundation pit excavation. It can be observed from Figure 7 that there were significant differences in settlement characteristics among the various monitoring points. However, the accumulated settlement values at each point ranged from 0.34 mm to 6.18 mm, indicating that the overall settlement deformation of the buildings was small. For each settlement monitoring point, no significant fluctuations in settlement values were observed before the foundation pit excavation. When excavation began, the settlement values at each monitoring point showed significant changes. Subsequently, due to the construction of the internal support system, the settlement values became stable again. However, as the foundation pit was re-excavated, the settlement values at each monitoring point showed significant fluctuations and an overall increasing trend with time. After the completion of the excavation, the settlement values gradually stabilized again during the construction of the foundation pit base. It is evident that the construction procedures and sequencing of the excavation not only impact the deformation of the excavation itself but also have implications for the stress distribution and deformation of the surrounding structures. Therefore, it is crucial to pay attention to the impact of excavation construction procedures on the deformation of the surrounding structures during the excavation process.

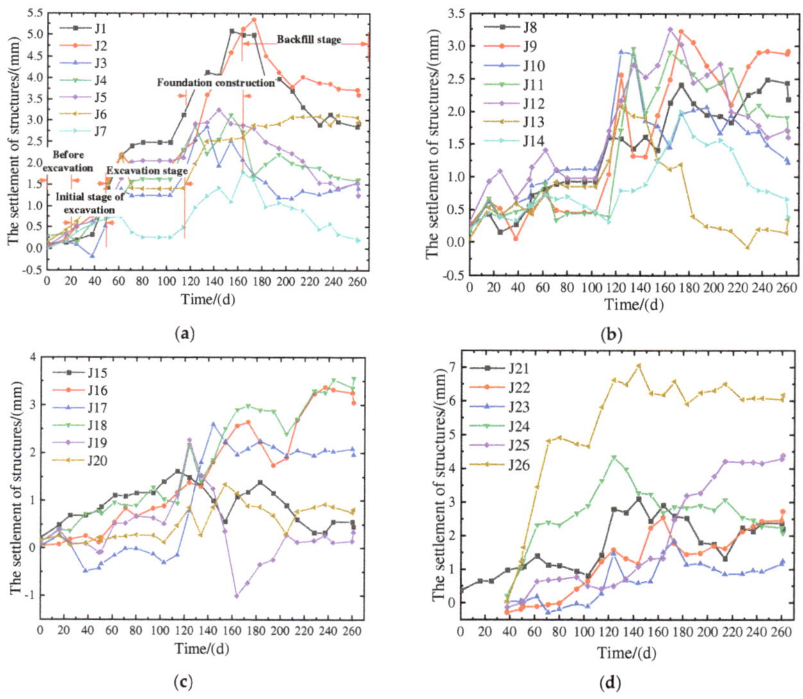

Figure 7. The curve of settlement variation in adjacent structures: (**a**) J1 to J7. (**b**) J8 to J14. (**c**) J15 to J20. (**d**) J21 to J26.

Furthermore, Figure 7 reveals significant variations in settlement values among monitoring points at different positions, which can be attributed to their varying distances from the excavation site. Taking Figure 7a as an example, it is evident that the settlement values of J1 and J2 are significantly greater than those of J3 to J7. This discrepancy can be attributed to the closer proximity of J1 and J2 to the excavation site, resulting in a more pronounced impact from the excavation process. Furthermore, the different underground structures of different structures are also one of the reasons for the differences in settlement values at different monitoring points.

(4) Settlement Analysis of Columns

The excavation consists of a total of 36 columns (L1–L36). Figure 8 depicts the variation in column settlements over time. Figure 8 shows that during the initial stage of excavation, the columns did not exhibit significant settlement, and the settlement rate remained relatively stable. However, as the excavation progressed further, noticeable settlement of the columns started to occur, and with increasing excavation depth, the settlement values of the columns exhibited a rapid increasing trend. Starting from day 100, the excavation phase of the basement was largely completed, and the subsequent focus shifted toward the construction of the basement floor. As a result, the settlement values of individual columns began to stabilize gradually after the 100th day, and with the progression of subsequent excavation activities, the variations in column settlement remained relatively small. Clearly, the settlement values of the columns exhibit significant variations during the excavation phase of the foundation pit while demonstrating relatively minor changes during the subsequent stages. Therefore, it is essential to place particular emphasis on monitoring the structural response and deformation of the internal bracing system during the excavation phase of a foundation pit. Moreover, enhancing the monitoring frequency during this stage is necessary.

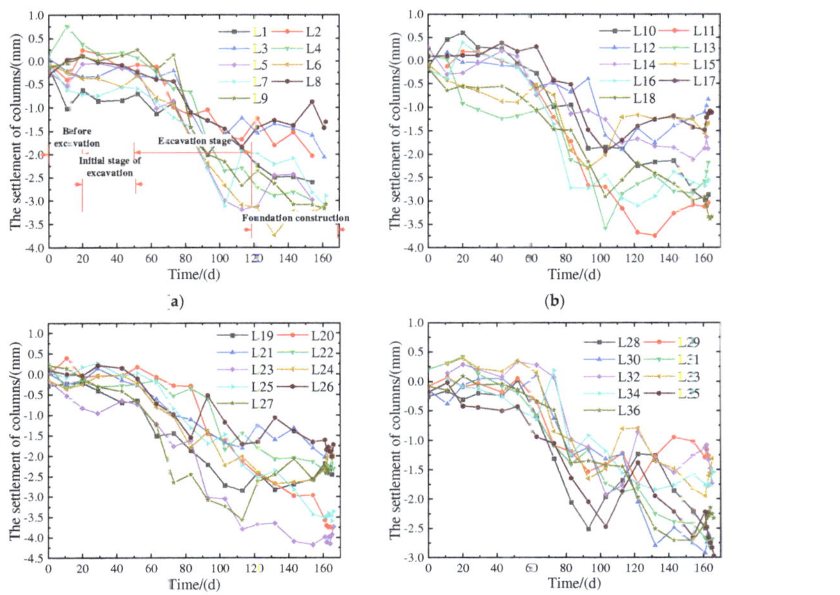

Figure 8. Column settlement change curves: (**a**) L1 to L9. (**b**) L10 to L18. (**c**) L19 to L27. (**d**) L28 to L36.

(5) Analysis of Displacement Monitoring for Surrounding Roads and Underground Pipelines

Fourteen displacement monitoring points (D1–D14) were installed for surrounding roads and underground pipelines. Monitoring data were collected starting 50 days prior to the excavation of the foundation pit. Figures 8 and 9 present the time-dependent curves of water level displacement and settlement values for each monitoring point. Based on Figures 9 and 10, it can be observed that prior to the excavation, no significant deformations were detected in the surrounding roads and underground pipelines. Both the horizontal displacement and settlement values were nearly zero. However, after 50 days, when the foundation pit excavation commenced, noticeable variations in horizontal displacement and settlement values were observed at each monitoring point. With the gradual increase in the depth of excavation of the foundation pit, noticeable increases were observed in the horizontal displacement and settlement values of the surrounding roads and underground pipelines. The increase in deformation was also positively correlated with the foundation pit's excavation depth. Upon completion of the excavation, as the construction of the basement floor and backfilling of the foundation pit took place, the horizontal displacement and settlement values at each monitoring point gradually stabilized. It can be observed that the deformations of the surrounding roads and underground pipelines were primarily concentrated during the excavation phase of the foundation pit. After the excavation, the deformations of the surrounding roads and underground pipelines gradually reached a state of stability, with minimal influence from the construction of the basement floor and backfilling of the foundation pit on the increase in deformation. Furthermore, the accumulated horizontal displacement values at each monitoring point range from 3.70 to 13.86 mm, while the accumulated settlement values range from 3.91 to 24.52 mm, all of which meet the design requirements.

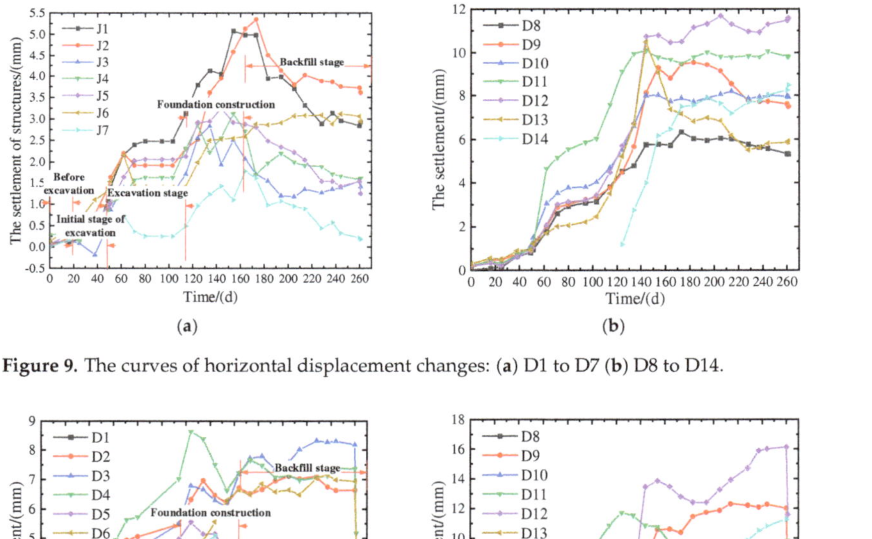

(**a**) (**b**)

Figure 9. The curves of horizontal displacement changes: (**a**) D1 to D7 (**b**) D8 to D14.

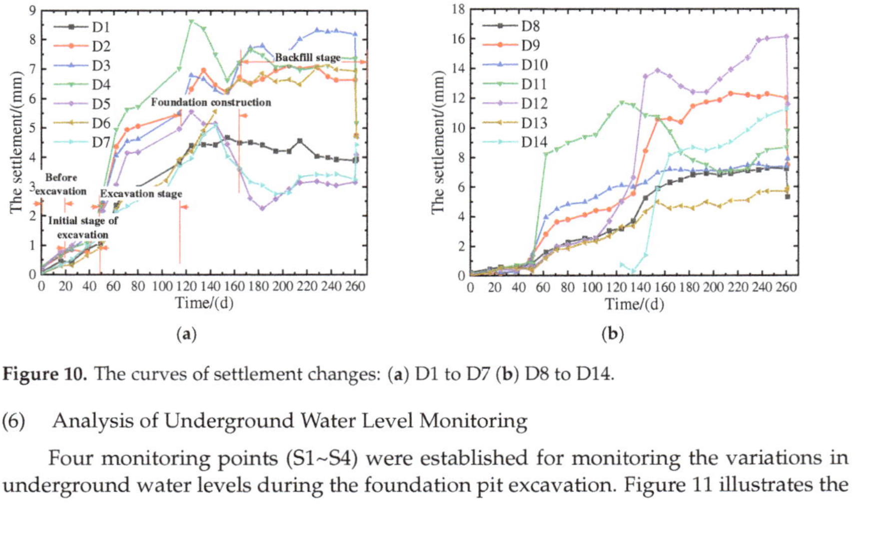

(**a**) (**b**)

Figure 10. The curves of settlement changes: (**a**) D1 to D7 (**b**) D8 to D14.

(6) Analysis of Underground Water Level Monitoring

Four monitoring points (S1~S4) were established for monitoring the variations in underground water levels during the foundation pit excavation. Figure 11 illustrates the changes in groundwater level with respect to the excavation. Based on Figure 11, it is evident that during the excavation phase of the foundation pit, the groundwater levels at each monitoring point exhibit a fluctuating trend. The maximum cumulative variation in groundwater levels ranges from 0.812 to 1.856 m. Importantly, during the excavation of the foundation pit and the subsequent construction of the underground structure, monitoring points S1 and S3 recorded cumulative variations that exceeded the predefined threshold of 1000 mm, indicating an alert level. This can be primarily attributed to significant rainfall events occurring during the initial pit excavation and the subsequent underground construction process, resulting in substantial fluctuations in groundwater levels at these specific monitoring points. According to the monitoring results, the initial buried depth of the groundwater level before foundation pit excavation was between 3.670~4.870 m and the maximum accumulated variation in groundwater level around the foundation pit excavation and basement construction was 0.812~1.856 m. Table 3 shows the observation data for the groundwater level around the foundation pit.

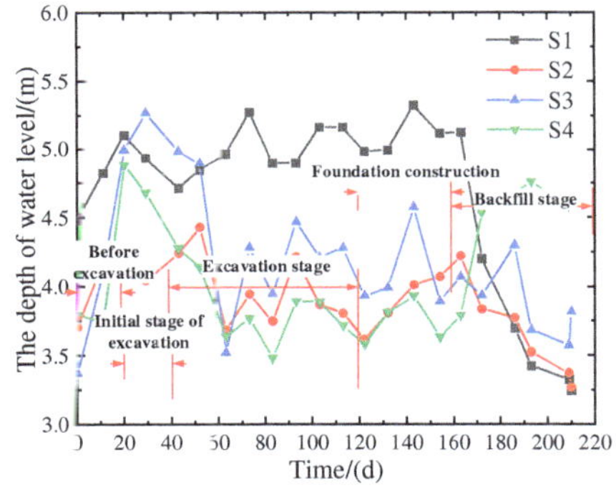

Figure 11. The curve of groundwater level changes.

Table 3. Groundwater level observation data around foundation pit.

Monitoring Point Number	S1	S2	S3	S4
Buried depth of initial water level (m)	4.54	3.67	4.87	3.95
Maximum water level buried depth (m)	5.624	4.482	5.272	4.882
Buried depth of minimum water level (m)	3.244	3.325	3.014	3.074
Maximum cumulative variation (m)	1.296	0.812	1.856	0.932
Buried depth of final water level (m)	3.244	3.268	3.816	4.681

(7) Stress analysis of inner support structure

As shown in Figure 12, the stress of the inner support at the Z1 and Z4 monitoring points was relatively small at the early stage before excavation, and Z1 was a positive value. This is because the internal bracing structure was mainly affected by the weight of the structure at the initial stage of excavation and shows the characteristics of tension. With the start of foundation pit excavation, the structure was gradually subjected to the effect of the foundation pit slope and became compressed; at this time, the stress became negative. With continuous excavation, the pressure on the internal support structure became larger and larger; at this time, the reinforcement stress continued to increase. When the excavation was completed, the stress of the inner support gradually decreased during the backfilling of the foundation pit.

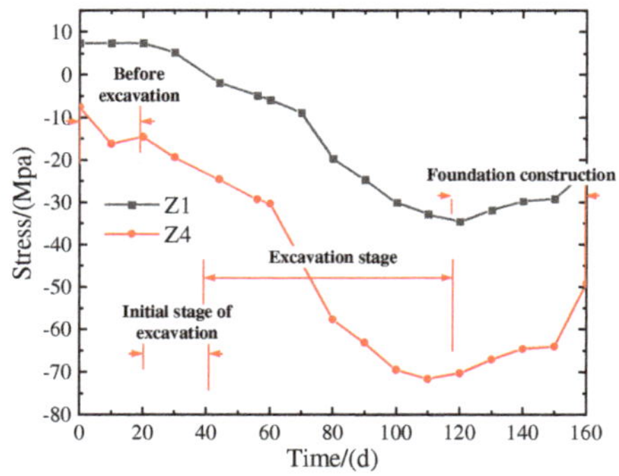

Figure 12. Stress variation in internal support structure.

(8) Summary of data

During the whole excavation process, the maximum horizontal displacement of the deep soil was 14.46~41.74 mm. The cumulative horizontal displacement of slope top was 9.53~27.22 mm; the cumulative settlement of the slope top was 7.03~22.48 mm; and the accumulated settlement of the columns ranged from −3.75 mm to −0.84 mm. During excavation, the maximum cumulative displacement of the horizontal displacement monitoring point C11 of deep soil was 41.74 mm, exceeding the allowable design value (40 mm), and the cumulative change and change rate of other monitoring points did not exceed the design warning value. After the foundation pit bottom was sealed, the deformation rate of the retaining structure tended to stabilize. The settlement and inclination rates of adjacent buildings were normal, all within the allowable design value range, the accumulated settlement of the adjacent buildings was 0.34~6.18 mm, and the accumulated inclination change rate of the adjacent buildings was −0.03 ‰~0.03 ‰, all within the allowable design value range. The cumulative horizontal displacement of surrounding roads and underground pipelines was 3.70~24.64 mm, and the cumulative settlement of surrounding roads and underground pipelines was 3.91~50.04 mm. The maximum cumulative variation in groundwater level was 0.812~1.856 m. During the foundation pit excavation and basement construction, the maximum cumulative variation in monitoring points S1 and S3 exceeded the warning value (1000 mm). Combined with the deformation results of the surrounding roads, it shows that the change in water level has great influence on the surrounding environment. Reinforcing the bar stress of the ring beam and supporting beam: According to the observation results, the accumulative change from reinforcing the bar stress of the ring beam and supporting beam was −163.84~150.11 MPa (a negative value "−" indicates compression). All were within the allowable design range.

It is worth noting that different foundation pit projects have their unique characteristics, and this paper only analyzes a single foundation pit. In order to understand the soft-soil area more comprehensively, the response of ring support combined with SMW pile support to deep foundation pit excavation needs more analysis. In addition, the monitoring method in this paper is also more traditional, and is affected by climate, construction methods, and other factors. In the future, intelligent monitoring methods can be used to analyze responses to the excavation of foundation pits.

5. Conclusions

This study was conducted in the context of an actual engineering project to investigate the excavation response of a bottomless foundation pit supported by a combination of SMW piles in soft-soil areas. The research primarily focuses on analyzing the displacement of the combined SMW piles, the settlement of columns, the deformation of surrounding structures, and the changes in groundwater levels. It also summarizes the specific deformation of the supporting structure in a construction method combining SMW piles in a soft-soil area and of the surrounding environment by analyzing the deformation during the foundation pit excavation process at different locations, and gives some suggestions to guide the design and construction of foundation-pit-supporting structures in soft-soil areas. Based on the analysis of these parameters, the following significant conclusions are obtained:

(1) The deformation of the combined circular internal bracing and SMW support structure in excavation projects exhibits distinct spatial distribution characteristics. Generally, the deformation lessens as the distance from the excavation corner increases, while the middle sections of the excavation walls experience more significant deformations. The excavation can also increase the deformation of the support structure. Therefore, in practical engineering, it is advisable to enhance the monitoring points or monitoring frequency in the middle section of the excavation walls and near existing structures near the top of the excavation to ensure the safety of the excavation.

(2) During the construction phase, the deformations of the internal support combined with the SMW support structure exhibit a significant time effect, meaning that the deformation characteristics of the support structure vary at different excavation stages. Typically, during the initial excavation and construction phases of the internal support structure, the deformation growth rate of the support structure is relatively low. During the excavation construction phase, as the foundation pit's excavation depth increases, the support structure's deformation growth rate significantly increases. However, during the construction of the foundation base and backfilling of the excavation pit, the deformation growth rate of the support structure decreases significantly. Therefore, to ensure the safety of excavation construction, it is essential to closely monitor the deformation of the internal support system composed of SMW piles, particularly during the excavation phase. This will help prevent sudden excessive deformations of the support structure.

(3) The deformations of the surrounding structures also demonstrate a notable 'time effect'. Based on the analysis of monitoring data from the surrounding structures, it is evident that the settlements of adjacent buildings and the deformations of surrounding roads and underground pipelines are primarily concentrated during the excavation phase of the foundation pit. Subsequently, upon completion of the excavation, the deformations of the surrounding roads and underground pipelines gradually stabilize, with minimal impact from the construction of the basement floor and backfilling of the foundation pit. Furthermore, the closer the structures are to the foundation pit, the greater their deformations. Therefore, to ensure the safety of the surrounding structures during the excavation of the foundation pit, it is crucial to pay close attention to the deformations of the structures.

(4) Throughout the entire period of construction of the foundation pit, the groundwater level exhibited a fluctuating deformation trend. Apart from the influence of rainfall, the variations in groundwater level were insignificant, indicating the effective water-sealing performance of the support system formed by the combination of internal supports and SMW piles.

Author Contributions: Conceptualization, B.T. and J.Z.; methodology, S.Y.; validation, M.S. and B.T.; formal analysis, B.T.; investigation, J.Z.; resources, S.Y.; data curation, M.S.; writing—original draft preparation, S.Y.; writing—review and editing, S.Y; supervision, B.T.; project administration, J.Z.; funding acquisition, S.Y. All authors have read and agreed to the published version of the manuscript.

Funding: This work was supported by the Collaborative Innovation Platform Project of Fuzhou-Xiamen-Quanzhou National Self-Innovation Zone (Grant No. 3502ZCQXT2022002), the Science and Technology plan of Fujian Province (Grant No. 2022I0014), the Construction Technology Project of Xiamen Municipal Construction Bureau (Grant No. XJK2022-1-12), and the Fundamental Research Funds for the Central Universities (Grant No. ZQN-1012).

Data Availability Statement: Due to legal issues and trade secrets, the monitoring data in this article cannot be provided free of charge. Nevertheless, all concepts and data are explained in the submitted studies, and some studies are available upon request.

Conflicts of Interest: The authors declare no conflict of interest.

References

1. Ou, C.Y.; Hsieh, P.G.; Chiou, D.C. Characteristics of ground surface settlement during excavation. *Can. Geotech. J.* **1993**, *30*, 758–767. [CrossRef]
2. Wong, I.H.; Poh, T.Y.; Chuah, H.L. Performance of excavations for depressed express Engineering in Singapore. *J. Geotech. Geoenviron. Eng.* **1997**, *123*, 617–625. [CrossRef]
3. Huang, M.S.; Wang, W.D.; Zheng, G. A review of recent advances in the underground engineering and deep excavations in soft soils. *China Civ. Eng. J.* **2012**, *45*, 146–161.
4. Ye, S.H.; Huang, A.P.; Gao, S. Deformation characteristics and environmental impact analysis of deep foundation pit excavation in a subway station in Lanzhou. *Water Resour. Hydropower Eng.* **2020**, *51*, 12–22.
5. Lee, F.H.; Yong, K.Y.; Quan, K.; Chee, K.T. Effect of corners in strutted excavations: Field monitoring and case histories. *J. Geotech. Geoenviron. Eng.* **1998**, *124*, 339–349. [CrossRef]
6. Cheng, J.L.; Li, Y.Z.; Liu, X.Z.; Wen, G.X. Application of many concentric ring supports with large diameter in super deep foundation pits. *Chin. J. Geotech. Eng.* **2014**, *36* (Suppl. 1), 122–126.
7. Li, Y.; Gu, K.Y.; Lin, J. Design method for bracing of asymmetric cylindrical foundation pits. *Chin. J. Geotech. Eng.* **2013**, *35* (Suppl. 2), 888–891.
8. Liu, G.; Guo, J.; Li, M.; Qin, T.; Huang, P. Measured behaviors of an oversized irregular basement excavation and its surrounding responses in thick soft clay. *Arab. J. Geosci.* **2019**, *13*, 3. [CrossRef]
9. Wang, J.H.; Xu, Z.H.; Wang, W.D. Wall and ground movements due to deep excavations in Shanghai soft soils. *J. Geotech. Geoenviron. Eng.* **2020**, *136*, 985–994. [CrossRef]
10. Finno, R.J.; Arboleda-Monsalve, L.G.; Sarabia, F. Observed performance of the one museum park west excavation. *J. Geotech. Geoenviron. Eng.* **2015**, *141*, 04014078. [CrossRef]
11. Ye, S.; Zhao, Z.; Wang, D. Deformation analysis and safety assessment of existing metro tunnels affected by excavation of a foundation pit. *Undergr. Space* **2021**, *6*, 421–431. [CrossRef]
12. Ye, S.H.; Zhou, J. Study on stress and deformation of shield tunnel plate under unloading of foundation pit excavation. *Arab. J. Geosci.* **2021**, *14*, 1–14. [CrossRef]
13. Wu, J.; Ye, S.H.; Wang, Z.Q.; Yang, D. Application and automatic monitoring and analysis of hybrid support structure in ultra-deep foundation pit engineering in the Lanzhou area under complex environmental conditions. *Water* **2023**, *15*, 1335. [CrossRef]
14. Ye, S.H.; Li, D.P. Monitoring and simulation analysis of deep and large foundation pit excavation in complex environment. *China Civ. Eng. J.* **2019**, *52* (Suppl. 2), 117–126.
15. Tan, Y.; Lu, Y.; Wang, D. Deep excavation of the gate of the orient in Suzhou stiff clay: Composite earth-retaining systems and dewatering plans. *J. Geotech. Geoenviron. Eng.* **2018**, *144*, 05017009. [CrossRef]
16. Ou, C.Y.; Shiau, B.Y.; Wang, I.W. Three-dimensional deformation behavior of the Taipei National Enterprise Center (TNEC) excavation case history. *Can. Geotech. J.* **2000**, *37*, 438–448. [CrossRef]
17. Ou, C.Y.; Hsieh, P.G.; Lin, Y.L. Performance of excavations with cross walls. *J. Geotech. Geoenviron. Eng.* **2010**, *137*, 94–104. [CrossRef]
18. Chen, W.; Cui, W. Study on monitoring of deep foundation pit with SMW engineering method plus anchor cable retaining structure. *AIP Conf. Proc.* **2018**, *1944*, 020016. [CrossRef]
19. Feng, Z.Y.; Xu, Q.; Xu, X.Y.; Tang, Q.; Li, X.D.; Liao, X. Deformation Characteristics of Soil Layers and Diaphragm Walls during Deep Foundation Pit Excavation: Simulation Verification and Parameter Analysis. *Symmetry* **2022**, *14*, 254. [CrossRef]
20. Chen, S.R.; Cui, J.F.; Liang, F.Y. Case Study on the Deformation Coupling Effect of a Deep Foundation Pit Group in a Coastal Soft Soil Area. *Appl. Sci.* **2022**, *12*, 6205. [CrossRef]
21. Kiet, H.N.; Phien-wej, N. Advanced soil parameters determination for Ho Chi Minh city soft clay to predict ground movements in deep excavations and tunneling. In Proceedings of the 4th International Conference on Geotechnics for Sustainable Infrastructure evelopment (GEOTEC HANOI), Hanoi, Vietnam, 28–29 November 2019; pp. 497–504.
22. Panchal, J.P.; McNamara, A.M.; Stallebrass, S.E. A new approach to modelling excavations in soft soils. In Proceedings of the 9th International Conference on Physical Modelling in Geotechnics (ICPMG), London, UK, 17–20 July 2018; pp. 1445–1450.
23. Ding, Y.; Wang, P.; Yu, S. A new method for deformation monitoring on H-pile in SMW based on BOTDA. *Measurement* **2015**, *70*, 156–168. [CrossRef]
24. Xian, Q.; Wang, Z.; Liu, X.; Ma, S.; Xiao, Z. Site Measurement Study on Mechanical Properties of SMW Piles of Building Structures in Sandy Soil Areas. *Buildings* **2022**, *12*, 1733. [CrossRef]
25. Chen, W.Z.; Gao, Q.C.; Jiang, J.S. Simulation experimental research for the influence of embedded ratio on SMW's working traits. *Adv. Mater. Res.* **2011**, *250–253*, 2322–2326. [CrossRef]
26. Finno, R.J.; Voss, F.T.; Rossow, E.; Blackburn, J.T. Evaluating damage potential in buildings affected by excavations. *J. Geotech. Geoenviron. Eng.* **2005**, *10*, 131. [CrossRef]
27. Son, M.; Cording, E.J. Estimation of building damage due to excavation-induced ground movements. *J. Geotech. Geoenviron. Eng.* **2005**, *131*, 162–177. [CrossRef]

28. Asker, K.; Fouad, M.T.; Bahr, M.; El-Attar, A. Numerical analysis of reducing tunneling effect on viaduct piles foundation by jet grouted wall. *Min. Miner. Depos.* **2021**, *15*, 75–86. [CrossRef]
29. Sharma, J.S.; Hefny, A.M.; Zhao, J.; Chan, C.W. Effect of large excavation on deformation of adjacent mrt tunnels. *Tunn. Undergr.Space Technol.* **2001**, *16*, 93–98. [CrossRef]
30. Huang, X.; Schweiger, H.F.; Huang, H. Influence of deep excavations on nearby existing tunnels. *Int. J. Geomech.* **2013**, *13*, 170–180. [CrossRef]

water

MDPI

Article

Geothermal Water Exploration of the Maoyanhe Hot Spring Scenic Spot in Zhangjiajie Using the Natural Electric Field Frequency Selection Method

Yulong Lu [1,2], Haiyang Ding [1], Tianchun Yang [1,*] and Yang Liu [1]

[1] School of Earth Sciences and Spatial Information Engineering, Hunan University of Science and Technology, Xiangtan 411201, China; 1010106@hnust.edu.cn (Y.L.); 22012002005@mail.hnust.edu.cn (H.D.); 1020143@hnust.edu.cn (Y.L.)

[2] Institute of Geological Survey of Hunan Province, Changsha 410083, China

* Correspondence: ytc6803@hotmail.com; Tel.: +86-130-1715-5739

Abstract: Natural electric field frequency selection method was proposed by Chinese scholars in the 1980s on the basis of imitating the field observation method of the magnetotelluric method (MT). It can only measure the magnetotelluric field components of one or several frequencies on the surface to determine the existence of underground geological bodies. This method has played an important role in shallow groundwater exploration. This paper mainly discusses the application of frequency selection method in the exploration of underground hot water in the Maoyanhe Scenic spot, Zhangjiajie City, Hunan Province, in order to illustrate the effectiveness of the frequency selection method in water exploration. According to the situation of the construction site, nearly 20 geophysical prospecting survey lines of varying lengths were laid flexibly within the red line of the Maoyan River Scenic Spot. Firstly, three-frequency (170 Hz, 67 Hz and 25 Hz) measurements were carried out on each survey line to preliminarily determine the possible horizontal location of underground hot water. Secondly, in the vicinity of the low potential anomaly of the three-frequency curve, the fine measurement by using multi-frequency bipolar profile method was further carried out. The specific distribution of underground hot water was determined based on the principle of frequency domain sounding and the static effect characteristics of the electromagnetic method so as to provide a scientific basis for the drilling layout. Finally, the reliability of the frequency selection method is verified by two verification boreholes. The results indicate that the frequency selection method is one of the effective geophysical exploration methods in groundwater exploration.

Keywords: geophysics; static shift method; groundwater; electromagnetic field; magnetotelluric (MT)

Citation: Lu, Y.; Ding, H.; Yang, T.; Liu, Y. Geothermal Water Exploration of the Maoyanhe Hot Spring Scenic Spot in Zhangjiajie Using the Natural Electric Field Frequency Selection Method. *Water* **2023**, *15*, 3418. https://doi.org/10.3390/w15193418

Academic Editor: Guilin Han

Received: 2 August 2023
Revised: 11 September 2023
Accepted: 26 September 2023
Published: 28 September 2023

1. Introduction

Geothermal energy is a kind of green, sustainable, and renewable energy, with the development of society and the improvement of people's living standards. Making full use of geothermal energy can maintain and promote the health of the ecological environment and improve the regional energy utilization structure. Geothermal resources of hydrothermal type, which integrate heat energy and water resources, have the advantages of wide distribution, stability and reliability, simple development and utilization, and high economic benefits, and therefore hydrothermal type geothermal resources occupy a relatively large proportion in the geothermal development market [1]. China is one of the countries with large reserves of geothermal resources in the world, especially the medium and low thermal resources of hydrothermal type, which have huge development potential [2,3].

Geophysical exploration has always played an important role in the exploitation of geothermal resources. Because of the uniformly distributed underground hot water channels, the temperature [4–7], density [8–12], electrical conductivity [13–17], wave velocity [18,19], magnetic susceptibility [20–22] and other physical parameters of underground media, these differences in physical properties provide a physical basis for geophysical exploration of geothermal energy. The conductivity of underground media is one of the important parameters to characterize the geothermal environment, and it is also the method basis for the application of the electromagnetic method in geothermal exploration, and the electromagnetic method has also obtained many achievements in the exploration of geothermal resources [23–26].

Maoyanhe Town, Yongding District, Zhangjiajie City, Hunan Province, is located in the core area of the World Natural Heritage connecting line—Jiaoziya to Wentang Road. There are "Li River First Bay", "Love Lake" and other Internet popular scenic spots, and some tourist attractions such as Maoyanhe Lake Tour, Maoyanhe rafting, Tianquanshan National Forest Park, etc. in this area. With the further development of hot spring hotels, tourist characteristic blocks and other projects, Maoyanhe Town has become more prominent in the core position of Zhangjiajie West Route tourism. At the same time, with the gradual increase in the number of tourists, the water quantity of hot springs in the Maoyan River scenic spot can no longer meet the demand of tourism development. Therefore, on the basis of previous geological data and field geological investigation, the authors carried out the exploration of underground hot water resources within the area of the Maoyanhe scenic spot by using natural electric field frequency selection method.

2. Description of the Study Area

2.1. Physiographic Conditions

The research area belongs to the subtropical mountain prototype monsoon humid climate area, which is characterized by warm climate, four distinct seasons, a rainy spring and summer, and often summer drought from July to September. The annual mean temperature is about 16.8 °C and the extreme maximum temperature is 40.7 °C. The annual average sunshine duration is 1449.6 h, the dominant wind direction was east wind, and the frequency of quiet wind was 47.7%. The average annual rainfall is 1336.1 mm [27–30].

Maoyan River tourist attractions near the cliffs are very dangerous for Zhangjiajie City and Xiangxi city boundaries. The width of the river varies greatly, and the widest part is located at the edge of Maoyanhe Town, about 250 m. The narrowest part is located in the 600 m section of the river downstream of the current scenic hot spring Q07 which is 30~40 m wide. On both sides of this section, there are steep walls tens of meters high, shaped like a throat Once encountered heavy rain, the river is blocked here, the river level can rise several meters or even more than ten meters in a short time.

Maoyan River is 186~202 m above sea level, and Hanshe—Ganxi is 196~330 m above sea level. It is bounded by the Shenghe River and the Ganxi River, with sand shale low hill-hills and cone-shaped peaks in the north and west. On the SE side, there are low hills composed of carbonate. In the middle of the hill, the slope is gentle, the top of the hill is rounded, and there are many conical peaks on both sides. On the whole, the terrain is low in the middle, and it is undulating to both sides. With the Maoyan River as the boundary, the central valley basin is on both sides of the low mountains—hills. The terrain on the whole NE side is slightly higher, and the SW side is low.

According to the origin of the geomorphology, the geomorphology of the exploration area can be divided into three types: dissolution tectonic geomorphology, denudation tectonic geomorphology, river erosion accumulation geomorphology, and the dissolution tectonic geomorphology can be divided into two sub-types: low platform dissolution depression and low mountain-hill cluster valley.

2.2. Hydrogeological Condition

The strata in and around the exploration area include Paleozoic Ordovician and Silurian, which is a NW trending monoclinal structure. Generally speaking, the strata change from old to new from SE to NW. The Ordovician are mainly composed of Marine carbonate rocks, mainly composed of argillaceous banded limestone, limestone and dolomite. The

Silurian system is distributed in the north and west of the area and is composed of shallow Marine clastic rocks. No magmatic rock outcrop was observed in the area (Figure 1).

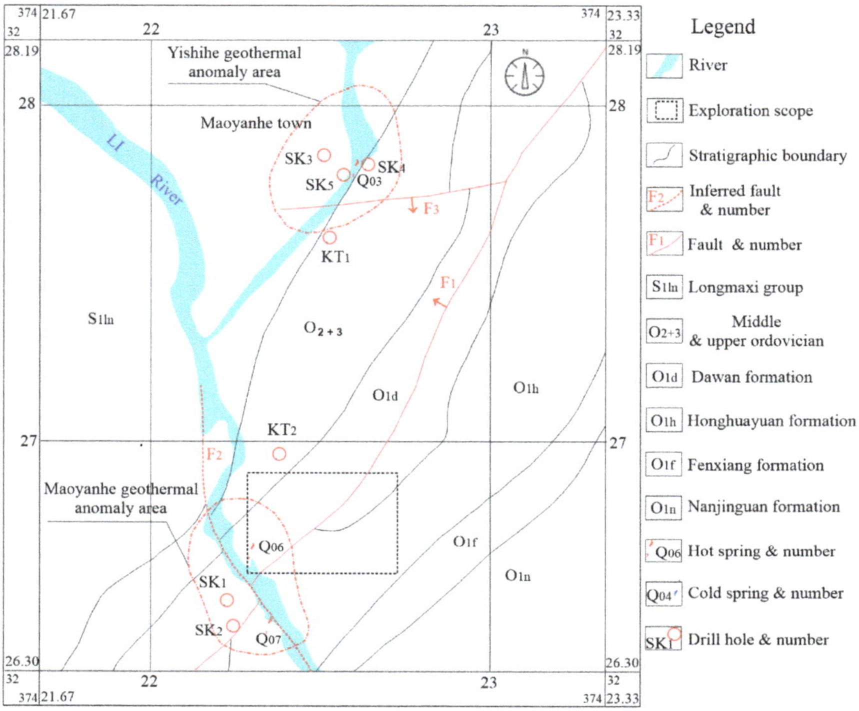

Figure 1. Geological scheme of the working area.

The exploration area is located in the NW of the regional deep fault Huayuan—Zhangjiajie—Cili fault belt, the NW wing of the ZhongHehu-Yaowan anticline, and the east wing of the Qinganping syncline. The structural traces in this area are mainly NE-trending and NNE-trending, and NW-trending and near-EW-trending are the second-order structures. The main structure of the exploration area and its surrounding area is fault F1, which has a trend of 30° east to the north and a dip to the NW with a dip Angle of 55~60°. The fault zone is argillaceous cementation and poor permeability, which is the main heat-controlling structure formed by hot spring water in the area. F2 fault strikes NW, spreads along the Maoyan River (Namely Lishui River), and leans NE, which is the main water channel for hot spring Q06 to emerge on the surface. F3 fault is spreading north 56° west and is a positive fault with a SW trend and a dip of about 75°, which is the main water-conducting structure of drill hole KT1.

In this area, carbonate rocks are mainly in Ordovician (O) limestone, argillaceous limestone, dolomite, and karst is relatively developed. According to the field investigation, the karst forms developed in the area mainly include dissolved trough (groove) and stone tooth, shaft, water cave, karst funnel, depression, blind valley, karst cave, etc. The anomaly distribution of hot spring water in the exploration area and surrounding area is mainly located in the Maoyan River anomaly area and the Shenghe anomaly area. Hot springs in the Maoyan River anomaly area include Q06, Q07 and Q08, among which Q07 and Q08 are located on the west bank of the Lishui River, within the scope of Yongshun County, Xiangxi Autonomous Prefecture, Hunan Province, not within the scope of this survey. Q06 is located in the Maoyan River scenic area, and the highest water temperature is 39.3 °C, which is obviously affected by rainfall. The lowest water temperature is 29.2 °C, and the maximum discharge is 11 L/s. In the past, the maximum water temperature of KT2 was 30.3~46.2 °C, and the water quantity was small. The specific size was unknown. Hot spring Q03 emerged from the anomaly area of Hanshe River. The highest water temperature of SK4, SK5 and KT1 drilled in the past was 28.4~43.7 °C. The hole depth of KT1 is about 305 m, and the self-flow rate of the hole is about 2300 m^3/d.

According to the lithology of water-bearing medium, the occurrence condition of groundwater, the water physical properties and the hydrodynamic characteristics, the groundwater in this area can be divided into two categories: clastic rock fissure water and carbonate rock fissure cave water. Carbonate fissure karst cave water can be divided into carbonate fissure karst cave water and carbonate fissure karst cave water with clastic rock. The carbonate fissure karst cave is rich in water, the carbonate fissure karst cave with clastic rock is medium rich in water, and the clastic fissure karst cave is poor in water. The groundwater in the working area is mainly replenished by atmospheric rainfall followed by lateral replenishment between different aquifers. In addition, since the water level difference between the Maoyan River and its tributaries can reach up to ten meters, when the river level is at a high level, it will recharge groundwater in reverse.

3. Methodology

3.1. Natural Electric Field Frequency Selection Method Technology

Limited by the work scope of the site, the geophysical exploration well can only be located within the scope of the Maoyan River Scenic spot. Therefore, the natural electric field frequency selection method (FSM) was adopted to carry out geophysical exploration in this survey, and the specific device form was used to observe the three-frequency method and two pole profile method [31–35]. Due to the abundant groundwater in the field, the underground fault-fissure structures are generally filled with water or mud, showing relatively low resistance characteristics. Therefore, there is an obvious conductivity difference between the target body and the surrounding rock in this work, which is mainly to search for water-bearing bodies with low resistance in relatively high resistance medium [36–38].

The natural electric field frequency selection method (FSM, Frequency selection method for short) is an extension and development of the audio telluric current method (AET) and the audio telluric field method (TEF). It is also a further improvement of these two methods. It is a kind of natural electric field method. The natural electric field frequency selection method was first proposed by Liang in 1976 [39]. He called it the electric pulse natural electric field method, and Yang called it the stray current method and used it in karst area [40]. All these methods are based on the work of Soviet geophysicist A.N. Tikhonov and Cagniard, L [41,42]. The theoretical basis of its application is essentially the use of the static effect in the natural electromagnetic method, that is, the static effect is used to identify the existence of abnormal bodies [43].

The natural electric field is mainly caused by the natural electrochemical action of the electronic conductor and the filtration or diffusion action of the ionic conductor in groundwater. Therefore, the natural electric field method is developed. The field source is a DC signal. The natural electric field frequency selection method is different from the conventional natural electric field method. Its field source is an AC signal, which mainly uses the magnetotelluric field [43–45]. The causes of the magnetotelluric field are very complex. The main sources are thunderstorm discharge, cosmic rays, slight variation of geomagnetic field and industrial stray current induced by the earth. Nearly all the telluric fields with frequencies over 1 Hz are generated by lightning, except for industrial currents at 50 Hz and high power high-frequency radio messages. On average, space lightning of large and small size can occur more than 100 times per second on the earth. Part of the huge energy of lightning flows into the ground along with the inductive current [46–49]. The electromagnetic field generated by distant lightning spreads through space and the Earth in the way of conduction wave, forming a relatively gentle and approximately uniform geodetic electric field. Therefore, in the surface of the solid Earth, in the atmosphere and in the oceans, there are electric currents flowing [50,51]. This natural electric field is called the magnetotelluric field, and its direction and strength change over time. Alternating electric fields are always accompanied by alternating magnetic fields, which are collectively referred to as Earth's magnetotelluric field.

The field sources of passive source method are all very complex, but according to the previous research results, the field source problem of the natural electric field frequency selection method can be summarized into three sources. First, it is believed that its field source is exactly the same as MT, mainly from outside the Earth; The second is the electromagnetic wave emanating upward from the Earth's inner mantle asthenosphere and the Earth's magnetic field below. Third, the field source is industrial stray current. Regardless of whether the field source of frequency selection method comes from outside the Earth, inside the Earth, or industrial stray current, the exploration depth of current application of the frequency selection method in underground target exploration is mostly shallow (general <400 m), so the three field sources can be classified as far-field sources. Therefore, it can be considered that the field source of the frequency selection method is the comprehensive effect of the above three sources and other unrecognized electromagnetic field sources.

The natural field frequency selection method measures the horizontal electrical component generated by the telluric electromagnetic field in the audio segment. Its field sources are mainly the three natural alternating electromagnetic fields mentioned above, which is the same with the magnetotelluric method (MT). At a distance from the field source, the electric field and magnetic field components of the three natural alternating electromagnetic fields are horizontal components. Therefore, the field source of natural field frequency selection method is similar to the MT method. The electromagnetic field can be regarded as a plane wave incident perpendicular to the ground (see Figure 1), and its field characteristics obey Maxwell's equations:

$$\mathrm{rot}H = j + \partial D/\partial t \quad (1)$$

$$\mathrm{div}B = 0 \quad (2)$$

$$\mathrm{rot}E = -\partial B/\partial t \quad (3)$$

$$\mathrm{div}D = q \quad (4)$$

In the formula, q is the density of free charge; t is time; j is current density; E is electric field intensity; H is magnetic field strength; B is magnetic induction intensity; D is electric displacement.

When the electromagnetic wave propagates in the medium, its amplitude advances 1/distance along the Z-axis perpendicular to the surface, and the amplitude decays 1/e times (about 37%) of the surface. It is customary to take the distance $\delta = 1/b$ as the skin depth (or penetration depth) of the electromagnetic wave. In a non-magnetic medium, it can be approximated.

$$\delta = 1/b \approx 503.3 \times \sqrt{\rho/f} \quad (5)$$

In formula δ is the penetration depth; ρ is resistivity; f is the frequency of electromagnetic wave, b is the attenuation coefficient of electromagnetic wave; e is a constant that is the base of the natural logarithm function.

As can be seen from the above, electric field intensity E decays negatively with the absorption coefficient b. That is, the penetration depth of electromagnetic wave increases with the increase in dielectric resistivity and decreases with the increase in electromagnetic wave frequency, which is the frequency domain characteristic of electromagnetic signal. At the same observation point, the resistivity of the formation is constant, and the purpose of detecting different depths is achieved by choosing different working frequency methods. Therefore, the profile detection result of frequency selection method has the function of both profile and sounding, the low potential and high potential anomalies can be detected above the low resistance body or high resistance body, respectively.

At the same time, because the natural electric field frequency selection method belongs to the category of the natural electromagnetic method, the detection frequency is only different from that of traditional methods, and its detection results also have obvious static effect, which makes the relative anomalies of low or high resistance bodies near the surface more obvious (Figure 2). Static effect is unavoidable in magnetotellus observation. Usually, because MT is mostly detected by deep targets, static migration is generally regarded as interference to suppress or eliminate. As in the early days of seismic exploration, Rayleigh waves were always used as interference signals. To this end, Yang Tianchun et al. recently made a special study on the causes of profile anomalies of the natural electric field frequency selection method and concluded that the profile anomalies were mainly caused by the static effect of electromagnetic method [27,28].

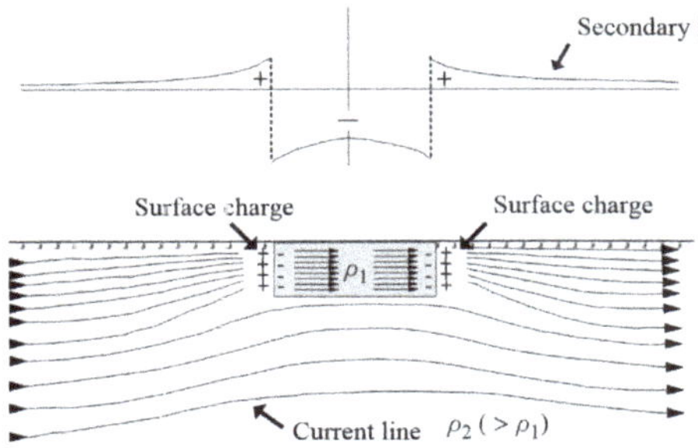

Figure 2. Schematic diagram of TM mode electric field response of near-surface low-resistivity anomalous body.

3.2. Layout of Field Survey Line

Since the Maoyan River Scenic Area Development Company is limited to looking for underground hot springs within the perimeter of the scenic area, which is basically completed, construction facilities, ground hardening, relief of terrain, strong artificial interference and other factors bring great difficulties to the selection of geophysical exploration methods and layout of survey lines (Figures 3 and 4). The line laying of the geophysical prospecting method can only be applied flexibly.

Figure 3. Site conditions and working photos: (**a**) By the river (**b**) In the bamboo forest by the river (**c**) Roadside scenic spot.

Figure 4. Schematic diagram of field line laying.

A total of 18 surveying lines, namely L1~L18, were laid around the scenic spot in this geophysical exploration work (Figure 4). Since the Maoyan River Scenic Area Development Company limited the final drilling to its red line, survey lines L5, L6, L14 and L15 outside the scenic area are not shown in Figure 4. They are located in the north of the area shown in Figure 4.

4. Results

4.1. Frequency-Selected Exploration Results

The construction of this method is similar to the audio frequency electromagnetic telluric method (AMT), but it only observes one or several horizontal electric field components of different frequencies in the telluric signal along the measured line. For example, the sampling frequency of the selected TC300 frequency selector is 40 frequency points within the range of 12~5000 Hz, and the sampling frequency of the TC500 frequency selector is 56 frequency points within the range of 8~5000 Hz.

During the field exploration, the observation pole distance MN is 10 m, and the observation point distance varies from 1 to 5 m according to the actual situation. Figure 5a shows the three-frequency detection result curve of L7 survey line by frequency selection method. The survey line is located in the parking lot of the scenic spot, and the distance between observation points is 5 m. The horizontal coordinate represents the distance of the profile line, and the vertical coordinate axis represents the potential difference. As can be seen from the results of the detection curve of a (0 in Figure 5), there is an overall relative trend of low potential near 40 m in the middle of the measurement line, but the relative amplitude of the anomaly is not very large. It is speculated that there is a small amount of groundwater under the anomaly point.

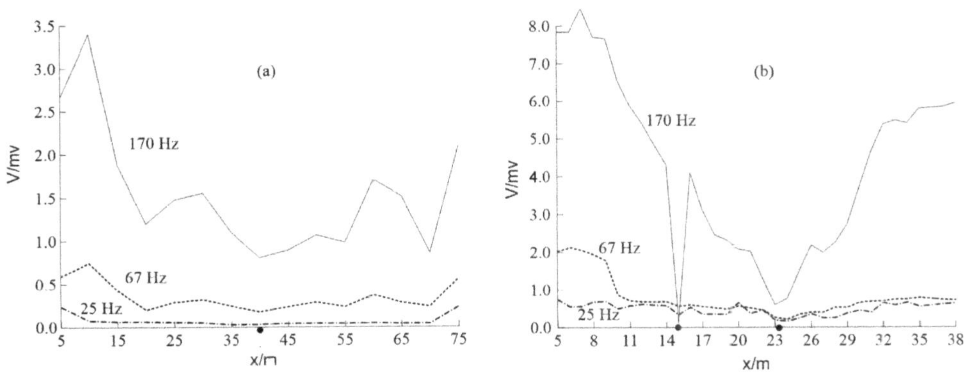

Figure 5. Results of three-frequency detection by the frequency selection method on line L7 (**a**) and line L16 (**b**).

Figure 5b shows the three-frequency detection result curve of frequency selection method on the L16 survey line. The survey line is located on the roadside on the east side of the heart-shaped pool in the scenic spot, and the distance between the observation points is 1 m. As can be seen from the curve results in Figure 5b, there are two obvious relatively low potential anomalies at 15 m and 23–24 m in the middle of the measurement line. The anomaly at 15 m has the nature of a single point with poor reliability, while the anomaly curve at 23–24 m has regular shape, wider and slower anomaly range, and greater reliability. In addition, compared with the anomaly on L7, the low potential anomaly on L16 is much more obvious, and it is speculated that the groundwater will be more abundant. Therefore, the relatively low potential anomaly at 23–24 m on the measuring line L16 should be the focus of the next step.

In order to further determine the anomalies near 40 m of line L7 and 24 m of line L16, bipolar profile detection was carried out near these two anomalies.

Figure 6 shows the observation results of bipolar profile method near 40 m of L7 survey line. Figure 6a,b, respectively, show the potential difference between the profile curve and pseudo-sectional diagram. There are 40 observed frequencies at each measuring point, corresponding to 40 profile curves in Figure 6a. Due to the density of the curves, the size of each observed frequency is not marked in the figure. In Figure 6b, the abscissa represents the profile distance, and the ordinate represents the apparent depth obtained after depth inversion. The observation instrument is TC300, and the observation point distance is 2 m. According to the detection results of the bipolar profile in Figure 6, it can be seen that 45 m of profile is a relatively low potential area, and the well formation location can be chosen there. As can be seen from Figure 6b, the potential difference ΔV curve near 45 m of measurement line 7 appears as the "hanging noodle" phenomenon. The contours in depths 0~125 m generally bend downward, the contours in depths 125~200 m generally bend upward, and the contours in depths above 200 m tend to bend downward. According to the results of pseudo section map Figure 6b, it is inferred that there is groundwater in the shallow part of the anomaly within 120 m, and the depth of the deep water may be 200 m later.

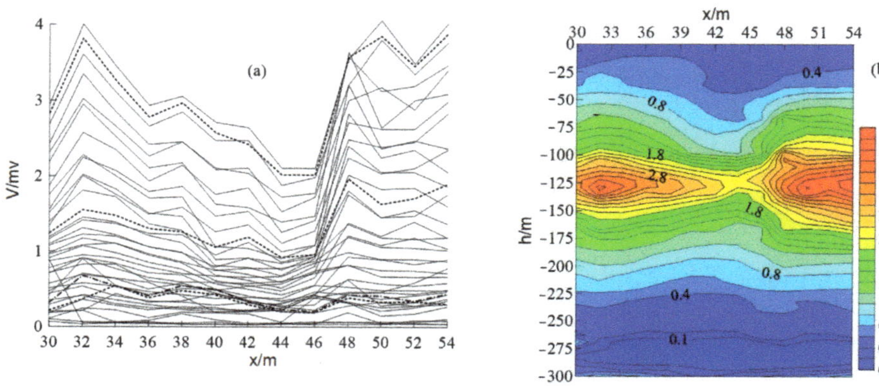

Figure 6. (**a**) bipolar profile curve and (**b**) pseudo-sectional diagram on L7. (dash line in (**a**) means potential abnormal profiles).

Figure 7 shows the observation results of the bipolar section method near 23 m of L16 survey line. Figure 7a,b, respectively, show the potential difference profile curve and pseudo-sectional diagram. There are 56 frequencies observed at each measuring point, corresponding to 56 profile curves in Figure 7a. The e instrument used in the field is TC500, and the observation point distance is 1 m. According to the results of the profile curve in Figure 7a, there is indeed a very obvious relatively low potential anomaly at 23 m of the measuring line. This exception is clearer than the one in Figure 6, and the relative size of the exception is much more obvious. In the potential difference pseudo Figure 7b, the curve near the surface is dense, which is caused by the interpolation of test results and does not affect the analysis of the results. It can be inferred from Figure 7b that the abnormal low potential at 23 m extends from the shallow part to the deep part, and the deeper the burial depth is, 270 m later, the abnormal range of low potential becomes wider, and it is speculated that the deep water may be caused by the development of underground cracks, abundant groundwater and good connectivity. The anomaly characteristics at 23 m on the pseudo-sectional map are obvious reflections of the static effect of electromagnetic method. It is expected that the anomaly at 23 m of line L16 is reliable and can be used as a favorable position for drilling. It is estimated that there is a small amount of groundwater in the depth of 100 m, and there may be very rich groundwater in the depth of 320~470 m.

4.2. Borehole Verification

Drilling holes ZK1 and ZK2 were arranged to verify the geophysical anomaly of L7 and L16 survey lines, respectively (see Figure 4). ZK1 is located in the parking apron of the Maoyan River Scenic spot, that is, 45 m of Line L7. With a drilling depth of 250 m, water-bearing cracks exist only near the depth of about 80 m and 200 m. The well had a water yield of about 100 m^3/d and a water temperature of about 22 °C in a 20 cm thick muddy pack near a depth of about 218 m. ZK2 is located at 23 m of line L16, with a well completion depth of 500 m and a water yield of about 1500 m^3/d. The shallow 200 m water is less and all of it is cold water. A large amount of groundwater was buried after 250 m, and the temperature increased significantly with the increase in depth. Finally, the temperature after mixing cold and hot water in the well was 34 °C. With the pumping test of ZK2 drilling, it can be seen that the water at the dewpoint of Q06 hot spring is significantly reduced, which also proves that ZK2 has drilled into the hot spring zone. The water temperature in the ZK2 well is expected to rise if the shallow water is isolated.

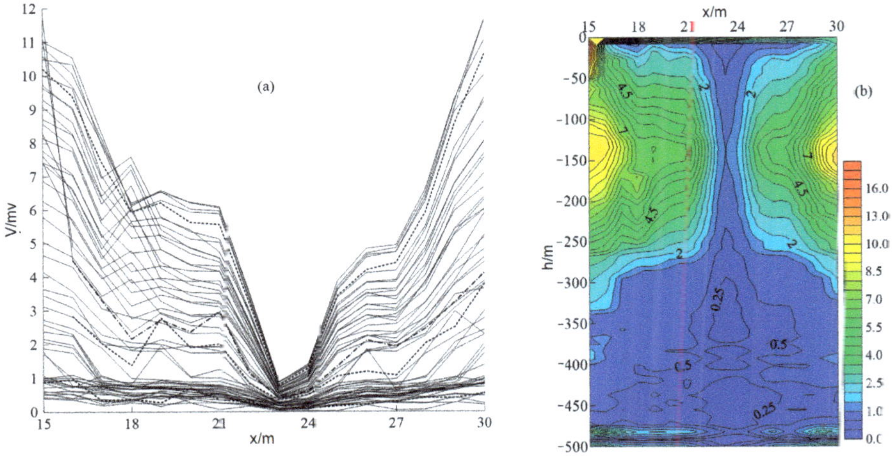

Figure 7. (**a**) bipolar profile curve and (**b**) pseudo-sectional diagram on L16. (dash line in (**a**) means potential abnormal profiles).

5. Discussion

The results of three-frequency detection by frequency selection method and the bipolar profile curve are shown in this research. The site selection was based on the results we have obtained. When underground water exists, the potential difference ΔV profile curve will appear obvious abnormal characteristics of low potential. On ΔV pseudo section map the isoline will appear as the "hanging noodles" phenomenon, which is an obvious reflection of the static effect of electromagnetic method. It can be seen from the verification of drilling results that the more abundant groundwater is, the more obvious the static effect phenomenon is, and the more significant the abnormal characteristics of low potential are on the profile curve. Therefore, the natural electric field frequency selection method is essentially the use of the electromagnetic method of static effect phenomenon and can be called static effect method. The reliability of the frequency selection method is verified by two verification boreholes.

Although the static effect phenomenon of the electromagnetic method can perfectly explain the cause of profile anomaly of the natural electric field frequency selection method, according to the author's years of practical application and theoretical research, the static effect cannot perfectly explain the abnormal causes of FSM sounding, nor can it reasonably explain the dynamic effect phenomenon of FSM. These problems need to be further studied in the future.

6. Conclusions

This paper mainly discusses the application of the frequency selection method in the exploration of underground hot water in the Maoyanhe Scenic spot, Zhangjiajie City, Hunan Province, in order to illustrate the effectiveness of the frequency selection method in water exploration. The reliability of the frequency selection method is verified by two verification boreholes. The results indicate that the frequency selection method is one of the effective geophysical exploration methods in groundwater exploration.

The natural electric field frequency selection method cannot directly observe the water temperature of groundwater. Hydrology can only affect the conductivity of underground media, and thus has an indirect effect on the exploration of electrical methods, but this effect is almost negligible for the observation of the frequency selection method. Therefore, the temperature anomaly observation of groundwater has to rely on other methods. But the frequency selection method as a simple and portable shallow groundwater exploration method is worth further promotion in the future exploration of water resources.

Author Contributions: Data curation, H.D.; writing—original draft preparation, Y.L. (Yulong Lu); writing—review and editing, Y.L. (Yang Liu); project administration, T.Y. All authors have read and agreed to the published version of the manuscript.

Funding: This research was funded by the National Natural Science Foundation of China (No. 42074219), the National Key R&D Program of China (No. 2018YFC0603900), the Natural Science Foundation of Hunan Province (Grant No.2022JJ30244), the Research Project of Teaching Reform of Hunan Province (Grant No. HNJG-2022-0790) and the Education Department of Hunan Province (No. 16K031).

Data Availability Statement: No new data were created or analyzed in this study. Data sharing is not applicable to this article.

Conflicts of Interest: The authors declare no conflict of interest.

References

1. Han, Y.H.; Shen, X.L.; Li, B.; Xu, D.-C.; Jia, Z.-G.; Wu, D.-L.; Wang, W.; Lyu, J. Target area prediction and drilling verification of the tectonic fissure-hosted geothermal water in Meixian County, Guanzhong Plain based on the integrated geophysical exploration. *Geophys. Geochem. Explor.* **2023**, *47*, 65–72.
2. Yin, Z.; Liu, Y.G.; Zhang, X.R.; Li, Y.S.; Feng, J.X. An analysis of the endowment characteristics and geneses of geothermal resources in the Zhangye Basin. *Hydrogeol. Eng. Geol.* **2023**, *50*, 168–178.
3. Zhou, Z.Y.; Liu, S.L.; Liu, J.X. Study on the characteristics and development strategies of geothermal resources in China. *J. Nat. Resour.* **2015**, *30*, 1210–1221.
4. Liu, Y.; Bouazza, A.; Gates, W.P.; Rowe, R.K. Hydraulic performance of geosynthetic clay liners to sulfuric acid solutions. *Geotext. Geomembr.* **2015**, *43*, 14–23. [CrossRef]
5. Wang, R.G.; Liu, S.Y.; Shangguan, D.H.; Radic, V.; Zhang, Y. Spatial Heterogeneity in Glacier Mass-Balance Sensitivity across High Mounta in Asia. *Water* **2019**, *11*, 776. [CrossRef]
6. Zhao, B.F.; Wang, Q.N.; Guan, D.W.; Chen, T.G.; Fang, W. Temperature field characteristics and thermal control factors of banded reservoirs geothermal field: An example of Reshui-town geothermal field, Hunan Province. *Geol. Bull. China* **2022**, *41*, 2035–2046.
7. Deng, R.J.; Shao, R.; Ren, B.Z.; Hou, B.L.; Tang, Z.E.; Hursthouse, A. Adsorption of Antimony(III) onto Fe(III)-Treated Humus Sludge Adsorbent: Behavior and Mechanism Insights. *Pol. J. Environ. Stud.* **2019**, *28*, 577–586. [CrossRef]
8. Zhang, Y.; Ren, B.Z.; Hursthouse, A.; Deng, R.J.; Hou, B.L. Leaching and Releasing Characteristics and Regularities of Sb and As from Antimony Mining Waste Rocks. *Pol. J. Environ. Stud.* **2019**, *28*, 4017–4025. [CrossRef]
9. You, B.; Xu, J.X.; Shi, S.L.; Liu, H.Q.; Lu, Y.; Liang, X.Y. Treatment of coal mine sewage by catalytic supercritical water oxidation. *Fresenius Environ. Bull.* **2020**, *29*, 497–502.
10. Liu, Y.; He, B.; Xie, J.; Lu, Y.; Zhang, L. Compatibility of geosynthetic clay liners at different temperatures. *J. Environ. Prot. Ecol.* **2021**, *22*, 2295–2306.
11. Yu, S.Q.; Mo, Q.F.; Chen, Y.Q.; Li, Y.W.; Li, Y.X.; Zou, B.; Xia, H.P.; Jun, W.; Li, Z.A.; Wang, F.M. Effects of seasonal precipitation change on soil respiration processes in a seasonally dry tropical forest. *Ecol. Evol.* **2020**, *10*, 467–479. [CrossRef] [PubMed]
12. Wang, Z.H.; Huang, W.; Wang, C.; Li, X.S.; He, Z.H.; Hursthouse, A.; Zhu, G.C. Variation of Al species during water treatment: Correlation with treatment efficiency under varied hydraulic conditions. *AQUA—Water Infrastruct. Ecosyst. Soc.* **2021**, *70*, 891–900. [CrossRef]
13. Zhou, Y.B. Application of vibroseis vehicle in geothermal exploration in newly developed urban area. *Geotech. Investig. Surv.* **2021**, *49*, 67–72.
14. Wu, J.W.; Hu, X.Y.; Huang, G.S.; Peng, R.H.; Zhou, W.L. Status and prospects of electromagnetic method used in geothermal resources exploration. *Acta Geosci. Sin.* **2023**, *44*, 191–199.
15. Chongo, M.; Christiansen, A.V.; Tembo, A.; Banda, K.E.; Imasiku, A.N.; Larsen, F.; Peter, B.G. Airborne and ground-based transient electromagnetic mapping of groundwater salinity in the Machile-Zambezi Basin, southwestern Zambia. *Near Surf. Geophys.* **2015**, *13*, 383–395. [CrossRef]
16. Zhou, W.; Hu, X.; Yan, S.; Guo, H.; Chen, W.; Liu, S.; Miao, C. Genetic Analysis of Geothermal Resources and Geothermal Geological Characteristics in Datong Basin, Northern China. *Energies* **2020**, *13*, 1792. [CrossRef]
17. Zhao, Y.L.; Luo, S.L.; Wang, Y.X.; Wang, W.J.; Zhang, L.Y.; Wan, W. Numerical analysis of karst water inrush and a criterion for establishing the width of water-resistant rock pillars. *Mine Water Environ.* **2017**, *36*, 508–519. [CrossRef]
18. Yang, Z.C. The least-square Fourier-series model-based evaluation and forecasting of monthly average water-levels. *Environ. Earth Sci.* **2018**, *77*, 328. [CrossRef]
19. Jin, C.S.; Liu, Y.X.; Li, Z.W.; Gong, R.Z.; Huang, M.; Wen, J.J. Ecological consequences of China's regional development strategy: Evidence from water ecological footprint in Yangtze River Economic Belt. *Environ. Earth Sci.* **2018**, *77*, 328. [CrossRef]
20. Zhou, J.Z.; Zhang, M.Q.; Ji, M.T.; Wang, Z.H.; Hou, H.; Zhang, J.; Huang, X.; Hursthouse, A.; Qian, G.R. Evaluation of heavy metals stability and phosphate mobility in the remediation of sediment by calcium nitrate. *Water Environ. Res.* **2020**, *92*, 1017–1026. [CrossRef]

21. Chen, B. Application of CSAMT and microtremor sounding in a geothermal resource exploration area, Zhangzhou. *Coal Geol. China* **2021**, *33*, 64–69.
22. Tang, L.; Tang, X.W.; Liu, Y.; Qu, S.X. Prediction of pore size characteristics of woven slit-film geotextiles subjected to unequal biaxial tensile strains. *Geotext. Geomembr.* **2020**, *48*, 724–734. [CrossRef]
23. Liu, Z.; Zeng, Z.F.; Li, J.; He, R.Q.; Huo, Z.J.; Lin, J.Y. Inversion calculation of Curie depth based on dual magnetic interface and variable susceptibility model. *Glob. Geol.* **2019**, *38*, 795–804.
24. Spichak, V.; Manzella, A. Electromagnetic sounding of geothermal zones. *J. Appl. Geophys.* **2009**, *68*, 459–478. [CrossRef]
25. Deng, R.J.; Jin, C.S.; Ren, B.Z.; Hou, B.L.; Hursthouse, A.S. The Potential for the Treatment of Antimony-Containing Wastewater by Iron-Based Adsorbents. *Water* **2017**, *9*, 794. [CrossRef]
26. Ilyushin, Y.V.; Kapostey, E.I. Developing a Comprehensive Mathematical Model for Aluminium Production in a Soderberg Electrolyser. *Energies* **2023**, *16*, 6313. [CrossRef]
27. Wang, P.F.; Tan, X.H.; Zhang, L.Y.; Li, Y.J.; Liu, R.H. Influence of particle diameter on the wettability of coal dust and the dust suppression efficiency via spraying. *Process Saf. Environ. Prot.* **2019**, *132*, 189–199. [CrossRef]
28. Wang, T.; Huang, C.; Du, G.; Liu, Y.; Xie, J.; Li, H. Geochronology, geochemistry and zircon Hf-isotopes of the early Mesoproterozoic Yaopengzi dolerite in SW Yangtze block (Sichuan, SW China): Implications for the Columbia supercontinent breakup. *Geosci. J.* **2019**, *23*, 557–573. [CrossRef]
29. Liu, T.; Chen, Z.S.; Li, Z.X.; Fu, H.; Chen, G.L.; Feng, T.; Chen, Z. Preparation of magnetic hydrochar derived from iron-rich Phytolacca acinosa Roxb. for Cd removal. *Sci. Total Environ.* **2021**, *769*, 145–159. [CrossRef]
30. Meju, M.A. Geoelectromagnetic exploration for natural resources: Models, case studies and challenges. *Surv. Geophys.* **2002**, *23*, 133–206. [CrossRef]
31. Yang, T.C.; Shen, J.P.; Li, G.M.; Cao, Y.J.; Zhang, H. Application and analysis of natural electric field frequency selection method for water-filled karst investigation. *Coal Geol. Explor.* **2014**, *42*, 71–75.
32. Lu, Y.L.; Yang, T.C.; Tizro, A.T.; Yang, L. Fast recognition on shallow groundwater and anomaly analysis using frequency selection sounding method. *Water* **2023**, *15*, 96. [CrossRef]
33. Zhou, Y.Y.; Ren, B.Z.; Hursthouse, A.S.; Zhou, S.J. Antimony Ore Tailings: Heavy Metals, Chemical Speciation, and Leaching Characteristics. *Pol. J. Environ. Stud.* **2019**, *28*, 485–495. [CrossRef] [PubMed]
34. Hou, B.L.; Liu, X.; Li, Z.; Ren, B.Z.; Kuang, Y. Heterogeneous fenton oxidation of butyl xanthate catalyzed by iron-loaded sewage sludge. *Fresenius Environ. Bull.* **2022**, *31*, 4125–4131.
35. Liu, Y.; Hao, Y.; Lu, Y. Improved design of risk assessment model for PPP project under the development of marine architecture. *J. Coast. Res.* **2018**, *83*, 74–80. [CrossRef]
36. Tang, J.T.; Zhou, C.; Xiao, X. Selection of minimum transmit-receive distance of CSAMT on complicated media. *Chin. J. Nonferrous Met.* **2013**, *23*, 1681–1693.
37. Shi, X.Y.; Ren, B.Z.; Hursthouse, A. Source identification and groundwater health risk assessment of PTEs in the stormwater runoff in an abandoned mining area. *Environ. Geochem. Health* **2022**, *44*, 3555–3570. [CrossRef]
38. Yuan, Q.; Zhu, G.C. A review on metal organic frameworks (MOFs) modified membrane for remediation of water pollution. *Environ. Eng. Res.* **2021**, *26*, 190435. [CrossRef]
39. Liu, Y.; Gates, W.P.; Bouazza, A. Acid induced degradation of the bentonite component used in geosynthetic clay liners. *Geotext. Geomembr.* **2013**, *36*, 71–80. [CrossRef]
40. Liang, B.H. Introduction of electric pulse natural electric field water search. *Guangxi Karst Geol.* **1976**, *3*, 1–4.
41. Yang, J. Experimental results and theoretical research of dissipative current method in karst area. *Geophys. Geochem. Explor.* **1982**, *1*, 41–54.
42. Tikhonov, A.N. On determining electric characteristics of the deep layers of the Earth's crust. *Dokl. Akad. Nauk. SSSR* **1950**, *73*, 295–297.
43. Cagniard, L. Basic theory of the magneto-telluric method of geophysical prospecting. *Geophysics* **1953**, *18*, 605–635. [CrossRef]
44. Yang, T.C.; Gao, Q.S.; Li, H.; Fu, G.H.; Hussain, Y. New Insights into the Anomaly Genesis of the Frequency Selection Method: Supported by Numerical Modeling and Case Studies. *Pure Appl. Geophys.* **2023**, *180*, 969–982. [CrossRef]
45. Pershin, I.M.; Papush, E.G.; Kukharova, T.V.; Utkin, V.A. Modeling of Distributed Control System for Network of Mineral Water Wells. *Water* **2023**, *15*, 2289. [CrossRef]
46. Ilyushin, Y.V.; Asadulagi, M.-A.M. Development of a Distributed Control System for the Hydrodynamic Processes of Aquifers, Taking into Account Stochastic Disturbing Factors. *Water* **2023**, *15*, 770. [CrossRef]
47. Liu, Y.; Gates, W.P.; Bouazza, A. Impact of acid leachates on microtexture of bentonites used in geosynthetic clay liners. *Geosynth. Int.* **2019**, *26*, 136–145. [CrossRef]
48. Yang, X.; Liu, Y.; Yang, C. Research on the slurry for long-distance large-diameter pipe jacking in expansive soil. *Adv. Civ. Eng.* **2018**, *9*, 47–54. [CrossRef]
49. Ikard, S.; Pease, E. Preferential groundwater seepage in karst terrane inferred from geoelectric measurements. *Near Surf. Geophys.* **2019**, *17*, 43–53. [CrossRef]
50. Liu, Y.; Li, X.; Tu, Y.; Lu, Y. Mining Leachates Effect on the Hydraulic Performance of Geosynthetic Clay Liners under Different Temperatures. *Water* **2023**, *15*, 1132. [CrossRef]

51. Liang, J.; Wei, Q.F.; Hong, J.; Zheng, S.Y.; Qin, Y.C.; Yan, F.S.; Feng, Y.X. Application of self-potential method to explore water in karst area. *Geotech. Investig. Surv.* **2016**, *44*, 68–78.

water

MDPI

Article

A Simplified Method for Effective Calculation of 3D Slope Reliability

Juxiang Chen [1,*], Dayong Zhu [2] and Yalin Zhu [1]

1 School of Civil Engineering, Hefei University of Technology, Hefei 230009, China; zhuyalin@hfut.edu.cn
2 School of Civil Engineering and Architecture, Ningbo Tech University, Ningbo 315100, China
* Correspondence: chenjx@hfut.edu.cn

Abstract: Traditional 3D slope reliability analysis methods have high computational costs and are difficult to popularize in engineering practice. Under the framework of the limit equilibrium method with 3D slip surface normal stress correction, the critical horizontal acceleration coefficient K_c, which is equivalent to the safety factor F_s, is selected to characterize the slope stability. The limit state function uses the difference between K_c and the known critical value K_{c0}. A simplified method for calculating the reliability of 3D slope is proposed. Through two typical slope examples, the 3D reliability calculation results of six methods after coupling two limit state functions and three reliability algorithms are compared. The results show that this method is reliable and effective, and the method coupled with subset simulation (SS) is the one with good calculation accuracy and efficiency. In the case of long slopes, 2D analysis results may underestimate the probability of slope instability, and 3D reliability of the slope must be analyzed.

Keywords: 3D slope reliability; limit equilibrium method; slip surface normal stress correction; critical horizontal acceleration coefficient; limit state function

1. Introduction

Slope stability evaluation is one of the important and popular research topics in geotechnical engineering. Due to the existence of complex loads and geological conditions, as well as the inherent variability of soil physical and mechanical parameters, slope failure is actually uncertain [1]. The influence of these uncertain factors on slope stability cannot be considered by a single safety factor [2]. Slope reliability analysis is a useful supplement to the deterministic method of slope stability [3].

In the past few decades, the research on slope reliability analysis methods has developed rapidly and achieved many meaningful results. However, most researchers simplified the slope stability analysis problems into two-dimensional plane strain problems, ignoring the three-dimensional effects of actual slopes, which will overestimate or underestimate their stabilities [4–6]. Especially for slopes with obvious longitudinal changes in soil properties along the slope surface, concentrated loads acting on the surface, irregular potential failure surfaces, or short slopes with non-negligible boundary conditions, their 3D effects are particularly significant, and 3D reliability analysis must be carried out [7].

Three-dimensional slope reliability analysis requires higher computational costs than two-dimensional simulation, and sometimes the calculation time is even more than one month [8]. This is one of the reasons why 3D slope reliability analysis cannot be widely used in practical engineering. However, in the preliminary design of highway engineering, it is often necessary to quickly evaluate the stability of slopes of hundreds of sections [9]. In the study of 3D slope reinforcement measures, it is also necessary to quickly judge the effects of different reinforcement measures [10,11]. In these cases, there is an urgent need for a method that can quickly perform 3D slope reliability analysis without significant loss of accuracy.

check for updates

Citation: Chen, J.; Zhu, D.; Zhu, Y. A Simplified Method for Effective Calculation of 3D Slope Reliability. *Water* **2023**, *15*, 3139. https://doi.org/10.3390/w15173139

Academic Editor: Roberto Greco

Received: 18 July 2023
Revised: 25 August 2023
Accepted: 29 August 2023
Published: 1 September 2023

In recent years, the research on 3D slope reliability analysis methods has gradually increased with the improvement of computer performance. The existing 3D slope reliability analysis methods mainly include the analytical method [12,13], stochastic finite element method (SFEM) [1,14], stochastic limit equilibrium method (SLEM) [15], and stochastic limit analysis method (SLAM) [16]. The research of these methods is carried out under the framework of combining the deterministic method of 3D slope stability and the reliability calculation method. The limit state performance function generally adopts the expression form of the difference between safety factor F_s and one. The obvious differences in accuracy and computational efficiency of various methods are mainly derived from the efficiency of the solution method of safety factor and the reliability algorithm. The analytical method has a small amount of calculation, but it is difficult to apply in engineering practice because the sliding surface must be a special combination shape with a cylindrical surface in the middle and a vertical plane or an ellipsoid surface at both ends. Based on the finite element method and strength reduction technology, the stochastic finite element method has the advantage of automatically obtaining the critical sliding surface. However, when the probability of slope failure is very small, its computational cost is very high [17]. The stochastic limit analysis method constructs the static allowable stress field or the maneuvering allowable velocity field according to the large-scale geotechnical parameter samples, based on the lower bound theorem or the upper bound theorem. The safety factor of the slope is solved by means of mathematical programming. If the number of soil parameter samples is large, the calculation efficiency is low. Therefore, the stochastic limit equilibrium method with a clear concept and high computational efficiency is still the most widely used method in engineering practice.

The conventional 3D slope limit equilibrium methods are obtained by extending the 2D slice methods. According to the different assumptions of the inter-column force, they can be divided into the 3D simplified Bishop method [18,19], 3D simplified Janbu method [18,19], 3D Morgenstern–Price method [20,21], and so on. Safety factor expressions are usually complex nonlinear implicit expressions, which require multiple numerical iterations to solve, and sometimes encounter non-convergence problems. Some scholars used the stochastic response surface method (SRSM) [22], intelligent response surface method (IRSM) [23], genetic algorithm (GA) [8], support vector machine (SVM) [4], and other surrogate models to simplify complex nonlinear implicit performance functions into approximate equivalent explicit expressions, which effectively solves the difficult problem of 3D slope reliability calculation. However, these methods require a large amount of data to generate enough samples to obtain surrogate models. For highly nonlinear and small failure probability problems, it is difficult to fit the surrogate models.

Sarma [24] proposed a good idea to replace the safety factor with a critical horizontal acceleration coefficient K_c for slope stability analysis. In the process of solving K_c, no numerical iteration occurs, and no non-convergence problem will be encountered. However, the formula of the Sarma method is cumbersome and the solving process is not very convenient. Zhu [25] proposed an explicit solution for the 3D slope safety factor by correcting the normal stresses on the slip surface. His method does not need to assume the inter-column force, so it does not belong to the "column method". The calculation principle of the method is simple without iteration and non-convergence problems, and the calculation results are reliable. In this paper, a new method for calculating the reliability of the 3D slope was proposed by using the critical horizontal acceleration coefficient as the criterion of slope stability under the mode of the 3D limit equilibrium method with normal stress correction of the sliding surface.

Among the reliability algorithms, the Monte Carlo simulation (MCS) method is the most widely used because of its simple concept. As long as there are enough samples, the approximate accurate reliability calculation results can be obtained. However, in the case of slopes with small failure probability, the calculation is very time-consuming. The subset simulation method (SS) reduces the sample requirements and is an efficient method for solving the reliability analysis of slopes with small failure probability [26]. In this paper, the subset simulation method is used to calculate the reliability of the 3D slope.

In the author's previous work [27], it has been proved that the limit state performance function using the critical horizontal acceleration coefficient expression can indeed significantly improve the computational efficiency probability of 2D slope reliability. In this paper, we extend the previous work to the 3D slope and comprehensively study the accuracy and efficiency of the obtained algorithm.

The research framework of this paper is as follows. Firstly, based on the 3D limit equilibrium method with normal stress correction on the sliding surface, four main equilibrium equations including the critical horizontal acceleration coefficient and three correction parameters were established. Secondly, according to the definition of the critical horizontal acceleration coefficient, its explicit expression was derived. Thirdly, a simplified method for calculating the reliability of the three-dimensional slope is proposed by using the difference expression between the critical horizontal acceleration coefficient and the critical value and combining it with the reliability algorithm. Finally, the effectiveness and reliability of the proposed method are verified by two slope examples.

2. Three-Dimensional Limit Equilibrium Method Based on the Modification of Normal Stresses over Slip Surface

2.1. Fundamental Assumptions

A 3D slope with a general shape sliding surface and a coordinate system are shown in Figure 1a. It is assumed that the horizontal projections of the sliding direction of each point on the sliding surface are parallel to each other and opposite to the positive direction of the x-axis. The slope surface is represented by $g(x, y)$, and the sliding surface is represented by $s(x, y)$. The slope is divided into many soil columns. As shown in Figure 1b, one of the soil columns has a width of dx and a length of dy. The horizontal projection area of the soil column is dA, and the angles with the x and y axes are α_x and α_y, respectively. The angle between the outer normal direction and the z-axis is γ_z.

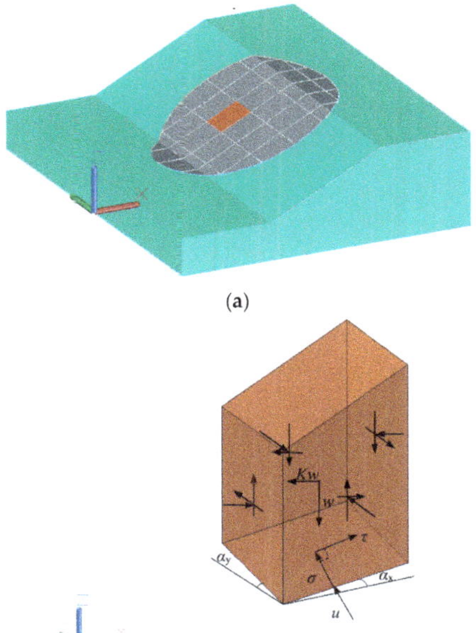

(a)

(b)

Figure 1. Three-dimensional slope, soil column, and forces. (**a**) Three-dimensional slope and sliding body. (**b**) Soil column and forces.

According to the geometric relation, dA can be calculated by Equation (1).

$$\mathrm{d}A = \Delta \mathrm{d}x\mathrm{d}y \tag{1}$$

where

$$\Delta = \sqrt{1 + s'^2_x + s'^2_y} \tag{2}$$

$$s'_x = \tan \alpha_x,\ s'_y = \tan \alpha_y \tag{3}$$

$$\cos \gamma_z = \frac{1}{\sqrt{1 + s'^2_x + s'^2_y}} = \frac{1}{\Delta} \tag{4}$$

Assuming that the soil column weight is $w(x,y)$ and the horizontal acceleration coefficient is K, then the horizontal seismic force is $Kw(x,y)$, and the position of the action point is (x, y, z_c).

There are normal stress and shear stress at the bottom of the column, which are expressed by $\sigma(x,y)$ and $\tau(x,y)$, respectively. The pore water pressure is expressed by $u(x,y)$. The cohesion and effective internal friction angle of the soil at the sliding surface are $c(x,y)$ and $\varphi(x,y)$, respectively.

The direction cosines of the normal force and shear force at the bottom of the column are $\left(n^x_\sigma, n^y_\sigma, n^z_\sigma\right)$ and $\left(n^x_\tau, n^y_\tau, n^z_\tau\right)$, respectively. As the sliding direction is parallel to the x-axis, $n^y_\tau = 0$. From the geometrical relation, it follows that

$$\left(n^x_\sigma, n^y_\sigma, n^z_\sigma\right) = \left(-\frac{s'_x}{\Delta}, -\frac{s'_y}{\Delta}, \frac{1}{\Delta}\right) \tag{5}$$

$$\left(n^x_\tau, n^y_\tau, n^z_\tau\right) = \left(\frac{1}{\Delta'}, 0, \frac{s'_x}{\Delta'}\right) \tag{6}$$

where

$$\Delta' = \sqrt{1 + s'^2_x} \tag{7}$$

In order to simplify the calculation, the initial normal stress distribution $\sigma_0(x,y)$ on the sliding surface adopts the 3D extended form of the Fellenius method [28].

$$\sigma_0(x,y) = \frac{w(x,y)}{1 + s'^2_x + s'^2_y} \tag{8}$$

The initial normal stresses over the slip surface need to be corrected to make the 3D sliding body meet the required equilibrium conditions. Let the correction function be $\xi(x,y)$, then

$$\sigma(x,y) = \sigma_0(x,y)\xi(x,y) \tag{9}$$

2.2. Three-Dimensional Limit Equilibrium Equations

For 3D slopes, only the reliability calculation results that rigorously meet the force balance in three directions and the moment balance conditions around three axes are accurate. The research shows that the quasi-rigorous solution is very close to the rigorous solution, and the difference can be ignored. The former is more suitable for practical engineering [25]. Therefore, in order to solve the problem of rapid analysis of 3D slope reliability, a quasi-rigorous simplified method of 3D slope reliability based on four main equilibrium equations of the sliding body is proposed.

When the 3D sliding body reaches the limit equilibrium state under the action of the force in three directions and the torque around the y-axis, the following equations can be listed (in order to simplify, "(x,y)" in all formulas are omitted.):

$$\iint \sigma \cdot n_\sigma^y \cdot \mathrm{d}A = 0 \tag{10a}$$

$$\iint (\sigma \cdot n_\sigma^x + \tau \cdot n_\tau^x)\, \mathrm{d}A = K \iint w \mathrm{d}x\mathrm{d}y \tag{10b}$$

$$\iint (\sigma \cdot n_\sigma^z + \tau \cdot n_\tau^z)\, \mathrm{d}A = \iint w \mathrm{d}x\mathrm{d}y \tag{10c}$$

$$\iint (\sigma \cdot n_\sigma^z + \tau \cdot n_\tau^z) \cdot x \cdot \mathrm{d}A - \iint (\sigma \cdot n_\sigma^x + \tau \cdot n_\tau^x) \cdot s \cdot \mathrm{d}A = \iint w \cdot x \cdot \mathrm{d}x\mathrm{d}y - K \iint w \cdot z_c \cdot \mathrm{d}x\mathrm{d}y \tag{10d}$$

Substituting Equations (7) and (8) into Equation (10a–d) leads to

$$\iint \sigma \cdot s'_y \cdot \mathrm{d}x\mathrm{d}y = 0 \tag{11a}$$

$$\iint \left(-\sigma \cdot s'_x + \tau \cdot \frac{\Delta}{\Delta'}\right) \mathrm{d}x\mathrm{d}y = K \iint w \mathrm{d}x\mathrm{d}y \tag{11b}$$

$$\iint \left(\sigma + s'_x \cdot \frac{\Delta}{\Delta'} \cdot \tau\right) \mathrm{d}x\mathrm{d}y = \iint w \mathrm{d}x\mathrm{d}y \tag{11c}$$

$$\iint \left(\sigma + s'_x \cdot \frac{\Delta}{\Delta'} \cdot \tau\right) \cdot x \cdot \mathrm{d}x\mathrm{d}y - \iint \left(-\sigma \cdot s'_x + \tau \cdot \frac{\Delta}{\Delta'}\right) \cdot s \cdot \mathrm{d}x\mathrm{d}y = \iint w \cdot x \cdot \mathrm{d}x\mathrm{d}y - K \iint w \cdot z_c \cdot \mathrm{d}x\mathrm{d}y \tag{11d}$$

It is assumed that safety factor values of the entire sliding surface are equal, and the relationship between normal stress and shear stress obeys the Mohr–Coulomb failure criterion.

$$\tau = \frac{1}{F_s}[(\sigma - u)\tan\varphi - c] \tag{12}$$

Let

$$\psi = \tan\varphi \tag{13a}$$

$$c_u = c - u \cdot \tan\varphi \tag{13b}$$

$$\rho = \frac{\Delta}{\Delta'} \tag{13c}$$

Equation (11a–d) is rearranged as

$$\iint \sigma \cdot s'_y \cdot \mathrm{d}x\mathrm{d}y = 0 \tag{14a}$$

$$\iint \left(-s'_x + \frac{\rho\psi}{F_s}\right) \cdot \sigma \cdot \mathrm{d}x\mathrm{d}y = K \iint w \mathrm{d}x\mathrm{d}y - \frac{1}{F_s} \iint \rho \cdot c_u \mathrm{d}x\mathrm{d}y \tag{14b}$$

$$\iint \left(1 + \frac{\rho\psi s'_x}{F_s}\right) \cdot \sigma \cdot \mathrm{d}x\mathrm{d}y = \iint w \mathrm{d}x\mathrm{d}y - \frac{1}{F_s} \iint s'_x \cdot \rho \cdot c_u \mathrm{d}x\mathrm{d}y \tag{14c}$$

$$\iint \left[(x + s \cdot s'_x) + \frac{\rho\psi(x \cdot s'_x - s)}{F_s}\right] \cdot \sigma \cdot \mathrm{d}x\mathrm{d}y = \iint w \cdot x \cdot \mathrm{d}x\mathrm{d}y - K \iint w \cdot z_c \cdot \mathrm{d}x\mathrm{d}y - \frac{1}{F_s} \iint \rho (x \cdot s'_x - s) \cdot c_u \cdot \mathrm{d}x\mathrm{d}y \tag{14d}$$

In order to make Equation (14a–d) statically determinate and solvable, the correction function must contain three undetermined parameters and its general form can be taken as

$$\xi(x,y) = \xi_0 + \lambda_1\xi_1 + \lambda_2\xi_2 + \lambda_3\xi_3 \tag{15}$$

Substituting Equations (12) and (15) into Equation (14a–d), we can obtain

$$\lambda_1 D_1 + \lambda_2 D_2 + \lambda_3 D_3 = E \tag{16a}$$

$$\lambda_1\left(A_{11} + \frac{B_{11}}{F_s}\right) + \lambda_2\left(A_{12} + \frac{B_{12}}{F_s}\right) + \lambda_3\left(A_{13} + \frac{B_{13}}{F_s}\right) + KG_1 = A_{14} + \frac{B_{14}}{F_s} \tag{16b}$$

$$\lambda_1\left(A_{21} + \frac{B_{21}}{F_s}\right) + \lambda_2\left(A_{22} + \frac{B_{22}}{F_s}\right) + \lambda_3\left(A_{23} + \frac{B_{23}}{F_s}\right) + KG_2 = A_{24} + \frac{B_{24}}{F_s} \tag{16c}$$

$$\lambda_1\left(A_{31} + \frac{B_{31}}{F_s}\right) + \lambda_2\left(A_{32} + \frac{B_{32}}{F_s}\right) + \lambda_3\left(A_{33} + \frac{B_{33}}{F_s}\right) + KG_3 = A_{34} + \frac{B_{34}}{F_s} \tag{16d}$$

where

$$\begin{array}{ll}
D_i = \iint \sigma_0\xi_i s'_y \mathrm{d}x\mathrm{d}y (i = 1,2,3) & E = -\iint \sigma_0\xi_0 s'_y \mathrm{d}x\mathrm{d}y \\
A_{1i} = -\iint \sigma_0\xi_i s'_x \mathrm{d}x\mathrm{d}y (i = 1,2,3) & A_{14} = \iint \sigma_0\xi_0 s'_x \mathrm{d}x\mathrm{d}y \\
A_{2i} = \iint \sigma_0\xi_i \mathrm{d}x\mathrm{d}y (i = 1,2,3) & A_{24} = \iint w\mathrm{d}x\mathrm{d}y - \iint \sigma_0\xi_0 \mathrm{d}x\mathrm{d}y \\
A_{3i} = \iint \sigma_0\xi_i (x + s\cdot s'_x)\mathrm{d}x\mathrm{d}y (i = 1,2,3) & A_{34} = \iint [w\cdot x - \sigma_0\xi_0(x + s\cdot s'_x)]\, \mathrm{d}x\mathrm{d}y \\
B_{1i} = \iint \rho\psi\sigma_0\xi_i \mathrm{d}x\mathrm{d}y (i = 1,2,3) & B_{14} = -\iint \rho(c_u + \psi\sigma_0\xi_0)\, \mathrm{d}x\mathrm{d}y \\
B_{2i} = \iint \rho\psi\sigma_0\xi_i s'_x \mathrm{d}x\mathrm{d}y (i = 1,2,3) & B_{24} = -\iint \rho(c_u + \psi\sigma_0\xi_0)s'_x \mathrm{d}x\mathrm{d}y \\
B_{3i} = \iint \rho\psi\sigma_0\xi_i (x\cdot s'_x - s)\mathrm{d}x\mathrm{d}y (i = 1,2,3) & B_{34} = -\iint \rho(x\cdot s'_x - s)(c_u + \psi\sigma_0\xi_0)\, \mathrm{d}x\mathrm{d}y \\
G_1 = -\iint w\mathrm{d}x\mathrm{d}y & G_2 = 0 \\
G_3 = -\iint w\cdot z_c \mathrm{d}x\mathrm{d}y &
\end{array}$$

Assuming that safety factor F_s is a known parameter and the initial normal stress distribution σ_0 is determined, then $D_1 \sim D_3$, $A_{11} \sim A_{14}$, $A_{21} \sim A_{24}$, $A_{31} \sim A_{34}$, $B_{11} \sim B_{14}$, $B_{21} \sim B_{24}$, $B_{31} \sim B_{34}$, $G_1 \sim G_3$, and E, these parameters can be solved by a definite integral. Substituting the 31 parameters into Equation (16a–d), we can obtain the statically determinate solvable limit equilibrium equations with four unknowns λ_1, λ_2, λ_3 and K.

Equation (16a–d) can be rewritten in a matrix form

$$\begin{bmatrix} D_1 & D_2 & D_3 & 0 \\ A_{11} + \frac{B_{11}}{F_s} & A_{12} + \frac{B_{12}}{F_s} & A_{13} + \frac{B_{13}}{F_s} & G_1 \\ A_{21} + \frac{B_{21}}{F_s} & A_{22} + \frac{B_{22}}{F_s} & A_{23} + \frac{B_{23}}{F_s} & G_2 \\ A_{31} + \frac{B_{31}}{F_s} & A_{32} + \frac{B_{32}}{F_s} & A_{33} + \frac{B_{33}}{F_s} & G_3 \end{bmatrix} \begin{pmatrix} \lambda_1 \\ \lambda_2 \\ \lambda_3 \\ K \end{pmatrix} = \begin{pmatrix} E \\ A_{14} + \frac{B_{14}}{F_s} \\ A_{24} + \frac{B_{24}}{F_s} \\ A_{34} + \frac{B_{34}}{F_s} \end{pmatrix} \tag{17}$$

According to Cramer's rule, λ_1, λ_2, λ_3 and K can be solved.

$$\lambda_1 = \frac{\Delta_1}{\Delta_0} \tag{18a}$$

$$\lambda_2 = \frac{\Delta_2}{\Delta_0} \tag{18b}$$

$$\lambda_3 = \frac{\Delta_3}{\Delta_0} \tag{18c}$$

$$K = \frac{\Delta_4}{\Delta_0} \tag{18d}$$

where

$$\Delta_0 = \begin{vmatrix} D_1 & D_2 & D_3 & 0 \\ A_{11} + \frac{B_{11}}{F_s} & A_{12} + \frac{B_{12}}{F_s} & A_{13} + \frac{B_{13}}{F_s} & G_1 \\ A_{21} + \frac{B_{21}}{F_s} & A_{22} + \frac{B_{22}}{F_s} & A_{23} + \frac{B_{23}}{F_s} & G_2 \\ A_{31} + \frac{B_{31}}{F_s} & A_{32} + \frac{B_{32}}{F_s} & A_{33} + \frac{B_{33}}{F_s} & G_3 \end{vmatrix} \tag{19}$$

$$\Delta_1 = \begin{vmatrix} E & D_2 & D_3 & 0 \\ A_{14} + \frac{B_{14}}{F_s} & A_{12} + \frac{B_{12}}{F_s} & A_{13} + \frac{B_{13}}{F_s} & G_1 \\ A_{24} + \frac{B_{24}}{F_s} & A_{22} + \frac{B_{22}}{F_s} & A_{23} + \frac{B_{23}}{F_s} & G_2 \\ A_{34} + \frac{B_{34}}{F_s} & A_{32} + \frac{B_{32}}{F_s} & A_{33} + \frac{B_{33}}{F_s} & G_3 \end{vmatrix} \quad (20)$$

$$\Delta_2 = \begin{vmatrix} D_1 & E & D_3 & 0 \\ A_{11} + \frac{B_{11}}{F_s} & A_{14} + \frac{B_{14}}{F_s} & A_{13} + \frac{B_{13}}{F_s} & G_1 \\ A_{21} + \frac{B_{21}}{F_s} & A_{24} + \frac{B_{24}}{F_s} & A_{23} + \frac{B_{23}}{F_s} & G_2 \\ A_{31} + \frac{B_{31}}{F_s} & A_{34} + \frac{B_{34}}{F_s} & A_{33} + \frac{B_{33}}{F_s} & G_3 \end{vmatrix} \quad (21)$$

$$\Delta_3 = \begin{vmatrix} D_1 & D_2 & E & 0 \\ A_{11} + \frac{B_{11}}{F_s} & A_{12} + \frac{B_{12}}{F_s} & A_{14} + \frac{B_{14}}{F_s} & G_1 \\ A_{21} + \frac{B_{21}}{F_s} & A_{22} + \frac{B_{22}}{F_s} & A_{24} + \frac{B_{24}}{F_s} & G_2 \\ A_{31} + \frac{B_{31}}{F_s} & A_{32} + \frac{B_{32}}{F_s} & A_{34} + \frac{B_{34}}{F_s} & G_3 \end{vmatrix} \quad (22)$$

$$\Delta_4 = \begin{vmatrix} D_1 & D_2 & D_3 & E \\ A_{11} + \frac{B_{11}}{F_s} & A_{12} + \frac{B_{12}}{F_s} & A_{13} + \frac{B_{13}}{F_s} & A_{14} + \frac{B_{14}}{F_s} \\ A_{21} + \frac{B_{21}}{F_s} & A_{22} + \frac{B_{22}}{F_s} & A_{23} + \frac{B_{23}}{F_s} & A_{24} + \frac{B_{24}}{F_s} \\ A_{31} + \frac{B_{31}}{F_s} & A_{32} + \frac{B_{32}}{F_s} & A_{33} + \frac{B_{33}}{F_s} & A_{34} + \frac{B_{34}}{F_s} \end{vmatrix} \quad (23)$$

3. Simplified Method of 3D Slope Reliability Calculation

3.1. Critical Horizontal Acceleration Coefficient K_c

Sarma [24] defined a parameter called the critical horizontal acceleration coefficient K_c. When the horizontal seismic force is applied to make the sliding body reach the critical equilibrium state (that is, safety factor F_s is equal to one), the ratio of the maximum horizontal seismic force to the gravity can be expressed by the critical horizontal acceleration coefficient K_c. Later, Sarma [29] further elaborated on the relationship between safety factor F_s and critical horizontal acceleration coefficient K_c. Safety factor F_s refers to the coefficient of available shear strength and movable shear stress when the sliding body enters the equilibrium state. It can be seen from the definition of the two parameters that the effects of F_s and K_c are equivalent. K_c can also be used as an index to measure safety factor F_s.

According to the above ideas, $F_s = 1$ can be substituted into Equations (18d), (19), and (23) to obtain K_c.

$$K_c = \frac{\begin{vmatrix} D_1 & D_2 & D_3 & E \\ A_{11} + B_{11} & A_{12} + B_{12} & A_{13} + B_{13} & A_{14} + B_{14} \\ A_{21} + B_{21} & A_{22} + B_{22} & A_{23} + B_{23} & A_{24} + B_{24} \\ A_{31} + B_{31} & A_{32} + B_{32} & A_{33} + B_{33} & A_{34} + B_{34} \end{vmatrix}}{\begin{vmatrix} D_1 & D_2 & D_3 & 0 \\ A_{11} + B_{11} & A_{12} + B_{12} & A_{13} + B_{13} & G_1 \\ A_{21} + B_{21} & A_{22} + B_{22} & A_{23} + B_{23} & G_2 \\ A_{31} + B_{31} & A_{32} + B_{32} & A_{33} + B_{33} & G_3 \end{vmatrix}} \quad (24)$$

3.2. Limit State Performance Function

The seismic force acting on the sliding body is expressed by the horizontal acceleration coefficient K. Sarma found that the functional relationship between K and safety factor F_s is monotonically decreasing when the sliding body is in critical equilibrium, as shown in Figure 2. As F_s decreases, K increases. When $F_s = 1$, $K = K_c$. K_c is greater than zero. While K equals zero, F_s is greater than one. According to the above analysis, when the slope is stable, F_s is greater than one, or K_c is greater than zero.

Similar to the safety factor having a boundary value of one, K_c also has a boundary value. The critical value of K_c is represented by the known horizontal seismic coefficient K_{c0} that actually acts on the slope. There is no seismic force on the slope, $K_{c0} = 0$. With seismic force, K_{c0} is a known value greater than zero. In summary, the limit state performance function expression can be written as the difference between K_c and K_{c0} as shown in Equation (25). The determination condition of slope stability can be changed from conventional $F_s > 1$ to $K_c > K_{c0}$ [30].

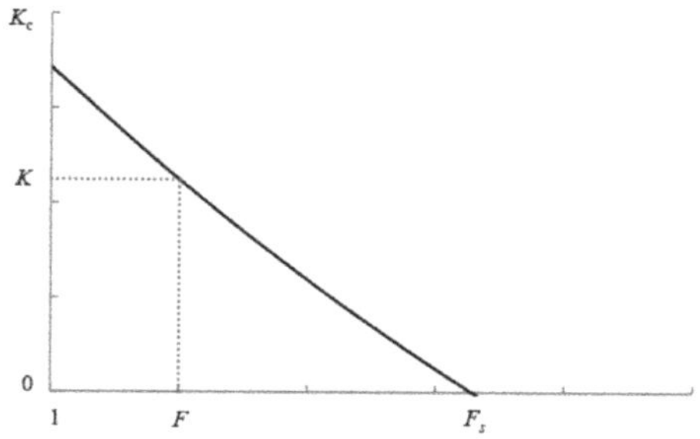

Figure 2. The relation curve of F_s and K.

In addition, it can be found from the expression composition of Equation (24) that K_c is an explicit function of soil strength parameters c and φ. The calculation method is simple and easy to understand and does not need iterative calculation, which will not lead to non-convergence problems. Therefore, using Equation (25) as the limit state performance function is bound to significantly improve the calculation efficiency of 3D slope reliability analysis.

$$Z = K_c - K_{c0} \tag{25}$$

3.3. Three-Dimensional Slope Reliability Calculation Method

According to the limit state performance function defined in Equation (25), the slope failure probability can be expressed as

$$P_F = P(K_c(x) < K_{c0}) \tag{26}$$

where $\{x : X_1, X_2, \cdots X_n\}$ denotes a random variable.

From Equation (24), K_c can be solved directly without iteration. When the distribution type, mean, and variance of random variables are known, the first-order second-moment method (FOSM) can be used to solve the approximate analytical solution of the reliability index [13,31].

$$\beta = \frac{\mu_Z}{\sigma_Z} = \frac{(K_c)_{X_i^*} - K_{c0}}{\sqrt{\sum_{i=1}^{n}\left(\frac{\partial Z}{\partial X_i}\sigma_{X_i}\right)^2}} \tag{27}$$

$$P_F = \Phi(-\beta) \tag{28}$$

The partial derivatives of the limit state performance function in Equation (27) can be approximated by the difference method in Equation (29).

$$\frac{\partial Z}{\partial X_i} = \frac{Z_i^+ - Z_i^-}{2\sigma_{X_i}} \tag{29}$$

where $Z_i^+ = Z(X_1^*, X_2^*, \cdots, X_i^* + \sigma_{X_i}, \cdots, X_n^*)$, $Z_i^- = Z(X_1^*, X_2^*, \cdots, X_i^* - \sigma_{X_i}, \cdots, X_n^*)$, X_i^* denotes the mean value of the i-th random variable.

The MCS method is recognized as a high-precision method in reliability algorithms. It is based on the law of large numbers, using the frequency of failure events in a large number of samples to approximate the failure probability. Only when the sample size N is large enough, the unbiased estimation of slope failure probability can be obtained.

However, computational inefficiency is the main drawback of the MCS method, especially for small failure probability problems.

$$\widehat{P}_F = \frac{1}{N}\sum_{i=1}^{N} I_F(x) \tag{30}$$

where $I_F(x)$ is the indicator function.

$$I_F(x) = \begin{cases} 1, K_c(x) < 0 \\ 0, K_c(x) \geq 0 \end{cases} \tag{31}$$

Au [26] proposed an improved MCS method called the subset simulation (SS) method, which can effectively improve the computational efficiency of small failure probability problems. The SS method divides the probability space into multiple sequence subsets by introducing intermediate failure events. The small failure probability is calculated by the product of a series of conditional probabilities.

Let $F = \{K_c(x) < K_{c0}\}$ be the target event of 3D slope reliability analysis, and the intermediate subset is $F_i = \{K_c(x) < K_{ci}, i = 1, 2, \cdots, m\}$, which conforms to the following relationship: $F_1 \supset F_2 \supset \cdots \supset F_m = F$ and $K_{c1} > K_{c2} > \cdots > K_{cm-1} > K_{cm} = K_{c0}$. That is,

$$F_k = \cap_{i=1}^{k} F_i (k = 1, 2, \cdots, m) \tag{32}$$

$$P_F = P(F) = P(F_m) = P(F_1)\prod_{i=2}^{m} P(F_i|F_{i-1}) \tag{33}$$

Let $\widehat{P}_1 = P(F_1)$, $\widehat{P}_i = P(F_i|F_{i-1}) = P(K_c(x) < K_{ci}|K_c(x) < K_{ci-1})(i = 1, 2, \cdots, m)$, then the unbiased estimate of Equation (34) is

$$\widehat{P}_F = \prod_{i=1}^{m} \widehat{P}_i \tag{34}$$

On the basis of the method in this paper, the specific steps of calculating the failure probability and reliability index of the slope by the subset simulation method are shown in Chen [27], which will not be repeated here.

4. Numerical Examples

By analyzing the calculation results of the 3D slope reliability of two classical examples, the feasibility of the proposed method is verified. The limit state performance function adopts the proposed K_c expression (denoted as K_c method) and F_s expression (denoted as F_s method, the specific calculation method is shown in Zhu [25]). The FOSM approximate analytical method, MCS, and SS methods were used to calculate the reliability. By coupling two limit state performance functions and three reliability algorithms, six methods are obtained, which are called the K_c + MCS method, K_c + SS method, F_s + MCS method, F_s + SS method, K_c + FOSM method, and F_s + FOSM method, respectively. The reliability index, failure probability, and calculation time of the six methods were compared. The calculation time is the CPU occupancy time of the program running on the same computer.

4.1. Example 1

Example 1 is derived from the homogeneous slope in Arai [32]. The geometric dimensions of the slope section are shown in Figure 3, and the soil parameters are listed in Table 1. The cohesion and internal friction angle of the soil are independent of each other and obey the normal distribution. Considering the two cases with and without seismic force on the slope, K_{c0} is 0 and 0.1, respectively. Both sliding surfaces are spherical, and the position and range are shown in Figure 4. The equations are $(x - 26.22)^2 + y^2 + (z - 46.86)^2 = 34.48^2$ (denoted by sliding surface 1, abbreviated as S1) and $(x - 29.529)^2 + y^2 + (z - 49.306)^2 = 35.667^2$ (denoted by sliding surface 2, abbreviated as S2). The results of the six methods are shown in Table 2.

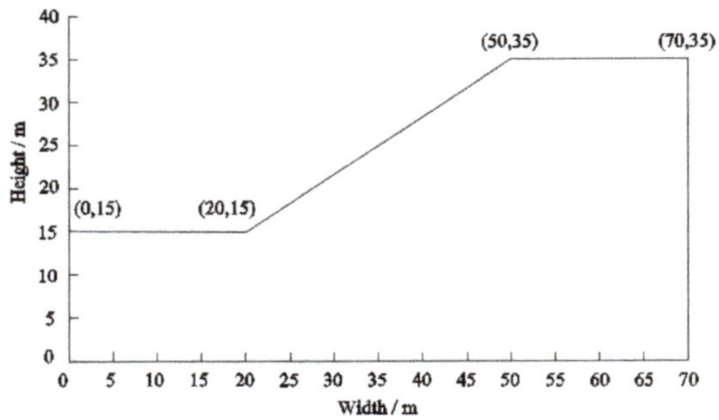

Figure 3. Cross-section of homogeneous slope in Example 1.

Table 1. Material parameters of the slope in Example 1.

Cohesion c (kPa)		Friction Angle φ (°)		Unit Weight γ (kN/m^3)
μ_c	σ_c	μ_ϕ	σ_ϕ	
41.65	8.00	15.00	3.00	18.82

Table 2. Comparison of reliability analysis results of Example 1.

Performance Functions	Reliability Methods	Horizontal Seismic Coefficient K_{c0}	Sampling Number N	β	P_f	Slip Surface Shape	Computation Time t (s)
F_s method	MCS	0	40,000	2.6376	0.0042	3D S1	21.70
K_c method	MCS	0	40,000	2.6606	0.0039	3D S1	12.45
F_s method	SS	0	40,000	2.6780	0.0037	3D S1	6.80
K_c method	SS	0	40,000	2.6438	0.0041	3D S1	5.45
F_s method	FOSM	0		2.6596	0.0039	3D S1	0.30
K_c method	FOSM	0		2.6937	0.0035	3D S1	0.13
Bishop	GA + FORM	0		2.6100	0.0045 [8]	3D S1	>3.8 h
Bishop (GeoStudio)	MCS	0	40,000	2.2422	0.0120	2D circular S1	
K_c method	MCS	0.1	40,000	1.8317	0.0335	3D S2	12.98
K_c method	SS	0.1	40,000	1.8633	0.0312	3D S2	5.33
K_c method	FOSM	0.1		1.8432	0.0326	3D S2	0.14
Bishop (GeoStudio)	MCS	0.1	40,000	1.1870	0.1176	2D circular S2	

MCS—Monte Carlo simulation; SS—subset simulation; GA—genetic algorithm; FORM—first-order reliability method; FOSM—first-order second-moment method.

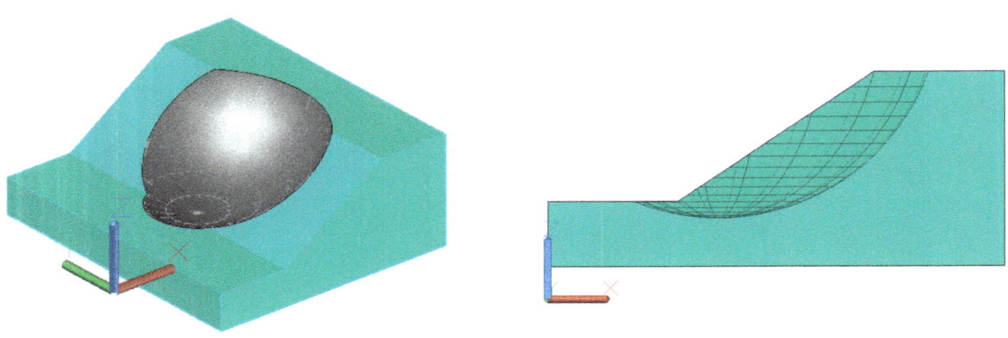

(a)

Figure 4. *Cont.*

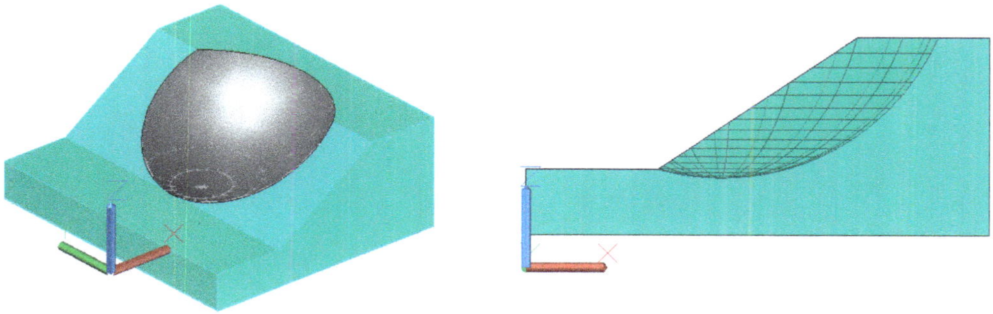

(**b**)

Figure 4. The sliding surface position and range of Example 1. (**a**) Three-dimensional slip surface at $K_{c0} = 0$. (**b**) Three-dimensional slip surface at $K_{c0} = 0.1$.

In the case of no seismic load on the slope (i.e., K_{c0} = 0.1) and the sliding surface is S1, the calculation results in Tun [8] are taken as the reference values. Among the six methods, the relative errors of the reliability index and failure probability of F_s + MCS method are the smallest, while that of the K_c + FOSM method is the largest. The relative error of the K_c + SS method is the second smallest. Except that the relative error of the failure probability of the K_c + FOSM method reaches 22.22%, those of other results are less than 20%. However, the computational efficiency of the six methods is very different. The calculation time of the F_s + MCS method is the longest, but it is much less than that in Tun [8]. Both the K_c + FOSM and F_s + FOSM methods have very short CPU time, even less than 0.5 s. The calculation time of the K_c + SS method is slightly less than that of the F_s + SS method, which is about half of the K_c + MCS method and one-fourth of the F_s + MCS method.

When there is a seismic load on the slope (i.e., K_{c0} = 0.1) and the slip surface is S2, the failure probability of the 3D slope increases and the reliability index decreases. The results of K_c + MCS, K_c − SS, and K_c + FOSM are very close. The maximum relative errors of the reliability index and failure probability are 3.80% and 7.37%, respectively. In Example 1, the reliability indexes of the 3D slope are slightly higher than that of the 2D slope.

From the above, it can be concluded that the K_c performance function can improve the efficiency of the 3D slope reliability analysis without an obvious loss of precision. The method coupled with the SS method can further reduce the amount of calculation.

4.2. Example 2

The second example is a slope with three soil layers, which is adapted from a slope stability test question of the Australian Computer Aided Design Association (ASCD) [33]. The distribution of the soil layer in the slope section is shown in Figure 5. The material parameter variables are normally distributed and are listed in Table 3.

In order to compare with the 2D results, several 3D slope reliability analyses with different lengths are carried out. Firstly, a 3D slope with a spherical sliding surface of 40 m long is designed. Then, keeping the shape of the cross-section of the slope unchanged, the length of the slope is expanded by an integer multiple (the slope length expansion multiple is denoted by η). The sliding surfaces of long slopes are all ellipsoids, and their longitudinal axis length increases with the slope length. The positions and ranges of two kinds of sliding surfaces are shown in Figure 6. Table 4 lists the results of the 3D slope reliability calculation with different methods under the condition of two kinds of sliding surfaces. The curves of the reliability index and failure probability with slope length are shown in Figure 7.

It can be seen from Table 4 that the failure probability of F_s + MCS, K_c + MCS, F_s + SS, and K_c + SS is basically the same as that of Tun [8] in the case of the 3D spherical slip surface, with the maximum relative error of reliability index at less than 5%. The calculated results of F_s + FOSM and K_c + FOSM are quite different from those of Tun [8], with the maximum relative error of failure probability at more than 50%. In terms of calculation efficiency, the calculation time of the six methods is still very different, but they are far less than that of Tun [8]. Among these methods, the K_c + SS method is the one with good calculation accuracy and efficiency.

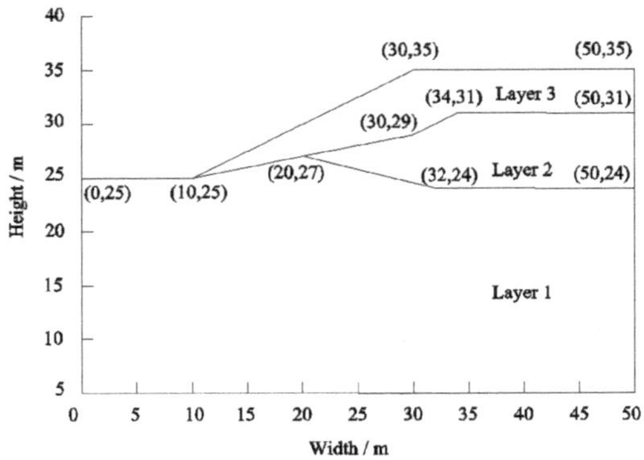

Figure 5. Cross-section of three-layer slope in Example 2.

Table 3. Material parameters of the slope in Example 2.

Soil Layer	Cohesion c (kPa)		Friction Angle φ (°)		Unit Weight γ (kN/m^3)
	μ_c	σ_c	μ_φ	σ_φ	
Layer 1	0	0	38	0.1	19.5
Layer 2	5.3	0.53	23	4.6	19.5
Layer 3	7.2	1.44	20	4.0	19.5

Table 4. Results of reliability analysis of Example 2 by different methods.

Performance Functions	Reliability Methods	Sampling Number N	β	P_f	Slip Surface Shape	Computation Time t (s)
F_s method	MCS	100,000	2.8576	0.0021	3D sphere	58.98
K_c method	MCS	100,000	2.8894	0.0019	3D sphere	31.48
F_s method	SS	100,000	2.8682	0.0021	3D sphere	18.23
K_c method	SS	100,000	2.8707	0.0020	3D sphere	16.89
F_s method	FOSM		2.7411	0.0031	3D sphere	0.80
K_c method	FOSM		2.7152	0.0033	3D sphere	0.44
Bishop	GA + FORM		2.8900	0.0019 [8]	3D sphere	~7 h
K_c method	MCS	100,000	2.5414	0.0055	3D ellipsoid (η = 7)	163.59
K_c method	SS	100,000	2.5360	0.0056	3D ellipsoid (η = 7)	56.92
K_c method	FOSM		2.4290	0.0076	3D ellipsoid (η = 7)	0.61
Bishop (GeoStudio)	MCS	100,000	2.5465	0.0054	2D circular	

MCS—Monte Carlo simulation; SS—subset simulation; GA—genetic algorithm; FORM—first-order reliability method; FOSM—first-order second-moment method.

When the slope length is increased by seven times (η = 7), the failure probabilities of the K_c + MCS and K_c + SS methods are very close and slightly larger than those of the 2D results. It can be seen that the results of the 2D slope reliability analysis are not always conservative.

Figure 7 shows the trend of increasing failure probability and decreasing reliability index with increasing slope length. When η = 6, the reliability results of the 3D slope are roughly equal to those of the 2D slope. If η is greater than 6, the failure probability of the 3D slope will exceed that of the 2D slope. For the case of long slopes, only the 2D reliability analysis does not necessarily provide a conservative result. It is consistent with the conclusions of Xiao [1] and Qi [34]. It is very necessary to analyze the 3D reliability of the actual long slope.

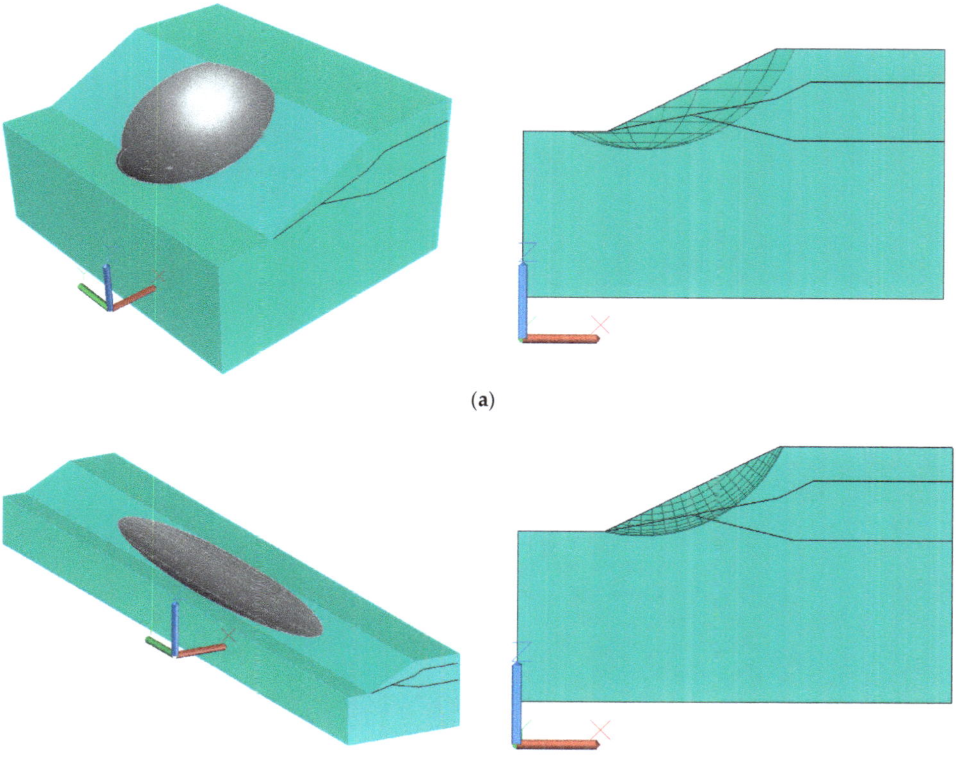

Figure 6. The slip surface position and range of the three-layer slope in Example 2. (**a**) Three-dimensional sphere slip surface. (**b**) Three-dimensional ellipsoid slip surface.

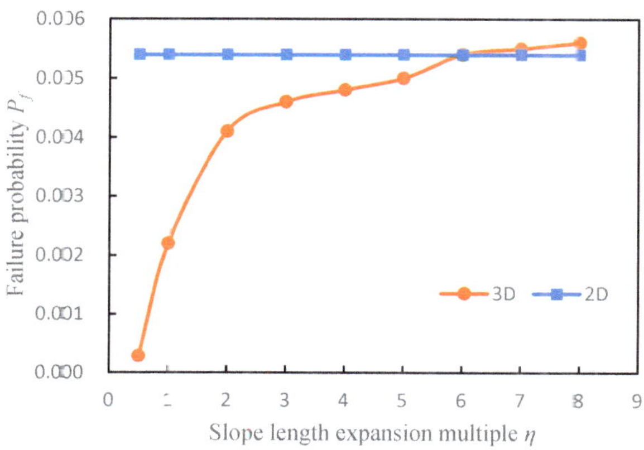

(**a**)

Figure 7. *Cont.*

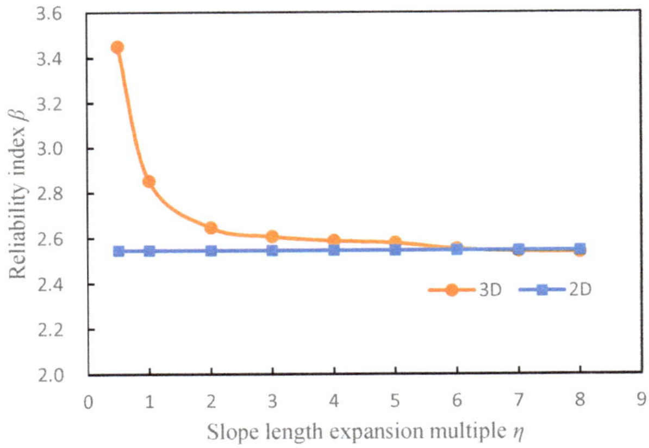

(**b**)

Figure 7. Relationship curves between reliability results and slope length. (**a**) The curve of failure probability with slope length. (**b**) The curve of reliability index with slope length.

5. Conclusions

The traditional 3D slope reliability analysis method has low computational efficiency. The reason is that the limit state function is generally expressed by the safety factor, which needs to be solved iteratively. Under the framework of the limit equilibrium method of the 3D slip surface normal stress correction, the critical horizontal acceleration coefficient K_c is used as an alternative to the safety factor F_s to measure the stability of the slope. Coupled with the reliability algorithm, a simplified method for calculating the reliability of the 3D slope is proposed.

By studying two 3D slope examples, the following conclusions can be drawn:

(1) This method has the advantages of simple calculation, no iterative convergence problem, and high calculation efficiency. Combined with the SS method, it can fully reflect the advantages of high accuracy and efficiency.

(2) By changing K_{c0} to a value greater than zero, this method can conveniently calculate 3D slope reliability under seismic loads without large-scale modification of the calculation program.

(3) In the case of a long slope, the results of 2D reliability calculation do not necessarily underestimate the stability of the slope, so it is necessary to carry out 3D slope reliability analysis.

This method only considers the force balance in three directions and moment balance in the y-axis direction and does not belong to the rigorous 3D limit equilibrium method. However, the quasi-rigorous solution is very close to the rigorous solution, and the difference can be ignored. At present, the spatial variability of soil parameters and the influence of groundwater on the reliability of 3D slopes are not considered in this method. Especially, the variability of soil parameters in the direction of slope length will make the results of 3D slope reliability analysis significantly different from those of the 2D slope. That will be the focus of the next research work on this method. In addition, on the basis of this method and a large number of practical 3D slope reliability studies, the design charts can be made for designers to use conveniently, which can improve the practical application value of this method.

Author Contributions: Conceptualization, J.C. and D.Z.; methodology, J.C. and D.Z.; software, J.C.; validation, D.Z. and Y.Z.; formal analysis, J.C., D.Z. and Y.Z.; investigation, J.C.; data curation, J.C.; writing—original draft preparation, J.C.; writing—review and editing, J.C., D.Z. and Y.Z.; visualization, J.C. All authors have read and agreed to the published version of the manuscript.

Funding: This work was funded by the National Natural Science Foundation of China (NSFC Grant No. 52079121).

Data Availability Statement: The data supporting the findings of this study are not publicly available due to privacy.

Conflicts of Interest: The authors declare no conflict of interest.

References

1. Xiao, T.; Li, D.-Q.; Cao, Z.-J.; Au, S.-K.; Phoon, K.-K. Three-dimensional slope reliability and risk assessment using auxiliary random finite element method. *Comput. Geotech.* **2016**, *79*, 146–158. [CrossRef]
2. Varkey, D.; Hicks, M.; Vardon, P. An improved semi-analytical method for 3D slope reliability assessments. *Comput. Geotech.* **2019**, *111*, 181–190. [CrossRef]
3. Duncan, J.M. Factors of Safety and Reliability in Geotechnical Engineering. *J. Geotech. Geoenviron. Eng.* **2000**, *126*, 307–316. [CrossRef]
4. Chen, C.-F.; Xiao, Z.-Y.; Zhang, G.-B. Time-variant reliability analysis of three-dimensional slopes based on Support Vector Machine method. *J. Central South Univ. Technol.* **2011**, *18*, 2108–2114. [CrossRef]
5. Al-Jeznawi, D.; Alzabeebee, S.; Shafiqu, Q.S.M.; Güler, E. Analysis of Slope Stabilized with Piles Under Earthquake Excitation. *Transp. Infrastruct. Geotechnol.* **2022**, 1–19. [CrossRef]
6. Tolun, M.; Un, B.; Emirler, B.; Yildiz, A. Stability analyses of a slope reinforced with piles subjected to static and dynamic loading conditions. *El-Cezerî J. Sci. Eng.* **2021**, *8*, 1360–1371. [CrossRef]
7. Auvinet, G.; González, J. Three-dimensional reliability analysis of earth slopes. *Comput. Geotech.* **2000**, *26*, 247–261. [CrossRef]
8. Tun, Y.W.; Llano-Serna, M.A.; Pedroso, D.M.; Scheuermann, A. Multimodal reliability analysis of 3D slopes with a genetic algorithm. *Acta Geotech.* **2019**, *14*, 207–223. [CrossRef]
9. Cheng, Y.M.; Li, L.; Liu, L.L. Simplified approach for locating the critical probabilistic slip surface in limit equilibrium analysis. *Nat. Hazards Earth Syst. Sci.* **2015**, *15*, 2241–2256. [CrossRef]
10. Xiao, S.; Zeng, J.; Yan, Y. A rational layout of double-row stabilizing piles for large-scale landslide control. *Bull. Eng. Geol. Environ.* **2017**, *76*, 309–321. [CrossRef]
11. Tang, H.; Hu, X.; Xu, C.; Li, C.; Yong, R.; Wang, L. A novel approach for determining landslide pushing force based on landslide-pile interactions. *Eng. Geol.* **2014**, *182*, 15–24. [CrossRef]
12. Vanmarcke, E.H. Reliability of Earth Slopes. *J. Geotech. Eng. Div.* **1977**, *103*, 1247–1265. [CrossRef]
13. Zhang, W.-S.; Luo, Q.; Jiang, L.-W.; Wang, T.-F.; Tang, C.; Li, Z.-T. Improved Vanmarcke analytical model for 3D slope reliability analysis. *Comput. Geotech.* **2021**, *134*, 104106. [CrossRef]
14. Hicks, M.; Spencer, W. Influence of heterogeneity on the reliability and failure of a long 3D slope. *Comput. Geotech.* **2010**, *37*, 948–955. [CrossRef]
15. Xu, Z.-X.; Zhou, X.-P. Three-dimensional reliability analysis of seismic slopes using the copula-based sampling method. *Eng. Geol.* **2018**, *242*, 81–91. [CrossRef]
16. Shu, S.; Ge, B.; Wu, Y.; Zhang, F. Probabilistic Assessment on 3D Stability and Failure Mechanism of Undrained Slopes Based on the Kinematic Approach of Limit Analysis. *Int. J. Geomech.* **2023**, *23*, 06022037. [CrossRef]
17. Cho, S. Probabilistic assessment of slope stability that considers the spatial variability of soil properties. *J. Geotech. Geoenviron. Eng.* **2010**, *136*, 975–984. [CrossRef]
18. Hungr, O.; Salgado, F.M.; Byrne, P.M. Evaluation of a three-dimensional method of slope stability analysis. *Can. Geotech. J.* **1989**, *26*, 679–686. [CrossRef]
19. Huang, C.-C.; Tsai, C.-C.; Chen, Y.-H. Generalized Method for Three-Dimensional Slope Stability Analysis. *J. Geotech. Geoenviron. Eng.* **2002**, *128*, 836–848. [CrossRef]
20. Chen, Z.; Mi, H.; Zhang, F.; Wang, X. A simplified method for 3D slope stability analysis. *Can. Geotech. J.* **2003**, *40*, 675–683. [CrossRef]
21. Zhou, X.; Cheng, H. Stability analysis of three-dimensional seismic landslides using the rigorous limit equilibrium method. *Eng. Geol.* **2014**, *174*, 87–102. [CrossRef]
22. Kasama, K.; Furukawa, Z.; Hu, L. Practical reliability analysis for earthquake-induced 3D landslide using stochastic response surface method. *Comput. Geotech.* **2021**, *137*, 104303. [CrossRef]
23. Song, L.; Xu, B.; Kong, X.; Zou, D.; Yu, X.; Pang, R. Reliability Analysis of 3D Rockfill Dam Slope Stability Based on the Copula Function. *Int. J. Géoméch.* **2021**, *21*, 04021001. [CrossRef]
24. Sarma, S.K. Stability analysis of embankments and slopes. *Géotechnique* **1973**, *23*, 423–433. [CrossRef]
25. Zhu, D.; Qian, Q. Rigorous and quasi-rigorous limit equilibrium solutions of 3D slope stability and application to engineering. *Chin. J. Rock Mech Eng.* **2007**, *26*, 1513–1528. [CrossRef]
26. Au, S.; Wang, Y. *Engineering Risk Assessment with Subset Simulation*; John Wiley & Sons: Singapore, 2014; pp. 157–202. [CrossRef]
27. Chen, J.; Zhu, D.; Zhu, Y. An efficient method for computing slope reliability calculation based on rigorous limit equilibrium. *Appl. Rheol.* **2023**, *33*, 20220147. [CrossRef]
28. Hovland, H.J. Three-Dimensional Slope Stability Analysis Method. *J. Geotech. Eng. Div.* **1977**, *103*, 971–986. [CrossRef]

29. Sarma, S. Stability analysis of embankments and slopes. Journal of Geotechnical and Geoenvironmental Engineering Division. *ASCE* **1979**, *105*, 1511–1524. [CrossRef]
30. Halatchev, R. Probabilistic stability analysis of embankments and slopes. In Proceedings of the 11th International Conference on Ground Control in Mining, Wollongong, NSW, Australia, 7–10 July 1992.
31. Zhu, Y. *Slope Reliability Analysis*; Metallurgical Industry Press: Beijing, China, 1993; pp. 218–274.
32. Arai, K.; Tagyo, K. Determination of Noncircular Slip Surface Giving the Minimum Factor of Safety in Slope Stability Analysis. *Soils Found.* **1985**, *25*, 43–51. [CrossRef]
33. Ji, J.; Low, B.K. Stratified Response Surfaces for System Probabilistic Evaluation of Slopes. *J. Geotech. Geoenviron. Eng.* **2012**, *138*, 1398–1406. [CrossRef]
34. Qi, S.; Ling, D.; Yao, Q.; Lu, G.; Yang, X.; Zhou, J.-W. Evaluating slope stability with 3D limit equilibrium technique and its application to landfill in China. *Eng. Geol.* **2021**, *280*, 105939. [CrossRef]

water

MDPI

Article

Study on the Analysis of Pile Foundation Deformation and Control Methods during the Excavation of Deep and Thick Sludge Pits

Dengqun Wang [1], Shuaihua Ye [2,*] and Liangliang Xin [2]

[1] HFUT Design Institute (Group) Co., Ltd., Hefei 230051, China
[2] School of Civil Engineering, Lanzhou University of Technology, Lanzhou 730050, China
* Correspondence: yeshuaihua@163.com

Abstract: This study aims to apply performance-based safety-assessment methods to the monitoring and numerical simulation of excavation engineering projects in order to comprehensively enhance engineering risk management and decision support. In this paper, a deep excavation project in Hefei with thick silty clay layers was studied. The analysis included the surface settlement, the deformation of support structures, the vertical and horizontal displacements of pile tops, axial forces in steel braces, settlement, and the horizontal displacement of a gravity retaining wall on the south side of the excavation using field-monitoring data. A refined three-dimensional finite element model was established to further analyze the distribution of uplift displacement at the bottom of the excavation, horizontal displacement, and bending moments of piles based on simulation results. The research findings indicate that phased excavation can reduce the spatial extent of disturbance to the surrounding soil caused by excavation. Additionally, the closer the location to the excavation and the thicker the underlying silty clay layer, the faster the rate of settlement change and the greater the surface settlement. The spatial structure formed by steel braces and pile foundations effectively reduced the horizontal displacement of the engineering piles. The study's use of field monitoring and finite element simulation provided valuable insights into the deformation of support structures and the response of the surrounding soil to excavation, confirming the rationality and applicability of the support structure in this paper. The proposed method can serve as a reference for similar complex stratum excavation design and construction. The performance-based safety assessment is introduced, and the monitoring data, numerical simulation results, and performance targets are comprehensively analyzed to provide a reliable scientific basis for engineering decision making.

Keywords: excavation of foundation pit; numerical simulation; field monitoring; silty clay layers; performance-based safety assessment

check for updates

Citation: Wang, D.; Ye, S.; Xin, L. Study on the Analysis of Pile Foundation Deformation and Control Methods during the Excavation of Deep and Thick Sludge Pits. *Water* **2023**, *15*, 3121. https://doi.org/10.3390/w15173121

Academic Editor: Yeshuang Xu

Received: 19 July 2023
Revised: 13 August 2023
Accepted: 28 August 2023
Published: 30 August 2023

1. Introduction

With the growth of the population and the advancement of science and technology, some areas need more construction land. Therefore, establishing engineering projects in complex geological formations and complicated surroundings has become inevitable. These projects often face significant risks during the excavation phase of deep excavations. A large excavation area, considerable depth, complex geometry, diverse support forms, and high construction risk coefficients are common characteristics of complex geological formations and complicated environmental conditions in deep excavation projects [1–6]. The structural performance of deep excavations is influenced by the physical and mechanical properties of the soil, spatial distribution patterns, support structure forms, and construction sequences [7,8]. Studying the deformation patterns of support structures and their impact on the surrounding environment caused by deep excavation in complex geological formations and complicated environmental conditions, as well as summarizing the successful experiences of special complex projects, is of great significance in guiding deep excavation projects in such challenging conditions.

Field monitoring, theoretical analysis, and numerical simulation are the primary research methods for studying the response of the surrounding soil and support structures for deep excavation [9–15]. In recent years, the combination of field monitoring and numerical simulation has become a prerequisite for studying complex deep excavation projects, thanks to advancements in field-monitoring technology, the availability of monitoring equipment, the rapid development of the computer industry, and the mature application of finite element algorithms. Field monitoring and numerical simulation are now the main methods for studying the deformation of support structures and the impact on the surrounding environment during complex deep excavation [1,16,17]. They provide real-time monitoring data and quantitative predictions, aiding in understanding the mechanical response of the support structure deformation and surrounding soil during excavation. The comprehensive use of field monitoring and numerical simulation allows for obtaining comprehensive information, optimizing excavation design and support schemes, and improving the safety and reliability of engineering projects. This integrated analysis method provides a scientific basis and practical guidance for the planning and construction of complex deep excavation projects. In the selection of deep excavation support schemes and the design of support structures, numerical simulation analysis serves as an important tool for designers and engineers in decision making. To accurately capture the main characteristics of excavation behavior, it is necessary to establish a refined finite element model that can accurately represent the main features of the actual project and the true mechanical response of the structures and surrounding soil.

Extensive research has been conducted on complex deep excavation, and researchers widely adopt the combination of field-monitoring data and numerical simulation [18–20]. Field-monitoring data can be analyzed in depth to provide timely feedback on the deformation development of deep excavations during construction. Analyzing the monitoring data enables the implementation of measures to proactively control construction risks and prevent accidents. Field monitoring offers real-time feedback to design professionals during the excavation process and provides valuable case histories for future reference [20]. Numerical simulation is a valuable tool for evaluating and analyzing the selection of support structures, design calculations, and the optimization of construction sequencing in complex site conditions, thereby reducing engineering risks.

There have been numerous studies on the influence of deep excavation on adjacent pile foundations, while research on the effects on engineering piles within the excavation is relatively limited [21]. The study of integrating support structures with the main underground structures in basement engineering is still in its early stages. Due to significant differences in loading and deformation characteristics between column piles and general engineering piles, there are complex engineering problems that require further investigation. Cui et al. [22] conducted a study on reinforcement engineering in a densely developed urban area. They used finite element simulation and field measurements to investigate the deformation characteristics of bored piles and inclined steel support retaining structures. They also examined the relationship between the measured lateral displacement at the top of bored piles and the excavation depth. Furthermore, they observed the influence of excavation on the lateral displacement of bored piles and ground movement at the corners of the excavation. Liu et al. [9] developed a Timoshenko beam model based on the Vlazov model to simulate pile–soil interaction. They derived explicit solutions using the finite difference method and validated the proposed method through numerical analysis and field case studies. Dong et al. [23] conducted a parameter study on the applicability of finite element models for complex excavation projects. They investigated the types of structural element models, the determination of appropriate material parameters, and simulation methods for soil–structure interface interaction. The results of this parameter study provided guidance for the development of finite element models that meet practical design requirements. Goh et al. [24] performed a series of two-dimensional (2D) and three-dimensional (3D) finite element analyses using the hardening-soil (HS) model. They studied the effects of soil properties, wall stiffness, excavation length, excavation depth, and clay thickness on excavation and wall deflection. Based on the finite element simulation results, they established an equation for wall deflection considering 3D effects and obtained an empirical formula for estimating the maximum wall deflection. Wu et al. [25] evaluated the excavation support scheme of a complex deep excavation project in Lanzhou using automatic monitoring technology and finite element simulation. They identified potential risks and weak links in the support system to ensure safe construction. Ye et al. [26] studied the influence of excavation on adjacent subway tunnels and underground pipelines based on a deep excavation project in Lanzhou. They combined automatic monitoring technology with the finite element analysis software PLAXIS 3D (2020 version) to simulate the entire excavation process. Guo et al. [27] analyzed the displacement and stress of deep excavation support structures considering asymmetric surcharge effects. They used the finite difference method (FDM) to analyze the horizontal displacement and bending moment of the underground diaphragm wall, as well as the bending moment and axial force of the piles, considering lateral ground movement. The FDM method was validated by comparing the results with field-monitoring data, showing good agreement in yield wall displacement and bending moment. Mei et al. [28] studied the deformation characteristics of the water table, axial force, surface settlement, and lateral displacement of retaining walls in an irregular deep excavation in soft clay in Hangzhou. They simulated the excavation process using Midas GTS NX software and conducted a comprehensive comparative analysis with field measurements. Despite the extensive research on the combined analysis of field-monitoring data and finite element simulation to study the response of surrounding soil and deformation of support structures during excavation, the current project site presents additional complexities. These complexities include significant stiffness variations in the subsoil layers beneath the foundation, a large excavation area, diverse surrounding support structures, and a high risk of pile inclination. The investigation of support schemes and construction methods for this excavation project holds practical significance for researchers and designers. A comprehensive examination of this specific project is necessary to provide valuable insights for similar excavation projects in comparable site conditions.

By introducing performance objectives and indicators, we have defined the requirements for engineering safety and stability, enabling us to quantify and measure the performance of the engineering in key aspects [29,30]. By utilizing displacement monitoring, settlement monitoring, and finite element numerical simulations, we collect and analyze monitoring data in real time and then compare them with performance objectives. Introducing performance-based safety assessment, we comprehensively analyze monitoring data, numerical simulation results, and performance objectives to provide reliable scientific foundations for engineering decision making. This paper presents a study conducted on a large excavation project situated at the edge of a "bowl-shaped" deep pit in Hefei, Anhui Province, China. The pit was previously a quarry that has been abandoned, and it contains thick layers of deep sludge. The sludge exhibits high water content, significant compressibility, and strong flowability, posing challenges in handling it. To improve the bearing capacity of the soil foundation and establish favorable construction conditions, deep soil-mixing piles were initially employed to reinforce the soil layers below a depth of 5 m beneath the excavation. To prevent the flow of sludge and water infiltration from the sludge into the excavation, a gravity retaining wall, measuring 10 m in depth and 6 m in width, was constructed on the south side of the excavation. During the excavation process, the flow of sludge exerts lateral forces on the pile foundation, which can lead to pile inclination and affect the load-carrying capacity of the pile foundation in the long term. To mitigate this issue, the design team proposed a layered excavation method with horizontal and inclined steel bracing between the pile foundations. This approach creates a spatial structure that connects the pile foundations with the steel bracing, aiming to reduce the horizontal displacement of the pile foundation. The steel bracing installed at the excavation's base remains in position and is seamlessly incorporated into the basement floor structure.

A comprehensive analysis was conducted using a combination of field monitoring and finite element simulation to examine various factors related to the large excavation project. These factors include ground settlement around the excavation, settlement and horizontal displacement of the gravity retaining wall on the south side of the excavation, settlement and horizontal displacement of the pile foundations within the excavation, and deep-level horizontal displacement and bending moments in the Y-direction of the pile foundation. By studying this unique engineering project, valuable insights were gained into the behavior of deep excavations in complex geological conditions and the surrounding environments. These findings serve as valuable references for the design and construction of similar complex excavation projects in challenging site conditions. Based on performance-based safety assessment methods, we do not confine ourselves to the collection of monitoring data, but we also closely integrate them with performance objectives. This integration enables us to effectively contrast the actual state of the engineering with the safety requirements. Through real-time monitoring technology and numerical simulation methods, we gain a deeper understanding of the engineering behavior. By employing performance-based safety assessment methods, we have conducted a more comprehensive evaluation of the engineering's safety and stability.

2. Project Overview

2.1. Introduction to Surrounding Environment and Site Geology

2.1.1. Project Introduction

The proposed site is located in the southwest corner of the intersection of Fanwa Road and Ningbei Road, Hefei City, Anhui Province. The site is bordered by Ningbei Road to the west, Stone Pond Park to the south, and residential and recreational facilities to the southwest. The planned land area for this project is 18,390.31 m^2, with a total construction area of 68,090.00 m^2. It includes the construction of two management service buildings, with 15 to 18 floors aboveground and 2 floors belowground. The foundation type is a pile foundation, and the superstructure consists of frame–shear-wall structure. Due to the complex geological conditions at the project site, the construction is divided into two phases. The first phase involves the construction of management service buildings, and the structural construction has been completed. The built project shown in Figure 1 represents the first phase of the project. A-B stands for east-west direction and C-D stands for north-south direction. In addition, there are plans to construct an administrative center with four floors aboveground and two floors belowground. The foundation will be a pile foundation, and the superstructure will be a frame structure. The excavation site is located on the edge of a large quarry, which contains a significant amount of silty soil with high water content and insufficient bearing capacity. As an artificial quarry, the inclination and dip angle of the weathered bedrock surface vary greatly and irregularly. Some locations of the bedrock surface at the pile's ends.

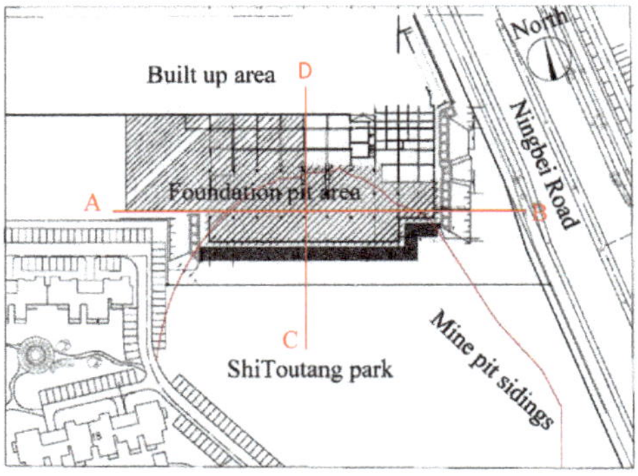

Figure 1. Schematic of the surrounding environment of the site.

2.1.2. Composition and Engineering Characteristics of Subsoil and Rock

Figure 2 illustrates the geological cross-section of the excavation project in the east–west direction, while Figure 3 presents the geological cross-section in the north–south direction. Table 1 shows the soil parameters from top to bottom in the field stratum. The upper fill thickness at the site exhibits significant variation and poor uniformity. The distribution of cohesive soil in the lower section is discontinuous. The overall assessment suggests that the soil and rock layers at the site lack uniformity and belong to non-uniform subsoil. The soil layers, from top to bottom, are as follows:

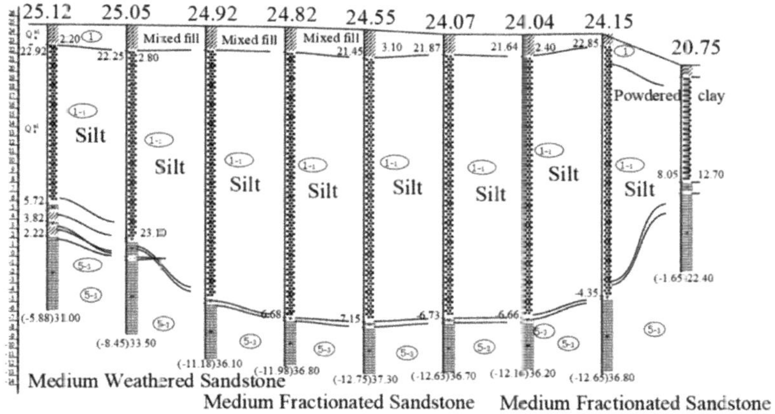

Figure 2. East-west geological profile diagram.

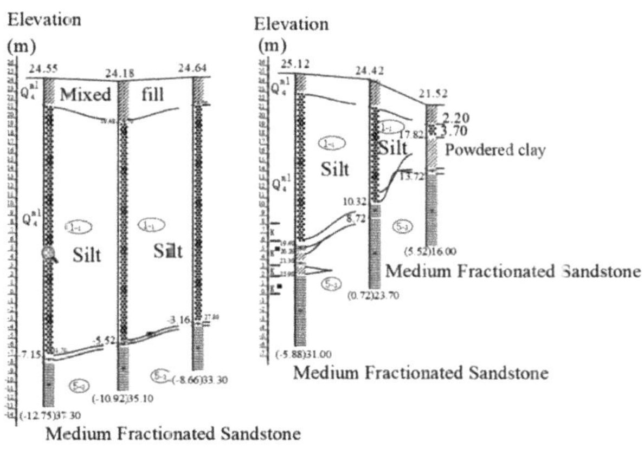

Figure 3. Geological profile at different sections in the north–south direction of the foundation pit.

Table 1. Soil Parameters of the Excavation Site.

Name of Soil	The Natural Unit Weight γ (kN/m^3)	Cohesive Strength C_k (kPa)	Internal Friction Angle φ_k()	Compressive Modulus $E_{S1\text{-}2}$ (MPa)	Permeability Coefficient k (cm/s)
① Miscellaneous fill soil	17.5	12.0	10.0	4.5	5.0×10^{-3}
② Silt	17.9	5.0	5.0	2.0	3.0×10^{-5}
③ Silty clay	19.2	42.0	12.2	7.0	3.0×10^{-6}
④ Clay	19.6	50.0	16.5	11.0	6.0×10^{-7}
⑤ Highly weathered sandy claystone	20.0	10.0	20.0	20.0	2.0×10^{-5}
⑥ Moderately weathered sandy claystone	22.5	30.0	40.0	500	2.0×10^{-6}

① Layer of miscellaneous fill soil (Q^{4ml}): gray-brown, wet, loose, predominantly composed of silt, containing a large amount of debris such as rubble and bricks. Widely distributed in the site area, with a thickness of 1.50–13.00 m and an average of 6.21 m.

② Layer of silty clay: gray-black, highly plastic, containing organic matter, mainly distributed on the south side of the site, with a thickness of 8.00–31.70 m, a bottom elevation of −24.04 to −8.28 m, and a burial depth of 17.50–33.20 m.

③ Layer of silty clay (Q^{4al+pl}): yellow-brown, in a plastic state, with a thickness of 1.90–7.00 m, a bottom elevation of −1.53 to 4.02 m, and a burial depth of 5.20–11.00 m.

④ Layer of clay (Q^{3al+pl}): yellow-brown, hard plastic, locally containing silt, with a thickness of 1.00–7.00 m, a bottom elevation of −3.72 to 4.10 m, and a burial depth of 4.50–13.00 m.

⑤ Layer of highly weathered sandstone (J): brown-red, locally weathered into soil-like material, with a thickness of 1.20–3.30 m, a bottom elevation of −5.56 to 2.60 m, and a burial depth of 6.00–15.00 m. The average SPT N-value from standard penetration tests is 58.

⑥ Layer of moderately weathered sandstone (J): gray-red, with a relatively fragmented rock mass, maximum exposed thickness of 7.00 m, characterized as relatively soft rock, and the degree of rock mass intactness is relatively fragmented.

3. Excavation Support Plan and Construction Plan

Excavation Support Plan

The foundation pit measures 82 m in length in the east–west direction and has an irregular shape in the north–south direction, with a maximum width of 40 m and a minimum width of 20 m. The project site presents complex geological conditions, with the southern side of the excavation bordering an abandoned quarry. The quarry consists of deep silt with a high water content, considerable compressibility, and low bearing capacity, necessitating reinforcement for construction purposes. To mitigate these challenges, the silt within the excavation undergoes treatment using deep soil-mixing piles before proceeding with pile foundation construction, thereby improving the bearing capacity of the silt at the bottom of the excavation. A gravity-type cemented soil wall, measuring 6 m in width and 10 m in depth, is erected on the southern side of the excavation. On the northern side of the excavation, there is a pre-existing underground basement, and the irregularly shaped outer wall of the basement serves as the existing support structure. Double-layered sheet pile support structures are employed on both the eastern and western sides of the excavation. Figure 4 illustrates the excavation site, and Figure 5 depicts the plan view of the support plan.

Figure 4. Foundation pit excavation site.

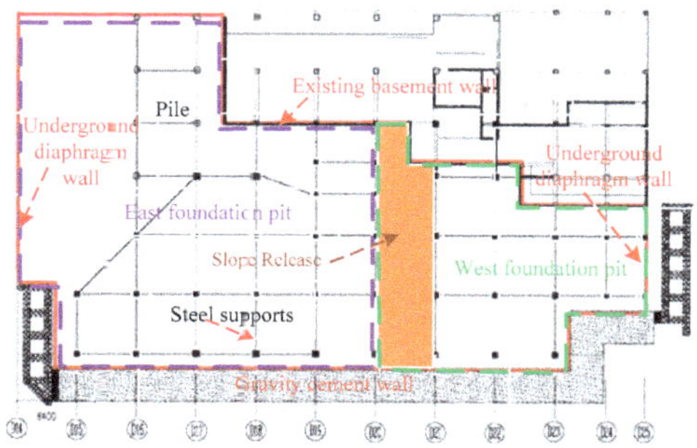

Figure 5. Layout of foundation pit supporting structure.

Once the mud soil at the bottom of the foundation pit has been reinforced, the construction of the pile foundation will begin. Each pile has a diameter of 1.5 m and lengths that vary between 35 m and 10 m. The piles are embedded into the moderately weathered sandy claystone rock to a depth of 3.5 m. The pile foundation employs concrete with a compressive strength of 30 MPa. Steel supports will be installed between the piles to prevent excessive bending moments and tilting of the pile foundation during the excavation of the foundation pit. The supports, made of Q235 steel, have a square cross-section measuring 350 mm on each side and a thickness of 8 mm.

The construction steps are as follows. Step 1: First, excavate the east area of the foundation pit for 5 m at a time, and then construct the steel support of the west area of the foundation pit at the bottom of the foundation pit, which is set above the excavation surface of the foundation pit, and the west area has only one steel support. Step 2: Apply the first steel support to the top of the pile in the east area of the foundation pit, and then carry out earthwork excavation in the west area of the foundation pit, first digging down for 3 m, and then digging for 2 m. The value shows that due to the deep silt in the foundation pit on the south side of the foundation pit, the first row of piles on the south side of the foundation pit is subjected to lateral horizontal force during the excavation process, which causes the pile foundation to tilt and affects the bearing capacity of the pile foundation in the later stage. Therefore, when excavating the west side of the foundation pit, it is necessary to leave the soil between the gravity cement soil wall on the south side of the foundation pit and the second row of piles on the south side of the foundation pit. See Figure 6 for the reserved earthwork location area. See Figure 7 for the reinforced soil at the bottom of the foundation pit.

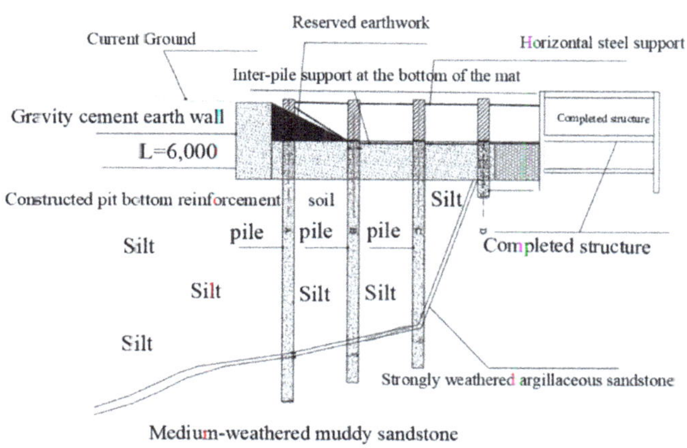

Figure 6. North–south elevation of foundation pit.

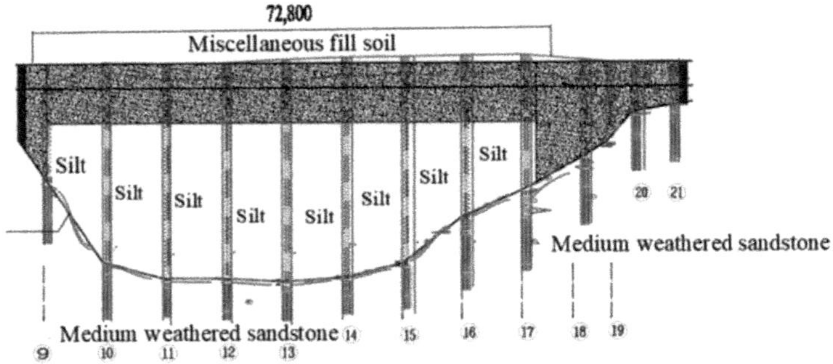

Figure 7. East-west elevation of foundation pit bottom.

Table 2 presents the construction sequence and duration of each construction activity for the foundation pit project. The subsequent sections will establish finite element models according to this construction sequence.

Table 2. Construction Process and Duration.

Stage	Construction Area	Construction Duration Days	Construction Content
Before excavation	North side of foundation pit	-	North building basement
	South side of the foundation pit	-	Gravity retaining wall construction
	The whole foundation pit area	-	Pile foundation construction
Under excavation	East and west sides of the foundation pit	-	Diaphragm wall construction
	East area of foundation pit	20	Excavate to 5 m.
	East area of foundation pit	30	Pile cutting-steel support construction
	East area of foundation pit	10	The first steel support construction
	East area of foundation pit	25	Excavation of the first layer of soil
	East area of foundation pit	25	Excavation of the second layer of soil
	East area of foundation pit	25	The second steel support construction
	East area of foundation pit	25	Demolition of steel support
	South side of foundation pit	10	Basement floor construction
After excavation	-	-	-

4. Engineering Monitoring Scheme and Monitoring Results Analysis

4.1. Monitoring Point Layout and Monitoring Content

Figure 8 depicts the layout plan of monitoring points, where G1–G19 represent the monitoring points for park settlement, D6–D16 and B1–B11 represent the monitoring points for road settlement along Ningbei Road, Z1–Z9 represent the displacement-monitoring points for soil between piles on the east side of the foundation pit, JC1–JC23 represent the settlement-monitoring points for soil between piles on the east side of the foundation pit, N1–N54 represent the axial force-monitoring points for the second row of steel supports on the west side of the foundation pit, N60–N85 represent the axial force-monitoring points for the second row of steel supports on the east side of the foundation pit, and w01–w45 represent the monitoring points for horizontal and vertical displacements at the tops of the piles. The identification numbers of displacement-monitoring equipment correspond to the numbering of pile foundations. For example, w01 refers to the displacement-monitoring equipment placed at the pile top of pile foundation #1. C1–C14 and 2-1-2-8 represent the monitoring points for building settlement. The specific arrangement of the monitoring equipment can be seen in Figures 8 and 9. For the requirements for the arrangement of monitoring points for excavation support structures and surrounding soil displacement, monitoring parameters, monitoring frequency, and performance indicators, refer to the standards [31,32].

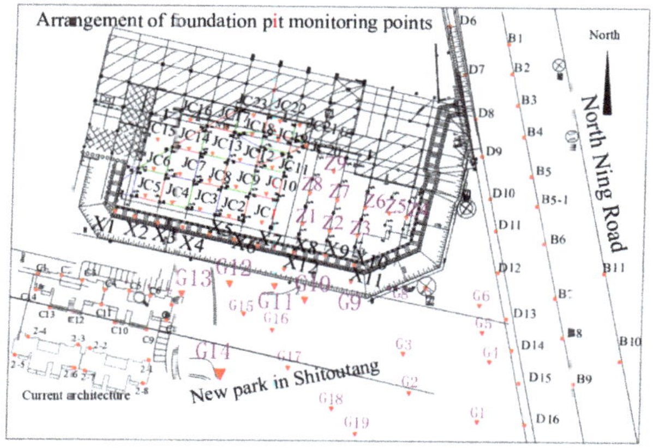

Figure 8. Layout plan of monitoring points.

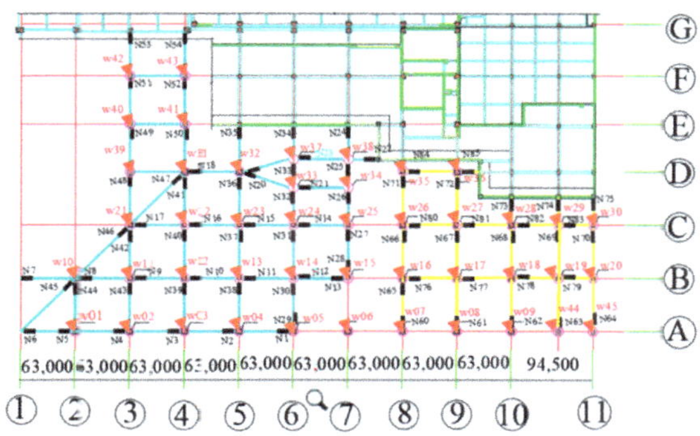

Figure 9. Layout of pile foundation displacement monitoring and steel support axial force monitoring.

4.2. Performance-Based Security Assessment Protocol

4.2.1. Background and Purpose

This agreement aims to ensure comprehensive monitoring, accurate assessment, and effective control of key factors such as the south-side gravity retaining wall, the settlement of the south-side park, surrounding building settlements, the settlement of the eastern access road, the deformation of pile foundations within the excavation pit, and axial forces on steel supports in the excavation project. By integrating performance-based safety-assessment methods, real-time monitoring techniques, and finite element numerical simulations, we will ensure the safety, stability, and sustainability of the engineering project.

4.2.2. Performance Objectives and Indicators

To achieve the safety and stability of the engineering project, we will establish clear performance goals and indicators to quantitatively assess standards. Specifically, we will focus on the following key indicators:

① South-side gravity retaining wall displacement: Limit the displacement of the retaining wall in the X- and Y-directions, ensuring its stability and staying within the specified maximum displacement limits. The maximum limit is 100 mm.

② Manage the park land subsidence to ensure it remains within the predefined limit and uphold the park's overall usability. The maximum limit is 200 mm.

③ Settlement of surrounding buildings: Monitor the settlement of neighboring buildings, ensuring their deformations remain within a safe range and preserving the structural integrity of the buildings. The maximum limit is 30 mm.

④ Settlement of eastern access road: Restrict road settlement to ensure safe traffic passage and prevent adverse impacts. The maximum limit is 60 mm.

⑤ Deformation of pile foundations within the excavation pit: Thoroughly monitor the deformation of the pile foundations within the excavation pit, ensuring they stay within specified limits, controlling pile head displacements and pile body inclinations and maintaining the later-stage bearing capacity of the piles. The maximum limit is 80 mm.

⑥ Axial forces on steel supports: Continuously monitor changes in axial forces on steel supports, ensuring they remain within acceptable ranges and guaranteeing the reliability of the support system. The maximum limit is 400 KN.

4.2.3. Monitoring Methods and Techniques

To achieve the performance goals, we will comprehensively utilize the following monitoring methods and technologies: ① Displacement-monitoring instruments: Install high-precision displacement-monitoring instruments at key locations to continuously record the displacements of the retaining wall, park, buildings, roads, and other structures. ② Settlement-monitoring system: Deploy settlement-monitoring instruments to conduct real-time monitoring and data collection of settlements in the park, buildings, and roads. ③ Finite element numerical simulation: Utilize finite element simulation technology to model the deformation process of the excavation project at various stages, providing validation and support for actual monitoring data. ④ Axial-force-monitoring instruments: Place axial-force-monitoring instruments on the steel supports of displacement piles to monitor changes in axial forces. By integrating these monitoring methods and technologies, we aim to ensure the performance goals are met, accurately assess the project's behavior, and take necessary actions for effective control.

4.2.4. Data Acquisition and Analysis

We will periodically collect the data generated by the monitoring instruments, carry out processing and analysis of data, and compare them with the performance goals and indicators. Through data analysis, we will be able to promptly identify any abnormal situations, thus providing a scientific basis for decision making.

4.2.5. Performance-Based Security Assessment

Since it will integrate monitoring data, numerical simulation results, and performance objectives, the basis of performance-based safety assessment will be crucial in evaluating the engineering project's safety and stability. Through this assessment, we can ascertain whether the actual performance of the project aligns with expectations and take appropriate measures in response.

4.2.6. Continuous Monitoring and Improvement

Throughout the entire lifecycle of the engineering project, we will sustain continuous monitoring activities and make timely adjustments and improvements based on monitoring results. Real-time monitoring, data analysis, and performance assessment will ensure our ability to promptly address potential risks and provide robust support for the successful implementation of the project.

4.2.7. Conclusions

Through the implementation of this agreement, we will comprehensively utilize performance-based safety-assessment methods, real-time monitoring technologies, and numerical simulations to comprehensively understand the project's risk profile, ensuring its stability and safety. Continuous monitoring, analysis, improvement, and learning will ensure our ability to promptly respond to potential risks. Through the collaboration outlined in this agreement, we offer a dependable safeguard for the smooth execution of the engineering project.

4.3. Analysis of Monitoring Results

4.3.1. Settlement Analysis of Park on the South Side of Foundation Pit

The vertical settlement-time–history curves of different monitoring points in the park on the south side of the foundation pit at different excavation stages are shown in Figure 10.

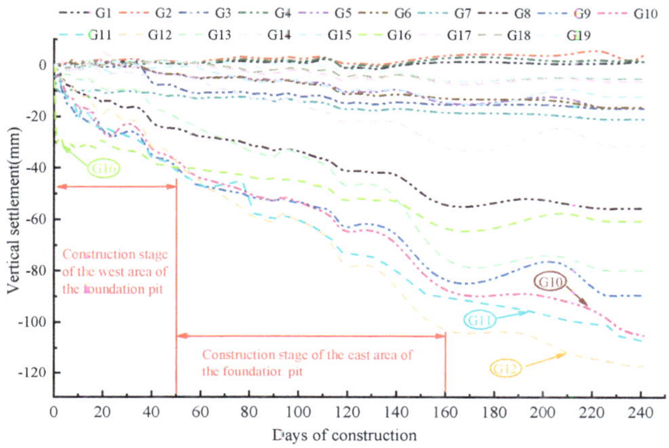

Figure 10. Settlement-time-history curve of monitoring points in the park.

Excavation of the foundation pit results in stress release in the surrounding soil, causing soil movement toward the interior of the pit. Settlement occurs on the surface of the park on the south side of the foundation pit, and the settlement increases with greater excavation depth. From Figure 8, it can be observed that during excavation in the eastern area of the foundation pit, the monitoring point G16 experiences the greatest settlement and the highest settlement rate, followed by G9, G10, and G11. This is because the location of G16 coincides with the center of the park, directly below which lies a deep layer of silt. Therefore, during the excavation of the soil in the foundation pit, this area undergoes the most significant deformation and experiences the greatest settlement. The positions of G9, G10, and G11 are located in the southern excavation area on the western side of the foundation pit, closer to the edge of the pit, resulting in relatively large settlements at these points. Prior to the excavation in the western area of the pit, the settlement at monitoring point G12 is relatively small. However, as construction progresses in the eastern area of the pit, the settlements at G12 and G13, being the closest to the excavation zone, increase. The maximum settlement at G12 reaches 117 mm. Therefore, Figure 10 illustrates that the excavation of the foundation pit has a spatial effect on the surrounding soil, with greater surface settlement occurring closer to the excavation area. Zone excavation can effectively mitigate the impact of large-scale excavation on the surrounding environment.

4.3.2. Analysis of Monitoring Results of Gravity Retaining Wall

During the entire excavation phase of the foundation pit, the time–history curves of vertical displacements at monitoring points X1–X12 of the gravity retaining wall on the south side of the pit can be seen in Figure 11. The locations of monitoring points X1–X12 are shown in Figure 8. During the excavation phase in the eastern area of the foundation pit, special attention was given to the vertical displacements at monitoring points X8, X9, X10, X11, and X12. Similarly, during the excavation phase in the western area of the foundation pit, focus was placed on the vertical displacements at monitoring points X3, X4, X5, X6, and X7. The displacement-monitoring results in the X-direction (east–west direction) at points X1–X12 can be seen in Figure 12, while the displacement-monitoring results in the Y-direction (north–south direction) of the gravity retaining wall can be seen in Figure 13. It should be noted that in the X-direction, positive values represent movement toward the east, while in the Y-direction, positive values represent movement toward the north.

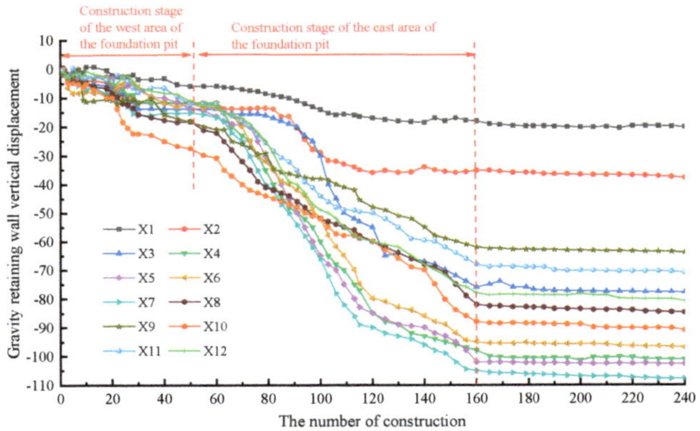

Figure 11. Settlement-time-history curves of vertical displacements at monitoring points of the gravity retaining wall.

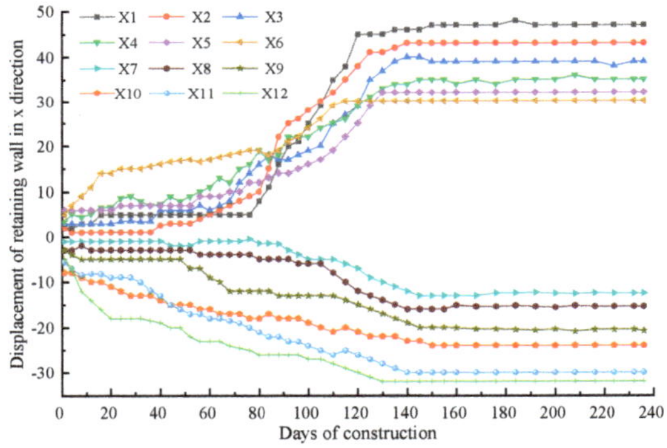

Figure 12. Time-history curve of displacement in the X-direction at monitoring point of gravity retaining wall.

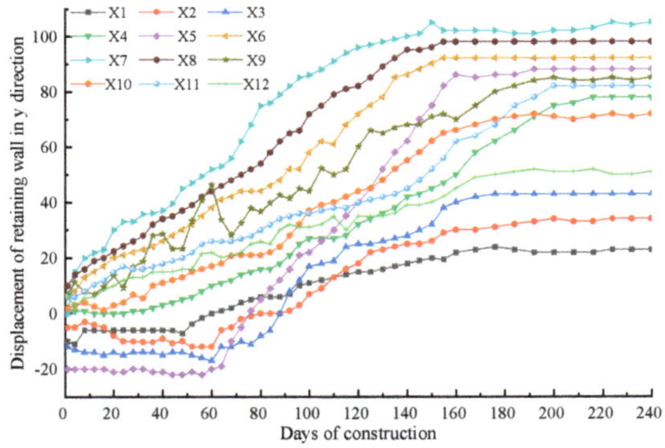

Figure 13. Time-history curve of the Y-direction displacement of monitoring point of gravity retaining wall.

According to Figure 11, during the excavation in the east area of the foundation pit, the maximum vertical settlement of the gravity retaining wall occurred at monitoring point X10, while after the excavation in the west area of the foundation pit, the maximum settlement of the south side retaining wall was observed at monitoring point X7. Throughout the excavation process, the minimum settlement was recorded at monitoring point X1.

Based on Figure 12, it can be observed that during different stages of the excavation process, the monitoring points moved in different directions. During the excavation in the east area of the foundation pit, monitoring points X1 to X6 moved eastward, with monitoring point X6 showing the largest eastward displacement and monitoring point X1 exhibiting the smallest eastward displacement. On the other hand, monitoring points X7 to X12 moved westward, with monitoring point X7 experiencing the smallest westward displacement and monitoring point X12 showing the largest westward displacement. After the excavation of the soil in the western area of the foundation pit, the eastward displacement increased for monitoring points X1 to X6, and the westward displacement increased for monitoring points X7 to X12. Ultimately, monitoring point X6 had the smallest eastward displacement, measuring 28 mm, while monitoring point X1 had the largest eastward displacement, measuring 46 mm. Monitoring point X7 had the smallest westward displacement, measuring 13 mm, while monitoring point X12 had the largest westward displacement, measuring 32 mm. In conclusion, during the excavation of the foundation pit, the horizontal displacement of the gravity retaining wall in the east–west direction followed a pattern of moving from both sides towards the center, with larger displacements observed at greater distances from the center and smaller displacements closer to the center in the X-direction of the retaining wall. This is due to the flow of silt beneath the gravity retaining wall toward the interior of the foundation pit after the excavation of the surrounding soil, resulting in the uplift of the foundation pit bottom and the settlement of the central portion of the gravity retaining wall, causing the eastward movement of monitoring points X1 to X6 and the westward movement of monitoring points X7 to X12.

According to Figure 13, throughout the entire excavation process of the foundation pit, monitoring point X5 initially moved southward and then moved northward. The monitoring points X6, X7, and X8 in the central portion of the retaining wall exhibited larger displacements, with monitoring point X7 showing the maximum northward displacement of 100 mm. This is attributed to the northward movement of the retaining wall induced by the soil pressure in the central portion of the wall, with the maximum displacement occurring at this section, which is consistent with the deformation pattern observed in typical excavation projects.

4.3.3. Analysis of Deformation Monitoring Results of Pile Foundation in the Foundation Pit

(1) The time–history curves of the settlement, X-direction displacement (east–west), and Y-direction displacement (north–south) at the top of selected pile foundations in the eastern area of the foundation pit are shown in Figures 14, 15, and 16, respectively. For the definition of positive directions, the eastward movement is considered positive in the X-direction, the northward movement is considered positive in the Y-direction, and the upward movement is considered positive in the Z-direction.

From Figure 14, it can be observed that during different stages of excavation, the pile foundations experience varying degrees of settlement. Among all the pile foundations in the eastern area of the foundation pit, Pile #7 exhibits the maximum settlement at the top of the pile, with a magnitude of 14.5 mm, while Pile #28 shows the minimum settlement, with a magnitude of 2.5 mm. The settlement of other pile foundations falls within the range of 2.5 mm to 14.5 mm.

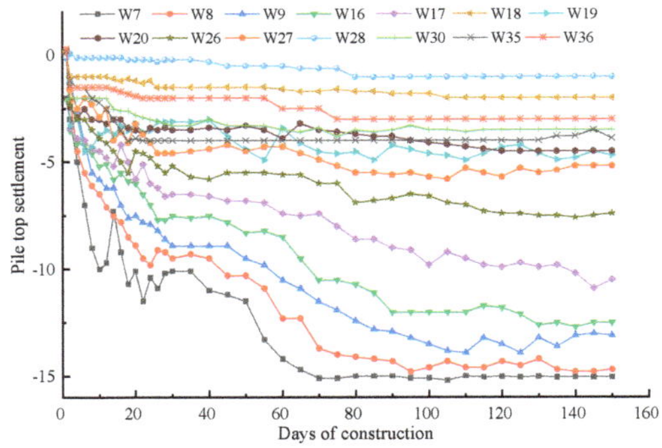

Figure 14. Time-history curve of pile top settlement in the east area of foundation pit.

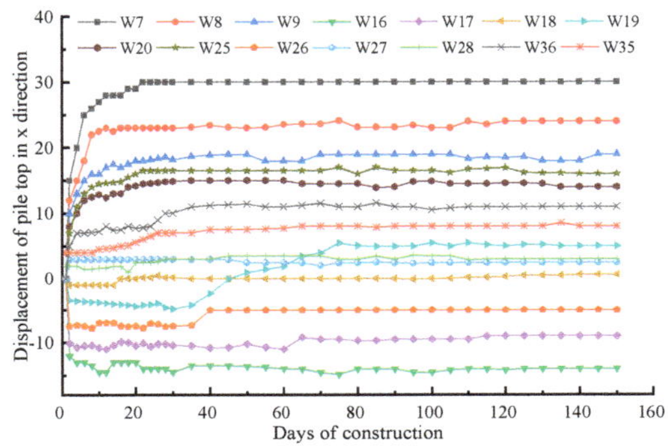

Figure 15. Time–history curve of displacement of pile top in the X-direction in the east area of foundation pit.

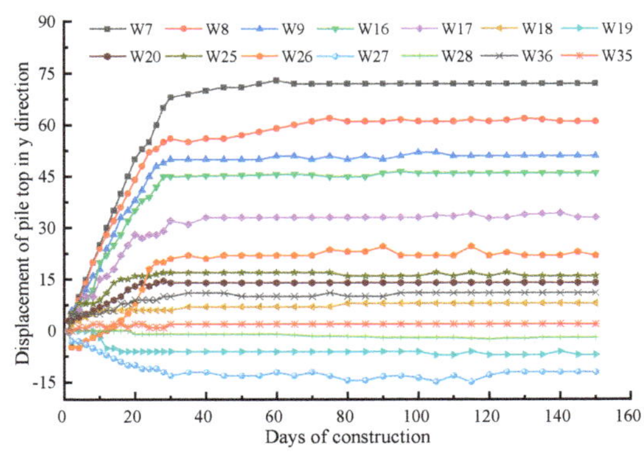

Figure 16. Time-history curve of displacement of pile top in Y-direction in the east area of foundation pit.

From Figure 15, it can be observed how the pile foundations in the eastern area of the foundation pit deform in the east–west direction. During the 25 days of excavation in the eastern area, the pile top's horizontal displacement of the eastern area's pile foundations shows significant variations. However, after the installation of the eastern steel supports, the pile foundations exhibit slower horizontal displacements. Piles #7, #8, and #9 experience notable eastward displacements, with Pile #7 showing the largest eastward displacement of 30 mm. On the other hand, Piles #16, #17, and #26 exhibit westward displacements, with Pile #16 exhibiting the largest westward displacement of 15 mm.

Figure 16 reveals the deformation behavior of the pile foundations in the north–south direction in the eastern area of the foundation pit. Prior to the installation of steel supports, the pile foundations in the western area of the pit, except for Piles #27, #28, #19, and #35, move in a southward direction, while the remaining pile foundations move in a northward direction. Combining this information with the pile foundation numbering in Figure 9, it can be observed that the pile foundations located in the central area of the pit move northward, whereas those near the perimeter of the pit eventually move southward.

Figures 15 and 16 demonstrate that, prior to the installation of steel supports, the pile foundations in the eastern area of the foundation pit experienced significant horizontal displacements at the pile tops. However, after the installation of steel supports, the rate of change in horizontal displacements decreased noticeably, indicating the effectiveness of the steel supports in reducing the horizontal displacements at the tops of the pile. This approach proves to be effective in mitigating pile top horizontal displacements and preventing pile foundation movements and tilting in excavations involving deep cohesive silt layers.

(2) The settlement, horizontal displacement in the X-direction (east–west direction), and horizontal displacement in the Y-direction (north–south direction) of the selected pile foundations in the western area of the foundation pit are presented in the time–history curves shown in Figures 17–19.

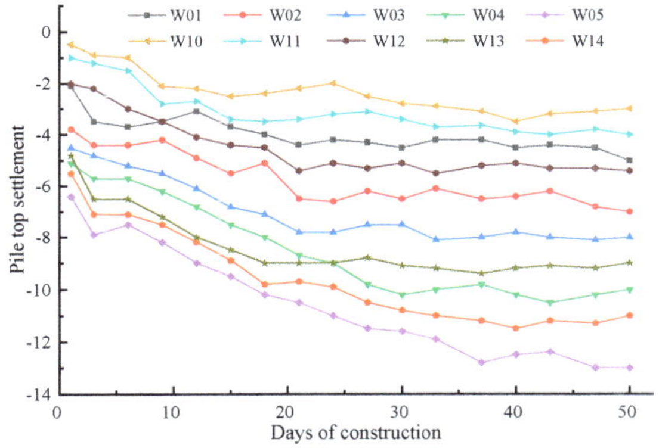

Figure 17 Time-history curve of pile top settlement in the west area of foundation pit.

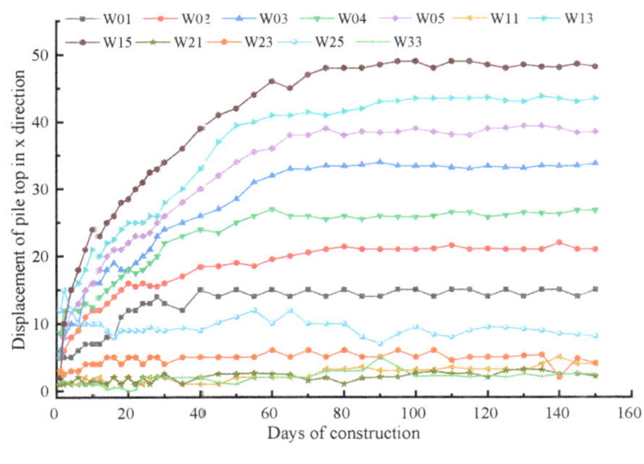

Figure 18. Time-history curve of displacement of pile top in the X-direction in the west area of foundation pit.

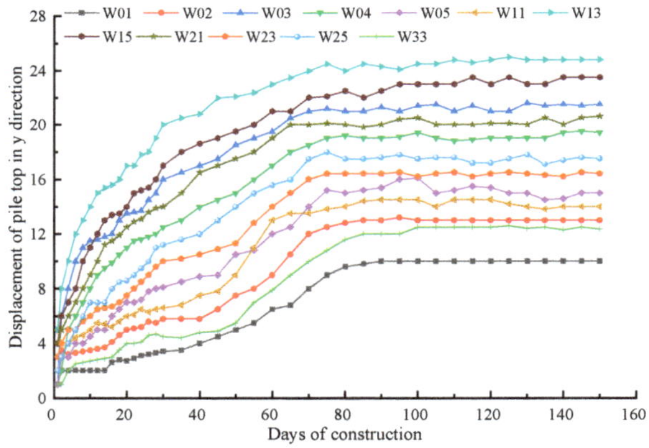

Figure 19. Time-history curve of displacement of pile top in the Y-direction in the west area of foundation pit.

From Figure 17, it can be observed that in the western area of the foundation pit, the maximum settlement is recorded at Pile #5, with a settlement of 13 mm, followed by Pile #15 with a settlement of 9 mm, and the minimum settlement is observed at Pile #10 with a settlement of 2.1 mm. Referring to the layout of the pile settlement monitoring points in Figure 9, it can be inferred that Piles #5, #13, and #4 are located in the central part of the foundation pit, with larger pile lengths and a thicker layer of underlying silt. Consequently, the excavation of surrounding soil results in varying degrees of settlement at the pile tops due to self-weight effects.

Figure 18 reveals that during the excavation of the western area of the foundation pit, Piles #15, #13, and #5 experience significant eastward displacement, with horizontal displacements of 49 mm, 46 mm, and 38 mm, respectively. Piles #21, #11, and #23, on the other hand, demonstrate smaller eastward displacement, measuring 1.5 mm, 1.8 mm, and 2 mm, respectively. Referring to Figure 9, it can be observed that in the central part of the foundation pit where Piles #15, #13, and #5 are located, there is a rapid increase in the pile top displacement before the implementation of steel support in the western area of the foundation pit. However, after the installation of steel support, the growth rate of the pile top's horizontal displacement decreases, exhibiting a diminishing trend. The presence of steel support between the piles effectively reduces the horizontal displacement of the piles.

Based on Figure 19, it can be observed that during the initial 50 days, there is a slow variation in the north–south displacement of the pile foundations in the western area of the foundation pit prior to excavation. However, once the construction in the western area of the foundation pit begins, the pile foundations in this area rapidly shift northward. Following the implementation of steel support in the western area, the northward displacement of the pile foundations gradually decreases and stabilizes. Ultimately, Pile #13 exhibits the largest displacement, moving 25 mm northward, followed by Piles #15 and #3 with displacements of 24 mm and 20 mm, respectively. Pile #1 demonstrates the smallest north–south horizontal displacement, with a movement of 10 mm northward. This behavior can be attributed to the construction of a steel support prior to the excavation in the western area of the foundation pit, resulting in minimal horizontal displacement of the piles during excavation.

4.3.4. Analysis of Monitoring Results of Axial Force of Steel Support

The direction of axial force is defined as positive tension and negative pressure. See Figure 9 for the number of steel supports and the specific location of axial force-monitoring points.

(1) Axial force of steel support in the east area of the foundation pit

From Figure 20, it can be observed that the steel supports N71, N77, and N73 at the bottom of the eastern area of the foundation pit are subjected to compressive forces, while the remaining steel supports experience tensile forces. Among them, the steel support N71 experiences the highest compressive force, measuring 300 KN, and the steel support N76 experiences a maximum compressive force of 135 KN. The steel support N74 experiences the smallest axial force. This is because the N74 steel support is located at the northeast corner of the foundation pit, where the horizontal displacement of the pile foundation is small, resulting in minimal force on the N74 steel support.

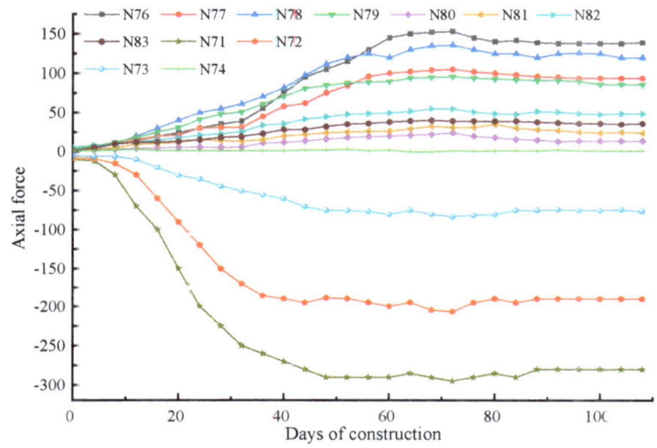

Figure 20. Axial force monitoring of steel support in the east area of foundation pit.

(2) Analysis of monitoring results of axial force of the first steel support in the west area of the foundation pit (upper steel support)

From Figure 21, it can be observed that after the construction of the upper steel supports in the western area of the foundation pit, a spatial structure is formed between the steel supports and the pile foundation. During the excavation of the soil in the western area of the foundation pit, the steel supports can reduce the horizontal displacement of the pile foundation, and the axial force gradually increases. Some steel supports experience compressive forces, while others experience tensile forces. Ultimately, the N18 steel support experiences the highest tensile force, with a magnitude of 70 KN, and the N09 steel support experiences the smallest compressive force, measuring 36 KN.

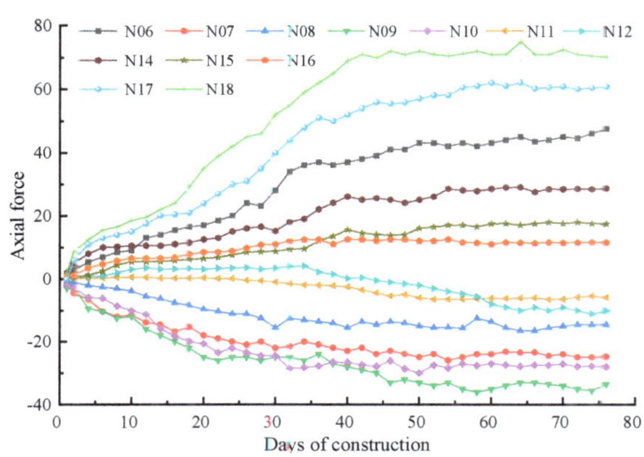

Figure 21. Axial force monitoring of steel support in the west area of the foundation pit.

(3) Analysis of monitoring results of the inclined steel support between gravity retaining wall and piles on the south side of the west area of the foundation pit

According to Figure 22, after the excavation of the soil in the western area, the inclined struts experience axial compression. Among them, the inclined strut between #6 and #15 experiences the highest axial compression, measuring 280 KN, while the inclined strut between #1 and #10 experiences the lowest axial compression, measuring 10 KN. Combining this information with Figure 9, it can be observed that the N1-1 steel support is located at the southwestern corner of the foundation pit, where the horizontal displacement between the retaining wall and the pile top is small, resulting in the smallest axial compression in the N1-1 steel support. On the other hand, the N6-1 steel support is located in the middle of the foundation pit, where the southern retaining wall undergoes significant deformation toward the interior of the foundation pit, leading to a higher axial compression in the N6-1 steel support compared to other inclined steel supports.

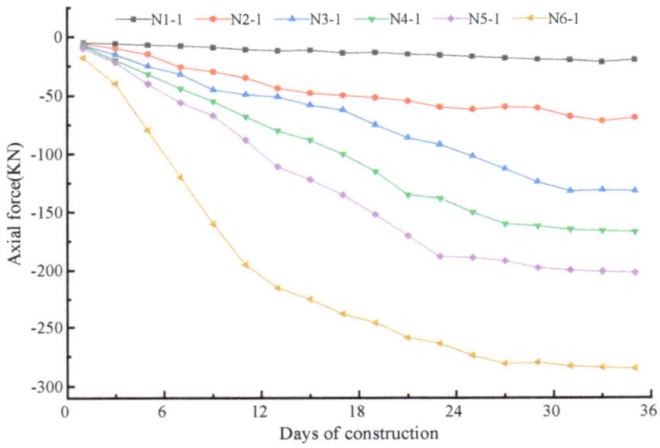

Figure 22. Axial force monitoring of inclined steel support.

4.3.5. The Eastern Side of the Excavation Site Is Adjacent to a Road

Monitoring points D06-D16 are positioned along the road's edge near the excavation site, while monitoring points B01-B11 are placed along the road's central axis, as illustrated in Figure 8. Figure 23 depicts the settlement-monitoring time–history curve along the road's edge, while Figure 24 displays the settlement-monitoring time–history curve along the road's central axis.

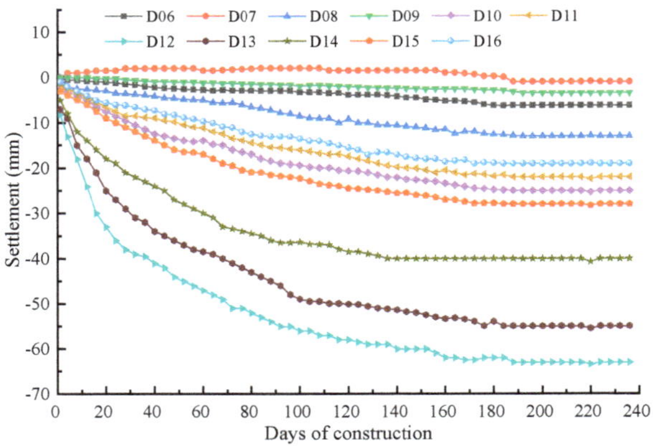

Figure 23. Settlement time-history curve along the road's edge.

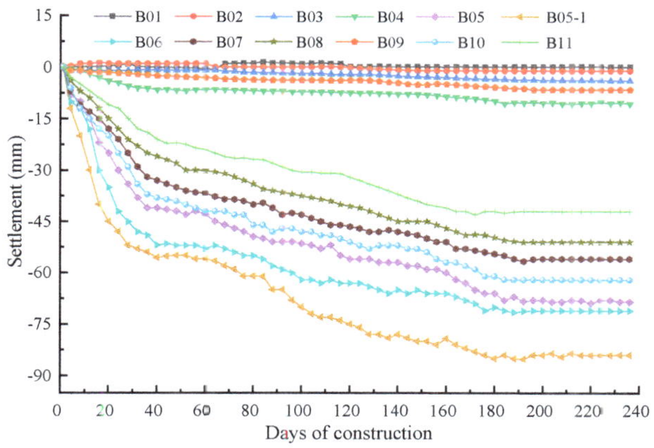

Figure 24. Settlement time-history curve along the road's center.

From Figure 23, it is evident that the settlement at monitoring point D07 is the smallest, with a stable value of 0.2 mm during the excavation of the west side of the excavation area. The maximum settlement occurs at monitoring point D12, with a stable settlement value of 60 mm. Following this, monitoring points D13 and D14 exhibit significant settlements, measuring 55 mm and 46 mm, respectively.

From Figure 24, it can be observed that along the central axis of the road, monitoring points B01, B02, B03, B04, and B09 exhibit relatively smaller settlements, likely due to their greater distance from the excavation edge of the pit. Conversely, monitoring points B05, B05-1, B06, and B07 show larger settlement values. Referring to the layout plan in Figure 8, it becomes apparent that monitoring points B05, B05-1, B06, and B07 are situated closer to the excavation pit's edge, leading to the larger settlements observed.

Comparing and analyzing Figures 22 and 23, it is evident that the maximum settlement displacement at the road's edge occurs at monitoring point D12, with a settlement value of 64 mm. On the road's central axis, the largest settlement is at monitoring point B05-1, measuring 81 mm. The reason for the larger settlement at monitoring point B05-1 compared to monitoring point D12 is attributed to the vehicular load on both sides of the road's central axis. At monitoring point D12, the smaller effect of the eastern pit support and the lesser vehicular load near the road's edge result in a smaller settlement value. This is why the settlement at monitoring point B05-1 is greater than that at monitoring point D12.

5. Finite Element Simulation Analysis

5.1. Establishment of Finite Element Model

The 3D model was established using the Midas GTS NX (2022 version) specialized finite element software to simulate the impact of excavation on the surrounding environment and the deformation response of the support structure during the excavation phase. To eliminate the influence of boundary effects, the model dimensions were set to 220 × 200 × 70 m in length, width, and height, respectively. According to the actual distribution of soil layers, the soil outside the pit was divided into four layers from top to bottom: 3 m of loose fill soil, 2 m of clay, 5 m of strongly weathered sandstone, and a bottom layer of rock with a thickness of 60 m. Inside the pit, the soil was divided into three layers: 3 m of loose fill soil, 2 m of clay, and an uneven thickness of silt layer, with a rock layer beneath it. The 3D finite element model is shown in Figure 25. The support structure model for the pit is shown in Figure 26, while the steel supports on the east and west sides of the pit are shown in Figure 27. See Figure 28 for the location of excavated slope on the west side and reserved soil on the east side of the foundation pit. The pile foundation model can be seen in Figure 29.

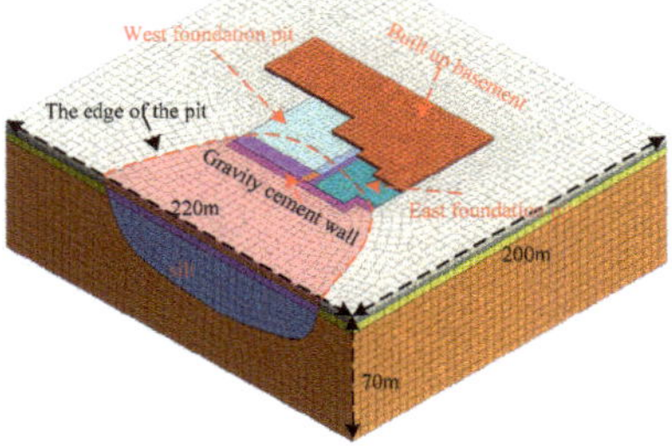

Figure 25. Three-dimensional finite element model.

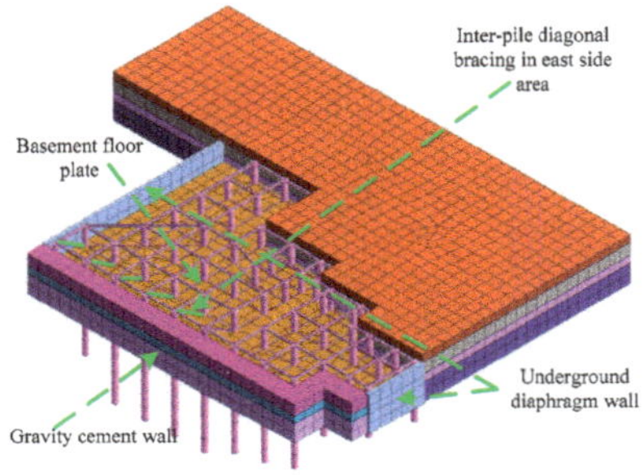

Figure 26. Model of foundation pit supporting structure.

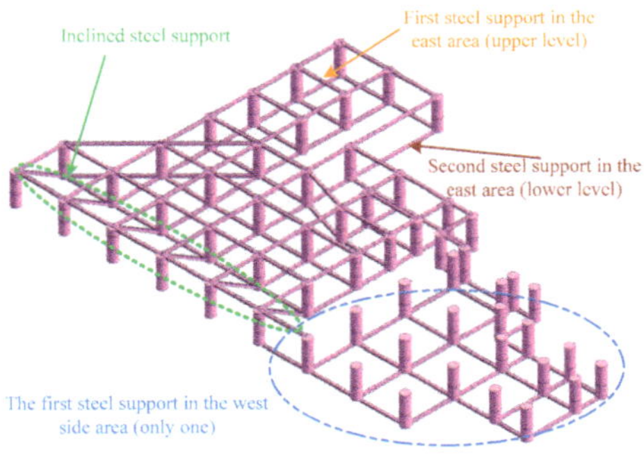

Figure 27. Steel support model of foundation pit.

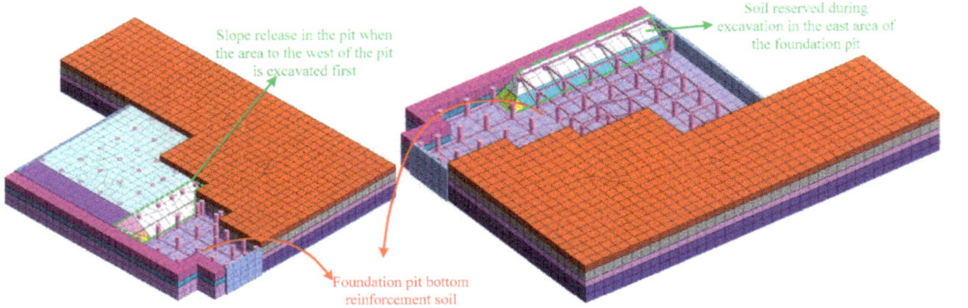

Figure 28. Location map of excavation slope on the west side of foundation pit and reserved soil on the east side.

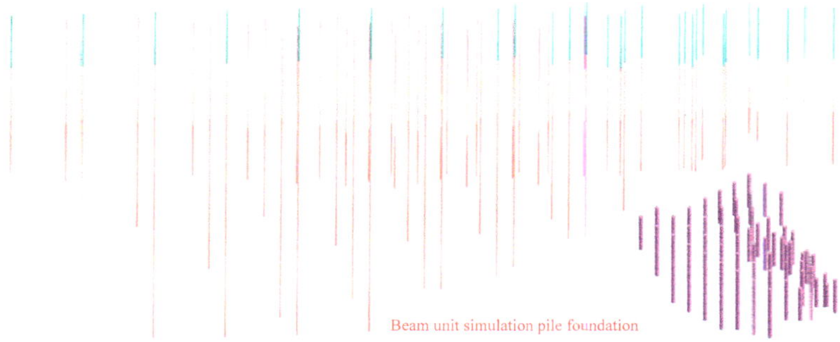

Figure 29. Beam element simulation pile foundation.

The pile foundation and steel supports were simulated using one-dimensional beam elements, while the underground retaining walls and basement floor on the east and west sides of the pit were simulated using two-dimensional plate elements. The remaining soil layers were simulated using three-dimensional solid elements. The constitutive model chosen for the soil was the Mohr–Coulomb model, while the structural elements were assumed to follow linear elastic behavior. The parameters for the soil and structural elements can be found in Table 3.

Table 3. Soil mass and structural parameters of finite element model.

Structure/Soil Mass	γ (10^3KN/m^3)	E (Mpa)	φ ()	v	Unit Type
Miscellaneous fill	18.5	20	15	0.40	Entity unit
Clay	19	30	20	0.35	Entity unit
Silt	18	6	5	0.45	Entity unit
Strongly weathered sandstone	26	7.0×10^4	45	0.31	Entity unit
Rock	28	1.8×10^5	50	0.30	Entity unit
Reinforcement soil at pit bottom	20	45	36	0.32	Entity unit
Gravity cement soil wall	19.5	50×10^4	40	0.30	Entity unit
Concrete diaphragm wall	25	4.5×10^4	-	0.30	Entity unit
Pile foundation/basement floor	25	3.25×10^4	-	0.28	Beam/plate element
Steel support	78	2.1×10^5	-	0.26	Beam element

Since a part of the pit is situated on an existing quarry, which has an irregular spatial surface, a three-dimensional surface was created based on the contour lines of the quarry using the software. The surface layers were established using the 3D solid segmentation function. The boundaries of the quarry and the extent of the silt layer can be observed in the model shown in Figure 25. The entire excavation area of the pit was divided into east and west regions, with the east region being excavated first, followed by the west region. In the east region, a 5 m excavation was conducted, and a slope with a gradient of 1:1.25 was formed at the junction between the east and west regions. In the east region, the soil between the southern retaining wall and the pile foundation was preserved during excavation, and the slope and reserved soil displacement can be seen in Figure 28.

5.2. Definition of Numerical Simulation Construction Stage

According to the actual construction sequence and excavation steps of the foundation pit, the finite element model is divided into the following 10 construction stages. Each stage aims to simulate the most critical locations under different working conditions, taking field-monitoring data into account. The study investigates the variation patterns of pile bending moment, horizontal displacement, steel support axial force, displacement of the southern cement soil wall, and deformation characteristics of the surrounding soil during different excavation stages.

(1) Stage 1: Initial stress equilibrium stage—Activate all soils, structures, and reset displacements to zero. (2) Stage 2: Construction of gravity retaining wall, continuous wall, and pile foundations—Reset displacements to zero. (3) Stage 3: Excavation of 5 m of eastern soil, with the eastern pile foundations reaching the bottom of the excavation. (4) Stage 4: Construction of steel support in the eastern area of the pit (below the bottom of the excavation after reaching the bottom with the eastern pile's foundations). This support is independent and not connected to the western area. (5) Stage 5: Construction of the first steel support in the western area of the pit. (6) Stage 6: Excavation of upper-level western soil, with a depth of 3 m, requiring the retention of soil. (7) Stage 7: Excavation of lower-level western soil, with a depth of 2 m, requiring the retention of soil. (8) Stage 8: Construction of the second steel support in the western area (including inclined braces on the southern side). The second steel support is at the same elevation as the eastern steel support. (9) Stage 9: Removal of the first steel support in the western area, trimming the pile foundations to the bottom, and excavation of the retained soil. (10) Stage 10: Construction of the basement floor in the eastern and western areas of the pit, applying a simulated vertical load of 6000 KN on the pile foundations to simulate deformations under the vertical load from the upper structure.

5.3. Analysis of Numerical Simulation Results

5.3.1. Vertical Displacement Analysis of Soil at the Bottom of Foundation Pit

Figure 30 shows the nephogram of vertical displacement in the ninth stage, and Figure 31 shows the position of monitoring points at the bottom of the foundation pit.

Figure 30. Cloud diagram of vertical displacement of foundation pit excavation.

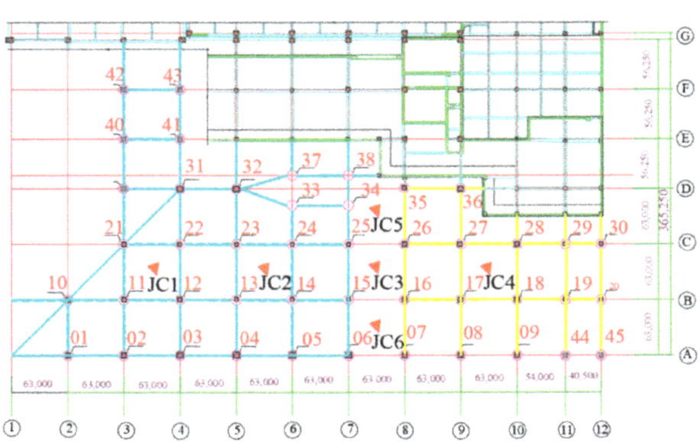

Figure 31. Numerical simulation monitoring points at the bottom of foundation pit.

From Figure 30, it can be observed that there is significant heave at the bottom of the excavation during Stage 9. Therefore, selected monitoring points are analyzed to assess the heave at the bottom of the excavation. The locations of the monitoring points can be found in Figure 31. Figure 32 presents the vertical displacements of the soil at the bottom of the excavation at these monitoring points. It can be observed from Figure 32 that monitoring points JC3 and JC6 exhibit larger vertical displacements, while JC1 and JC5 have smaller vertical displacements. Considering the positions of the monitoring points shown in Figure 32, it can be concluded that in the east–west direction, the soil at the bottom of the excavation in the central area experiences larger vertical displacements, while in the north–south direction, the closer the excavation is to the south side, the larger the vertical displacements of the soil at the bottom of the excavation. This can be attributed to the specific distribution of soil layers in the excavation. The central and southern parts of the excavation are characterized by deep layers of soft soil. After the excavation of the soil, the bottom of the excavation experiences rebound, and under the influence of the self-weight stress of the overlying soil in the southern part, the deep soft soil moves toward the excavation, resulting in uneven heave at the bottom of the excavation.

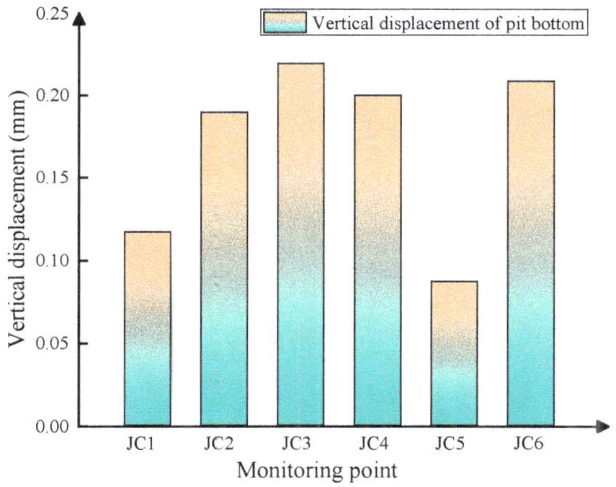

Figure 32. Vertical displacement of pit bottom.

5.3.2. Comparative Analysis of Axial Force of Steel Support and Numerical Simulation

(1) Comparative analysis of the monitoring value and numerical simulation of the axial force of steel support in the east area of the foundation pit.

From Figure 33, it can be observed that the numerical simulation results align well with the measured data at monitoring points N76, N77, and N78, which are located at the steel support in the eastern region of the excavation. This confirms the reliability of the monitoring data and the reasonableness of the numerical simulation results. Additionally, the accuracy of the numerical simulation results meets the engineering requirements. It is important to note that the available field-monitoring data only cover a period of 110 days, from the construction of the steel support in the eastern region to the construction of the basement floor. Based on the numerical simulation results, it is evident that significant changes occur in the axial forces of the steel support after the construction of the basement floor, and these changes occur in different directions.

(2) Comparative analysis of axial-force-monitoring value and numerical simulation of the first steel support on the west side of the foundation pit. The comparison between the monitored axial force of the steel support and numerical simulation results is shown in Figure 34.

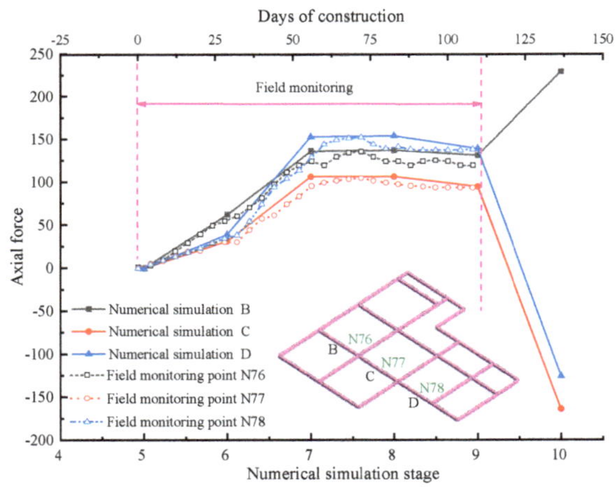

Figure 33. Comparison between monitoring value and numerical simulation of axial force of steel support on the east side of the foundation pit.

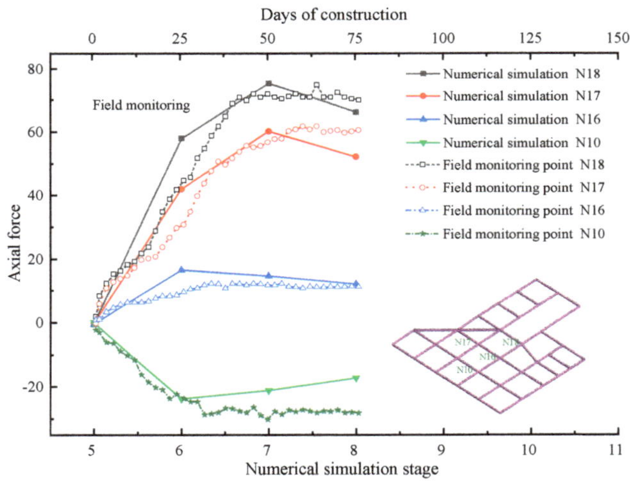

Figure 34. Comparison between monitoring value and numerical simulation of axial force of the upper steel support on the west side of foundation pit.

From Figure 34, the field-monitoring data for the upper steel support in the western region of the excavation cover a period of 75 days, corresponding to stages 5–8 of the numerical simulation. By comparing the measured data from four monitoring locations, namely N18, N17, N16, and N10, with the simulated values, it is observed that the numerical simulation results closely match the field measurements. This indicates that numerical simulation can be effectively combined with field monitoring to analyze and study complex excavation projects, providing a basis for design decisions. Additionally, it can serve as a supplement to data in cases where on-site monitoring is not conducted.

(3) Numerical simulation of axial force of lower-layer steel support on the west side of foundation pit.

In the entire excavation area, the steel supports at the bottom of the excavation are not removed. After the construction of the basement floor at the bottom of the excavation, there are no measured axial force data for the steel supports. Figure 35 shows the simulated axial force values of the steel supports under vertical loads from the upper main structural elements after the construction of the basement floor. From Figure 35, it can be observed that M9 experiences axial tension with a magnitude of 1700 kN, M9 experiences axial compression with a magnitude of 1400 kN, M4 experiences the smallest axial compression with a value of 210 kN, and M11 experiences the smallest axial tension with a value of 400 kN

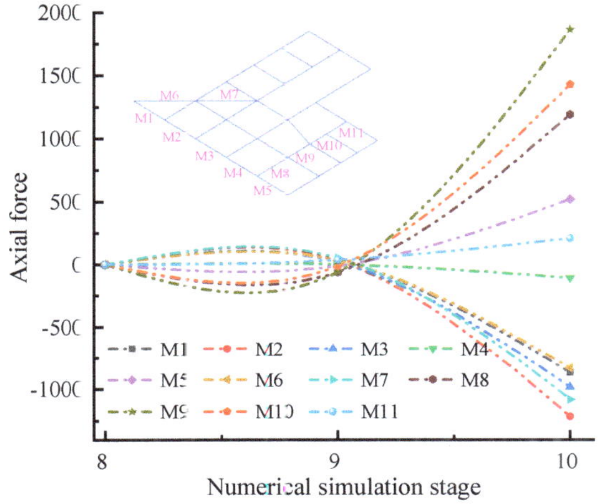

Figure 35. Numerical simulation value of the axial force of the lower-layer steel support on the west side of foundation pit.

5.3.3. Analysis of Displacement and Bending Moment of Pile Foundation

Figure 36 shows the relationship curve between the displacement of the #6 pile in the middle of the first row of piles on the south side of the excavation and the X-direction Figure 37 depicts the relationship curve between the displacement of the #6 pile and the Y-direction. Lastly, Figure 38 illustrates the relationship curve between the #6 pile and the bending moment in the Y-direction. Please refer to Figure 9 for the specific location of the #6 pile in the pile foundation.

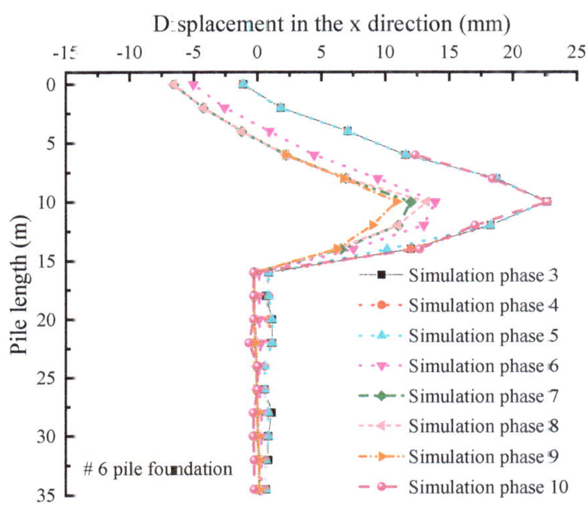

Figure 36. Displacement of pile body in the X-direction.

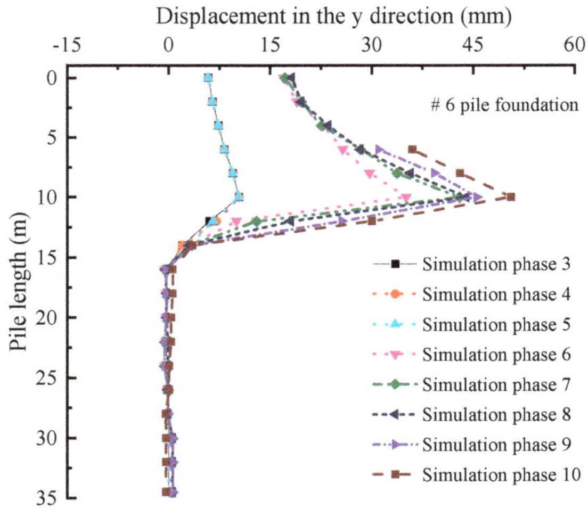

Figure 37. Displacement of pile in the Y-direction.

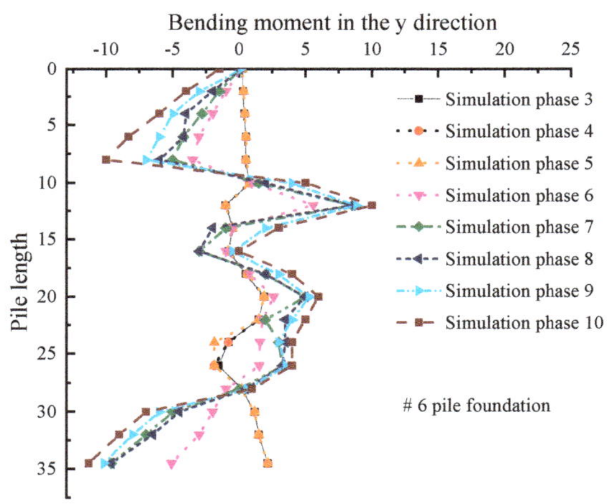

Figure 38. Bending moment in Y-direction of the pile body.

From Figures 36 and 37, it can be observed that before the pile top is cut off at 5 m, the #6 pile exhibits maximum horizontal displacement in both the X- and Y-directions at a depth of 10 m below the pile top. Below 15 m, the deep horizontal displacements in the X- and Y-directions of the pile foundation are close to zero. After the pile is cut off in the ninth stage, the pile top displacements are 12 mm in the X-direction and 35 mm in the Y-direction. At a depth of 5 m below the bottom of the excavation, the horizontal displacements of the pile foundation in the X- and Y-directions reach their maximum values, with 26 mm displacement in the X-direction and 52 mm displacement in the Y-direction.

From Figure 38, it can be observed that the pile bending moment varies at different construction stages. In stages 3–5, the pile bending moment is relatively small, while in stages 6–10, the bending moment increases. The maximum bending moment occurs at 8 m and at the pile tip, with a magnitude of 10 kN·m. During the construction of the basement slab, the steel support is cast in place together with the pile top, resulting in a non-zero bending moment at the top of the pile.

After integrating the principles of performance-based risk assessment, the authors conclude that both the deformation of the excavation support structure and the deformation of the foundation piles within the pit remain within the specified performance limits. This indicates the rationality of the engineering design and construction plan, confirming the feasibility of measures taken to control pit deformation.

Key performance indicators that were closely monitored include the displacement in the Y-direction of the gravity retaining wall, the horizontal displacement of foundation piles within the pit, and the axial forces in the steel supports. The monitoring results demonstrate that the maximum displacement in the Y-direction for the gravity retaining wall is 100 mm. Additionally, for the foundation piles within the pit, the maximum horizontal displacements are 30 mm and 70 mm in the X- and Y-directions, respectively, for the eastern area, and 49 mm and 25 mm for the X- and Y-directions, respectively, for the western area. The maximum settlement in the southern park area of the pit is 117 mm. The axial force in the steel supports on the eastern side of the pit is 300 KN, while the upper-layer steel supports on the western side have a maximum axial force of 70 KN and the lower-layer steel supports have a maximum axial force of 280 KN. Notably, the maximum values of the key monitoring parameters do not exceed the performance indicators established based on the principles of performance-based assessment.

In this study, we have successfully integrated the concept of performance-based safety assessment into the monitoring and numerical simulation of excavation projects, providing a more comprehensive and scientifically grounded assurance for the engineering's safety and stability. The objective of this paper is to ensure a thorough monitoring, accurate assessment, and effective control of multiple critical aspects of the engineering project, including the performance of the south-side gravity retaining wall, the settlement of the south-side park within the excavation, the settlement of the road on the east side, the deformation of the foundation piles within the excavation, and the axial forces in the steel supports. By combining on-site monitoring data, numerical simulation results, and performance-based safety assessment, our intention is to provide a more reliable foundation for engineering decision-making and risk management. The incorporation of performance targets and indicators enables us to measure the engineering's safety and stability with greater precision. Building on the foundation of monitoring and numerical simulation, we integrate real-world data with performance standards, providing a more comprehensive and accurate understanding of the actual conditions of the engineering. By comprehensively considering measured data, numerical simulation, and performance requirements, we can better comprehend the engineering's actual performance and potential risks. Supported by real-time monitoring technology and numerical simulation methods, the performance-based safety assessment approach offers us a more precise and comprehensive means of engineering analysis.

6. Conclusions

The performance-based safety-assessment method offers new perspectives and approaches to our research, ensuring a more reliable guarantee for the safety and stability of excavation engineering projects. In future engineering practices, we will continue to promote the application of this method, continually refining our methodologies and processes and making further contributions to safety management and risk control in the field of engineering.

(1) Layered and zonal excavation can effectively reduce the impact of internal stress release in the soil due to the excavation of surrounding structures and the minimization of the spatial disturbance. This method is an effective approach to mitigate the effects on the surrounding environment during excavation in complex geological conditions and complex excavation projects. When there are differences in elevation or varying construction sequences at the boundaries between different regions, slope construction needs to be implemented in different excavation areas within the excavation pit.

(2) Controlling the tilting and displacement of pile foundations is a challenging aspect of this excavation project. This paper proposes the solution of applying steel bracing between pile foundations in different layers. The study reveals that the steel bracing experiences both compression and tension and that the axial forces vary with the depth of excavation. By introducing steel bracing between the engineering piles, the pile's foundations are interconnected, forming a spatial skeletal structure. This approach effectively reduces the horizontal displacement of the pile foundations and resolves the issue of pile tilting caused by lateral forces during excavation due to the flow of silt. The proposed solution in this paper can serve as a reference for similar excavation projects.

(3) In cases where the bearing capacity requirements or construction conditions of the soil at the bottom of the excavation pit are not met, it is possible to reinforce the soil at the bottom of the pit before excavation. In this paper, prior to pile construction and soil excavation, deep mixing piles are constructed in the silt layer below 5 m from the bottom of the pit to strengthen the pit soil. Practice and research have shown that using deep mixing piles to reinforce the silt at the bottom of the excavation pit can effectively reduce the deep horizontal displacement of pile foundations. Additionally, it improves the bearing capacity of the soil at the bottom of the pit and reduces the seepage of water in the silt, creating favorable and dry construction conditions.

(4) Field monitoring provides real-time and dynamic responses to the surrounding environment during excavation, while finite element analysis is an effective method for analyzing and evaluating construction plans and sequences. The data obtained from finite element simulations serve as a valuable supplement to data that cannot be monitored or are difficult to monitor in the field. The combination of numerical simulation and field monitoring is an effective approach for analyzing and studying complex excavation projects. By utilizing both field monitoring and finite element simulation, comprehensive information can be obtained, leading to optimized pit design and support solutions, thereby improving the safety and reliability of the project. This integrated analysis approach provides scientific guidance and practical insights for the planning and construction of complex excavation projects.

(5) This study adopted an integrated research methodology involving field monitoring and numerical simulation to analyze the deformation behavior of pile foundations and support structures within a complex geological and environmental context during the excavation of deep pits in locations characterized by substantial and thick silt layers. The study explored the deformation patterns of pile foundations and support structures within the excavated pit, proposed effective strategies for managing pile-foundation deformation, and validated the soundness and practicality of these methods through empirical validation. Nevertheless, with the anticipated rise in environmental complexity, exclusive dependence on numerical simulation and monitoring approaches remains inherently restrictive. The development of theoretically robust analytical methods would enable comprehensive analyses and calculations for excavations in analogous site pits if realized. The fusion of these theoretical analytical methods with numerical simulation tools can furnish robust safety assurances and elevate the overall quality of pit-construction endeavors. Thus, the establishment of a robust theoretical analysis method becomes imperative. Going forward, a theoretical analysis approach could be adopted to investigate how pit excavation affects the internal pile foundation of the excavation site. This entails examining forces and deformations and formulating a comprehensive and pragmatic methodology for analyzing and mitigating pile foundation deformations.

Author Contributions: D.W.: Conceptualization, Methodology, Software, Validation, Formal analysis, Writing—review and editing. S.Y.: Conceptualization, Methodology, Supervision, Writing—review and editing, Funding acquisition, Project administration. L.X.: Methodology, Validation, Data curation, Writing—review and editing, Software. All authors have read and agreed to the published version of the manuscript.

Funding: This research was funded by National Natural Science Foundation of China (No. 52168050).

Data Availability Statement: Due to legal issues and trade secrets, the monitoring data in this article cannot be provided free of charge. Nevertheless, all concepts and data are explained in the submitted studies, and some studies are available upon request.

Acknowledgments: The authors gratefully acknowledge the financial support provided by National Natural Science Foundation of China (No. 52168050).

Conflicts of Interest: The authors declare that they have no known competing financial interests or personal relationships that could have appeared to influence the work reported in this paper.

References

1. Whittle, A.J.; Hashash, Y.M.A.; Whitman, R.V. Analysis of Deep Excavation in Boston. *J. Geotech. Eng.* **1993**, *119*, 69–90. [CrossRef]
2. Schäfer, R.; Triantafyllidis, T. Modelling of Earth and Water Pressure Development during Diaphragm Wall Construction in Soft Clay: Pore pressure development during wall construction. *Int. J. Numer. Anal. Methods Geomech.* **2004**, *28*, 1305–1326. [CrossRef]
3. Likitlersuang, S.; Surarak, C.; Wanatowski, D.; Oh, E.; Balasubramaniam, A. Finite Element Analysis of a Deep Excavation: A Case Study from the Bangkok MRT. *Soils Found.* **2013**, *53*, 756–773. [CrossRef]
4. Li, D.; Li, Z.; Tang, D. Three-Dimensional Effects on Deformation of Deep Excavations. *Proc. Inst. Civ. Eng. Geotech. Eng.* **2015**, *168*, 551–562. [CrossRef]
5. Lee, F.-H.; Hong, S.-H.; Gu, Q.; Zhao, P. Application of Large Three-Dimensional Finite-Element Analyses to Practical Problems. *Int. J. Geomech.* **2011**, *11*, 529–539. [CrossRef]
6. Chheng, C.; Likitlersuang, S. Underground Excavation Behaviour in Bangkok Using Three-Dimensional Finite Element Method. *Comput. Geotech.* **2018**, *95*, 68–81. [CrossRef]
7. Tan, Y.; Wang, D. Characteristics of a Large-Scale Deep Foundation Pit Excavated by the Central-Island Technique in Shanghai Soft Clay. I: Bottom-Up Construction of the Central Cylindrical Shaft. *J. Geotech. Geoenviron. Eng.* **2013**, *139*, 1875–1893. [CrossRef]
8. Li, Z. Displacement Monitoring during the Excavation and Support of Deep Foundation Pit in Complex Environment. *Adv. Civ. Eng.* **2021**, *2021*, 5715306. [CrossRef]
9. Liu, J.; Shi, C.; Cao, C.; Lei, M.; Wang, Z. Improved Analytical Method for Pile Response Due to Foundation Pit Excavation. *Comput. Geotech.* **2020**, *123*, 103609. [CrossRef]
10. Zhang, W.; Goh, A.T.C.; Xuan, F. A Simple Prediction Model for Wall Deflection Caused by Braced Excavation in Clays. *Comput. Geotech.* **2015**, *63*, 67–72. [CrossRef]
11. Arai, Y.; Kusakabe, O.; Murata, O.; Konishi, S. A Numerical Study on Ground Displacement and Stress during and after the Installation of Deep Circular Diaphragm Walls and Soil Excavation. *Comput. Geotech.* **2008**, *35*, 791–807. [CrossRef]
12. Zdravkovic, L.; Potts, D.M.; St John, H.D. Modelling of a 3D Excavation in Finite Element Analysis. *Géotechnique* **2005**, *55*, 497–513. [CrossRef]
13. Finno, R.J.; Harahap, I.S. Finite Element Analyses of HDR-4 Excavation. *J. Geotech. Eng.* **1991**, *117*, 1590–1609. [CrossRef]
14. Song, G.; Xu, J.; Marshall, A.M. Numerical Study on the Effect of Protective Wall Depth in Reducing Structure Deformations Caused by Tunnelling. *Comput. Geotech.* **2023**, *158*, 105374. [CrossRef]
15. Sugimoto, M.; Sramoon, A. Theoretical Model of Shield Behavior During Excavation. I: Theory. *J. Geotech. Geoenviron. Eng.* **2002**, *128*, 138–155. [CrossRef]
16. Finno, R.J.; Blackburn, J.T.; Roboski, J.F. Three-Dimensional Effects for Supported Excavations in Clay. *J. Geotech. Geoenviron. Eng.* **2007**, *133*, 30–36. [CrossRef]
17. Finno, R.J.; Bryson, S.; Calvello, M. Performance of a Stiff Support System in Soft Clay. *J. Geotech. Geoenviron. Eng.* **2002**, *128*, 660–671. [CrossRef]
18. Orazalin, Z.Y.; Whittle, A.J.; Olsen, M.B. Three-Dimensional Analyses of Excavation Support System for the Stata Center Basement on the MIT Campus. *J. Geotech. Geoenviron. Eng* **2015**, *141*, 05015001. [CrossRef]
19. Tan, Y.; Wei, B. Observed Behaviors of a Long and Deep Excavation Constructed by Cut-and-Cover Technique in Shanghai Soft Clay. *J. Geotech. Geoenviron. Eng.* **2012**, *138*, 69–88. [CrossRef]
20. Liu, G.B.; Ng, C.W.; Wang, Z.W. Observed Performance of a Deep Multistrutted Excavation in Shanghai Soft Clays. *J. Geotech. Geoenviron. Eng.* **2005**, *131*, 1004–1013. [CrossRef]
21. Pile Foundation Analysis and Design: Poulos, H G; Davis, E H New York: Wiley, 1980, 397P. *Int. J. Rock Mech. Min. Sci. Geomech. Abstr.* **1981**, *18*, 100. [CrossRef]
22. Cui, X.; Ye, M.; Zhuang, Y. Performance of a Foundation Pit Supported by Bored Piles and Steel Struts: A Case Study. *Soils Found.* **2018**, *58*, 1016–1027. [CrossRef]
23. Dong, Y.P.; Burd, H.J.; Houlsby, G.T. Finite Element Parametric Study of the Performance of a Deep Excavation. *Soils Found.* **2018**, *58*, 729–743. [CrossRef]
24. Goh, A.T.C.; Zhang, F.; Zhang, W.; Zhang, Y.; Liu, H. A Simple Estimation Model for 3D Braced Excavation Wall Deflection. *Comput. Geotech.* **2017**, *83*, 106–113. [CrossRef]
25. Wu, J.; Ye, S.; Wang, Z.; Yang, D. Application and Automatic Monitoring and Analysis of Hybrid Support Structure in Ultra-DEEP Foundation Pit Engineering in the Lanzhou Area under Complex Environmental Conditions. *Water* **2023**, *15*, 1335. [CrossRef]

26. Ye, S.; Zhao, Z.; Wang, D. Deformation Analysis and Safety Assessment of Existing Metro Tunnels Affected by Excavation of a Foundation Pit. *Undergr. Space* **2021**, *6*, 421–431. [CrossRef]
27. Guo, P.; Gong, X.; Wang, Y. Displacement and Force Analyses of Braced Structure of Deep Excavation Considering Unsymmetrical Surcharge Effect. *Comput. Geotech.* **2019**, *113*, 103102. [CrossRef]
28. Mei, Y.; Wang, L.; Zhou, D.; Fu, L. Displacement Characteristics of a Deep Excavation in Hangzhou Soft Clay. *Adv. Civ. Eng.* **2022**, *2022*, 5469471. [CrossRef]
29. Huang, Y.; Zhu, C.Q. Safety Assessment of Antiliquefaction Performance of a Constructed Reservoir Embankment. I: Experimental Assessment. *J. Perform. Constr. Facil.* **2017**, *31*, 04016101. [CrossRef]
30. Zhu, C.Q.; Huang, Y. Safety Assessment of Antiliquefaction Performance of a Constructed Reservoir Embankment. II: Numerical Assessment. *J. Perform. Constr. Facil.* **2017**, *31*, 04016102. [CrossRef]
31. *GB 50497-2019*; Technical Standard for Monitoring of Building Excavation Engineering. National Standardization Administration of China: Beijing, China, 2019.
32. *JGJ 106-2014*; Technical Specification for Testing of Building Foundation Piles. Ministry of Housing and Urban-Rural Development of the People's Republic of China: Beijing, China, 2014.

water

MDPI

Article

Risk Reduction Measures and Monitoring Analysis of Deep Foundation Pit with Water in a Metro Station in Hefei

Dengqun Wang [1], Shuaihua Ye [2,*] and Jun Zhang [2]

[1] HFUT Design Institute (Group) Co., Ltd., Hefei 230051, China
[2] School of Civil Engineering, Lanzhou University of Technology, Lanzhou 730050, China
* Correspondence: yeshuaihua@163.com

Abstract: The construction of an urban metro will inevitably involve deep excavation. Risk assessment before deep excavation, risk reduction measures, and real-time monitoring during excavation can effectively ensure the safety of deep excavation. Taking the deep excavation pit of Lingbi Road Station of Hefei Rail Transit Line 8 as the research object, this paper first analyses and evaluates the self-risk, groundwater risk, and surrounding environmental risk of the deep excavation pit, and gives the corresponding measures to reduce the risk of the deep excavation pit. Then, the monitoring content of the excavation process is determined according to the environment of the excavation, the hydrogeological conditions, and the type of supporting structure, and the monitoring scheme is designed. Finally, the entire excavation process is monitored in real time. By analyzing the monitoring data of 13 projects, such as horizontal displacement of the wall top, axial support force, groundwater level, etc., it is found that the monitoring values of 13 projects do not exceed the control value. This proves that the composite internal bracing structure of the underground diaphragm wall is suitable for deep foundation pit support in the Hefei area, as the selection of the water-bearing deep foundation pit support structure, the value of the support structure parameters, and the design of the foundation pit dewatering scheme are all reasonable. The study of this paper also serves as a case reference for the support design of water-bearing deep excavation of subway station in Hefei area.

Keywords: subway station; water-bearing deep foundation pit; composite internal bracing of diaphragm wall; risk reduction measures; monitoring and analysis

check for updates

Citation: Wang, D.; Ye, S.; Zhang, J. Risk Reduction Measures and Monitoring Analysis of Deep Foundation Pit with Water in a Metro Station in Hefei. *Water* **2023**, *15*, 3007. https://doi.org/10.3390/w15163007

Academic Editor: Chin H Wu

Received: 19 July 2023
Revised: 8 August 2023
Accepted: 16 August 2023
Published: 21 August 2023

1. Introduction

In the last few years, China's accelerating urbanization has led to a massive increase in the urban population, which leads to the shortage of urban land and great pressure on traffic. To this end, many cities began to vigorously develop underground space, and use underground space to build garages, commercial streets, and subways [1,2]. On the one hand, the construction of underground garages, commercial streets, and subways can greatly alleviate the pressure of urban land use. On the other hand, the construction of underground garages, commercial streets, and subways can promote the development of a city [3–5]. When building an underground car park, a shopping street or a metro, it is inevitable that you will come across a foundation pit. The excavation depth of underground station foundation pits tends to be relatively large and the environment relatively complex, which makes the design of underground station foundation pit support more difficult than the design of general foundation pit support.

At present, the support structure forms of foundation pits in different cities are also different due to the different geological conditions, hydrogeological conditions, and environment of foundation pits in different cities [6–9]. The research on the excavation process of foundation pit can clarify the change law between the displacement of support structure, groundwater level and surface settlement, and the excavation depth of foundation pit. Sun et al. [10] investigated the deep horizontal displacement and deformation of the pile body during the excavation of the deep foundation pit of the subway transfer station by means of field monitoring in combination with finite element simulation. Yang et al. [11] used the finite element software Midas GTS NX to simulate the construction of deep foundation pits in the vicinity of subway stations in areas with soft soil. The effect of constructing deep foundation pits on the structural deformation and stress of the subway station is investigated, and the influence of the pit supporting structure on the station is analyzed, based on geotechnical investigation, building information modeling, and multi-source sensor technology. Hong et al. [12] proposed an integrated intelligent method for the monitoring and management of deep foundation pits in subway stations. And real-time and high-precision monitoring of the foundation pit can be achieved using this method. Yang et al. [13] collected the monitoring data of deep foundation excavations in soft soil of six railway lines in Fuzhou, and obtained the deformation law of large and deep foundation excavations in Fuzhou's soft soil layers through comparison with the research results of similar projects in domestic and foreign countries. Wang et al. [14] proposed a method suitable for deep foundation pit support of swampy metro stations by analyzing and optimizing the parameters of grouting reinforcement, which can significantly reduce the amount of grouting. Liu et al. [15] used field observation combined with numerical simulation to study the stress and deformation of the support structure of the deep excavation of an underground railway, and made some proposals on the monitoring and construction of the deep excavation of the underground railway according to the analysis results. Based on 3D FEA software, Zhang et al. [16] conducted a deep excavating and drainage simulation for a T-shaped metro station, and analyzed deformation characteristics during excavating. Relying on a metro station in Xiamen, Zhao et al. [17] investigated the deformation response of an underground foundation pit with diaphragm wall composite internal reinforcement using the method of field observation combined with numerical simulation. Risk assessment of the trench before it is excavated can predict the potential safety hazards of the trench and reduce the likelihood of accidents. Wu et al. [18] established a risk assessment index system on the basis of the support structure, the asymmetric earth pressure of the subway station, the foundation pit condition, the surroundings, and the safety and security management on construction sites. And the system can reasonably and effectively evaluate the asymmetric risk of building a deep foundation for a metro station. Ye et al. [19] used PLAXIS 3D FES to model the excavation of foundation pits adjacent to subway tunnels. The influence of the excavation of the foundation pit due to the deformation of the adjacent tunnel is analyzed, and the supporting structure of the excavation is optimized in accordance with the numerical simulation results to keep the excavations. Shen et al. [20] established a three-level fuzzy comprehensive assessment model on the premise of analyzing the factors involved in building subway stations. And this model can comprehensively assess the risk involved in constructing subway station excavations. Zhou et al. [21] established an intelligent model based on the machine learning model for risk prediction of deep foundation excavations in metro stations. And the importance assessment function of this model can help site engineers determine the cause of safety risks.

From the above research, it can be found that the current research on the deep foundation pit of the subway station is mainly focused on the safety of the foundation pit itself and its impact on the surrounding buildings, and there is little research on the potential risk assessment of the foundation pit before the foundation pit excavation. In particular, there is almost no risk assessment and monitoring of the excavation of the foundation pit for the Hefei metro station. This paper first analyzes and assesses the self-risk, the ground-water risk, and the surrounding environmental risk of the deep excavation of the Lingbi Road Station of Hefei Railway No. 8, and then provides the corresponding measures to reduce the risk of the deep excavation. Then, the entire excavation process is monitored in real time, and the safety of the excavation process and the rationality of the choice of support structure and parameters are revealed by analyzing the monitoring data from 13 projects. The study of this paper can provide a case reference for the support design of water-bearing deep foundation pit of subway station in Hefei area.

2. Project Overview

2.1. Site Profile

Lingbi Road Station is the second station of the first phase of Rail Transit Line 8 in Hefei City, Anhui Province, China. The station is located at the intersection of Fuyang North Road and Lingbi Road, and is built along Fuyang North Road. The station is an open-cut underground three-story double-column and three-span island station. The east side of the station is Fuyang North Road viaduct. The northeast quadrant of the station is the dormitory of Hefei Prospecting Machinery Factory. The southeast quadrant of the station is Anhui Survey and Design Institute. The southwest quadrant of the station is the second dormitory of the Municipal Public Security Bureau. The northwest quadrant of the station is the warehouse of Hefei Prospecting Machinery Factory. The red line width of Fuyang North Road in this site is 60 m, which is the main road of the city. The viaduct of Fuyang North Road is laid along the road, and the deck width of the standard section of the viaduct is 25.5 m. The red line of Lingbi Road is 24 m wide and is a side road of the city. The surrounding plan of the foundation pit is shown in Figure 1.

Figure 1. Surrounding environment plan of foundation pit.

There are two groups of wind pavilions, four entrances and exits, and two safety exits at Lingbi Road Station. The standard section excavation depth is approximately from 22.8 m to 23.6 m, the small mile end shaft excavation depth is approximately 24.1 m, and the large mile end shaft excavation depth is approximately 25.1 m. The total standard section width is 21.9 meters, the effective platform width is 13 meters, and the total station length is 155 meters. The two ends of the station are built using the shielding technique, with the small mileage end of the station equipped with a shielded launch pad and the large mileage end equipped with a shielded receiving pad. The station's main body is constructed by the open-cut method.

2.2. Engineering Geology Overview

The landform unit of Lingbi Road Station is a first-class terrace, the site is flat and open, with a natural slope of about 3–5° and an elevation of 23–27 m. The site is mainly the main road of urban traffic and residential area. Due to the long-term construction, most of the original landform has been changed. Table 1 shows the physical and mechanical parameters of the soil layer at Lingbi Road Station.

Table 1. The physical and mechanical parameters of each soil layer in Lingbi Road Station.

Soil code	Name of soil layer	Status	Floor level (m)	Natural Heavy $\gamma/(kN/m^3)$	Cohesion $c/(kPa)$
1	Fill soil	Loose to slightly dense	20.28~26.29	18	5
2-1	Clay soil	Malleable	12.28~21.29	19.4	35
2-2	Clay soil	Hard plastic	5.25~10.50	19.7	40
3-1	Silt	Medium secret	0.19~4.08	20.1	20
3-2	Silty sand	Medium secret	−1.91~1.28	20.1	1
4	Fully weathered argillaceous sandstone	The core is in the form of an earth column	−4.34~−1.11	20	30
5	Strongly weathered argillaceous sandstone	The core is massive	−8.34~−3.24	21	90
6	Moderately weathered argillaceous sandstone	The core is columnar	Not penetrated	23	140

Soil code	Name of soil layer	Internal friction Angle $\varphi/(°)$	Permeability coefficient (m/d)	Standard value of limit resistance of soil on the side of pile (kPa)	Standard value of limit resistance of soil at the end of pile (kPa)
1	Fill soil	7	0.1~5	—	—
2-1	Clay soil	17.4	0.0035	60	—
2-2	Clay soil	17	0.0024	85	—
3-1	Silt	23.1	0.1~0.5	45	—
3-2	Silty sand	24	1~2	45	—
4	Fully weathered argillaceous sandstone	16.2	0.1~0.2	85	1300
5	Strongly weathered argillaceous sandstone	22	0.1	160	1800
6	Moderately weathered argillaceous sandstone	28	0.05	200	2800

2.3. Hydrogeological Conditions

There are no rivers, reservoirs, and other water bodies in the proposed project area. The groundwater in the proposed project area is mainly perched water, micro-confined water, bedrock pore water, and fissure water. The perched water mainly occurs in the artificial fill, the water quantity is poor, the distribution is discontinuous, and the buried depth of the water level is 2.4~5.2 m. The confined water mainly occurs in silt (3-1) and silty sand (3-2). The buried depth at the top of silt (3-1) is 15.7–20.6 m, and the elevation of the top is 5.25–9.97 m. The buried depth at the top of silty sand (3-2) is 21.4~25.5 m, and the elevation of the top of the layer is 1.5~4.1 m. Silt (3-1) layer and silty sand (3-2) layer are generally rich in water, and the amount of groundwater gushing is small and has micro-pressure bearing property. The pore water of bedrock mainly occurs in fully weathered argillaceous sandstone (4), which is rich in water and weak in permeability. Bedrock fissure water mainly occurs in strongly weathered argillaceous sandstone (5) and moderately weathered argillaceous sandstone (6), which is rich in water and weak in permeability. In this survey section, the corrosion grade of groundwater to concrete structure is micro, and the corrosion grade of steel bar in concrete structure is micro. The corrosion grade of surface water to concrete structure is micro, and the corrosion grade of steel bar in concrete structure is micro.

3. Risk Characteristics and Countermeasures of Foundation Pit Engineering

There are various risks involved in the excavation of the foundation pit at Lingbi Road Subway Station. In order to reduce the likelihood of excavation accidents, it is necessary to evaluate the various risks of the excavation project and formulate countermeasures before excavating the foundation pit.

3.1. Own Risks and Countermeasures

3.1.1. Risk Situation

The excavation depth of the station standard section is 22.8~23.6 m, and the excavation width is 21.9 m. The excavation depth of the small mileage end well foundation pit is about 23.8 m, the excavation depth of the long mileage end well foundation pit is about 24.8 m, and the excavation width of the final pit is approximately 26.45 m. The excavation method of the foundation pit is the open-cut method, and the surrounding environment is normal. Soils within the pit excavation include clay (2-1), clay (2-2), and silt (3-1), and there is micro-confined water in silt (3-1) and silt (3-2). After a comprehensive evaluation, the risk level of the foundation pit itself is level one.

3.1.2. Countermeasures

(1) The ground in a certain area around the foundation pit is hardened to cut off the recharge of surface water and atmospheric precipitation to groundwater, and at the same time, the drain water on the ground and in the pit in time. (2) During the construction of foundation pit, it is necessary to ensure layered excavation, timely installation of bracing, strict prohibition of over excavation, and the steel bracing has anti-loosening and pre-stressing additional measures. (3) Monitoring of the foundation pit and observation and treatment of the diving level during construction. (4) To formulate emergency measures to deal with unfavorable weather, such as sudden heavy rainfall and instantaneous rainstorm, during construction. The foundation pit's own residual risk level after the countermeasures is level three.

3.2. Risks of Foundation Pit Dewatering and Its Countermeasures

3.2.1. Risk Situation

Confined water mainly occurs in silt (3-1) and silt (3-2). According to the on-site exploration, the water-rich situation of this layer is general, the amount of groundwater gushing is small, and it has micro-pressure bearing property. The risk level of foundation pit dewatering is level two.

3.2.2. Countermeasures

(1) The foundation pit adopts an underground diaphragm wall with a thickness of 1000 mm, and the bottom of the wall enters the moderately weathered rock layer to isolate the confined water layer. (2) Pressure-relief wells and water-collecting open drains are installed in the foundation pit. The type of pressure-relief wells is tube well with a diameter of 600 mm, and the pipe well is 6 m deeper than the bottom of foundation pit. Observation wells and emergency backup wells are installed outside the foundation pit. The risk level of foundation pit dewatering after the countermeasures is level three.

3.3. Environmental Risks and Countermeasures

3.3.1. Risk Situation of Underground Pipeline

(1) The DN1000 rainwater pipe is buried at a depth of about 2.0 m, which is arranged along Lingbi Road and is located on the west side of the main body of the station. (2) The DN700 sewage pipe is buried at a depth of about 1.5 m and is located on the west and south sides of the main foundation pit. (3) The D200 gas pipe is buried at a depth of about 1.4 m and is located on the west and south sides of the main foundation pit. (4) The DN700 heat pipe is buried at a depth of about 2.1 m and is located on the east side of the main foundation pit. The foundation pit risk level caused by underground pipelines is level two.

3.3.2. Countermeasures

(1) Before the construction of the main foundation pit, prepare a special emergency plan for risk events, such as excessive pipeline deformation and pipeline leakage. (2) The relocation distance of the pipeline meets the requirements of the code, and at the same time, it is analyzed that the deformation caused by the excavation of the foundation pit must meet the deformation protection requirements. (3) Strengthen the monitoring and protection of the pipeline. After the pipeline is relocated, direct measuring points should be arranged and the formation deformation at the location of the pipeline after the relocation should be monitored. Once the alarm value is exceeded, take the corresponding control measures, such as grouting and reinforcement. The residual risk level of the underground pipeline after the countermeasures is level three.

3.3.3. Risk Situation of Fuyang North Road Viaduct and Buildings around the Foundation Pit

Fuyang Road viaduct (see Figure 2), the main bridge of Fuyang Road viaduct, is laid along the north-south direction of Fuyang North Road, and is located on the east side of the main body of the station. The bridge adopts pile foundation + cap foundation, the pile diameter is 1.5 m, the pile length is 39~41 m, and the bridge deck width is 25.5 m. The bridge piles of the main bridge are about 20.66 m closest to the main foundation pit of the station. The on-ramp bridge is laid along the north-south direction of Fuyang North Road and is located on the east side of the main body of the station. The ramp bridge adopts pile foundation + cap foundation, the pile diameter is 1.2 m, the pile length is 32~35 m, and the bridge deck width is 8.5 m. The nearest distance between the ramp bridge pile and the main foundation pit of the station is about 4.31 m. There is an abandoned civil air defense passage at the north end of the station and near the entrance and exit on the east side of the station, running east-west and north-south. The connecting channel is a brick arch structure with a buried depth of about 10 m, a width of about 2 m, and a height of about 2.5 m. The risk level of the foundation pit caused by the Fuyang North Road viaduct is level one.

Figure 2. Fuyang Road viaduct and buildings around the foundation pit.

The buildings around the foundation pit are shown in Figure 2. (1) The guard room of the factory building of Hefei Prospecting Machinery Factory is a brick structure on the first floor, about 25 m away from the main foundation pit of the station. (2) The private house is a one-story brick structure, about 43 m away from the main foundation pit of the station. (3) The factory building of Hefei Prospecting Plant is a one-story brick structure, about 39 m away from the main foundation pit of the station. (4) The No. 1 residential building in the second living area of Hefei Public Security Bureau is a seven-story concrete structure, and the closest distance to the main foundation pit of the station is about 32 m. (5) The No. 5 residential building in the second living area of the Hefei Public Security Bureau is a six-story concrete structure, and the closest distance to the main foundation pit of the station is about 29 m. (6) The outpatient building is a four-story concrete structure, about 14 m away from the main foundation pit of the station. The risk level of the foundation pit caused by the surrounding buildings is level two.

3.3.4. Countermeasures

(1) The foundation pit support adopts underground diaphragm wall + internal bracing system, the thickness of the ground connection wall is 1000 mm, the first bracing adopts concrete support, and the second, third, and fourth bracing positions are steel bracings. The length of the underground continuous wall is 31.10~34.39 m, and the embedded depth is 10.36~10.49 m. The section size of the first reinforced concrete bracing is 1000 mm × 1000 mm, the diameter of second and third steel bracings is 800 mm, and the wall thickness is 20 mm. The diameter of fourth steel bracing is 800 mm and the wall thickness is 16 mm The axial force servo system is added within the influence range of the viaduct piles, and the deformation of the foundation pit and the surrounding environment is controlled by strengthening the rigidity of the support structure. The section view of the support structure is shown in Figure 3. (2) Before construction, the third party shall evaluate the current quality and allowable deformation of the building, and propose the limit deformation value, warning value, and early warning value of the foundation allowed by the current building. The early warning value is 70% of the ultimate deformation value of the foundation, and the warning value is 80% of the ultimate deformation value of the foundation. (3) During the construction of the foundation pit, strengthen the monitoring of viaduct pile foundations and other buildings. According to the monitoring results, the construction plan and method can be dynamically adjusted in a timely manner, and the grouting can be tracked in real time. The residual risk level of the surrounding environment after the countermeasures is level three.

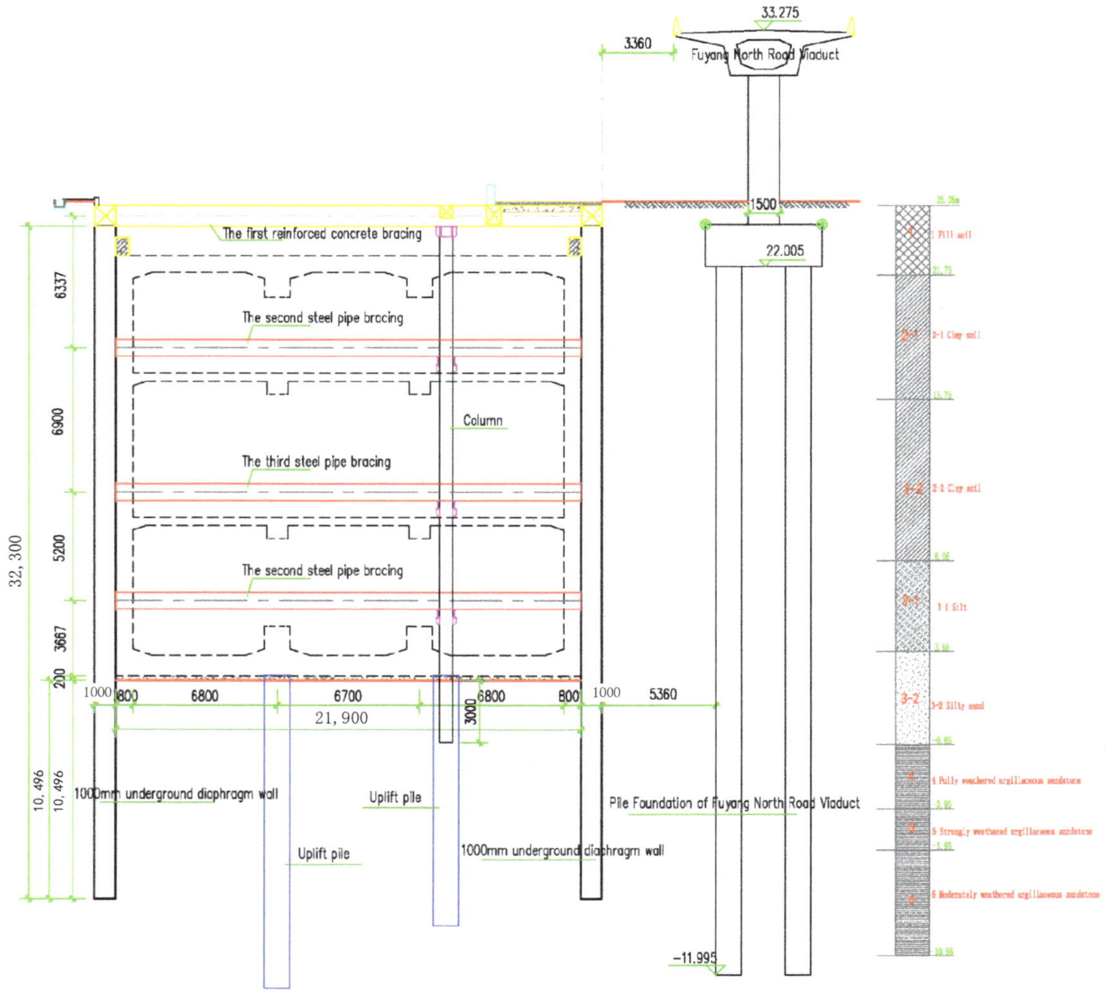

Figure 3. The section view of the support structure. Field monitoring and data analysis.

4. Field Monitoring and Data

According to the surrounding environment, hydrogeological conditions, and supporting structure types of the foundation pit of Lingbi Road Station, the monitoring objects of this project mainly include open-cut foundation pit, cover-excavated roof, surrounding important buildings, surrounding ground roads, surrounding important underground pipelines, and surrounding rocks and soils. For the above monitoring objects, the specific monitoring items of this foundation pit project mainly include horizontal displacement at the top of wall, vertical displacement at the top of wall, horizontal displacement in the deep layer of wall, axial force of the internal bracing, settlement of the columns, horizontal displacement of the columns, settlement of the building, groundwater level, surface settlement, pipeline settlement, settlement of the bridge abutment, horizontal displacement of the bridge abutment, and tilt of the bridge abutment. Figure 4 shows the plan view of the monitoring points in foundation pit, and the number of monitoring points are shown in Table 2.

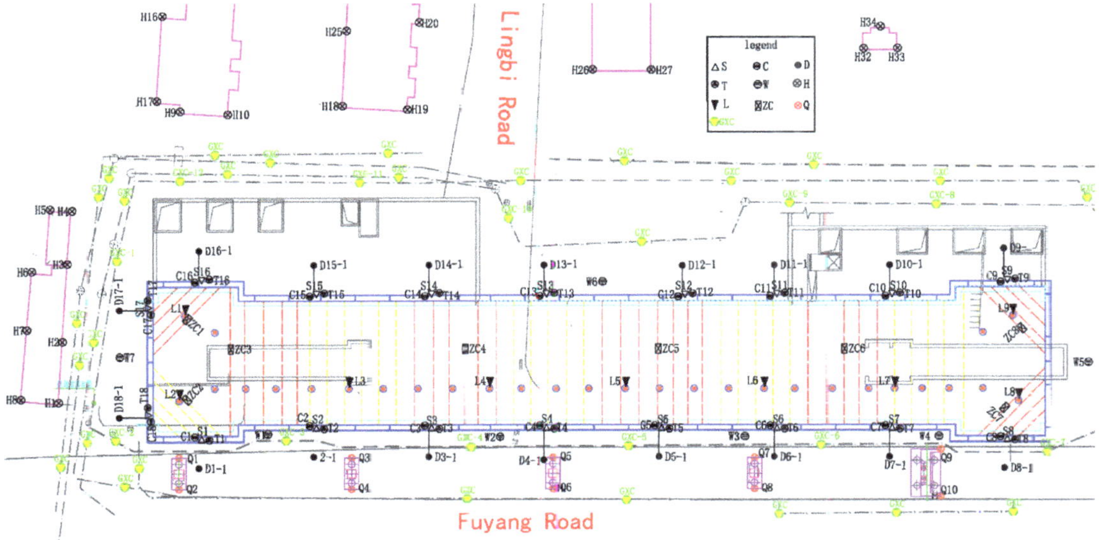

Figure 4. Plan view of monitoring points in foundation pit.

Table 2. Number of monitoring points.

Monitoring Project	Number of Monitoring Points
Horizontal displacement at the top of wall	18
Vertical displacement at the top of wall	18
Horizontal displacement in the deep layer of wall	18
Axial force of the internal bracing	32
Settlement of the columns	9
Horizontal displacement of the columns	9
Settlement of the building	37
Groundwater level	7
Surface settlement	74
Pipeline settlement	35
Settlement of the bridge abutment	60
Horizontal displacement of the bridge abutment	60
Tilt of the bridge abutment	60

4.1. Horizontal Displacement at the Top of Wall

The construction process of the underground diaphragm wall is shown in Figure 5. The horizontal displacement monitoring points at the top of the retaining wall are arranged in an array-like pattern along the perimeter of the foundation pit, with one measurement point every 20 to 40 m. It is advisable to place the monitors in the middle of each side of the trench, at the sun angles and near the shelter, and to place two monitors at each end-well on either side of the station. A total of 18 monitoring points are in place, with numbers from S1 to S18. The instrument used to monitor the horizontal displacement at the top of the wall is a Leica TS09 total station, and the monitoring map of the field is shown in Figure 6. Figure 7 shows the results of monitoring the horizontal displacement at the top of the wall.

Figure 5. Construction picture of underground diaphragm wall.

Figure 6. Field monitoring map of horizontal displacement at the top of wall.

As shown in Figure 7, before 1 August 2021, the horizontal cumulative displacement of each monitoring point is gradually increasing, and after 1 August 2021, the cumulative horizontal displacement of each monitoring point decreases sharply. This is due to the fact that before 1 August 2021, the excavation of the foundation pit begins after the construction of the diaphragm wall and the first reinforced concrete bracing is completed, and the cumulative value of the horizontal displacement at the top of the wall increases with the increase in the excavation depth. After 1 August 2021, the second steel pipe bracing was constructed, and the second steel pipe bracing played a certain role in controlling the deformation of the foundation pit; therefore, the cumulative value of horizontal displacement at the top of the wall decreased sharply. After 15 November 2021, each monitoring point's cumulative horizontal displacement remains basically unchanged. This is due to the fact that the foundation pit has been excavated to the bottom of the pit on 15 November 2021, after which the foundation construction and the construction of the main structure of the subway station begin. When the depth of the foundation pit no longer changes, the horizontal displacement at the top of the wall changes a little. It can also be seen from Figure 7 that the horizontal cumulative displacement value of S14 monitoring point is the largest, and the displacement value is 2.34 mm. In the whole process of foundation pit excavation, the monitoring values of the horizontal displacement monitoring points at the top of the wall did not exceed the control value (15 mm) set by the code [22], which meets the design requirements.

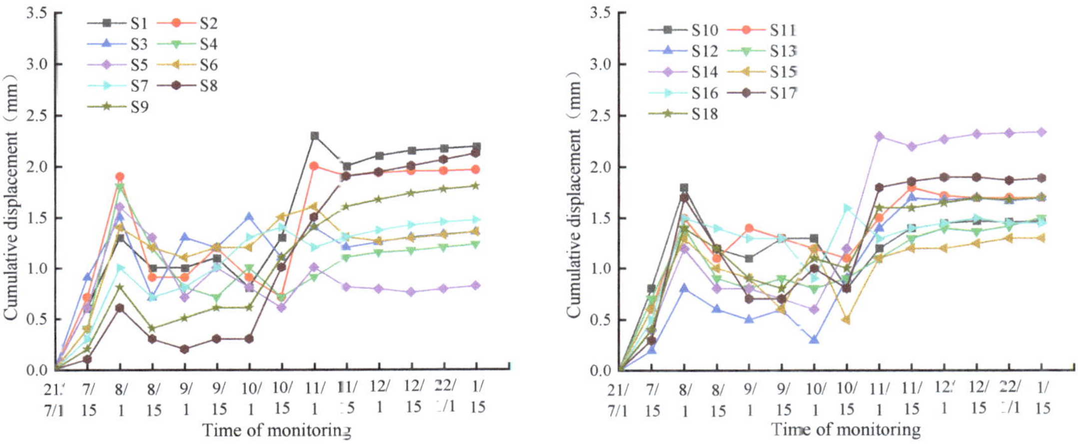

Figure 7. Time history curve of horizontal displacement at the top of wall.

4.2. Vertical Displacement at the Top of Wall

The vertical displacement monitoring point at the top of the enclosure wall and the horizontal displacement monitoring point at the top of enclosure wall are common points. The vertical displacement monitoring instrument at the top of wall is Trimble DINI03 electronic level, and the field monitoring map is shown in Figure 8. The horizontal displacement monitoring results at the top of wall are shown in Figure 9.

Figure 8. Field monitoring map of vertical displacement at the top of wall.

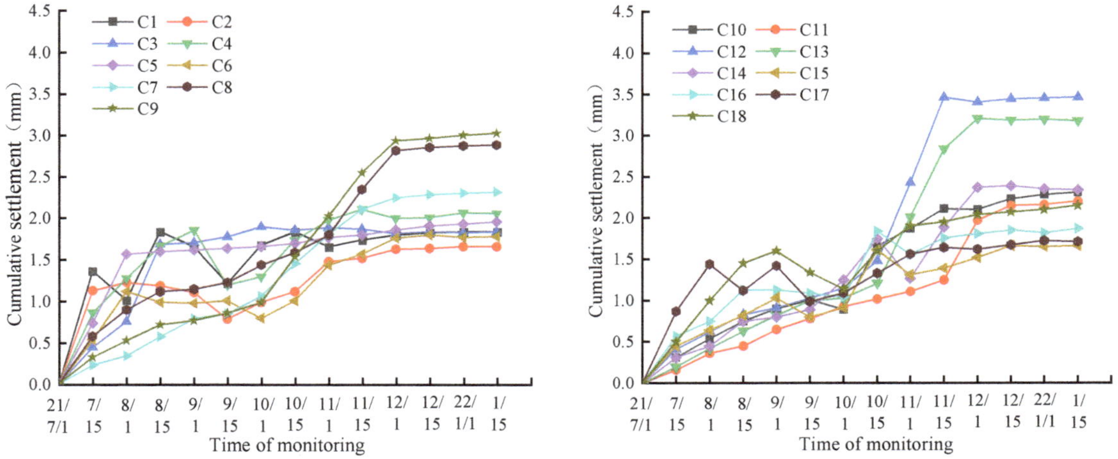

Figure 9. Time history curve of vertical displacement at the top of wall.

As shown in Figure 9, before 1 December 2021, the cumulative settlement of each monitoring point showed an overall increasing trend, but the cumulative settlement of some monitoring points showed a decreasing trend within the range of the small area. After 1 December 2021, the cumulative settlement of each monitoring point has changed very little. It can also be seen from Figure 5 that the cumulative settlement of the C12 monitoring point is the largest, with a settlement of 3.47 mm, and the cumulative settlement of the C2 monitoring point is the smallest, with a settlement of 1.65 mm. During the entire excavation process of the foundation pit, the monitoring values of each vertical displacement monitoring points at the top of the wall did not exceed the control value (10 mm) stipulated in the code [22], which meets the design requirements.

4.3. Horizontal Displacement in the Deep Layer of Wall

The layout of the monitoring points for the deep horizontal displacement of the enclosure wall is the same as that of the horizontal displacement monitoring points for the top of the enclosure wall. The inclinometer pipe is fixed on the steel cage of the wall by direct binding, and after the steel cage is put into the groove, concrete is poured, as shown in Figure 10. A total of 18 monitoring points are in place, with numbers from T1 to T18. The deep horizontal displacement monitoring instrument of the wall is the CX-3C foundation pit inclinometer, and the field monitoring map of horizontal displacement in the deep layer of wall (see Figure 11). According to the importance of the location of the monitoring points, the monitoring results of the deep horizontal displacement of the four monitoring points T2, T4, T6, and T12 with the largest displacement are shown in Figure 12.

Figure 10. The actual picture of the laying of the inclinometer pipe.

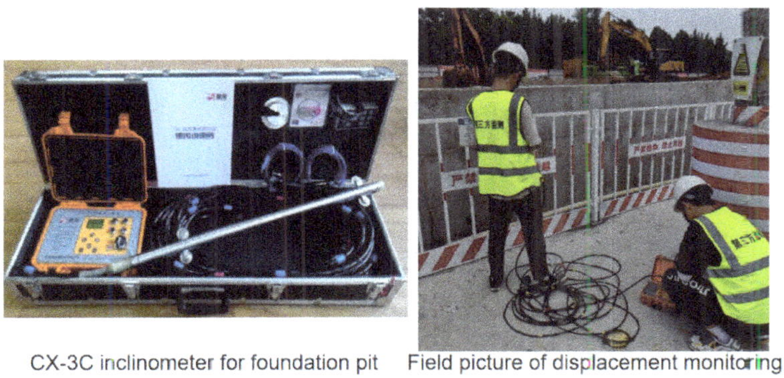

Figure 11. Field monitoring map of horizontal displacement in the deep layer of wall.

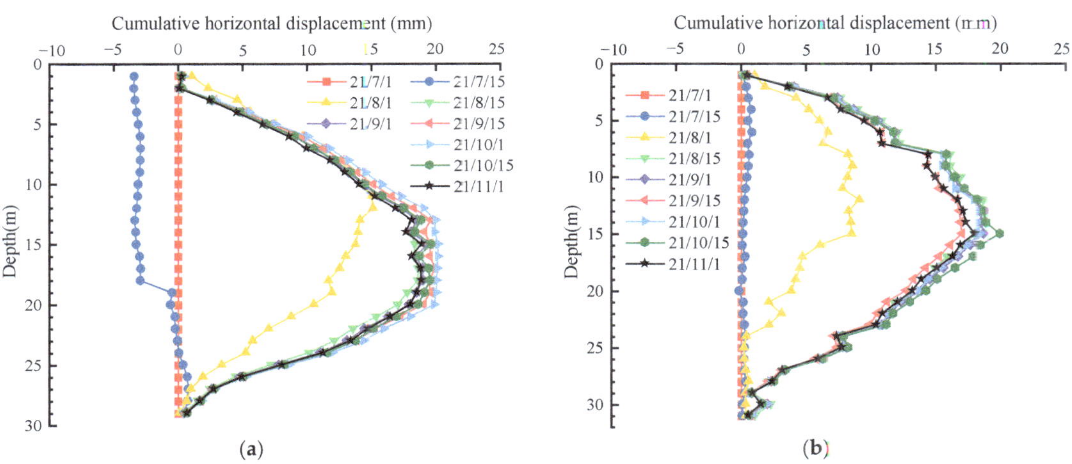

Figure 12. *Cont.*

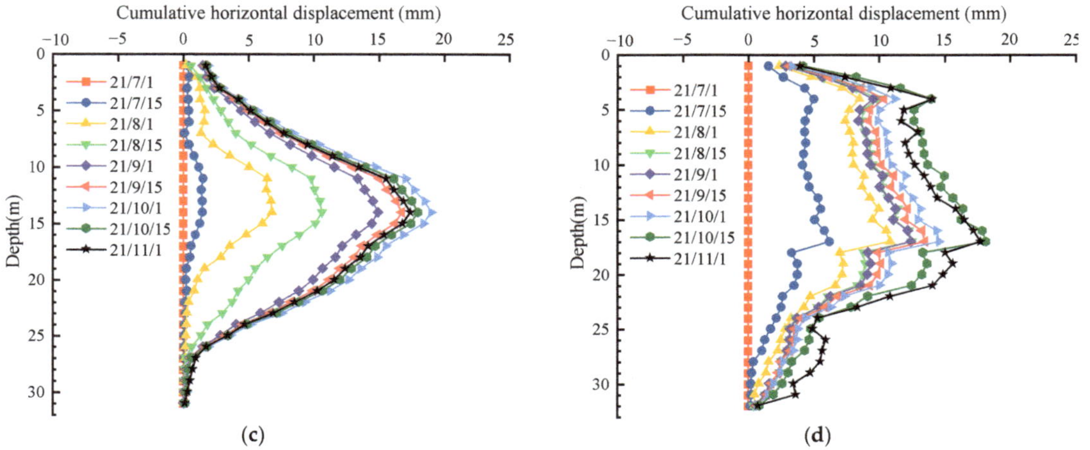

Figure 12. Time history curve of horizontal displacement in the deep layer of wall. (**a**) Time history curve of horizontal displacement in the deep layer of T2. (**b**) Time history curve of horizontal displacement in the deep layer of T4. (**c**) Time history curve of horizontal displacement in the deep layer of T6. (**d**) Time history curve of horizontal displacement in the deep layer of T12.

As shown in Figure 12, except for 1 July and 15 July 2021, the deep horizontal displacements of T2, T4, T6, and T12 monitoring points all showed a trend that is large in the middle and small at the ends for the rest of the time. The above rules also fully demonstrate that in the design of foundation pit support, the middle and lower parts of the foundation pit are the focus of the design. From Figure 12a, it can be seen that the maximum displacement of T2 monitoring point occurred at the buried depths of 15 m and 16 m on 1 October 2021, and the maximum displacement was 20.20 mm. From Figure 12b, it can be seen that the maximum displacement of T4 monitoring point occurred at a buried depth of 15 m on 15 October 2021, and the maximum displacement was 19.98 mm. From Figure 12c, it can be seen that the maximum displacement of T6 monitoring point occurred at the buried depth of 14 m on 1 October 2021, and the maximum displacement was 19.06 mm. From Figure 12d, it can be seen that the maximum displacement of T12 monitoring point occurred at a buried depth of 17 m on 15 October 2021, and the maximum displacement was 18.2 mm. Throughout the excavation of the foundation pit, the monitoring values of the deep horizontal displacement monitoring points of each wall did not exceed the control value (30 mm) specified in the code [22], which meets the design requirements.

4.4. Axial Force of the Internal Bracing

The construction process of the first reinforced concrete internal bracing is shown in Figure 13. The number of monitoring points for each layer of bracing is not less than 10% of the number of each layer of bracing, and not less than three. Eight monitoring points are arranged for the first reinforced concrete bracing, numbered from ZC1 to ZC8. Eight monitoring points are arranged for the second steel bracing, numbered from ZC2-1 to ZC2-8, and the axial force meter of bracing is shown in Figure 14. The bracing axial force monitoring instrument is CTY-202 vibrating wire frequency reading instrument, and the field picture of bracing axial force (see Figure 15). Considering the importance of the bracing position, only the monitoring data of the first bracing and the second bracing are analyzed. Due to the damage of the second bracing monitoring point during the construction process, there are only four monitoring points that can obtain monitoring data. The monitoring results of the bracing axial force are shown in Figure 16.

Figure 13. Construction process of the first reinforced concrete internal bracing.

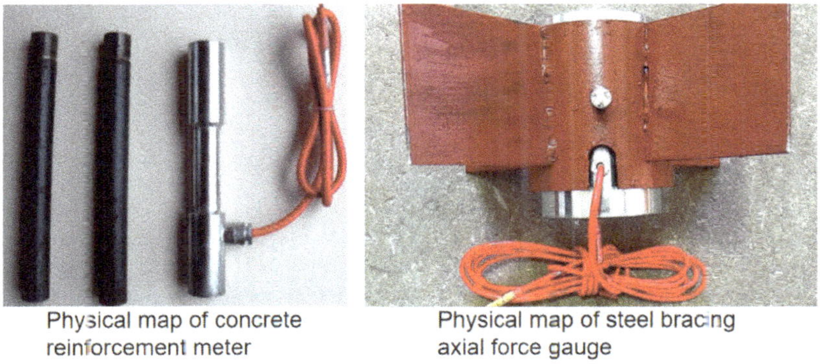

Figure 14. The axial force meter of bracing.

Figure 15. Field picture of bracing axial force.

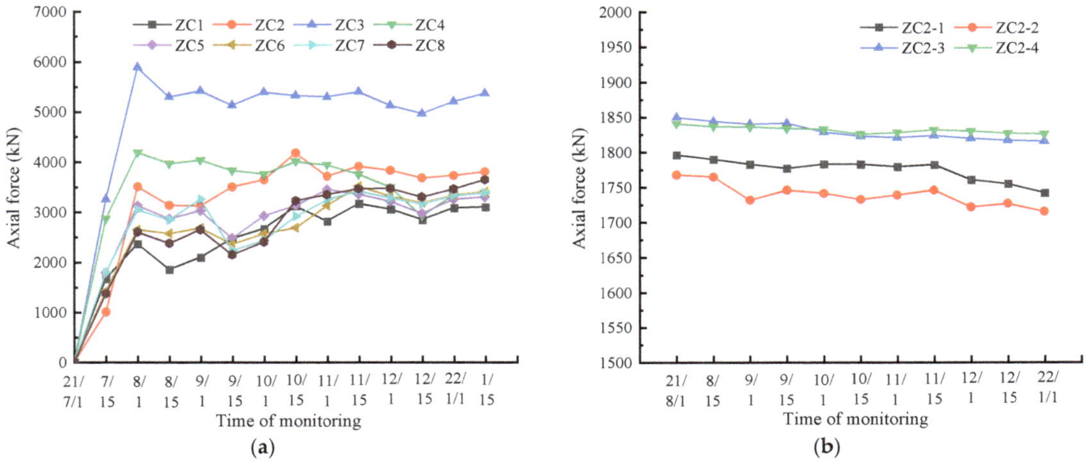

Figure 16. Axial force variation curve of internal bracing. (**a**) Axial force variation curve of first bracing. (**b**) Axial force variation curve of second bracing.

As shown in Figure 16a, before 1 August 2021, the axial force of each monitoring point on the first reinforced concrete bracing gradually increases, and after 1 August 2021, the axial force of each monitoring point Axial force is slightly reduced and then remains stable. This is due to the fact that before 1 August 2021, the excavation of the foundation pit earthwork will start after the construction of the first reinforced concrete bracing. As the excavation depth increases, the earth pressure that the first bracing needs to bear will increase, thus the axial force value is also gradually increasing. After 1 August 2021, the second steel pipe bracing was constructed. The second steel pipe bracing shared part of the earth pressure and reduced the earth pressure borne by the first bracing; therefore, the axial force of the first bracing decreased rapidly. It can be seen from Figure 16b that the axial force of the second steel bracing is gradually decreasing, but the decreasing range is small. This is due to the fact that after the construction of the second steel bracing, the third steel bracing and the fourth steel bracing will be constructed, and the third steel bracing and the fourth steel bracing have reduced the earth pressure on the second steel bracing. As a result, the axial force of the second steel bracing has been showing a decreasing trend. Among the four bracings, the first reinforced concrete bracing has the largest axial force at the monitoring point ZC3, with an axial force value of 5893.52 kN. Throughout the excavation of the foundation pit, the monitoring values of the bracing axial force monitoring points did not exceed the control value (14,000 kN) specified in the code [22], which meets the design requirements.

4.5. Settlement of the Columns

The construction picture of the column is shown in Figure 17. The monitoring number of column settlement shall not be less than 5% of the total number of columns, and not less than three. The column settlement monitoring point is arranged in the middle of the foundation pit and at the intersection of multiple bracings. The layout method is to paste the reflector to the top side of the column exposed by excavation. A total of nine monitoring points are in place, with numbers from L1 to L9, and the column settlement monitoring instrument is TS09 total station. Due to the destruction of some monitoring points in the process of construction, there are only four monitoring points that can read the data. The monitoring results of column settlement are shown in Figure 18.

Figure 17. The construction picture of the column.

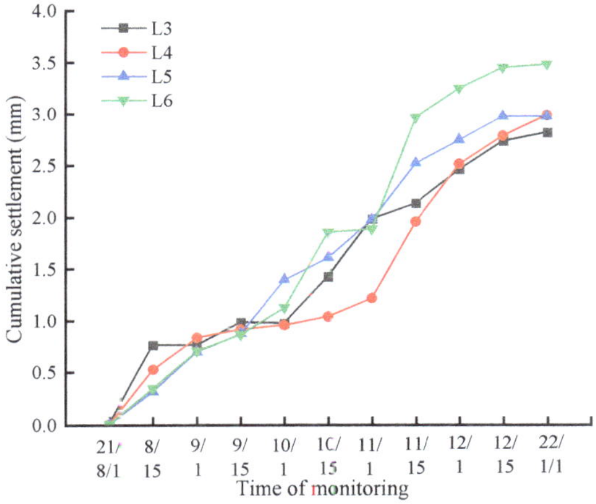

Figure 18. Time history curve of column settlement.

As shown in Figure 18, the cumulative settlement of the column increases gradually. This is due to the fact that in addition to the construction of the first reinforced concrete bracing, three steel bracings must be applied along with the excavation of the foundation pit. The bracing itself also has weight, the weight will increase as the number of supports increases, and the cumulative settlement of the column will increase accordingly. Moreover, it can be seen from Figure 18 that among the four monitoring points where monitoring data can be obtained, the cumulative settlement of L6 is the largest, with a settlement of 3.48 mm. Throughout the excavation of the foundation pit, the monitoring values of the column settlement monitoring points did not exceed the control value (10 mm) stipulated in the code [22], which meets the design requirements.

4.6. Horizontal Displacement of the Columns

The horizontal displacement monitoring point and the settlement monitoring point of the column are common points. The settlement monitoring instrument of the column is TS09 total station, and the monitoring results of horizontal displacement of the column are shown in Figure 19.

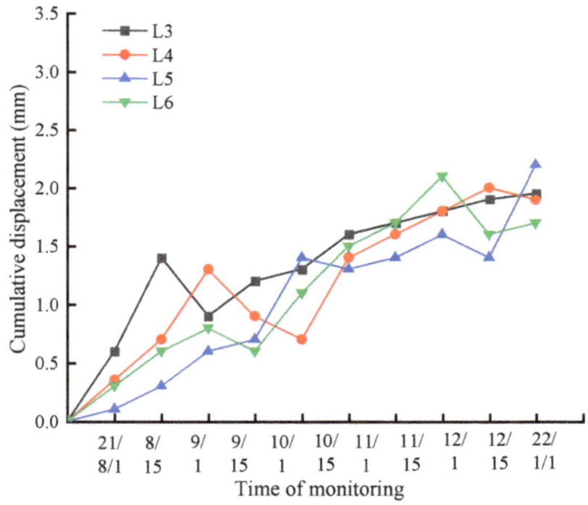

Figure 19. Time history curve of horizontal displacement of column.

As shown in Figure 19, the horizontal cumulative displacement of the column generally shows a tendency to increase, but the cumulative settlement of some monitoring points shows a decreasing trend in the small interval. Among the four monitoring points where monitoring data can be obtained, the horizontal cumulative displacement of L5 monitoring point is the largest, and the displacement value is 2.2 mm. In the whole process of foundation pit excavation, the monitoring value of the horizontal displacement monitoring point of the column did not exceed the control value (10 mm) set by the code [22], which meets the design requirements.

4.7. Settlement of the Building

Points are arranged around the building, and each building has no less than four monitoring points. A total of 37 monitoring points are set up, numbered as H1 to H37, and the building settlement monitoring instrument is Trimble DINI03 electronic level. Considering the importance of the location of the monitoring point, the monitoring point with large settlement is selected for analysis. The monitoring results of building settlement are shown in Figure 20.

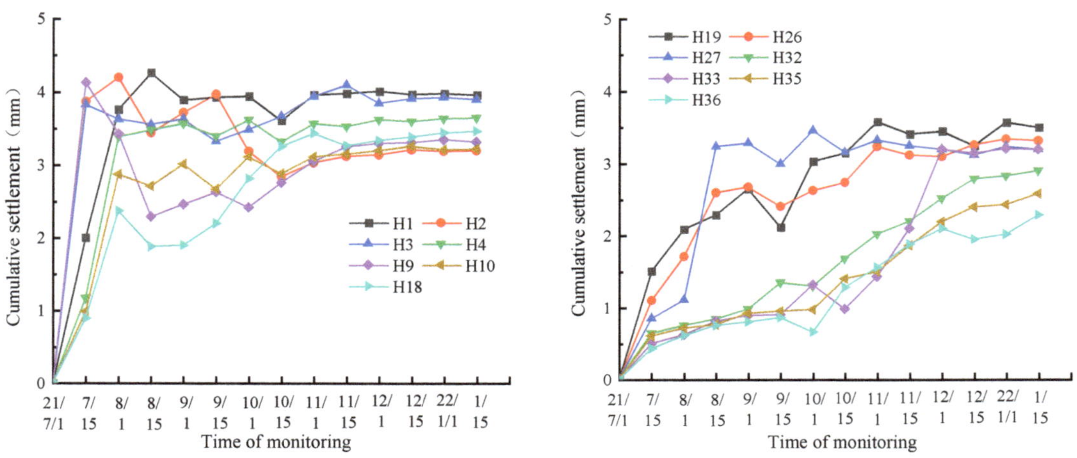

Figure 20. Time history curve of building settlement.

According to Figure 20, the cumulative settlement of the 14 monitoring points will first increase, then fluctuate within a certain range, and finally tend to stabilize. This is mainly related to the construction conditions of the foundation pit. the foundation pit begins to be excavated after the completion of the first bracing construction, and the settlement of the surrounding buildings will increase rapidly with the increase in the depth of the foundation pit. Then, the second, third, and fourth steel bracings are constructed to make the settlement of the surrounding buildings fluctuate within a certain range. When excavated to the bottom of the pit for foundation construction and the construction of the main structure of the station, the settlement of the surrounding buildings will remain stable. It can also be seen from Figure 20 that the closer the building is to the foundation pit, the greater the cumulative settlement. Among the 14 monitoring sites, the cumulative settlement of H1 is the largest, and the settlement is 4.26 mm. In the whole process of foundation pit excavation, the monitoring value of the horizontal displacement monitoring point of the column did not exceed the control value (10 mm) set by the code [22], which meets the design requirements.

4.8. Groundwater Level

The groundwater level monitoring points are arranged around the foundation pit, and at least one groundwater level observation hole is arranged on each side of the foundation pit. A total of seven monitoring points are in place, with numbers from W1 to W7. The SWJ-90 steel ruler water level meter is used to monitor the groundwater level, and the field picture of groundwater level is shown in Figure 21. Due to the destruction of two monitoring points during construction, there are only five monitoring points that can read the data. The monitoring results of building settlement are shown in Figure 22.

Figure 21. Field picture of groundwater level.

As shown in Figure 22, the groundwater table has been on a downward trend until 15 July 2021. This is due to the fact that in order to reduce the influence of groundwater on foundation pit construction during foundation pit excavation, dewatering wells are used to dewater the foundation pit. After 15 July 2021, the water table has risen. This is due to the fact that after the dewatering of the foundation pit, it is replenished by the surrounding water sources; therefore, the water level of the foundation pit rises slightly. After 1 August 2021, the groundwater level remains basically unchanged, and the height of the groundwater level is below the bottom of the foundation pit, which does not affect the construction of the foundation pit.

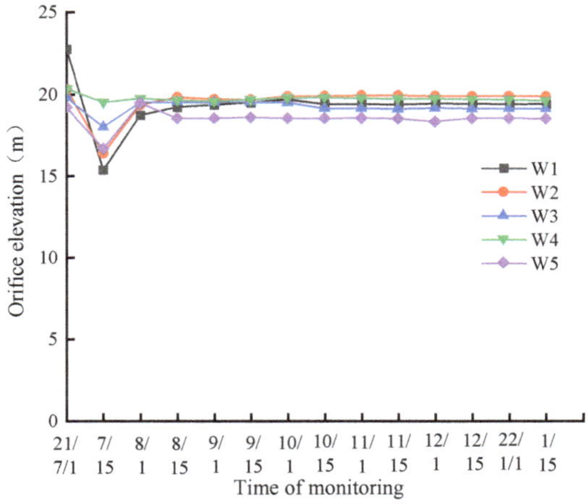

Figure 22. Time history curve of groundwater level.

4.9. *Surface Settlement*

A transverse monitoring section perpendicular to the boundary line of the foundation pit is arranged along the surface spacing around the foundation pit every 20 to 40 m, with four monitoring points on each side, with a total of 74 monitoring points. The surface subsidence monitoring instrument is TrimbleDINI03 electronic level, and the field picture of surface settlement is shown in Figure 23. Considering the importance of the location of the monitoring points, nine monitoring points close to the foundation pit are selected for analysis. The monitoring results of surface settlement are shown in Figure 24.

Figure 23. Field picture of surface settlement.

As shown in Figure 24, the surface subsidence shows an increasing trend. The cumulative settlement of each monitoring point increases rapidly before 1 November 2021, and slows down after 1 November 2021. This is due to the fact that on 1 November 2021, the excavation of the foundation pit has been completed, the depth of the foundation pit is no longer increased, and the impact on surface settlement is reduced. Among nine monitoring sites, the cumulative settlement of D18-1 is the largest, and the settlement is 10.95 mm. In the whole process of foundation pit excavation, the monitoring value of the surface settlement monitoring point did not exceed the control value (30 mm) set by the code [22], which meets the design requirements.

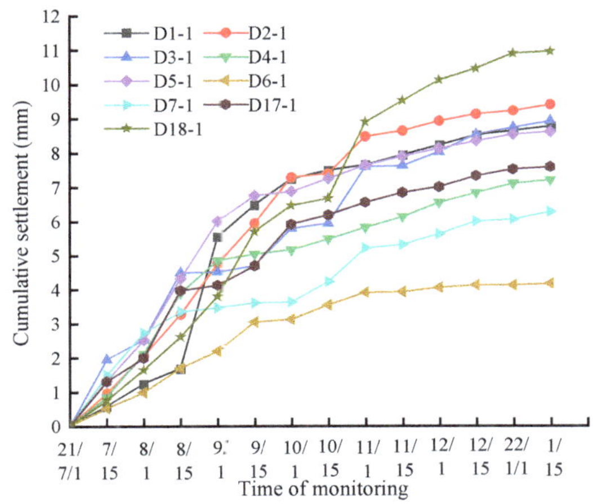

Figure 24. Time history curve of surface settlement.

4.10. Pipeline Settlement

The underground pipeline located in the main influence area shall set up one monitoring point every 5 m to 15 m, and the underground pipeline in the secondary influence area shall set up one monitoring point every 1500 m, with a total of 35 monitoring points. The surface subsidence monitoring instrument is Trimble DINI03 electronic level. Considering the importance of pipeline location, the monitoring points close to the foundation pit are selected for analysis. The monitoring results of pipeline settlement are shown in Figure 25.

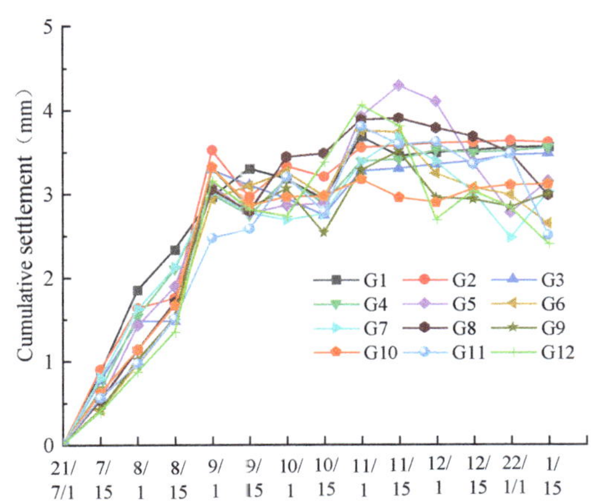

Figure 25. Time history curve of pipeline settlement.

As shown in Figure 25, before 1 September 2021, the cumulative settlement at each monitoring point continued to increase. After 1 September 2021, the cumulative settlement of each monitoring point fluctuated to a small extent, but after 1 November 2021, the cumulative settlement of G1, G2, G3, and G4 remained basically unchanged. Among 12 monitoring sites, the cumulative settlement of G5 monitoring site is the largest, and the settlement is 4.29 mm. In the whole process of foundation pit excavation, the monitoring value of surface settlement monitoring point did not exceed the control value (10 mm) set by the code [22], which meets the design requirements.

4.11. Settlement of the Bridge Abutment

One settlement monitoring point is arranged on one side of the bridge abutment and one side along the bridge, and two settlement monitoring points are set up on each bridge abutment, a total of 60 monitoring points are set up. The instrument for monitoring the settlement of bridge abutment is Trimble DINI03 electronic level. Due to the importance of the bridge abutment position, the monitoring point close to the foundation pit is selected for analysis. The monitoring results of settlement of the bridge abutment are shown in Figure 26.

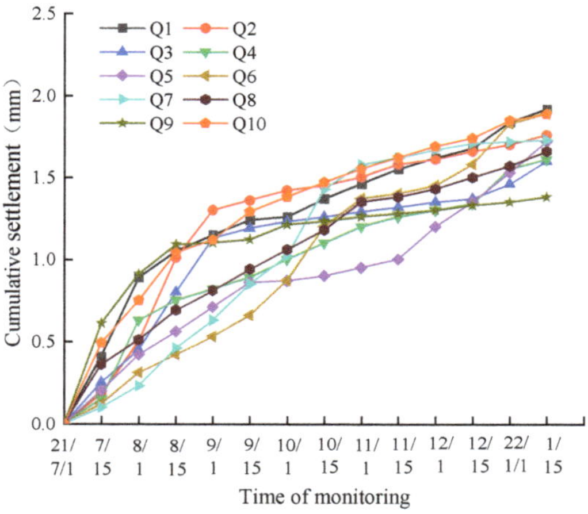

Figure 26. Time history curve of settlement of the bridge abutment.

As shown in Figure 26, with the increase in the excavation depth of the foundation pit, the settlement of the bridge abutment continues to increase. The cumulative settlement of each monitoring point increases rapidly before 1 September 2021, and slows down after 1 September 2021. Among 10 monitoring sites, the cumulative settlement of Q1 monitoring site is the largest, and the settlement is 1.92 mm. In the whole process of foundation pit excavation, the monitoring value of bridge abutment settlement monitoring points did not exceed the control value (5 mm) set by the code [22], which meets the design requirements.

4.12. Horizontal Displacement of the Bridge Abutment

One horizontal displacement monitoring point is arranged on one side of the bridge abutment and one horizontal displacement monitoring point on the side along the bridge, and two horizontal displacement monitoring points are set up on each bridge abutment, with a total of 60 monitoring points. The horizontal displacement monitoring points are arranged by directly pasting Leica reflective paste. The instrument for monitoring the horizontal displacement of bridge abutment is TS09 total station. Due to the importance of the position of the bridge abutment, the monitoring point closer to the foundation pit is selected for analysis. The monitoring results of the horizontal displacement of the bridge abutment are shown in Figure 27.

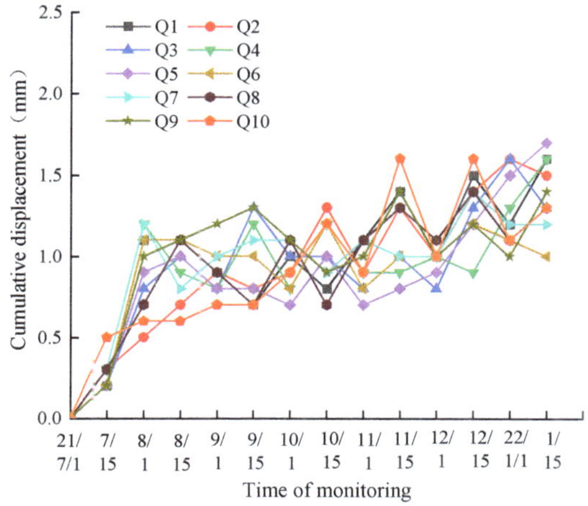

Figure 27. Time history curve of horizontal displacement of the bridge abutment.

According to Figure 27, the horizontal displacement of the bridge abutment is generally increasing, but it decreases in some time periods. Before 1 August 2021, the horizontal cumulative displacement of bridge abutment is increasing, but after 1 August 2021, the horizontal cumulative displacement of bridge abutment decreases sharply. This is due to the fact that the second bracing is applied after 1 August 2021, which controls the deformation of the foundation pit and reduces the horizontal cumulative displacement of the bridge abutment. Among the 10 monitoring points, the horizontal cumulative displacement of Q5 monitoring point is the largest, and the displacement value is 1.7 mm. In the whole process of foundation pit excavation, the monitoring value of the horizontal displacement monitoring point of the bridge abutment did not exceed the control value (5 mm) set by the code [22], which meets the design requirements.

4.13. Tilt of the Bridge Abutment

One tilt monitoring point is arranged on one side of the bridge abutment and one tilt monitoring point on the side along the bridge, and two tilt monitoring points are arranged on each bridge abutment, with a total of 60 monitoring points. The tilt monitoring points are arranged with L-shaped monitoring prisms, and one L-shaped monitoring prism is arranged on the transverse direction of the pier and the side along the bridge. Two prisms on the same side of the pier are a group of tilt monitoring points. The instrument for monitoring the inclination of bridge abutment is TS09 total station. Due to the importance of the location of the bridge abutment, the monitoring points close to foundation pit are selected for analysis. The monitoring results of the tilt of bridge abutment are shown in Figure 28.

As shown in Figure 8, before 1 August 2021, the tilt of bridge abutment is increasing, and after 1 August 2021, the tilt of bridge abutment is basically horizontal. However, after 15 August 2021, the cumulative tilt of bridge abutment decreases sharply. This shows that before the application of the second support, the foundation pit excavation has a great influence on the tilt of the bridge abutment. After the application of the second support, the influence of excavation on the tilt of the pier becomes smaller. Among the 10 monitoring points, the cumulative tilt of Q2 monitoring point is the largest, and the displacement value is 0.0519 mm. In the whole process of foundation pit excavation, the monitoring value of the bridge abutment tilt monitoring point did not exceed the control value (0.2 mm) set by the code [22], which meets the design requirements.

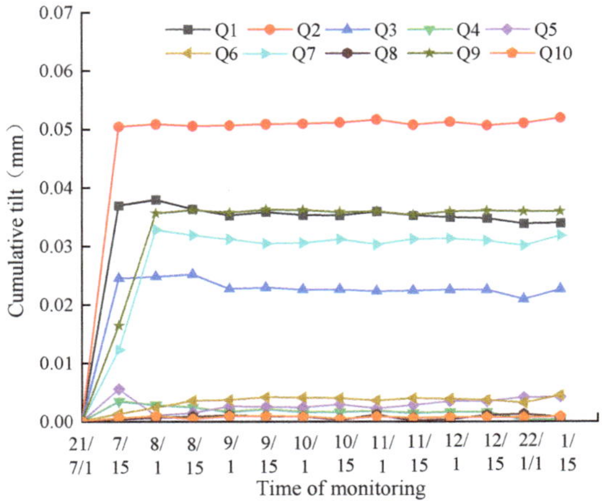

Figure 28. Time history curve of tilt of the bridge abutment.

5. Conclusions

The foundation pit of Lingbi Road Station of Hefei Rail Transit Line 8 not only has a large excavation depth, but also contains groundwater in the excavation depth. In the view of environmental risk and groundwater risk around the foundation pit, it is determined that the support structure of the foundation pit is the composite internal bracing of the underground diaphragm wall, and the dewatering measures are to set up a pressure-reducing well and a water-collecting open drain in the pit, and set up an observation well and an emergency backup well outside the pit. According to the characteristics of the foundation pit, a monitoring plan was set up. During the construction of the foundation pit, 13 items such as the horizontal displacement at the top of the wall, the axial force of the support, and the groundwater level were monitored in real time. The following conclusions were drawn by analyzing the monitoring data of the 13 items.

(1) The risk analysis of the foundation pit before the excavation of the water-containing deep foundation pit at Lingbi Road Station is very necessary, and the countermeasures given can greatly reduce the risk level of the foundation pit.
(2) During the monitoring process, the monitoring data of 13 monitoring items such as the horizontal displacement at the top of wall, the axial force of the bracing, and the groundwater level did not exceed the control value, and the construction process of the foundation pit was safe.
(3) The selection of support structure, the selection of support structure parameters, and the design of dewatering scheme for water-containing deep foundation pit at Lingbi Road Station are all reasonable, and the composite internal bracing structure of underground diaphragm wall is suitable for deep foundation pit support in Hefei area.
(4) The foundation pit of Lingbi Road Station is a typical foundation pit project with complex surrounding environment, large excavation depth, and water within the excavation depth range. The study in this paper also provides a case reference for the design of deep foundation pit support for subway stations in Hefei.

Author Contributions: Conceptualization, D.W. and S.Y.; methodology, D.W., S.Y. and J.Z.; validation, D.W. and S.Y.; formal analysis, D.W. and J.Z.; investigation, S.Y. and J.Z.; resources, D.W. and S.Y.; data curation, D.W. and J.Z.; writing—original draft preparation, D.W. and J.Z.; writing—review and editing, S.Y. and J.Z.; supervision, D.W.; project administration, D.W.; funding acquisition, D.W. and S.Y. All authors have read and agreed to the published version of the manuscript.

Funding: This research was funded by the National Natural Science Foundation of China (No. 52168050).

Data Availability Statement: Due to legal issues, the data in the article cannot be provided free of charge.

Conflicts of Interest: The authors declare no conflict of interest.

References

1. Han, K.H.; Su, D.; Hong, C.Y.; Chen, X.S.; Lin, X.T. An analytical model for face stability of tunnels traversing the fault fracture zone with high hydraulic pressure. *Comput. Geotech.* **2021**, *140*, 104467. [CrossRef]
2. Ye, S.H.; Li, D.P. Monitoring and simulation analysis of deep and large foundation pit excavation in complex environment. *China Civ. Eng. J.* **2019**, *52*, 117–126.
3. Shakya, K.M.; Saad, A.; Aharonian, A. Commuter exposure to particulate matter at underground subway stations in Philadelphia. *Build. Environ.* **2021**, *186*, 107322. [CrossRef]
4. Kim, C.; Kim, S.W.; Kang, H.J.; Song, S.M. What Makes Urban Transportation Efficient? Evidence from Subway Transfer Stations in Korea. *Sustainability* **2017**, *9*, 2054. [CrossRef]
5. An, P.Z.; Liu, Z.M.; Jia, B.X.; Zhou, Q.; Meng, F.L.; Wang, Z.X. Comparison and Economic Envelope Structure Schemes for Deep Foundation Pit of Subway Stations Based on Fuzzy Logic. *Comput. Intell. Neurosci.* **2022**, *2022*, 1148856. [CrossRef] [PubMed]
6. Pronozin, Y.A.; Bragar, E.P. Changes in soil properties at unloading of base of deep foundation pit. In Proceedings of the International Scientific-Technical Conference on Geotechnical Fundamentals and Applications in Construction-New Materials, Structures, Technologies and Calculations (GFAC), Saint Petersburg, Russia, 6–8 February 2019; pp. 290–295.
7. Wu, J.; Ye, S.H.; Wang, Z.Q.; Yang, D. Application and automatic monitoring and analysis of hybrid support structure in ultra-deep foundation pit engineering in the Lanzhou area under complex environmental conditions. *Water* **2023**, *15*, 1335. [CrossRef]
8. Gotman, A.L.; Gotman, Y.A. Numerical Analysis of the Shorings of Deep Foundation Pits with Regard for the Soil Solidification. *Soil Mech. Found. Eng.* **2019**, *56*, 225–231. [CrossRef]
9. Morozovskiy, P.; Babanina, A.; Ziiaeva, K.; Shulzhenko, S. Optimization of parameters of the bearing structures of deep excavations with minimization of environmental impacts. In Proceedings of the International Scientific Conference on Business Technologies for Sustainable Urban Development (SPbWOSCE), Saint Petersburg, Russia, 10–12 December 2019; p. 01002.
10. Sun, W.; Chen, H.J.; Li, B.; Liu, Y.F. Monitoring and finite element analysis of deep horizontal displacement of foundation pit enclosure pile of a subway transfer station during construction. In Proceedings of the 2nd International Conference on Big Data and Artificial Intelligence and Software Engineering (ICBASE), Zhuhai, China, 24–26 September 2021; pp. 319–323.
11. Yang, T.; Xiong, S.Y.; Liu, S.L.; Liu, Y.; Zhao, H.; Li, Y.W. Numerical Analysis of the Influence of Deep Foundation Pit Construction on Adjacent Subway Stations in Soft Soil Areas. *Adv. Civ. Eng.* **2022**, *2022*, 6071868. [CrossRef]
12. Hong, C.Y.; Zhang, J.Y.; Chen, W.B. An Integrated Intelligent Approach for Monitoring and Management of a Deep Foundation Pit in a Subway Station. *Sensors* **2022**, *22*, 8737. [CrossRef] [PubMed]
13. Yang, J.H.; Kong, D.Y. Deformation of deep and large foundation pit in soft soil of Fuzhou Subway. *Arab. J. Geosci.* **2020**, *13*, 36. [CrossRef]
14. Wang, W.; Han, Z.; Deng, J.; Zhang, X.Y.; Zhang, Y.F. Study on soil reinforcement param in deep foundation pit of marshland metro station. *Heliyon* **2019**, *11*, e02836. [CrossRef] [PubMed]
15. Liu, H.F.; Li, K.Z.; Wang, J.Q.; Cheng, C.X. Numerical Simulation of Deep Foundation Pit Construction under Complex Site Conditions. *Adv. Civ. Eng.* **2021**, *2021*, 6669466. [CrossRef]
16. Zhang, Y.G.; Yi, L.X.; Zhang, L.; Yang, Y.F.; Hao, X.; Li, H.; Ma, H.F. Causation Identification and Control Measures of Deformation by Integrated Dewatering-Excavation Process Simulation of a T-Shaped Deep Foundation Pit. *Water* **2022**, *14*, 535. [CrossRef]
17. Zhao, J.P.; Tan, Z.S.; Yu, R.S.; Li, Z.L.; Zhang, X.R.; Zhu, P.C. Deformation responses of the foundation pit construction of the urban metro station: A case study in Xiamen. *Tunn. Undergr. Sp. Tech.* **2022**, *128*, 5527–5538. [CrossRef]
18. Wu, B.; Zhao, R.; Meng, G.W.; Chen, H.H.; Huang, W.; Liu, J.L.; Cheng, Y. Safety risk assessment of asymmetric construction of subway deep foundation pit based on fuzzy theory. *J. Civ. Environ. Eng.* **2022**, *44*, 8–15.
19. Ye, S.H.; Zhao, Z.F.; Wang, D.Q. Deformation analysis and safety assessment of existing metro tunnels affected by excavation of a foundation pit. *Undergr. Space* **2021**, *6*, 421–431. [CrossRef]
20. Shen, Y.S.; Wang, P.; Li, M.P.; Mei, Q.W. Application of Subway Foundation Pit Engineering Risk Assessment: A Case Study of Qingdao Rock Area, China. *KSCE J. Civ. Eng.* **2019**, *23*, 4621–4630. [CrossRef]
21. Zhou, Y.; Li, S.Q.; Zhou, C.; Luo, H.B. Intelligent Approach Based on Random Forest for Safety Risk Prediction of Deep Foundation Pit in Subway Stations. *J. Comput. Civ. Eng.* **2019**, *33*. 05018004. [CrossRef]
22. *GB 50497-2019*; Technical Standard for Monitoring of Building Excavation Engineering. China Planning Press: Beijing, China, 2019.

water

MDPI

Article

Monitoring Analysis of a Deep Foundation Pit with Water Supported by Cast-in-Place Pile and Internal Bracing in a Soft Soil Area of Fuzhou

Bingxiong Tu [1,*], Jinhuo Zheng [2], Minglong Shen [2] and Weilong Ni [3,*]

1 College of Civil Engineering, Huaqiao University, Xiamen 361021, China
2 Fujian Provincial Institute of Architectural Design and Research Co., Ltd., Fuzhou 350001, China; ak9206@126.com (J.Z.); m15392008809@163.com (M.S.)
3 Xiamen Holsin Engineering Testing Co., Ltd., Xiamen 361027, China
* Correspondence: tubingxiong@hqu.edu.cn (B.T.); 18106936279@163.com (W.N.)

Abstract: In addition to selecting an effective support structure to control deformation, precipitation and water stopping should also be considered when designing a support scheme for water-bearing foundation pits in soft soil areas. This paper presents a detailed description of the foundation pit support scheme, the precipitation and water-stopping scheme, and the monitoring scheme of the foundation pit project of Taijiang Square in Fuzhou. During the construction of the foundation pit, the monitoring data of 12 items such as the deep horizontal displacement of the enclosure pile, the horizontal displacement at the top of the foundation pit, the settlement at the top of the foundation pit, the axial force of the internal bracing, and the axial force of the enclosure pile were obtained through 12 months of monitoring. The analysis of the monitoring data for each item led to the following two main findings. The first finding is that, during the construction of the pit, the monitoring values of the 12 monitoring items did not exceed the alarm values, which proves that the support scheme of the cast-in-place pile enclosure structure and internal bracing can meet the design requirements of deep foundation pits in soft soil areas. The second finding is that tube-well dewatering is an effective way to lower the groundwater level in water-containing deep foundation pits in soft soil areas, and double-wheel deep-mixing water-stopping curtain walls can effectively control the infiltration of groundwater outside the water-containing deep foundation pits in soft soil areas. This foundation pit project is representative, and it provides a good reference case for the design of water-bearing deep foundation pit projects in soft soil areas.

Keywords: soft soil areas; cast-in-place pile and internal bracing; deep foundation pit with water; monitoring analysis

check for updates

Citation: Tu, B.; Zheng, J.; Shen, M.; Ni, W. Monitoring Analysis of a Deep Foundation Pit with Water Supported by Cast-in-Place Pile and Internal Bracing in a Soft Soil Area of Fuzhou. *Water* **2023**, *15*, 3008. https://doi.org/10.3390/w15163008

Academic Editor: Akbar Javadi

Received: 24 July 2023
Revised: 14 August 2023
Accepted: 15 August 2023
Published: 21 August 2023

1. Introduction

Soft soil is a kind of soft plastic–flow plastic clayey soil with high moisture content, strong compressibility, and low bearing capacity, and it is mainly distributed in coastal areas, plain areas, inland lake basin areas, and other areas [1–4]. Soft soil in China is widely distributed in the Pearl River Delta, Yangtze River Delta, Bohai Bay, Zhejiang, Fujian, Shanghai, and other coastal areas [5–8]. The construction of buildings in soft soil areas inevitably involves foundation excavation engineering, which is often complex and difficult due to the nature of soft soil. At present, the main forms of foundation pit supporting structures in soft soil areas are steel sheet pile, steel sheet pile–internal bracing, underground diaphragm wall, underground diaphragm wall–internal bracing, cast-in-place pile, cast-in-place pile–internal bracing, and so on [9–13]. The existing research methods of deep foundation pits in soft soil areas are mainly finite element simulation, field monitoring, and finite element simulation combined with field monitoring. In terms of finite element simulation, Yuan et al. [14] used MIDAS/GTS software to simulate the water-bearing deep foundation pit of a subway station and study the influence of seepage on the deformation of deep foundation pit in the process of deep foundation pit construction. Harahap et al. [15] used two-dimensional finite element software to study the variation characteristics of diaphragm walls and surfaces with time during soft soil deep foundation pit excavation. Bal et al. [16] proposed a numerical model for analyzing soft soil excavation using single excavation tools and multiple excavation tools that can simulate tool–soil interaction in the process of excavation. Uribe-Henao et al. [17] used a new finite element model to study the interaction between the support structure and soil of soft clay deep foundation pits and revealed the mechanism of the support structure on the soil. Based on the deep foundation pit of a complex soft soil in Wenzhou, Sun et al. [18] studied the deformation characteristics of underground diaphragm walls under different working conditions by using the methods of numerical simulation. In the aspect of field monitoring, Wu et al. [19] studied the application of a hybrid retaining structure in an ultra-deep foundation pit in a complex environment and monitored and analyzed the construction process of the whole foundation pit. Dmochowski et al. [20] studied the impact of deep excavation of foundation pits on the surrounding building walls, estimated the displacement of the surrounding building walls during the excavation of foundation pits through continuous monitoring and calculation, and provided a solution for the safety of building walls. Chen et al. [21] used a monitoring system to observe the deformation characteristics of deep foundation pits and the interaction between adjacent foundation pits. Rybak et al. [22] summarized the types of damage that could easily occur during the construction of deep foundation pits' retaining walls and in the process of foundation pit excavation, and they took the supporting wall of a steel sheet pile foundation pit as an example to analyze the damage caused by foundation pit over-excavation and its influence on the surrounding buildings. Yang et al. [23] performed a statistical analysis of monitoring data from 15 deep foundation pits in Suzhou and revealed the notable impact of spatial effects on pit deformation. Sun et al. [24] analyzed the site monitoring data of a deep foundation pit supported by pile–bracing–anchor in a soft soil area and derived the variation law of the displacement of the supporting structure and the axial force of the anchor cable. As regards finite element simulation combined with field monitoring, Chen et al. [25] took an offshore underground diaphragm wall–internal bracing foundation pit as the research object and studied the performance of supporting structure under wave, tide, vibration, and unbalanced load via numerical simulation and site monitoring. Czajewska [26] studied the displacement changes in the surrounding ground and structures during the excavation of the foundation pit of a subway station and established the vertical displacement prediction equation of the tunnel near the deep foundation pit. Feng et al. [27] used the numerical simulation method to study the deformation law of supporting structure and surrounding soil during the excavation of soft soil foundation pits on the basis of verifying the reliability of the numerical simulation with the site monitoring data. Panchal et al. [28] developed a new process and equipment for simulating the excavation of deep foundation pits in soft soil; this excavation process can ensure the stability of foundation pits and reduce the disturbance to surrounding buildings. Kiet et al. [29] measured the soil hardening model (HSM) parameters of Ho Chi Minh City's soft clay through experimental means, which laid a foundation for the numerical simulation of a local deep foundation pit. Nguyen et al. [30] analyzed the deformation law and groundwater-level change law of a deep foundation pit of a subway station in Ho Chi Minh City by using numerical simulation combined with field monitoring methods.

In summary, the research on foundation pit support in soft soil areas is more mature, but most of the research focuses on underground diaphragm wall–internal bracing, and there is less research on water-bearing deep foundation pits supported by pile–bracing in Fuzhou. In this paper, based on the foundation pit engineering of Taijiang Square, 12 items, namely the deep horizontal displacement of the enclosure pile, the horizontal displacement at the top of the foundation pit, the settlement at the top of the foundation pit, the axial force of internal bracing, the axial force of the enclosure pile, the settlement of the column, the groundwater level, the layered settlement of soil, the settlement of surrounding buildings, the horizontal displacement of surrounding buildings, the tilt of surrounding buildings, and the settlement of surrounding surfaces and underground pipelines are monitored while considering the complex environment around the foundation pit and the influence of groundwater on foundation pit engineering. The monitoring data were analyzed to determine the changes in the deep horizontal displacement of the enclosure pile, the horizontal displacement at the top of the foundation pit, the settlement at the top of the foundation pit, the axial force of the internal bracing, and the axial force of the enclosure pile with the construction of the foundation pit, and to verify the reasonableness of the support scheme. This foundation pit project provides a good reference case for the design of water-bearing deep foundation pit projects in soft soil areas.

2. Project Overview

2.1. Site Overview

Fuzhou is located in the southeastern coastal area of China, as shown in Figure 1, and it is the capital city of Fujian Province. The Taijiang Square project is situated at the southeastern corner of the intersection between Wuyi South Road and Yingzhou Road in Fuzhou, as shown in Figure 2. The development plan comprises a 36-story mixed-use commercial and office tower, as well as a 5-story commercial podium. The height of the mixed-use tower above ground level is approximately 130 m, while the commercial podium stands at 22.5 m. The upper structure of the mixed-use tower is proposed to be a framed tube system, with individual columns capable of withstanding a maximum load of 45,000 kN. The upper structure of the commercial podium is designed as a framed system, with individual columns capable of supporting a maximum load of 10,000 kN. The project includes three basement levels, with each level having a clear height of 4.8 m, 5.3 m, and 5.3 m from top to bottom. The site elevation is 6.90 m, and the floor slab is buried at a depth of approximately 20 m below the designed outdoor ground level, with an elevation of −13.10 m.

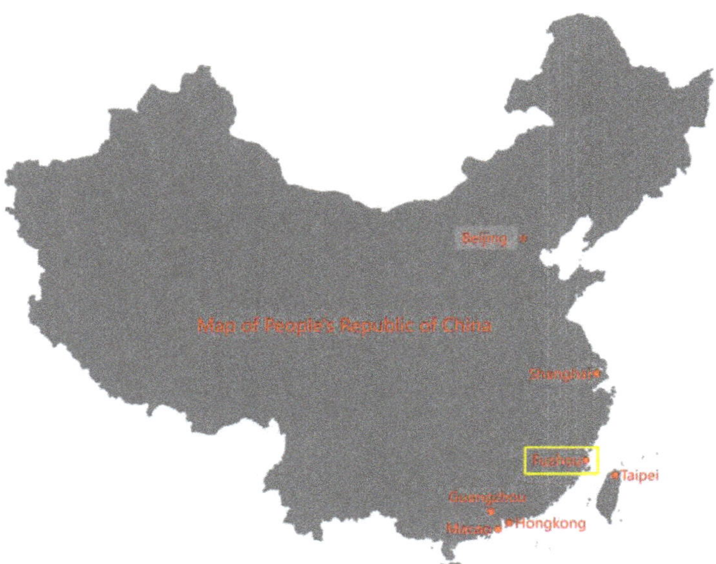

Figure 1. The general location map of Fuzhou.

Figure 2. An image of the surrounding area of the foundation pit.

An image of the surrounding area of the foundation pit is shown in Figure 2. The northern boundary of the site's basement is located approximately 4.5 m from the edge of Yingzhou Road, with a distance of approximately 22 m from the two-story wooden structures of Rongcheng Ancient Street, which have shallow foundations. On the western side of the site, the basement boundary is roughly 8.5 m away from the edge of Wuyi South Road. On the east side, the distance between the basement boundary and the existing five-story factory building, constructed as a masonry and reinforced concrete structure supported by a raft foundation, is approximately 15.0 m. The east side of the site is situated around 30.5 m away from the Yingzhou River. This project holds a Level I significance classification.

2.2. Geotechnical Conditions

Based on the geotechnical investigation conducted on the site, the sequence of geological layers within the influence range of the excavation of the foundation pit is as follows: Listed from top to bottom are miscellaneous fill (1-1), mucky fill (1-2), silt sand mixed with muck (2), muck mixed with sand (3), silt sand mixed with muck (4), muck mixed with sand (5), silt sand mixed with muck (6), mucky soil mixed with sand (7), silty sand (8), silty sand (9), and pebble (10). The physical and mechanical parameters of each geological layer can be found in Table 1.

Table 1. Physical and mechanical parameters of each geotechnical layer.

Soil Layer	Name of Soil Layer	Status	Thickness (m)	Natural Heavy γ/(kN m^3)	Cohesion c/(kPa)	Internal Friction Angle φ/(°)
1-1	Miscellaneous fill	Slightly denser	1.50 to 8.30	18.0	8.0	14.0
1-2	Mucky fill	Flow plastic	1.20 to 5.30	17.0	10.5	11.0
2	Silt sand mixed with muck	Slightly denser to medium density	3.90 to 11.0	18.0	3.0	21.0
3	Muck mixed with sand	Flow plastic	1.60 to 8.20	16.2	8.9	10.5
4	Silt sand mixed with muck	Medium density	2.70 to 11.3	18.2	5.0	23.0
5	Muck mixed with sand	Flow plastic	1.80 to 9.0	16.6	11.0	11.0
6	Silt sand mixed with muck	Medium density to dense	5.90 to 9.85	18.4	5.0	26.0
7	Mucky soil mixed with sand	Flow plastic	1.52 to 3.87	16.9	12.0	13.0
8	Silty sand	Medium density to dense	2.60 to 10.0	18.6	3.0	30.0
9	Silty sand	dense	0.50 to 9.80	18.7	3.0	33.0
10	Pebble	Medium density	4.29 to 5.80	21.0	0	38.0

2.3. Hydrogeological Conditions

The maximum excavation depth of this project is about 17 m, and the groundwater within the excavation area of the foundation pit is mainly as follows: The first type of groundwater is pore subsoil water in a loose layer, which mainly occurs in miscellaneous fill layer (1-1). This water-bearing medium is mostly filled with cohesive soil, with poor connectivity, weak water permeability, small aquifer thickness, and weak water-rich property. The second kind of groundwater is pore-confined water in a loose layer, which occurs in silt sand mixed with muck (2), silt sand mixed with muck (4), silt sand mixed with muck (6), silty sand (8), silty sand (9), and pebble (10). The medium structure of the aquifer above is relatively loose, the pore connectivity is good, and the water permeability is moderate to strong. The silty sand (8), silty sand (9), and pebble (10) aquifers have direct hydraulic connections. The layer with silt sand mixed with muck (2) has a buried depth of 3.40 to 3.50 m at static water level, the layer with silt sand mixed with muck (4) has a static water level buried depth of 7.30 to 7.60 m, and the layer with silt sand mixed with muck (6) has a static water level buried depth of 10.00 to 10.50 m. The initial groundwater level observed in the boreholes within the site ranges approximately from 2.94 to 4.16 m below the surface, whereas the depth of the mixed groundwater level ranges approximately from 2.78 to 3.88 m. The water level experiences an annual variation of around 0.5 to 1.0 m. The underground basement boundary on the southeastern side of the site is situated approximately 30.5 m away from the Yingzhou River, which spans about 16 m in width and flows in a northeasterly to southwesterly direction. The water depth in the river fluctuates between 1.0 and 5.0 m, influenced by the tidal movements of the Min River. The pore subsoil water of the loose layer on the surface of the site receives the supply of this water, and the surface water has a certain impact on the excavation of the basement foundation pit of this project. Protective measures, such as the construction of retaining walls and slope reinforcement, have been implemented along both banks of the river to ensure the current stability of the site.

3. Design Scheme of Foundation Pit Support and Construction Stage of Foundation Pit

3.1. Form of Enclosure Structure

Based on the analysis of the surroundings of the foundation pit, the foundation pit is supported by cast-in-place piling and internal bracing. To reduce the impact on ground traffic and to control the deformation of the surrounding ground, the piles on the east side of the foundation pit near the plant are 1.1 m in diameter, 1.4 m in spacing, and 33 m long, and the piles in the remaining locations are 1.0 m in diameter, 1.3 m in spacing, and 32 m long. To ensure the water retention effect and reduce the impact of precipitation on the surrounding environment, in this foundation pit project, an 800 mm thick double-wheel deep-mixing water-stopping curtain wall is used on the outside of the enclosure pile to retain water; the curtain wall's depth is 34 m. The profile view of the supporting structure is shown in Figure 3.

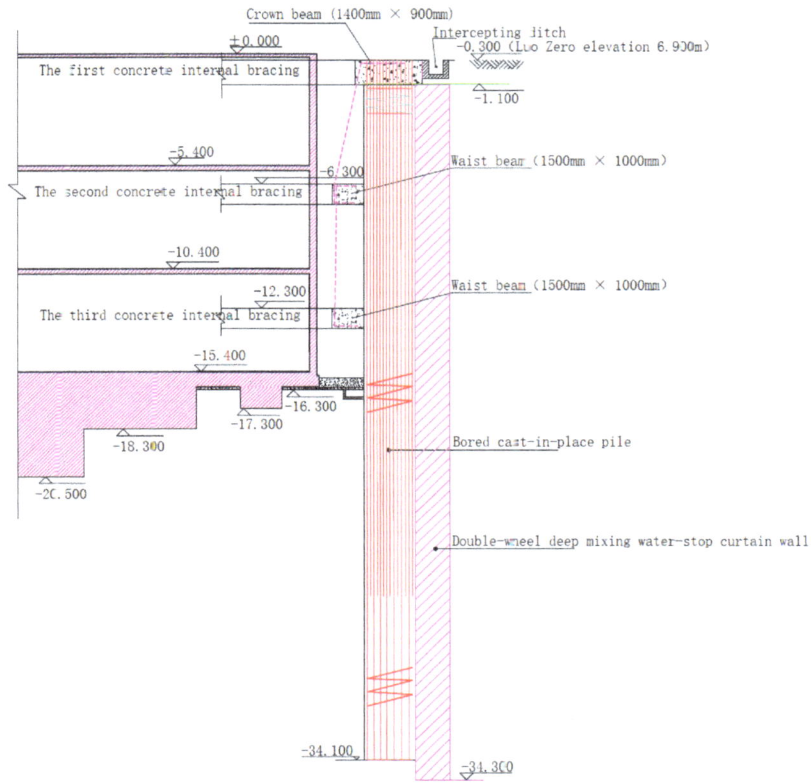

Figure 3. The profile view of the supporting structure.

3.2. *Foundation Pit Precipitation and Drainage*

This foundation pit project uses tube wells for rainfall and catchment trenches at the top of the slope. A 400 mm × 400 mm collection trench is constructed around the perimeter of the pit, and a 600 mm × 600 mm collection well is constructed at 30 m intervals. A submersible pump is used to extract groundwater from the catchment pit and discharge it into the cut-off trench at the top of the slope, which is then discharged into the municipal pipeline.

3.3. *Construction Stage of Foundation Pit*

The construction process of the foundation pit is as follows: First, the construction of the cast-in-place pile and crown beam begins before foundation pit excavation, and the cast-in-place pile and crown beam are maintained after construction. When the strength of the cast-in-place pile and crown beam meets the design requirements, the foundation pit excavation begins. When the foundation pit is excavated to the bottom elevation of the first concrete internal bracing, the construction of the first concrete internal bracing begins. When the strength of the first internal bracing meets the design requirements, the foundation pit excavation begins. When the foundation pit is excavated to the bottom elevation of the second concrete internal bracing, the construction of the second concrete internal bracing begins. When the strength of the second internal bracing meets the design requirements, the foundation pit excavation begins. When the foundation pit is excavated to the bottom elevation of the third concrete internal bracing, the construction of the third concrete internal bracing will begin. When the strength of the third internal bracing meets the design requirements, the excavation of the foundation pit begins. When the foundation pit is excavated to the bottom elevation of the platform and bottom plate, the construction of the platform and bottom plate begins. When the strength of the platform and bottom plate meets the design requirements, the construction of the shear wall in the basement begins. Until the construction of the basement is completed, the construction of the entire foundation pit will be completed. The construction stage of the foundation pit is shown in Table 2.

Table 2. The construction stage of the foundation pit.

Time (Months)	Construction Stage
0–2.5	Cast-in-place pile construction
2.5–3	The first internal bracing construction
3–4	Earthwork excavation
4–4.5	The second internal bracing construction
4.5–5	Earthwork excavation
5–5.5	The third internal bracing construction
5.5–6	Earthwork excavation
6–6.5	Platform construction
6.5–7	Bottom plate construction
7–12	Shear wall construction

4. Analysis of Site Monitoring Data

Given the extensive excavation area, considerable depth, and complex geological conditions of this foundation pit project, the selection of monitoring items should be based on careful consideration of the engineering and hydrogeological conditions, pit classification, characteristics of the support structure, and deformation control requirements. In addition to conventional inspection through visual inspection and inspection with the aid of other tools, the main instrumentation monitoring items are as follows: (1) the monitoring of the deep horizontal displacement of the enclosure pile; (2) the monitoring of the horizontal displacement at the top of foundation pit; (3) the monitoring of the settlement at the top of the foundation pit; (4) the monitoring of the axial force of the internal bracing; (5) the monitoring of the axial force of the enclosure pile; (6) the monitoring of the settlement of the column; (7) the monitoring of the groundwater level; (8) the monitoring of the layered settlement of the soil; (9) the monitoring of the settlement of the surrounding buildings; (10) the monitoring of the horizontal displacement of the surrounding buildings; (11) the monitoring of the tilt of the surrounding buildings; and (12) the monitoring of settlement of the surrounding surface and underground pipelines. The plan layout of the foundation pit monitoring points is shown in Figure 4, and the number of monitoring points, monitoring methods, and instrument names are shown in Table 3.

Table 3. Summary of monitoring points.

Monitoring Content	Monitoring Methods	Instrument Name	Number of Monitoring Points
Deep horizontal displacement of enclosure pile	Level survey (2nd class)	Su Yiguang DSZ2 automatic leveling level and FS1 micrometer	9
Horizontal displacement at the top of the foundation pit	Small angle method (Theodolite)	Su Yiguang DT302L electronic theodolite	13
Settlement at the top of the foundation pit	Inclinometer	American SINCO digital sliding inclinometer	13
The axial force of internal bracing	Rebar stress gauges	Jinyuan civil engineering CTY-202 frequency meter	6

Table 3. *Cont.*

Monitoring Content	Monitoring Methods	Instrument Name	Number of Monitoring Points
The axial force of the enclosure pile	Rebar stress gauges	Jinyuan civil engineering CTY-202 frequency meter	4
Settlement of the column	Level survey (2nd class)	Su Yiguang DSZ2 automatic leveling level and FS1 micrometer	23
Groundwater level	Water level meter	Jinyuan civil engineering SJ-92 water level gauge	6
Layered settlement of soil	Level survey (2nd class)	Su Yiguang DSZ2 automatic leveling level and FS1 micrometer	3
Settlement of surrounding buildings	Small angle method (Theodolite)	Su Yiguang DT302L electronic theodolite	19
Horizontal displacement of surrounding buildings	Point Casting Method (Theodolite)	Su Yiguang DT302L electronic theodolite	19
The tilt of surrounding buildings	Level survey (2nd class)	Su Yiguang DSZ2 automatic leveling level and FS1 micrometer	3
Settlement of surrounding surface and underground pipelines	Level survey (2nd class) and Stratified sedimentation meter	Su Yiguang DSZ2 automatic leveling level, FS1 micrometer and Wuhan foundation deep stratified sedimentation meter	9

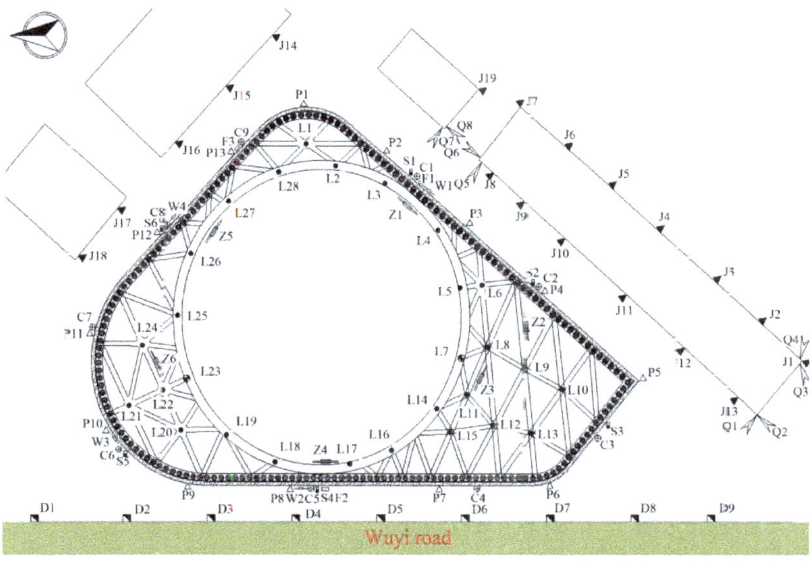

Figure 4. The layout plan of monitoring points of the foundation pit.

4.1. Monitoring of the Deep Horizontal Displacement of the Enclosure Pile

Along the key parts around the foundation pit, nine holes for monitoring the deep horizontal displacement of the retaining piles were arranged, numbered C1 to C9. The height of the inclinometer hole was equivalent to the height of the ground, and the hole depth was 34 m. The maximum horizontal displacement of the deep layer of the enclosure pile at each monitoring point is shown in Table 4.

Table 4. Maximum horizontal displacement of enclosure pile in the deep layer at each monitoring point.

Monitoring Point	C1	C2	C3	C4	C5	C6	C7	C8	C9
Maximum horizontal displacement (mm)	13.36	6.78	40.96	41.40	37.70	44.22	46.52	52.60	36.22

It can be seen from Table 3 that the cumulative displacement of the piles at the C1 and C2 monitoring points is relatively small. This is because the surrounding buildings on one side of C1 and C2 points are only one-story staff dormitory and concierge, which have little impact on the foundation pit. The cumulative displacement of the piles from C3 to C9 monitoring points is larger because there are both roads and multi-story concrete buildings on the side of C3 to C9, coupled with the disturbance of vehicles, which has a great impact on the foundation pit. In the process of earthwork excavation and foundation construction of the foundation pit, the cumulative displacement of the piles from C3 to C9 monitoring points exceeded the design alarm value (30 mm), but the deformation was stable. The cumulative displacement change rate of the piles from C3 to C9 monitoring points did not exceed the alarm value.

4.2. Monitoring of the Horizontal Displacement at the Top of the Foundation Pit

At the top of the foundation pit, 13 horizontal displacement monitoring points were arranged along the periphery, numbered P1 to P13. The observation mark was driven into the top of the crown beam with a measuring nail. Theodolites were used to monitor the horizontal displacement of these 13 points at the top of the foundation pit. The monitoring results are shown in Figure 5.

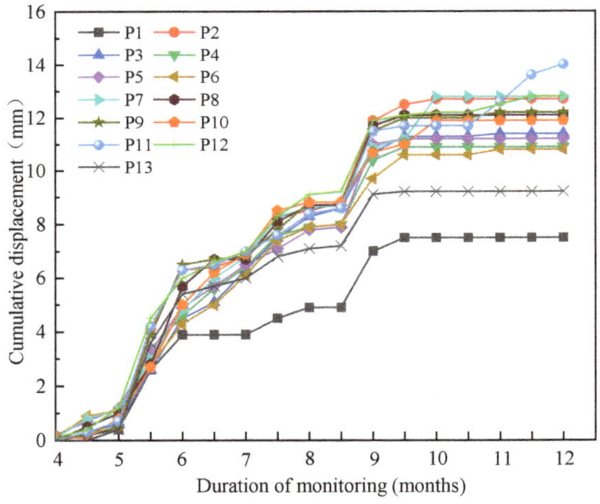

Figure 5. The time history curve of the horizontal displacement at the top of the foundation pit.

It can be seen from Figure 5 that, among the 13 monitoring points, the point with the largest cumulative horizontal displacement was the monitoring point P11, with a cumulative displacement of 14.0 mm. The point with the smallest cumulative horizontal displacement was the monitoring point P1, with a displacement of 7.5 mm. It can also be seen from Figure 5 that when the earthwork of the foundation pit was excavated, the horizontal displacement at the top of the foundation pit gradually increased with the increase in the excavation depth. The displacement accumulation and displacement increase rate of each monitoring point in the excavation process did not exceed the design alarm value. During the construction of the basement floor and shear walls after the completion of earth excavation, the horizontal displacement at the top of the foundation pit did not increase significantly, and the horizontal deformation at the top of the foundation pit tended to be stable. During the entire earthwork excavation and foundation construction process of the foundation pit, the accumulated horizontal displacement value and deformation rate at the top of the foundation pit did not exceed the design alarm value, thus meeting the requirements of the design and specifications.

4.3. Monitoring of the Settlement at the Top of the Foundation Pit

At the top of the slope, 13 settlement monitoring points were arranged along the periphery, numbered P1to P13. The observation mark was driven into the top of the crown beam with a measuring nail. Levels were used to monitor these 13 points at the top of the foundation pit. The monitoring results are shown in Figure 6.

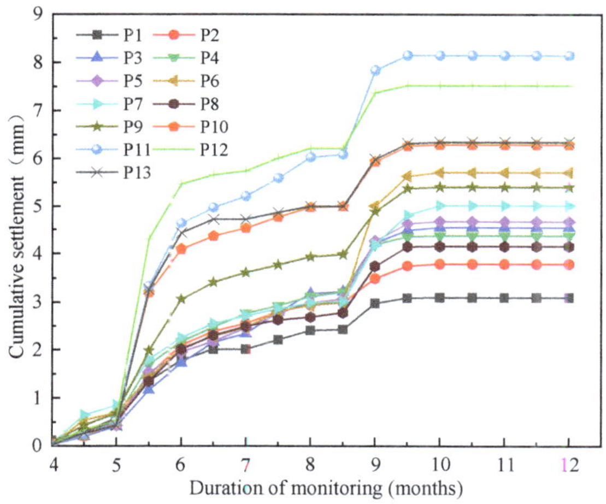

Figure 6. The time history curve of the vertical displacement at the top of the foundation pit.

It can be seen from Figure 6 that, among the 13 monitoring points, the point with the maximum cumulative value of vertical displacement was the monitoring point P11, with a displacement value of 8.15 mm. The point with the minimum cumulative value of vertical displacement was the monitoring point P1, with a displacement value of 3.10 mm. It can also be seen from Figure 6 that when the earthwork of the foundation pit was excavated, the cumulative vertical displacement at the top of the foundation pit gradually increased with the increase in the excavation depth. The settlement accumulation and settlement rate in the excavation process did not exceed the design requirement alarm value. During the construction of the basement floor and shear walls after the excavation of the earth, the vertical displacement at the top of the foundation pit did not increase significantly, and the vertical displacement at the top of the foundation pit tended to be stable. During the entire earthwork excavation and foundation construction process of the foundation pit, the accumulated vertical displacement and deformation rate at the top of the foundation pit did not exceed the design alarm value, which met the requirements of the design and specifications.

4.4. Monitoring of the Axial Force of the Internal Bracing

Considering the importance of the location of internal bracing, six monitoring points were arranged for the first reinforced concrete internal bracing, numbered Z1 to Z6. A rebar axial dynamometer was placed on the beam surface and the beam bottom at each point, and the layout method was welding the rebar axial dynamometer and the main bar in series. The monitoring results are shown in Figure 7.

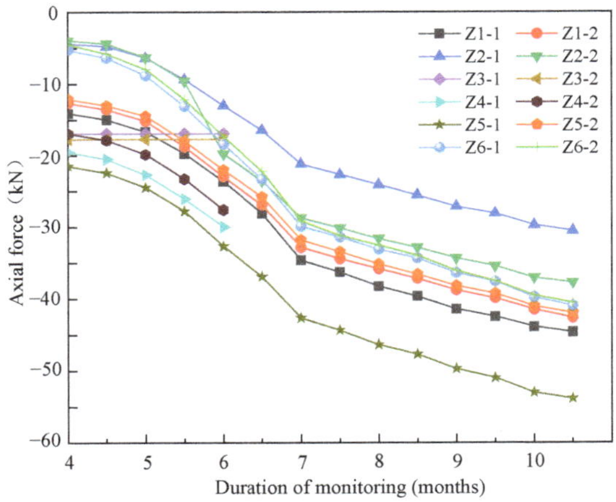

Figure 7. The axial force variation curve for the first internal bracing.

As shown in Figure 7, the Z3 and Z4 monitoring points on the first reinforced concrete internal bracing had no monitoring data after the third month, which is mainly because the Z3 and Z4 monitoring points were destroyed during the construction process in the third month. From the variation curves of the Z1, Z2, Z5, and Z6 monitoring points, it can be seen that, before the 7th month, the axial force value of the first reinforced concrete internal bracing gradually increased with the excavation depth of the foundation pit, and the increase rate was faster. After the 7th month, when the construction of the shear wall began after the excavation of the foundation pit, the axial force of the first reinforced concrete internal bracing also increased gradually, but the increase rate slowed down noticeably. In the process of monitoring, the axial force of the first reinforced concrete internal bracing did not exceed the alarm value, thus meeting the design requirements.

4.5. Monitoring of the Axial Force of the Enclosure Pile

A total of four monitoring points were arranged on the enclosure pile, numbered W1 to W4. A rebar axial dynamometer was arranged at each point, and the layout method was welding the rebar axial dynamometer in series with the main reinforcement of the enclosure pile. The monitoring results are shown in Figure 8.

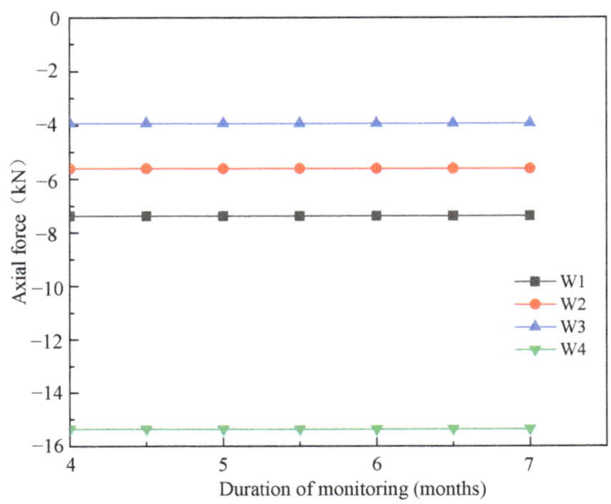

Figure 8. The axial force variation curve of the enclosure pile.

It can be seen from Figure 8 that the axial force value of the enclosure pile remains basically unchanged with the increase in the excavation depth of the foundation pit. Among the four monitoring points, the axial force value at point W4 was the largest, and the axial force value was 15.35 kN. The axial force value of the W3 point was the smallest, and the axial force value was 3.92 kN. During the monitoring process, the axial force data of each point of the enclosure pile did not exceed the alarm value, and the axial force of the pile body was within the design allowable range and therefore met the design requirements.

4.6. Monitoring of the Settlement of the Column

There were 28 settlement monitoring points at the top of the column, numbered L1 to L28. The observation mark was driven into the top of the column with a measuring nail. The level was used to monitor the 28 column-settlement monitoring points, and the monitoring results of column settlement are shown in Figure 9.

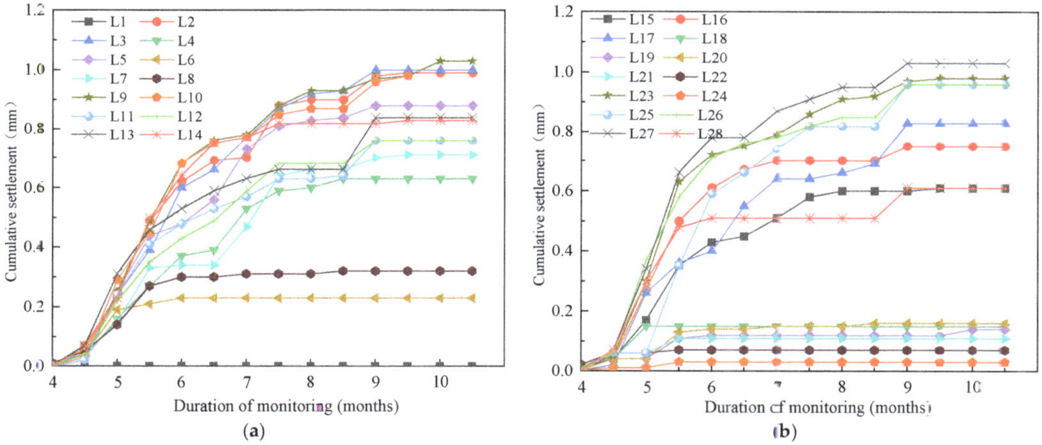

Figure 9. Time history curve of vertical displacement of the column: (**a**) time history curve of vertical displacement of L1 to L14 columns; (**b**) time history curve of vertical displacement of L15 to L28 columns.

It can be seen from Figure 9 that the cumulative settlement of the L1 monitoring point is 0. Except for the L1 monitoring point, the cumulative settlement of other monitoring points increased at first and then remained basically unchanged. The settlement of L6, L8, L18, L19, L20, L21, L22, and L24 monitoring points remained basically unchanged after the 7th month, and the settlement of the other monitoring points remained basically unchanged after the 9th month. The maximum cumulative settlement occurred at the L10 monitoring point, and the displacement was 1.14 mm, which did not exceed the alarm value and therefore met the design requirements. This shows that there was no obvious settlement of the column in the whole process of foundation pit excavation and foundation construction, and the cumulative settlement and settlement rate did not exceed the alarm value in the whole process, thus meeting the requirements of the design.

4.7. Monitoring of the Groundwater Level

Six groundwater level observation holes were arranged along the key parts within 2 m outside the periphery of the foundation pit, numbered S1 to S6. The monitoring results of the groundwater level are shown in Figure 10.

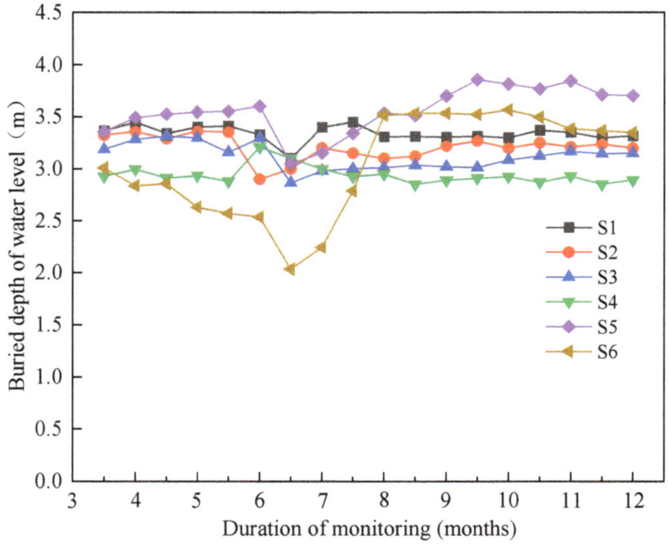

Figure 10. The time history curve of groundwater level monitoring in the foundation pit.

It can be seen from Figure 10 that the groundwater level of the six monitoring points changes in a very small range, and the groundwater level outside the pit did not markedly change during the whole process of foundation pit excavation and foundation construction. This shows that the water-stopping effect of the enclosure structure of the foundation pit project is good. At the same time, the whole process of foundation pit dewatering had no obvious impact on the surrounding environment and therefore met the design requirements.

4.8. Monitoring of the Layered Settlement of the Soil

The layered settlement of the soil was monitored by laying a layered settlement magnetic ring and using a layered settlement meter combined with a leveling method for measurement. In this project, a total of three monitoring points for layered soil settlement were arranged in representative parts close to the protected objects, numbered F1 to F3. Five settlement magnetic rings were arranged vertically and equidistantly in the settlement tube at each point, the settlement tube depth was 40 m, and the vertical spacing of each magnetic ring was 5 m. A stratified sedimentation meter was used to monitor the three monitoring points of layered settlement of soil, and the monitoring results are shown in Figure 11.

It can be seen from Figure 11 that the cumulative settlements of the three monitoring points all showed a law of first increasing and then remaining unchanged. The largest accumulative settlement occurred in F3 monitoring point, and the settlement was 3.17 mm, which did not exceed the design alarm value. This shows that, during the whole process of foundation pit excavation and foundation construction, the soil did not significantly settle, and the cumulative settlement and settlement rate did not exceed the alarm value, thus meeting the design requirements.

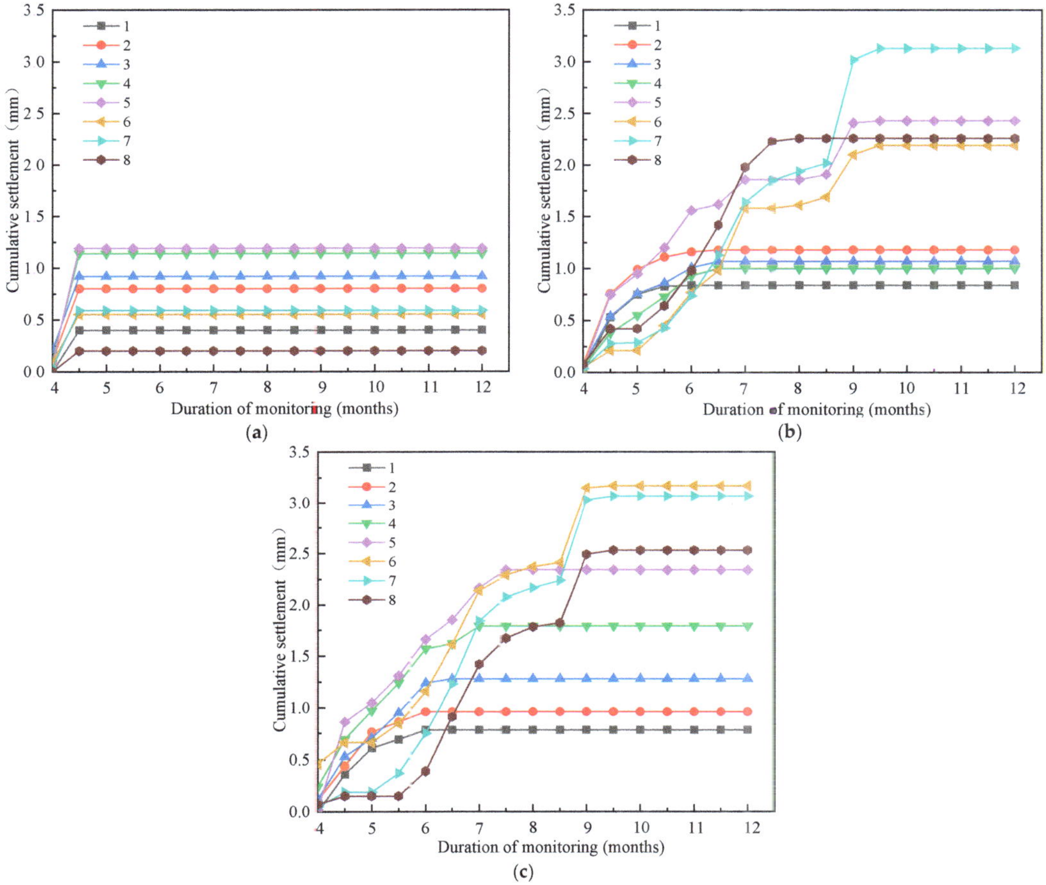

Figure 11. The time history curve of vertical displacement of each monitoring point in soil stratification: (**a**) the time history curve of F1 vertical displacement; (**b**) the time history curve of F2 vertical displacement; (**c**) the time history curve of F3 vertical displacement.

4.9. Monitoring of the Settlement of the Surrounding Buildings

Nineteen settlement monitoring points were set up in the corner of the outer wall and other key deformation parts of the Rongcheng ancient street building on the north side of the foundation pit and the factory building on the east side of the foundation pit, numbered J1 to J19. The observation sign was fixed on the wall or column by using the existing settlement observation point of the building or with expansion bolts. The level was used to monitor the settlement of these 19 points, and the monitoring results are shown in Figure 12.

It can be seen from Figure 12 that, before the 8th month, the cumulative settlement of each monitoring point increased with the increase in the excavation depth of the foundation pit. From the initiation of the project to the 4.5th month, the cumulative settlement of the monitoring points increased at a large rate. From the 4.5th month to the 8th month, the cumulative settlement growth rate of the monitoring points decreased significantly. After the 8th month, the cumulative settlement of each monitoring point remained basically unchanged. The point of the maximum cumulative settlement was the J7 monitoring point, with a deformation value of 12.42 mm. The minimum cumulative settlement occurred at the J17 monitoring point, with a deformation value of 7.05 mm. This shows that the whole foundation pit excavation and foundation construction process of the project did not have a significant impact on the surrounding buildings, and the monitoring data did not exceed the alarm value, thus meeting the design requirements.

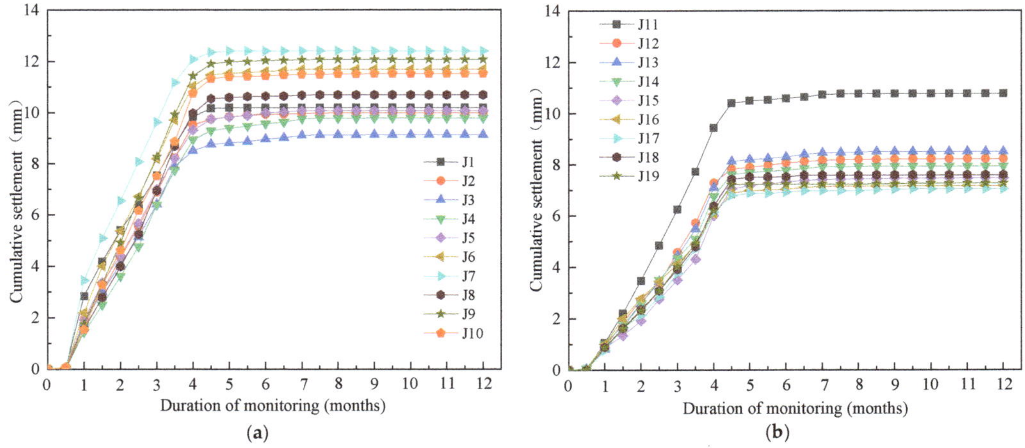

Figure 12. The time history curve of cumulative settlement of surrounding buildings: (**a**) the time history curve of cumulative settlement at J1 to J10 monitoring points; (**b**) the time history curve of cumulative settlement at J11 to J19 monitoring points.

4.10. Monitoring of the Horizontal Displacement of the Surrounding Buildings

Nineteen horizontal displacement monitoring points were set up in the corner of the outer wall and other key deformation parts of the Rongcheng ancient street building on the north side of the foundation pit and the factory building on the east side of the foundation pit, numbered J1 to J19. Theodolites were used to monitor the horizontal displacement of these 19 points, and the monitoring results are shown in Figure 13.

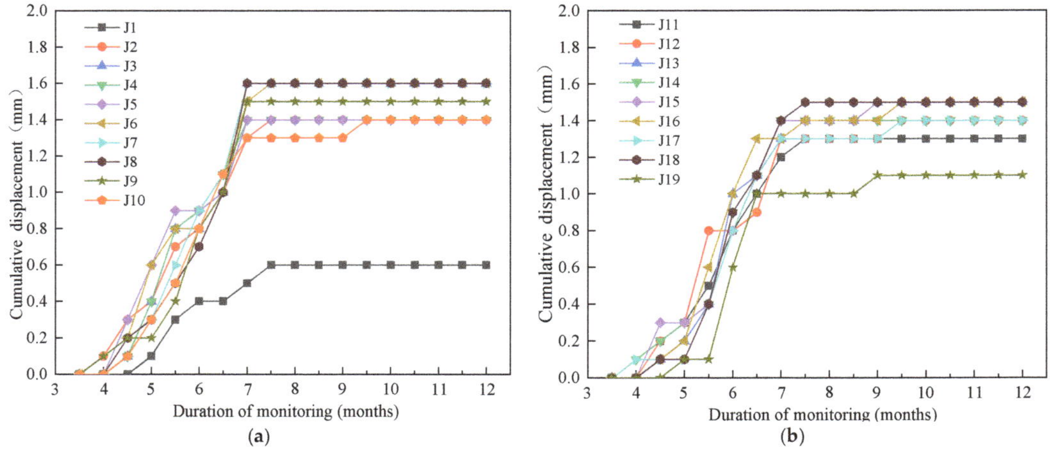

Figure 13. The time history curve of cumulative horizontal displacement of surrounding buildings: (**a**) the time history curve of cumulative horizontal displacement at J1 to J10 monitoring points; (**b**) the time history curve of cumulative horizontal displacement at J11 to J19 monitoring points.

It can be seen from Figure 13 that, before the 7th month, the cumulative horizontal displacement of each monitoring point increased with the increase in the excavation depth of the foundation pit. After the 7th month, the cumulative horizontal displacement of each monitoring point remained basically unchanged. The maximum cumulative horizontal displacement occurred at the J3 monitoring point, and the horizontal displacement was 1.6 mm. The minimum cumulative horizontal displacement occurred at the J1 monitoring point, and the horizontal deformation value was 0.6 mm. This shows that the earthwork excavation and foundation construction of the foundation pit project had very little impact on the surrounding buildings, and the monitoring data did not exceed the alarm value, which met the design requirements.

4.11. Monitoring of Tilt of the Surrounding Buildings

Tilt monitoring points were arranged on the key deformation parts of the building within two times the excavation depth of the foundation pit excavation boundary line, and a total of eight tilt monitoring points were arranged, numbered Q1 to Q8. Theodolites were used to monitor the tilt monitoring points of the surrounding buildings, and the tilt monitoring results of the surrounding buildings are shown in Figure 14.

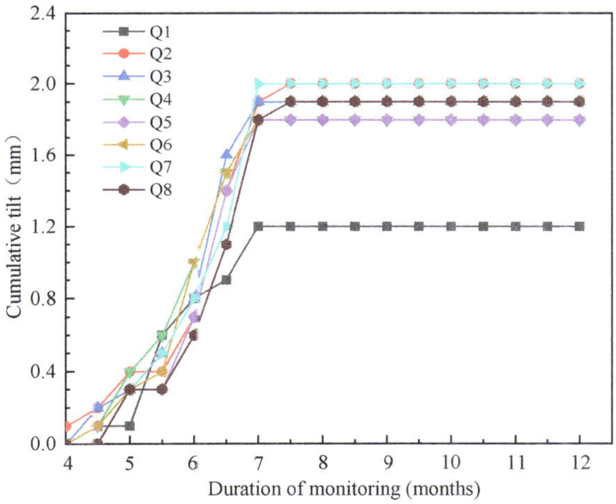

Figure 14. The time history curve of the cumulative tilt of the surrounding buildings.

It can be seen from Figure 14 that, before the 7th month, when the foundation pit was excavated, and the foundation was constructed, the cumulative tilt of the surrounding buildings increased with the excavation depth of the foundation pit. After the 7th month, when the foundation pit excavation and foundation construction were completed, and the basement shear wall construction started, the cumulative tilt of the surrounding buildings remained basically unchanged. During the whole monitoring process, the point with the maximum cumulative tilt was the Q2 monitoring point, and the tilt value was 2.0 mm. The point with the minimum cumulative tilt was the Q1 monitoring point, and the tilt value was 1.2 mm. This shows that the whole foundation pit excavation and foundation construction process of this project did not have a significant impact on the surrounding buildings, and none of the monitoring data exceeded the alarm value, which met the design requirements.

4.12. Monitoring of Settlement of the Surrounding Surface and Underground Pipelines

In the range of three times the excavation depth of the foundation pit from the boundary line of the foundation pit, nine settlement monitoring points were arranged on the surface and underground pipelines, numbered D1 to D9. The observation mark was driven into the ground with a measuring nail. The level was used to monitor the settlement of these nine points, and the monitoring results are shown in Figure 15.

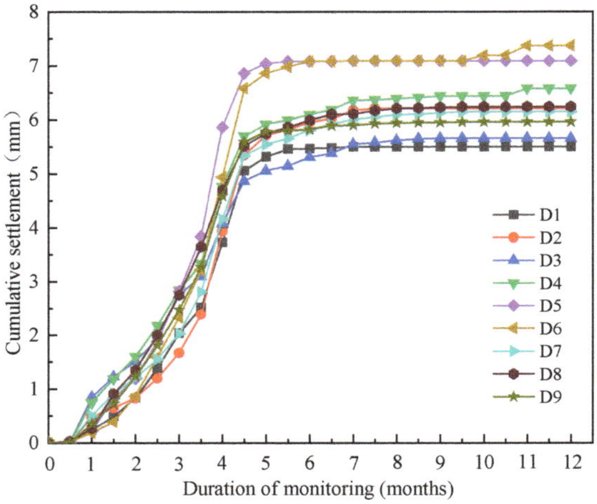

Figure 15. The time history curve of the cumulative settlement of the surrounding surface and underground pipelines.

It can be seen from Figure 15 that, before the 8th month, the cumulative settlement of each monitoring point increased with the increase in the excavation depth of the foundation pit. From the start of the project to the 4.5th month, the cumulative settlement growth rate of the monitoring point was relatively fast, and from the 4.5th month to the 8th month, the cumulative settlement growth rate of the monitoring point decreased significantly. After the 8th month, the cumulative settlement of each monitoring point remained basically unchanged. In the whole monitoring process, the maximum cumulative settlement occurred at the D6 monitoring point, and the settlement value was 7.37 mm. The minimum cumulative settlement occurred at the D1 monitoring point, and the settlement value was 5.50 mm. This shows that the whole process of foundation pit excavation and foundation construction had no obvious impact on the surrounding surface and underground pipelines, and the monitoring data of the project did not exceed the alarm value, which met the design requirements.

5. Conclusions

The surrounding area of the foundation pit of the Taijiang Square project in Fuzhou is complex, and the safety grade is high. The cast-in-place pile enclosure structure and three reinforced concrete internal bracing are adopted for support. In the whole process of foundation pit construction, 12 items such as the deep horizontal displacement of the enclosure pile, the horizontal displacement at the top of the foundation pit, the settlement at the top of the foundation pit, the axial force of internal bracing, and the axial force of the enclosure pile were monitored, and the following conclusions were drawn by analyzing the monitoring data of each item:

(1) During the construction of the foundation pit, the monitoring values of 12 monitoring items, including the deep horizontal displacement of the retaining pile, the horizontal displacement at the top of the foundation pit, the settlement at the top of the foundation pit, the axial force of the support, and the axial force of the retaining pile, did not exceed the alarm value, which met the design requirements.

(2) The support scheme of the cast-in-place pile enclosure structure and three internal bracing is reasonable and can effectively control the deformation of the foundation pit in soft soil areas.

(3) Tube-well dewatering can effectively lower the groundwater level below the bottom of the foundation pit, and the double-wheel deep-mixing water-stopping curtain wall can effectively control the infiltration of water outside the foundation pit.

(4) This foundation pit project is representative, and it provides a good reference case for the design of water-bearing deep foundation pit projects in the soft soil areas of Fuzhou.

Author Contributions: Conceptualization, B.T. and W.N.; methodology, J.Z. and M.S.; validation, B.T., W.N. and J.Z.; formal analysis, B.T.; investigation, M.S.; resources, B.T. and W.N.; data curation, J.Z. and M.S.; writing—original draft preparation, B.T.; writing—review and editing, W.N.; supervision, W.N.; project administration, M.S.; funding acquisition, B.T. and W.N. All authors have read and agreed to the published version of the manuscript.

Funding: This research was funded by the Collaborative Innovation Platform Project of Fuzhou-Xiamen-Quanzhou National Self-Innovation Zone under Grant No. 3502ZCQXT2022002; the Science and technology plan of Fujian Province under Grant No. 2022I0014; the Supported by Construction Technology Project of Xiamen Municipal Construction Bureau under Grant No. XJK2022-1-12; and the Supported by the Fundamental Research Funds for the Central Universities under Grant No. ZQN-1012.

Data Availability Statement: The monitoring data used for this research are not freely available due to legal concerns and commercial confidentiality. Nevertheless, all the concepts and procedures are explained in the presented research, and parts of the research may be available upon request.

Conflicts of Interest: The authors declare that they have no conflict of interest.

References

1. Rodriguez, J.A. Deep excavations for buildings in the Sabana Formation Bogota. In Proceedings of the 10th International Symposium on Geotechnical Aspects of Underground Construction in Soft Ground (IS-Cambridge), Cambridge, UK, 27–29 June 2022; pp. 145–152.
2. Fan, N.; Nian, T.K.; Guo, X.S.; Jiao, H.B. Piecewise strength model for three types of ultra-soft fine-grained soils. *Soils Found.* **2020**, *60*, 778–790. [CrossRef]
3. Gotman, A.L.; Gotman, Y.A. Numerical Analysis of the Shorings of Deep Foundation Pits with Regard for the Soil Solid-ification. *Soil Mech. Found. Eng.* **2019**, *56*, 225–231. [CrossRef]
4. Hettler, A.; Becker, P.; Borchert, K.M.; Kinzler, S. Report of the working group for recommendations on excavations: Outlook 6th edition EAB—Underpinnings, excavations in soft soils, head deformations of non-supported walls. *Bautechnik* **2019**, *96*, 785–792. [CrossRef]
5. Jun, S.H.; Kwon, H.J. Constitutive Relationship Proposition of Marine Soft Soil in Korea Using Finite Strain Consolidation Theory. *J. Mar. Sci. Eng.* **2020**, *8*, 429. [CrossRef]
6. Sun, Q.Q.; Dias, D.; Sousa, L.R.E. Soft soil layer-tunnel interaction under seismic loading. *Tunn. Undergr. Space Technol.* **2020**, *98*, 103329. [CrossRef]
7. Qiao, F.; Bo, J.S.; Wang, L.; Chang, C.Y.; Zhang, Z.P.; Qi, W.H. Survey of China's soft soil and its dynamic characteristics. *World Earthq. Eng.* **2019**, *35*, 150–161.
8. Ye, S.H.; Zhao, Z.F.; Wang, D.Q. Deformation analysis and safety assessment of existing metro tunnels affected by excavation of a foundation pit. *Undergr. Space* **2021**, *5*, 421–431. [CrossRef]
9. Takada, N.; Simono, K.; Oka, F.; Kimoto, S.; Higo, Y.A. Numerical evaluation of the deformation of earth retaining wall reinforced by soil buttress method during the excavation in soft soil. In Proceedings of the Computer Methods and Recent Advances in Geomechanics, Kyoto, Japan, 22–25 September 2014; pp. 1037–1042.
10. Zheng, G. Method and application of deformation control of excavations in soft ground. *Chin. J. Geotech. Eng.* **2022**, *44*, 1–36.
11. Song, X.G.; Wang, Z.Y.; Bai, W.W.; Wang, Z.; Xie, S.M.; Xia, K.Z. Study on engineering characteristics of large-scale deep soft soil in the central area of western Zhuhai. *Chin. J. Rock Mech. Eng.* **2019**, *38*, 1434–1451.
12. Zhang, Y.C.; Yang, G.H.; Hu, H.Y.; Chen, F.Q.; Huang, Z.X.; Chen, W.C. Some problems about retaining structures for shallow pits in deep and soft soil areas of Pearl River Delta. *Chin. J. Geotech. Eng.* **2014**, *36*, 1–11.
13. Ye, S.H.; Zhou, J. Study on stress and deformation of shield tunnel plate under unloading of foundation pit excavation. *Arab. J. Geosci.* **2021**, *14*, 2490. [CrossRef]
14. Yuan, C.F.; Hu, Z.H.; Zhu, Z.; Yuan, Z.J.; Fan, Y.X.; Guan, H.; Li, L. Numerical Simulation of Seepage and Deformation in Excavation of a Deep Foundation Pit under Water-Rich Fractured Intrusive Rock. *Geofluids* **2021**, *2021*, 6628882. [CrossRef]
15. Harahap, S.E.; Ou, C.Y. Finite element analysis of time-dependent behavior in deep excavations. *Comput. Geotech.* **2020**, *119*, 103300. [CrossRef]
16. Bal, A.R.L.; Dang, T.S.; Meschke, G. A 3D particle finite element model for the simulation of soft soil excavation using hypoplasticity. *Comput. Part. Mech.* **2020**, *7*, 151–172. [CrossRef]

17. Uribe-Henao, A.F.; Arboleda-Monsalve, L.G.; Mackie, K. Soil-Structure Interaction on Excavation-Induced Response of Moment-Resisting Frame Buildings. In Proceedings of the 2nd Geo Congress—Deep Foundations, Earth Retention, and Underground Construction, Charlotte, NC, USA, 20–23 March 2022; pp. 341–350.
18. Sun, F.X.; Liu, M.Q.; Zhu, Y.H.; Li, X.C.; Ge, G. Research on Numerical Simulation of Top-Down Construction Effect of Diaphragm Wall of Deep and Large under Different Working Conditions in Complex Stratum. *Adv. Civ. Eng.* **2022**, *2022*, 2576122. [CrossRef]
19. Wu, J.; Ye, S.H.; Wang, Z.Q.; Yang, D. Application and automatic monitoring and analysis of hybrid support structure in ultra-deep foundation pit engineering in the Lanzhou area under complex environmental conditions. *Water* **2023**, *15*, 1335. [CrossRef]
20. Dmochowski, G.; Szolomicki, J. Technical and Structural Problems Related to the Interaction between a Deep Excavation and Adjacent Existing Buildings. *Appl. Sci.* **2021**, *11*, 481. [CrossRef]
21. Chen, S.R.; Cui, J.F.; Liang, F.Y. Case Study on the Deformation Coupling Effect of a Deep Foundation Pit Group in a Coastal Soft Soil Area. *Appl. Sci.* **2022**, *12*, 6205. [CrossRef]
22. Rybak, J.; Ivannikov, A.; Kulikova, E.; Zyrek, T. Deep excavation in urban areas—Defects of surrounding buildings at various stages of construction. In Proceedings of the 9th International Scientific Conference on Building Defects (Building Defects 2017), Ceske Budejovice, Czech Republic, 23–24 November 2017; p. 02012.
23. Yang, T.; Liu, S.L.; Wang, X.Y.; Zhao, H.; Liu, Y.; Li, Y.W. Analysis of the Deformation Law of Deep and Large Foundation Pits in Soft Soil Areas. *Front. Earth Sci.* **2022**, *10*, 828354. [CrossRef]
24. Sun, L.; Mao, K.; Wang, Z.Z.; Ye, S.H.; Su, T.T.; Dai, G.L.; Xu, G.X.; Sun, J.L. Design and Field Monitoring of a Pile-Anchor-Brace Supporting System in a Soft Soil Area. *Water* **2022**, *14*, 3949. [CrossRef]
25. Chen, J.S.; Lin, C.; Liu, S.Z.; Mo, H.H. Study on Supporting Structure Performance of Deep Soft Soil Foundation Pit near Sea under Waves, Tides, Vibration, and Unbalanced Loads. *Adv. Civ. Eng.* **2020**, *2020*, 8830199. [CrossRef]
26. Czajewska, M.M. A study of displacements of structures in the vicinity of deep excavation. *Arch. Civ. Mech. Eng.* **2019**, *19*, 547–556. [CrossRef]
27. Feng, Z.Y.; Xu, Q.; Xu, X.Y.; Tang, Q.; Li, X.D.; Liao, X. Deformation Characteristics of Soil Layers and Diaphragm Walls during Deep Foundation Pit Excavation: Simulation Verification and Parameter Analysis. *Symmetry* **2022**, *14*, 254. [CrossRef]
28. Panchal, J.P.; McNamara, A.M.; Stallebrass, S.E. A new approach to modelling excavations in soft soils. In Proceedings of the 9th International Conference on Physical Modelling in Geotechnics (ICPMG), London, UK, 17–20 July 2018; pp. 1445–1450.
29. Kiet, H.N.; Phien-wej, N. Advanced soil parameters determination for Ho Chi Minh city soft clay to predict ground movements in deep excavations and tunneling. In Proceedings of the 4th International Conference on Geotechnics for Sustainable Infrastructure Development (GEOTEC HANOI), Hanoi, Vietnam, 28–29 November 2019; pp. 497–504.
30. Nguyen, B.P.; Ngo, C.P.; Tran, T.D.; Bui, X.C.; Doan, N.P. Finite Element Analysis of Deformation Behavior of Deep Excavation Retained by Diagram Wall in Ho Chi Minh City. *Indian Geotech. J.* **2022**, *52*, 989–999. [CrossRef]

water

Article

Influences of Underwater Shield Tunnelling on River Embankment Seepage Stability Considering Various Overburden Thickness

Wenyu Shu [1], Jingjing Ma [2], Ningning Geng [2], Yang Xiang [2], Shiyu Ma [2], Xian Li [1,*], Fang Tong [1] and Shisheng Fang [1]

[1] College of Civil Engineering, Hefei University of Technology, Hefei 230009, China; 2020110566@mail.hfut.edu.cn (W.S.)
[2] Hefei Rail Transit Group Limited Company, Hefei 230011, China
* Correspondence: xianli@hfut.edu.cn

Abstract: Underwater shield tunneling will disturb the soil near the river, especially in water-rich soft ground. This may cause a groundwater infiltration hydraulic gradient to exceed the critical value, leading to calamities such as unexpected flooding or submerged erosion. To ensure the security of construction and the stability of river embankment seepage, it is crucial to assess the safety of the underwater tunnel cover thickness. A shield tunnel project under a river in Hefei is used as an example. The numerical model established by the finite element method is used for calculating and analyzing the changes in the groundwater flow field and the stability state of embankment seepage induced by underwater shield tunneling under different overburden thickness conditions. The results show that the construction disturbance of the shield tunnel through the river is increased, the internal force environment of the embankment slope is destroyed, and the maximum seepage hydraulic gradient is increased. In the case study, the embankment keeps in a stable state of seepage when the cover thickness of the shield tunnel has 2.9 times its outer diameter. The findings of this study can serve as a scientific guide to assure seepage stability in an underwater shield tunneling project and to stop river embankment erosion.

Keywords: river-crossing tunnel; cover thickness; seepage stability; numerical simulation

check for updates

Citation: Shu, W.; Ma, J.; Geng, N.; Xiang, Y.; Ma, S.; Li, X.; Tong, F.; Fang, S. Influences of Underwater Shield Tunnelling on River Embankment Seepage Stability Considering Various Overburden Thickness. *Water* **2023**, *15*, 2346. https://doi.org/10.3390/w15132346

Academic Editors: Glen R. Walker and Dan Ma

Received: 5 May 2023
Revised: 2 June 2023
Accepted: 16 June 2023
Published: 25 June 2023

1. Introduction

In recent years, urban subway construction in China has grown significantly, and tunneling projects that cross rivers are becoming increasingly prevalent [1–4]. The shield technique is the most widespread tunneling build process [5]. When the shield method is utilized, the construction pace is fast and safe. However, the surrounding soil will inevitably be disturbed while the shield tunnels cut through the rivers. Furthermore, the physical and mechanical characteristics of the disturbed soil can be changed as a side effect of tunnel construction [6], weakening its ability to resist seepage [7] and significantly increasing the infiltration hydraulic gradient [8]. This increases the seepage erosion of the soil on the river embankment slope and causes significant soil loss, which in turn causes infiltration damage, such as flowing soil and pipe surges [9,10]. For example, in the construction of the river-crossing tunnel of Shanghai rail transit line 4, accidents occurred in the connecting channel flowing sand and gushing water, tunnel structures damaging, ground setting, and the direct cause of the accident was the sudden surge of pressurized water [11]. It was reported that sand and water surged at the end of the shield seal when one shield tunnel crossing the river was being built in Wuhan, China [12]. A rapid surge disaster happened during the construction of Tianjin Metro Line 2's shield structure beneath the river while traversing a thin layer of water-rich silt [13]. The Quaternary fluvial deposits, which mostly consist of clay, pulverized clay, dust, silt layers, and sand layers accumulated in the alluvial phase of ancient rivers, frequently predominate in the riverine strata [14]. These rising and falling are practically coordinated considering the hydraulic connection between the groundwater in the riverine strata and the stream of surface water [15]. There is a high likelihood that seepage channels will be created during tunnel construction, changing the seepage environment and harming the embankment slope's internal force environment [16–18]. The fine particles of the disturbed pore space rock and soil are continuously carried away by the groundwater flow under the influence of high permeability during high water levels in flood season and the sudden drop in water level. This further development leads to infiltration damage caused by seepage, which results in embankment slope instability [19]. To maintain the security of construction and river embankment safety and avoid infiltration damage, it is crucial to design the shield tunnel with the appropriate cover thickness.

Therefore, the minimum cover depth, as one of the most critical aspects to design underwater tunnels, must be addressed. The Japanese minimum water seepage method [20,21], the Norwegian empirical method [22], the mechanical equilibrium method [23], and the Chinese underwater mining empirical method [24] are popular methods for predicting the minimum cover depth, but it is still not certain to what capacity they can be used to estimate the safe thickness of shield tunnels. The level of the tunnel cover depth, together with additional elements, such as soil permeability [25–27] and dynamic disturbances [28–30], have significant effects on the tunnel and ground reaction.

Several researchers have developed models of the surface settlement caused by tunnel excavation while taking the effect of tunnel cover depth into consideration [31,32]. However, when contemplating realistic modeling of the seepage field for underwater tunnels, it is crucial to take the influence of the depth of cover into account. The seepage field variations in submerged tunnel excavation have been predicted and examined by numerous studies [33–36]. Additionally, it is beneficial to better understand the fluid-mechanical interaction that occurs in underwater tunneling [37–41]. These studies shed light on the hydraulic interaction between an underwater tunnel and groundwater, which is essential for ensuring the ideal design and construction management of underwater tunnels. Associated research have further shown that tunnel excavation has a stronger effect on the seepage stability of the formation and the region of seepage erosion of the formation is larger when the shield tunnel cover thickness is shallow [42,43]. In this study, the investigation of the maximum hydraulic gradient variation of the embankment slope serves as the primary methodology for estimating the minimum depth of cover for submerged tunnels.

To study the impact of various cover thickness conditions on the seepage stability of the embankment slope, the shield tunnel crossing the river in Hefei serves as an illustration, establishing a two-dimensional seepage model by means of the finite element software Geo-studio. Furthermore, the safety of the overburden thickness of the underwater shield tunnel for this project was also assessed.

2. Methods

To simulate the two-dimensional steady-state seepage and transient seepage field changes on the river embankment slope under the design flood level and the design flood level declining respectively, the seepage calculations in this study primarily use the finite element method of the Geo-studio software. The two-dimensional seepage differential control equations are [44]:

$$\frac{\partial}{\partial x}\left(K_x\frac{\partial h}{\partial x}\right)+\frac{\partial}{\partial z}\left(K_z\frac{\partial h}{\partial z}\right)=C\frac{\partial h}{\partial t} \tag{1}$$

$$C=\frac{\partial \theta}{\partial h} \tag{2}$$

where $h(x, y, z, t)$ is the head function to be sought (m), K_x and K_y are the permeability coefficient (cm·s^{-1}) in the direction of the x and z axis as the main axis, C is the water capacity, θ is the unit volume of water content.

The fixed solution conditions are as follows:

(1) Initial conditions:

$$h|_{t=0} = h_0(x, z, 0) \tag{3}$$

(2) Assume that the boundary is either $\Gamma = \Gamma_1$ or $\Gamma = \Gamma_2$. Where Γ_1 is the first class of boundary conditions, such as given head bounds and upstream and downstream water level boundary surfaces. Water separation boundaries and other well-known flow borders are examples of the second category of boundary conditions or Γ_2.

Type I boundary conditions:

$$h|_{t=0} = h_0(x, z, t) \tag{4}$$

Type II boundary conditions:

$$K_n \frac{\partial H}{\partial n}|_{\Gamma_2} = q(x, z, t) \tag{5}$$

where K_n is the normal permeability coefficient of the boundary surface (cm·s^{-1}), and n is the normal vector outside the boundary surface.

The variational approach is used to further get the finite element computation format:

$$[K]\{h\} + [M]\left\{\frac{\partial h}{\partial t}\right\} + [D]\{q\} = \{F_0\} \tag{6}$$

where $\{F_0\}$ is the nodal water vector created by sources or sinks inside the seepage field, $[K]$ is the unit seepage conduction matrix, $[M]$ is the matrix of water absorbed or released from the pore with unit head variation, $[D]$ is the matrix of water generated by unit flow variation at the flow boundary.

$$K_{ij} = \iiint\limits_{\Omega_e} \left(K_x \frac{\partial N_j}{\partial x}\frac{\partial N_i}{\partial x} + K_z \frac{\partial N_j}{\partial z}\frac{\partial N_i}{\partial z}\right) dxdz \tag{7}$$

$$M_{ij} = C \iiint\limits_{\Omega_e} N_i N_j dxdz \tag{8}$$

$$D_{ij} = C \iiint\limits_{\Omega_e \cap \Gamma_2} N_i N_j d\Gamma \tag{9}$$

where K_{ij}, M_{ij}, and D_{ij} are the relevant matrix elements and iterative calculation is used to determine the seepage field in the embankment slope.

3. Study Area

The Nanfei River is crossed by the metro shield tunnel project in Hefei, Anhui Province. Figure 1 shows the precise site of the project crossing. The shield tunnel cutting through the river is constructed by a single-circle shield method with prefabricated assembly-type single-layer lining. The thickness of the single-circle tunnel segment in the shield section is 300 mm, with an inner diameter of 5.4 m and an outer diameter of 6 m. The project is situated in the Nanfei River, also known as Shishui, Jindou River, which is a major tributary of the Chaohu Lake water system. It originates from the Jianghuai watershed and continues to extend southward from the great submerged mountain of Changgang. The main river channel has a length of 70 km, and the watercourse of crossing position is relatively straight. Moreover, according to the river's measured part, the project section's river embankment has an elevation of 12.6 m, the water's surface is about 80 m wide, and the river's minimum elevation is 6.0 m. The Nanfei River portion, where the project is located, currently has a 100-year flood control standard.

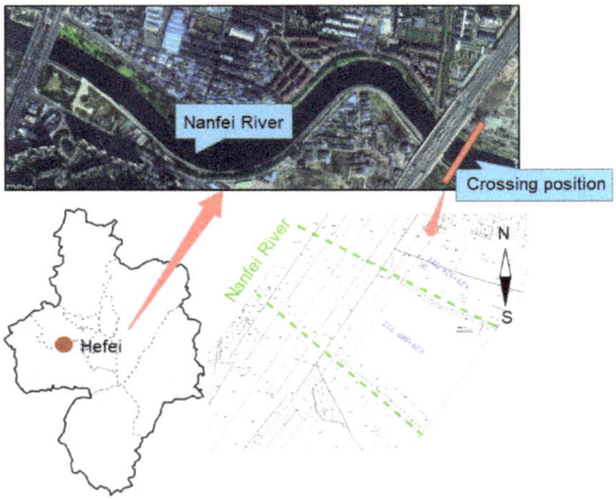

Figure 1. Project location map.

The Nanfei River's Second Bottom and river floodplain make up the terrain of the project river segment. The project river section's top layer is a layer of Quaternary artificial fill, mainly miscellaneous fill that is localized as plain fill. Layers of Quaternary Holocene alluvial clay, pulverized clay, dust, and silt-sand are found beneath the top layer. The deep soil layers are Quaternary alluvial clay, chalky clay, powder clay, and sandy soil layers of the alluvial phase of the ancient river of modest depth. According to the investigation region's geological survey report, the stratum at the project location is generalized in this study, and the stratum at the crossing section is made up of miscellaneous fill, clay, powdery clay, chalk, and fine sand in that order from top to bottom. The pertinent parameters of the stratum at the river section are shown in Table 1.

Table 1. Physical and mechanical parameters of the stratum section at the project river section.

Materials	E (MPa)	γ (kN·m^{-3})	c (kPa)	φ (°)	K (cm·s^{-1})	*Enhanced K* (cm·s^{-1})
Miscellaneous fill	10	19.5	0	9	5.0×10^{-5}	-
Clay	55	19.8	75	12.5	1.0×10^{-5}	-
Silty clay	18	20.1	70	12	2.0×10^{-5}	2.0×10^{-4}
Silt	19	20.5	67.5	22.5	3.0×10^{-5}	3.0×10^{-4}
Silty fine sand	20	21.0	0	27.5	1.0×10^{-4}	1.0×10^{-3}
Shield tunnel	3.0×10^{5}	25	-	-	0	-

The primary aquifers in the study region are water-bearing formations of rocks that are submerged (somewhat pressured), exhibit excellent permeability, receive recharge from atmospheric precipitation, and share fair recharge and discharge relationships with rivers. Additionally, there is no infiltration inside the underwater tunnel in seepage impact of interest. The two-dimensional modeling considered the shield tunnel layer's permeability coefficient, which is set to 0, in order to mimic the tunnel cross-section. The relevant parameters of the shield tunnel are also shown in Table 1.

4. Numerical Modeling

In this work, the underwater shield tunnel crossing the river embankment was modeled and examined using the finite element analysis program Geo-studio. A two-dimensional model was created for the semi-river profile at the shield tunnel crossing position based on the project profile. Model the slope's top elevation (12.6 m), the river's bottom elevation (6.0 m), and the shield tunnel's outer diameter (6 m). As mentioned in

Section 3, the shield tunnel layer permeability coefficient is configured as 0 to simulate the tunnel section in 2D modeling. While taking into account the shield tunnel's impact on the surrounding soil disturbance, designate the shield tunnel's outer diameter 2 m range as the troublesome zone, and increase the soil permeability coefficient value by one order of magnitude for the troublesome zone's permeability coefficient. The numerical model mesh of seepage is split as indicated in Figure 2 and the upper boundary condition of the model is the first kind of specified head boundary. The prior testing calculations show that the number and size of the grid could accurately define the river boundaries in the case of meeting the demands of numerical calculation accuracy.

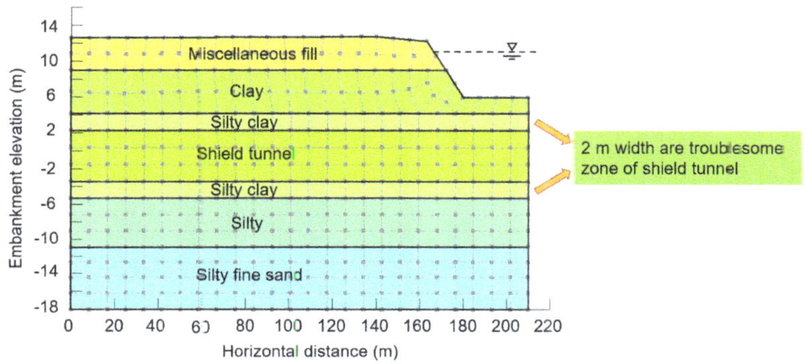

Figure 2. Diagram of the numerical model of seepage in the study area.

Two seepage calculation conditions, steady and unsteady seepage conditions, are used. The 100-year design flood level of 12.21 m is the steady seepage calculation condition, and the unstable seepage calculation is that the design flood level decreases sharply by 2 m in 24 h. The specific calculation circumstances are provided in Table 2.

Table 2. Seepage calculation conditions.

Initial Conditions	Descriptions
Stabilized seepage condition	100-year design flood level 12.21 m
Unsteady seepage condition	The 100-year design flood level plunges 2 m in 24 h

Along with the investigation of the region's engineering geology, the shield tunnel's diameter was held constant, and the overburden thicknesses were chosen to be 3.76 m, 9.07 m, and 17.51 m for the three overburden thicknesses to evaluate the impact of the various overburden thicknesses on the stability of river embankment seepage. Moreover, the top elevation of the embankment minus the minimum top elevation of the tunnel, which corresponds to the location at the burial depth of 17.51 m (the distance between the river bottom and the top of the tunnel), gives the maximum burial depth of the tunnel in the design scheme. Table 3 displays the various operating circumstances and the resulting overburden thicknesses.

Table 3. The cladding thickness of the underwater shield tunnel.

Numbers	Types of Strata Crossed by the Underwater Shield Tunnel	Cladding Thickness (m)
1	Silty clay	3.76
2	Silt	9.07
3	Silty fine sand	17.51

5. Results and Analysis

5.1. Analysis of the Impact of Underwater Shield Tunnel on River Embankment Seepage Stability under Design Flood Level

5.1.1. The Results of Seepage Head under Design Flood Level

From a general perspective, the seepage head is smaller the closer it is to the river embankment, and it gradually grows larger the farther it is from the river embankment. This is given as the numerical model makes the top of the slope a zero pressure hydraulic boundary, and the water flow path from the embankment to the river flow, which needs to conquer soil resistance causing the seepage head loss and enhances further the flow of seepage head loss. It is observed that the seepage head contours within the disturbance zone of the underwater shield tunnel with different burial depths are vertical. Consider Figure 3a as a case study, the seepage head within the disturbance zone of the underwater shield tunnel cutting through the powdered clay layer has almost no changes in the vertical direction and only gradually decreases in the horizontal direction, and the seepage head distribution below the disturbance zone changes drastically.

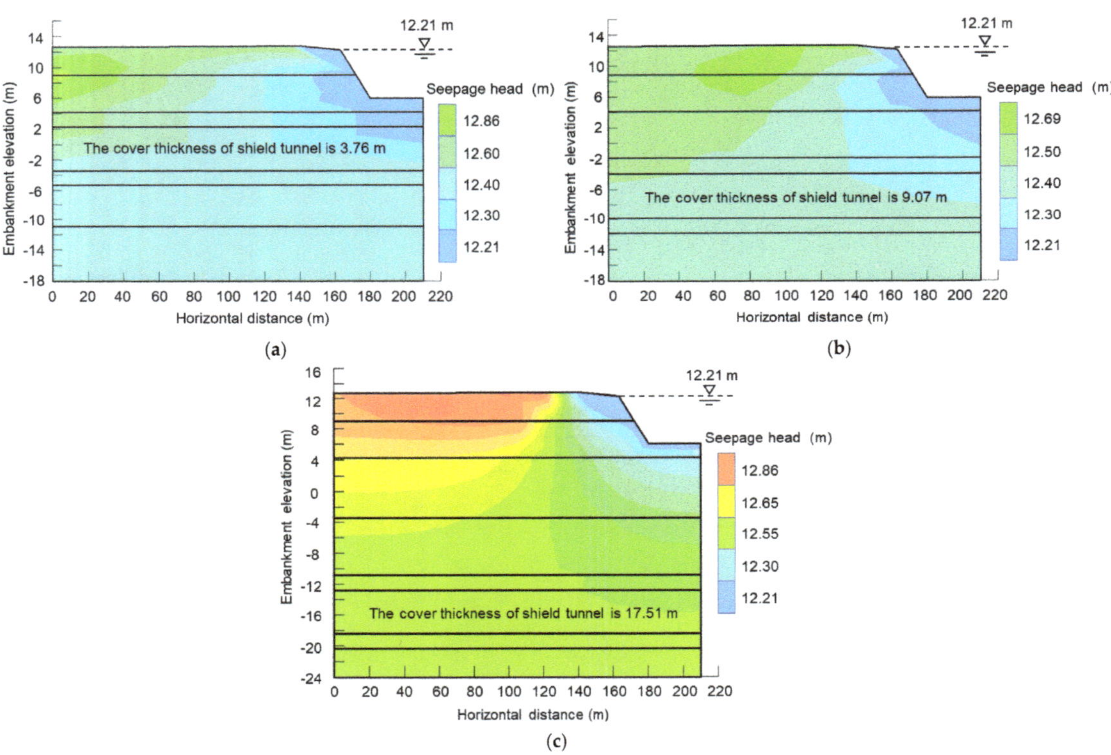

Figure 3. Seepage head contours under the design flood water level of: (**a**) 3.76 m of underwater tunnel overburden thickness; (**b**) 9.07 m of underwater tunnel overburden thickness; (**c**) 17.51 m of underwater tunnel overburden thickness.

According to Figure 4, the submerged shield tunnel cover thickness of 9.07 m has a maximum seepage head of 12.69 m, which is somewhat less than the maximum seepage head for the 3.76 m and 17.51 m submerged tunnel cover thicknesses. The seepage head contour in Figure 3a is nearly vertical within the soil near the top of the embankment slope, indicating that there is still a significant amount of horizontal seepage at this location. When the underwater tunnel overburden thickness is shallow, the tunnel has a greater impact on the embankment seepage field distribution. The analysis's findings index that the submerged shield tunnel influences the distribution of seepage head on the river embankment and that the impacts are larger on the seepage field the shallower the overburden thickness.

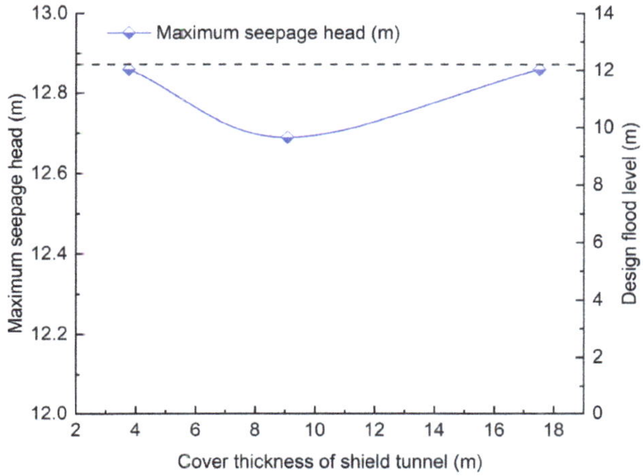

Figure 4. Maximum seepage head of embankment under the design water level.

5.1.2. The Results of Seepage Hydraulic Gradient under Design Flood Level

To examine the change in the infiltration hydraulic gradient of the river embankment under various overburden thickness excavation in the underwater shield tunnel at the design flood level condition (100-year event), simulation results from the numerical calculation model of steady seepage are utilized (Figure 5). The top of the river embankment, there remaining violent water flow, is that the infiltration damage occurs at a high level of the water, according to calculation findings, and it is safer to regulate the infiltration hydraulic gradient within 0.30 in accordance with the relevant specification [45]. The findings indicate that when excavating at 3.76 m, the maximum hydraulic gradient reaches 0.30 and there is a likelihood of infiltration damage. In contrast, the infiltration hydraulic gradient of the river embankment under other overburden thickness conditions is below the permitted hydraulic gradient, and there is no immediate risk of infiltration damage on the river embankment.

Figure 6 provides a detailed comparison and analysis of the results of river embankment infiltration hydraulic gradient for various shield tunnel cover thicknesses. The dotted line indicates the design flood level. (12.21 m). The maximum infiltration hydraulic gradient is 0.30 for underwater shield tunnel cover thickness of 3.76 m, 0.14 for underwater shield tunnel cover thickness of 9.07 m, and 0.09 for underwater shield tunnel cover thickness of 17.51 m. The comparison consequences reveal that the overburden thickness rises 141% when 3.76 m to 9.07 m, whereas the river embankment infiltration hydraulic gradient falls by 53.3% From 9.07 m to 17.51 m, the overburden thickness of the shield tunnel rises 93.1%, while the maximum infiltration hydraulic gradient of the river embankment falls 35.7%. The maximum hydraulic gradient under the design flood level gradually decreases when the burial thickness of the underwater shield tunnel rises

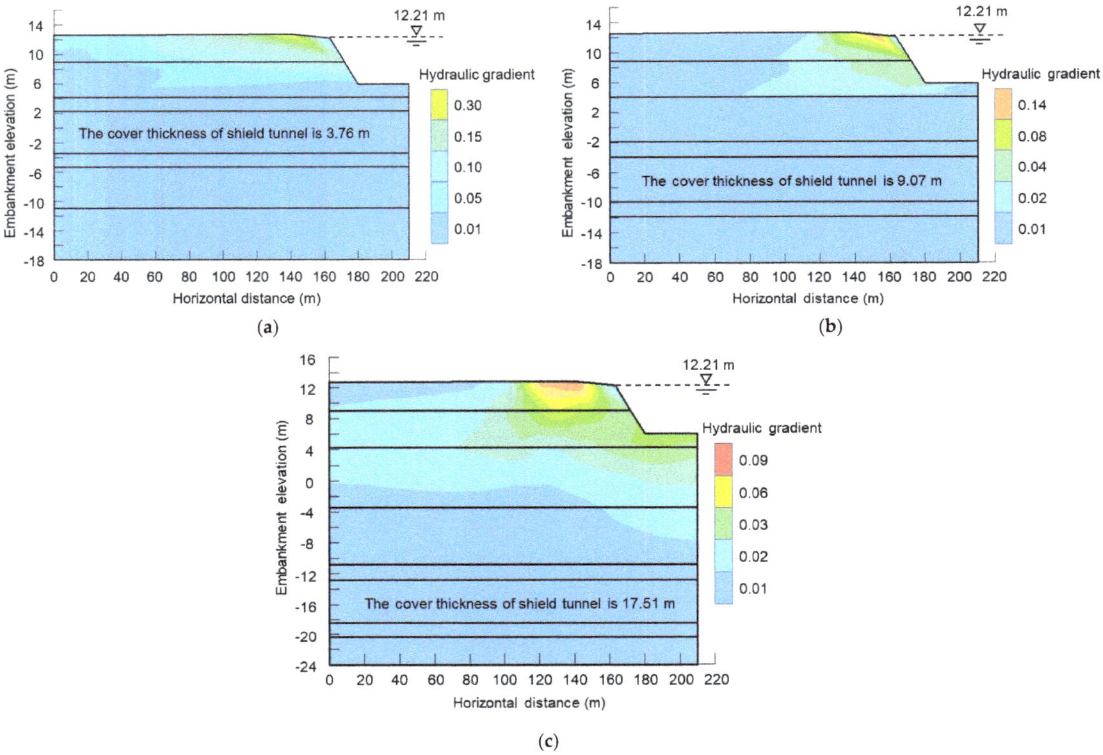

Figure 5. Seepage hydraulic gradient contours under design flood water level of: (**a**) 3.76 m of underwater tunnel overburden thickness; (**b**) 9.07 m of underwater tunnel overburden thickness; (**c**) 17.51 m of underwater tunnel overburden thickness.

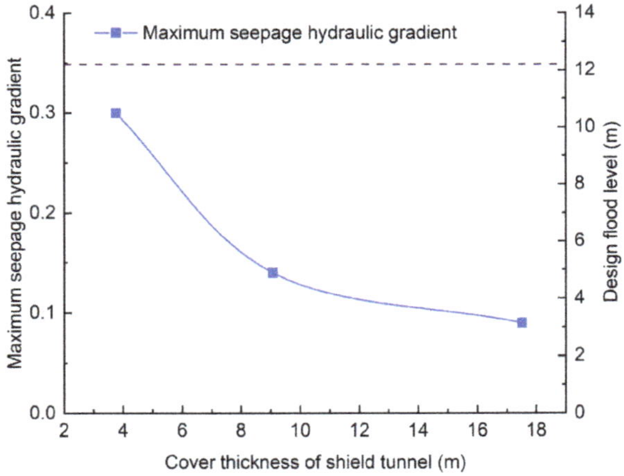

Figure 6. Maximum seepage hydraulic gradient of embankment under design water level.

5.2. Analysis of the Impact of Underwater Shield Tunnel on River Embankment Seepage Stability under Sudden Drop of Design Flood Level

5.2.1. The Results of Seepage Head under Sudden Drop of Design Flood Level

Based on the preliminary analysis of Figure 7, the maximum seepage head value at the underwater shield tunnel overburden thickness of 3.76 m is 13.67 m, the highest seepage head value at the underwater shield tunnel overburden thickness of 9.07 m is 12 77 m, and the maximum seepage head value is 12.86 m at the underwater shield tunnel overburden thickness of 17.51 m.

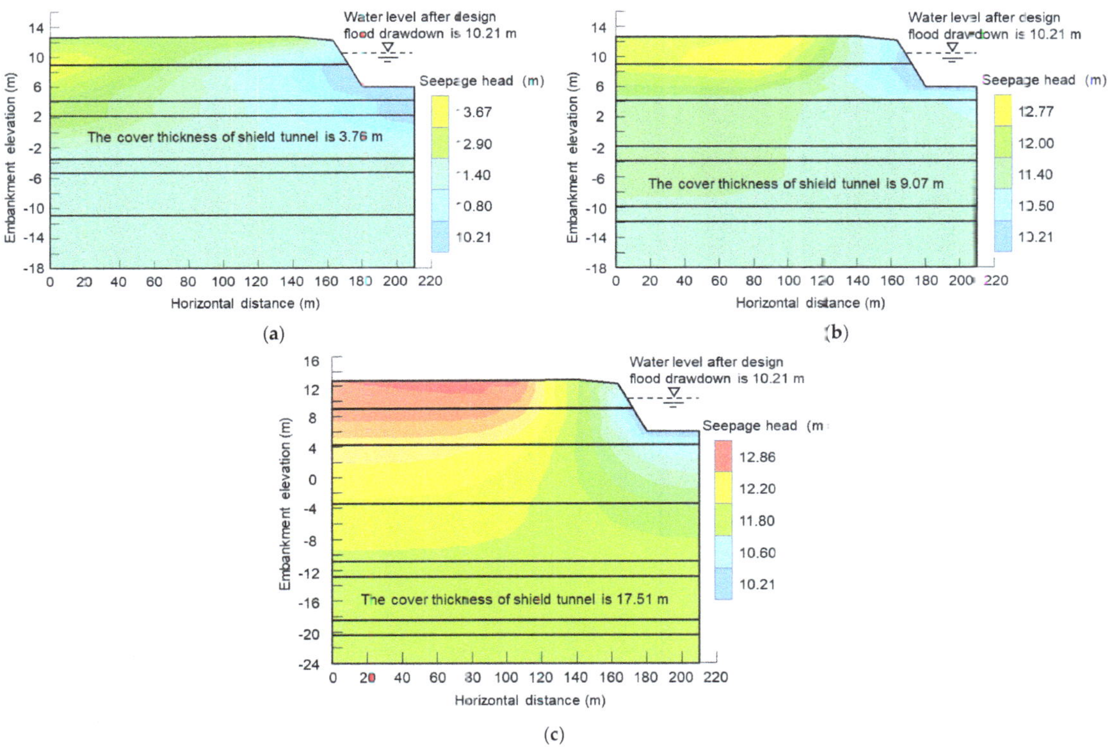

Figure 7. Seepage head contours under the sudden drop of design flood level of: (**a**) 3.76 m of underwater tunnel overburden thickness; (**b**) 9.07 m of underwater tunnel overburden thickness; (**c**) 17.51 m of underwater tunnel overburden thickness.

Figure 8 shows that the maximum seepage head values within the river embankment decrease by 6.6% when the underwater shield tunnel overburden thickness increases by 141% (from 3.76 m to 9.07 m). Moreover, Figure 8 shows that the maximum seepage head of the river embankment increases by 0.7% when the underwater shield tunnel overburden thickness increases by 93.1% (from 3.76 m to 9.07 m). The dotted line indicates the design flood level plunges by 2 meters in 24 hours. The findings demonstrate that the shallow submerged shield tunnel under unsteady seepage conditions influences both the maximum seepage head and the spatial distribution of the seepage head.

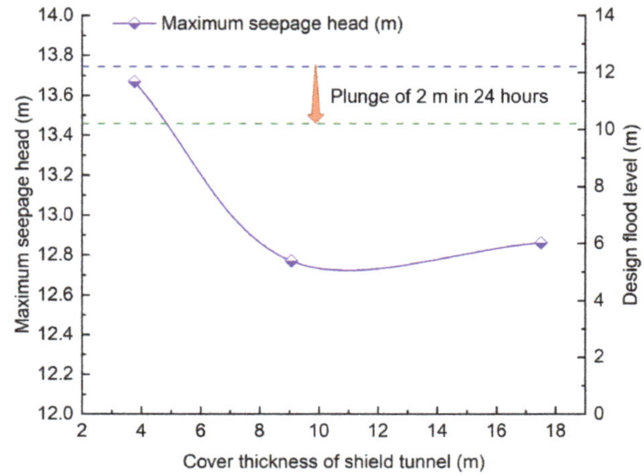

Figure 8. Maximum seepage head of the embankment under the sudden drop of design flood level.

Additionally, it compares the maximum seepage head values of the river embankment under various hydraulic conditions in Figure 9. When the underwater shield tunnel overburden thickness is 3.76 m, the maximum seepage head value under the sudden drop of design flood level is 6.3% higher than the maximum seepage head value under the design flood level. When the underwater shield tunnel overburden thickness is 9.07 m, the maximum seepage head value is 0.63% greater than the maximum head value under the design flood level. The maximum seepage head values for the sudden drop of design flood level and design flood level are the same for the underwater shield tunnel overburden thickness of 17.51 m. The findings demonstrate that when the shield tunnel overburden thickness is shallow, the maximum seepage head value under a sudden drop in design flood level is higher than that of the design flood level. According to the detailed analysis of the causes of the phenomenon, the reasons are that the maximum seepage head increases due to the rapid decline in water level and the low permeability of the soil in the river embankment, which prevents water from flowing out of the embankment in time to discharge.

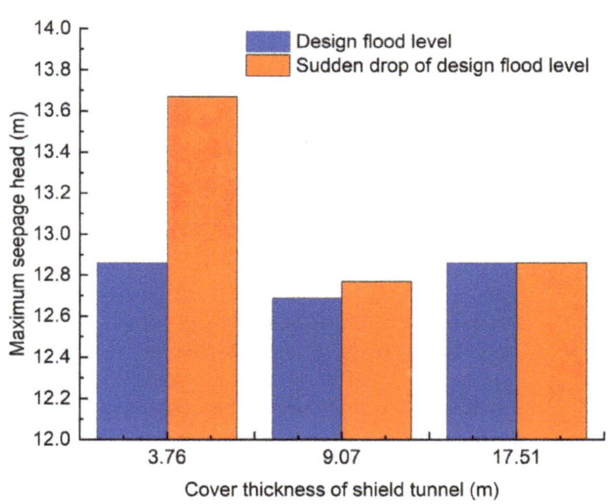

Figure 9. Comparison of maximum seepage head of river embankment.

5.2.2. The Results of Seepage Hydraulic Gradient under Sudden Drop of Design Flood Level

Analyze the changing laws of infiltration hydraulic gradient of the river embankment of the shield tunnel excavated with different overburden thickness under a sudden drop of design flood level (a sudden drop of design flood level by 2 m in 24 h), according to simulation results of the numerical calculation of unsteady seepage (Figure 10). When the underwater shield tunnel overburden thickness is 3.76 m, the maximum infiltration hydraulic gradient is 0.65 and the maximum hydraulic gradient is at the top of the river embankment. When the underwater shield tunnel overburden thickness is 9.07 m, the maximum infiltration hydraulic gradient is 0.40 and the maximum hydraulic gradient is at the intersection of the river channel and the river embankment. The greatest infiltration hydraulic gradient value is 0.24 when the shield tunnel overburden thickness is 17.51 m, and the maximum hydraulic gradient is also found at the embankment and river channel converge. The maximum hydraulic gradient values for the underwater shield tunnel overburden thickness of 3.76 m and 9.07 m are 0.65 and 0.40 respectively, which exceed the allowable slope drop of the code. Under these two work conditions, the river embankment may sustain infiltration damage.

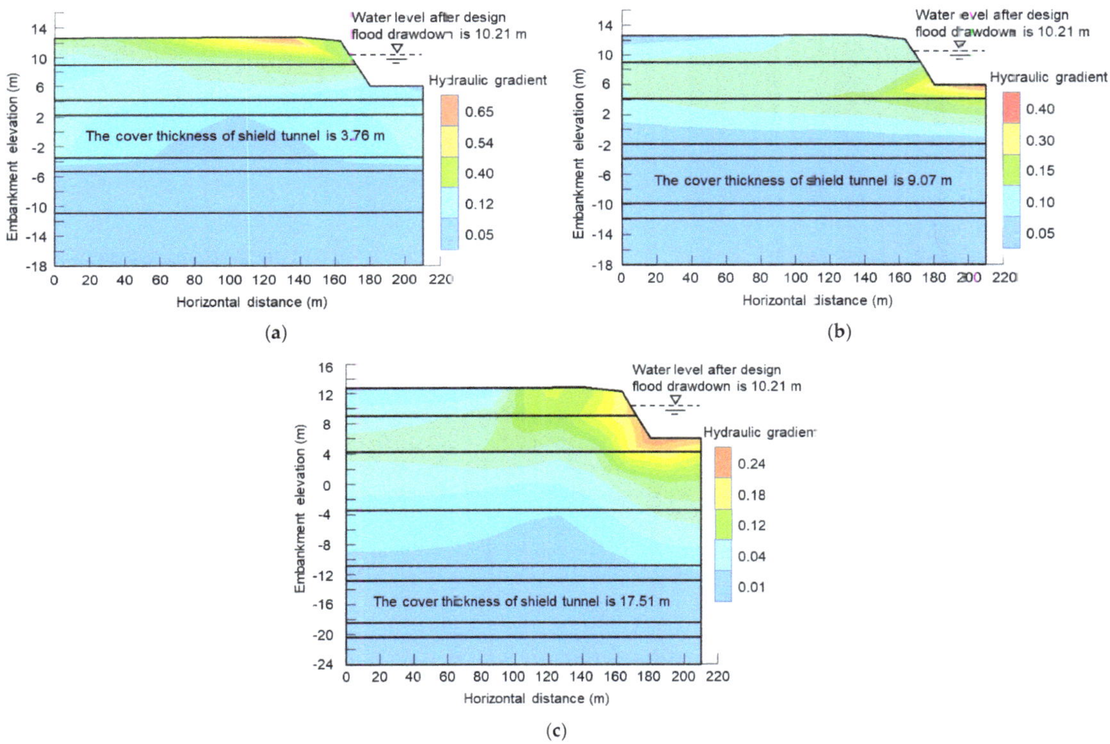

Figure 10. Seepage hydraulic gradient contours under the sudden drop of design flood level of: (**a**) 3.76 m of underwater tunnel overburden thickness; (**b**) 9.07 m of underwater tunnel overburden thickness; (**c**) 17.51 m of underwater tunnel overburden thickness.

The maximum infiltration hydraulic gradient value continually declines with the increasing underwater shield tunnel overburden thickness, and this rule is consistent with the changing trend of the maximum hydraulic gradient under the parameters mentioned above for the design flood level. Comparing and analyzing the infiltration hydraulic gradient results in Figure 11, it can be reported that the underwater shield tunnel overburden thickness increases from 3.76 m to 9.07 m, and its maximum hydraulic gradient decreases by 39.4%. While it increases from 9.07 m to 17.51 m, its overburden thickness increases by 93.1%, and the embankment's maximum infiltration hydraulic gradient decreases by 40%. And the dotted line indicates the design flood level plunges by 2 m in 24 h in Figure 11.

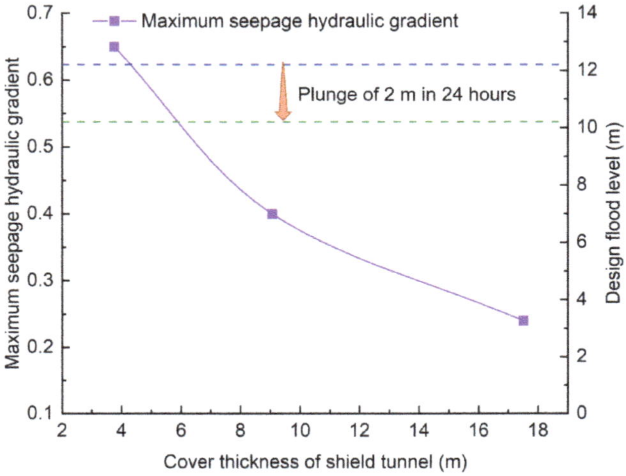

Figure 11. Maximum seepage hydraulic gradient of embankment under a sudden drop of design flood level.

To investigate the influences of various seepage circumstances and tunnel covering thickness on the maximum infiltration hydraulic gradient of the river embankment, the outcomes for the maximum hydraulic gradient are further compared (Figure 12). When the underwater shield tunnel overburden thickness is 3.76 m, the maximum infiltration hydraulic gradient under a sudden drop of the design flood level is 120% greater than the maximum infiltration hydraulic gradient under the design flood level. When the underwater shield tunnel overburden thickness is 9.07 m, the maximum infiltration hydraulic gradient under the sudden drop of the design flood level is 186% greater than the value under the design flood level. When the underwater shield tunnel overburden thickness is 17.51 m, the maximum infiltration hydraulic gradient under the design flood level is 186% greater than the value under the design flood level. This demonstrates that when the underwater shield tunnel overburden thickness increases, the maximum hydraulic gradient under the sudden drop of design flood level condition is significantly greater than the design flood level condition, which is caused by the flood level plunge creating a large hydraulic head difference between the groundwater and surface water body, resulting in the larger hydraulic gradients.

Moreover, when the underwater shield tunnel overburden thickness is 3.76 m, the maximum infiltration hydraulic gradient occurs at the top of both the river embankment near the shoulder under the design flood level condition and sudden drop of design flood level condition. When the underwater shield tunnel overburden thickness is 9.07 m and 17.51 m, the maximum infiltration hydraulic gradient occurs at the junction between the river channel and the river embankment under the sudden drop of design flood level condition, while the maximum hydraulic gradient always occurs at the top of the river embankment under the design flood level. The outcome reveals that when the underwater shield tunnel overburden thickness is modest, the maximum hydraulic gradient value appears in the top region surrounding the river embankment. When the underwater shield tunnel overburden thickness is large, the maximum hydraulic gradient appears at the junction of the river channel and the river embankment. The simulation results are consistent with the results of unsteady seepage physical tests performed by Duan et al. [46], i.e., they confirm that the maximum infiltration hydraulic gradient is generally near the water level of the surface water body. Thus, the maximum infiltration hydraulic gradient is easily at the junction location between the river channel and the river embankment when the water level drops abruptly.

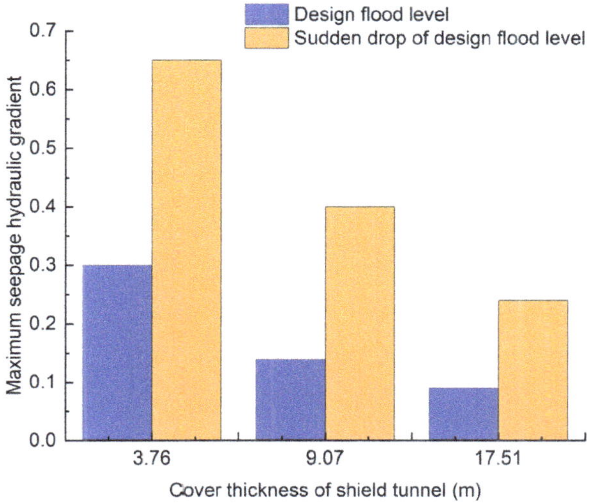

Figure 12. Comparison of maximum seepage hydraulic gradient of river embankment.

5.3. Analysis of the Effect of Shield Construction Disturbance on River Embankment Seepage Stability

The surrounding soil is readily disturbed to some level during shield tunnel construction, and the impacts of construction disturbance on the seepage stability of river embankment are modeled for varied overburden thicknesses in response to increased construction disturbance. In this section, the permeability coefficients of the disturbance zone are increased by raising orders of magnitude to generalize the growth of disturbance, and a model for calculating the seepage stability of river embankment with different overburden thickness under the design flood level and sudden drop of design flood level is established. The stratigraphic parameters used in the calculation are shown in Table 1.

Figure 13 illustrates the findings of infiltration hydraulic gradient under the design flood level and design flood level plunging circumstances respectively, and compares them in detail to the infiltration hydraulic gradient results of Figures 6 and 11. In comparison to the previous results for hydraulic gradient when the permeability coefficient of the disturbance zone is raised by one order of magnitude, the maximum hydraulic gradient rises when the permeability coefficient of the disturbance zone is increased by two orders of magnitude. Under design water level conditions, the underwater shield tunnel overburden thickness is 3.76 m, with the maximum hydraulic gradient increasing by 10%. The underwater shield tunnel's overburden thickness is 9.07 m, with the maximum hydraulic gradient increasing by 2.9%. With the underwater shield tunnel overburden thickness of 17.51 m, the maximum hydraulic gradient does not increase. However, the maximum hydraulic gradient increases by 15% when the shield tunnel overburden thickness is 3.76 m, 12.5% when the shield tunnel overburden thickness is 9.07 m and 8.3% when the shield tunnel overburden thickness is 17.51 m under the design flood level plunging scenario. The results show that when the permeability coefficient of the disturbed zone grows, the maximum hydraulic gradient increases, while the influence of shield disturbance on the permeability hydraulic gradient decreases as the overburden thickness increases.

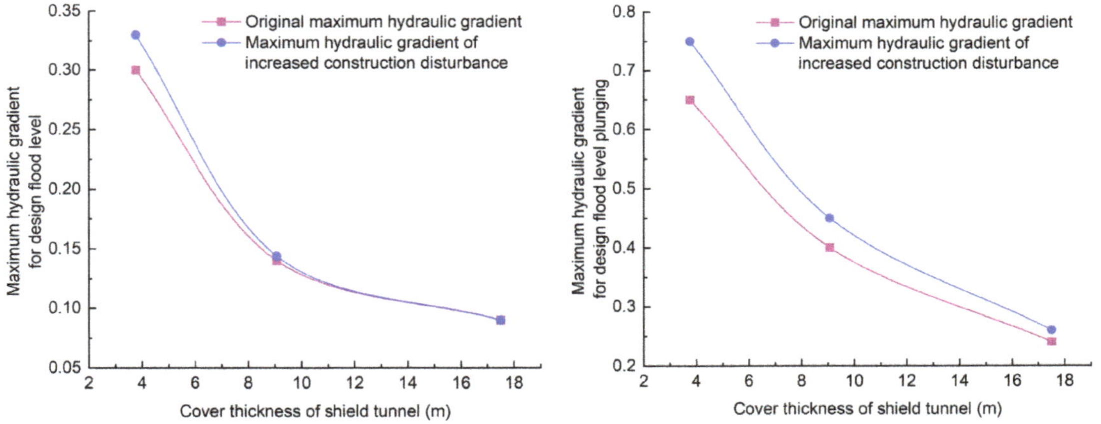

(**a**) (**b**)

Figure 13. Comparison for original maximum hydraulic gradient and maximum hydraulic gradient of increased construction disturbance: (**a**) Design flood level condition; (**b**) Sudden drop of design flood level condition.

5.4. Analysis of Underwater Shield Tunnel Cover Thickness

According to the findings of the previous investigation, when the underwater shield tunnel overburden thickness is 3.76 m under the design flood level condition, the maximum hydraulic gradient gets the permissible value, the river embankment seepage stability is in the critical state, and when the underwater shield tunnel overburden thickness is 9.07 m, the maximum hydraulic gradient value is within the permissible range. Under the design flood level plunging condition, the maximum hydraulic gradient for shield tunnel overburden thickness of 3.76 m and 9.07 m exceeds the code's allowable hydraulic gradient, and infiltration damages may occur on the river embankment under the two work conditions, whereas the maximum hydraulic gradient for shield tunnel overburden thickness of 17.51 m does not exceed the allowable value.

Considering the engineering geological background of this study, the river embankment is in seepage stability when the overburden thickness of the underwater shield tunnel is 1.5 times the tunnel outer diameter (9.07 m) under the design flood level condition (Figure 14). Figure 14 shows that the overburden thickness is safe under the design flood level plunge scenario, where the underwater shield tunnel overburden thickness is 2.9 times the tunnel outer diameter (17.51 m), the river embankment is in a seepage stable state, and no infiltration damage occurs.

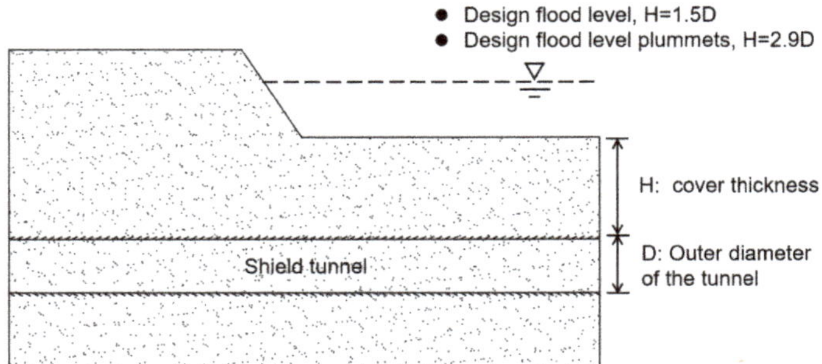

Figure 14. Diagram of shield tunnel cover thickness.

6. Conclusions

This study is conducted on the underground shield tunnel crossing the river project in Hefei. The effects of shield tunnel cover thickness on river embankment seepage stability are investigated under the design flood level and design flood level plunge conditions, as well as the effects of tunnel shield construction on river embankment seepage stability under different cover thickness, to evaluate the safety of the shield tunnel overburden thickness under engineering geology. The following conclusions were drawn:

(1) The maximum infiltration hydraulic gradient value of the river embankment gradually lowers as the thickness of the underwater shield tunnel overburden grows. When the underwater shield tunnel overburden thickness is thin, the maximum infiltration hydraulic gradient occurs at the top of the river embankment and the maximum seepage head values increase. When the underwater shield tunnel overburden thickness is thick, the maximum infiltration hydraulic gradient arises at the junction of the river channel and the river embankment.

(2) Shield tunnel construction has effects on river embankment seepage stability when the construction disturbance is large, i.e., the permeability coefficient of the disturbance zone increases, and the infiltration hydraulic gradient increases. However, as the underwater shield tunnel cover thickness increases, the effect of shield construction disturbance on the infiltration hydraulic gradient weakens, and the effect on river embankment seepage stability decreases.

(3) In the case of the underwater shield tunnel cover thickness reaching 1.5 times the tunnel outer diameter in the design flood level condition, the river embankment may keep seepage stability. When the design flood level plunges, an underwater shield tunnel cover thickness that remains 2.9 times the outer diameter of the tunnel can keep the river embankment in seepage stability.

Author Contributions: Conceptualization, X.L.; methodology, W.S.; software, W.S.; validation, W.S. and X.L.; formal analysis, X.L.; investigation, W.S.; resources, X.L.; data curation, W.S.; writing—original draft preparation, W.S.; writing—review and editing, X.L., J.M., N.G., Y.X., F.T., S.F. and S.M.; visualization, J.M., N.G., Y.X. and S.M.; supervision, X.L.; project administration, X.L.; funding acquisition, X.L. and W.S. All authors have read and agreed to the published version of the manuscript.

Funding: This research was funded by the National Natural Science Foundation of China (Nos. 4210071795, 42107082), and the Natural Science Foundation of Anhui Province, China (Nos. 2108085QD165) and the Foundation of Hefei Rail Transit Group Limited Company, China (Nos. 2021BFFBZ02689).

Data Availability Statement: All data, models, or code that support the findings of this study are available from the corresponding author upon reasonable request.

Conflicts of Interest: The authors declare no conflict of interest.

References

1. Liang, Y.; Chen, X.Y.; Yang, J.S.; Huang, L.C. Risk analysis and control measures for slurry shield tunneling diagonally under an urban river embankment. *Adv. Civ. Eng.* **2020**, *24*, 336–353. [CrossRef]
2. Ma, W.H.; Yang, C.Y.; Peng, H.; Bai, Y.; Cheng, L.; Gao, L.H.; Liu, Z.Y. Settlement control on retaining wall embankment affected by underneath traversing large-diameter slurry shield tunnels. *J. Hunan Univ.* **2020**, *47*, 44–53.
3. Hong, K. Typical underwater tunnels in the mainland of China and related tunneling technologies. *Engineering* **2017**, *3*, 871–879. [CrossRef]
4. Li, X.J.; Lu, D.D.; Sun, H.B.; Hu, J.L.; Li, J.L.; Wu, X.; Xu, J.B. Stratum deformation law of tidal river by shield tunneling under shallow overburden soil. *Sci. Technol. Eng.* **2023**, *23*, 1270–1277.
5. Zhang, H.C. Construction Deformation Control Technology of Large Diameter Shield Undercrossing Subway Tunnel in Operation. *Mod. Tunn. Technol.* **2022**, *59*, 934–940.
6. Guo, P.P.; Gong, X.N.; Wang, Y.X.; Lin, H.; Zhao, Y.L. Analysis of observed performance of a deep excavation straddled by shallowly buried pressurized pipelines and underneath traversed by planned tunnels. *Tunn. Undergr. Space Technol.* **2023**, *132*, 104946. [CrossRef]
7. Wu, S.M.; Lin, C.G.; Zhang, Z.M.; Wang, N. Risk analysis and control for slurry shield under-passing embankment. *J. Rock Mech. Eng.* **2011**, *30*, 1034–1042.

8. Kong, Q.Y.; Ma, X.L.; Wang, T.T. Seepage stability analysis of Yangtze River levee as tunnel traversing. *J. Water Resour. Archit. Eng.* **2012**, *10*, 74–77.
9. Sun, Y.S.; Li, Z.H.; Yang, K.; Wang, G.H.; Hu, R.L. Analysis of the influence of water level change on the seepage field and stability of a slope based on a numerical simulation method. *Water* **2023**, *15*, 216. [CrossRef]
10. Gu, J.Y.; Zhang, Q.; Lu, X.C.; Hu, J.; Zhu, J.W. Centrifugal model test study on deformation mechanism of reservoir bank landslide under reservoir water level variation. *J. Yangtze River Sci. Res. Inst.* 2022. Available online: https://kns.cnki.net/kcms/detail/42.1171.tv.20221122.1555.006.html (accessed on 4 May 2023).
11. Li, H.R.; Li, Q.M.; Lu, Y. Statistical analysis on regularity of subway construction accidents from 2002 to 2016 in China. *Urban Rapid Rail Transit* **2017**, *30*, 12–19.
12. Wu, S.X. Analysis and discussion of water gushing accident in a shield tunnel of a metro river-crossing interval. *Intell. City* **2017**, *3*, 118–119.
13. Li, X.Y.; Zhang, D.L.; Hou, Y.J.; Cao, L.Q.; Li, Q.Q. Analysis of ground and structure deformation characteristics during shield tunneling in Tianjin subway. *China Railw. Sci.* **2018**, *39*, 71–80.
14. Bomer, E.J.; Wilson, C.A.; Datta, D.K. An integrated approach for constraining depositional zones in a tide-influenced river: Insights from the Gorai River, southwest Bangladesh. *Water* **2019**, *11*, 2047. [CrossRef]
15. Jiang, B.H.; Yang, Y.P.; Peng, H.M.; Tian, H.R. Analysis on improving weak links and solutions of low-grade levees in middle and lower reaches of Changjiang River. *Yangtze River* **2021**, *52*, 127–130.
16. Qian, J.G.; Li, W.Y.; Yin, Z.Y.; Yang, Y. Influences of buried depth and grain size distribution on seepage erosion in granular soils around tunnel by coupled CFD-DEM approach. *Trans. Geotech.* **2021**, *29*, 100574. [CrossRef]
17. Wang, K.; Li, W.J.; Tang, K.; Liang, B. Influence of rainfall infiltration on slope stability of shallow-buried bias tunnel entrance. *Water Resour. Power* **2022**, *40*, 121–124.
18. Tao, Z.G.; Zhang, Q.Z.; Yang, X.J.; Zhao, F.F.; Cao, S.D.; Li, Y.P. Physical model test study on steady state effect of underpass tunnel excavation on loose deposit slope. *J. China Coal Soc.* **2022**, *47*, 61–76.
19. Li, X.F. Study on seepage characteristics and slope stability of Caipo accumulation body in Three Gorges Reservoir area under combined rainfall condition. *J. Water Resour. Water Eng.* **2019**, *30*, 194–200.
20. Zhang, L.W.; Tian, Z.N.; Li, S.C.; Wang, Y. Preventing water outburst in excavation in subsea tunnel. *Proc. Int. Conf. Transp. Eng.* **2009**, *2009*, 2102–2107.
21. Chen, S.W. Estimation of Minimum Rock Cover for a Sub-Water Tunnel. In Proceedings of the 3rd ISRM Young Scholars Symposium on Rock Mechanics, Xi'an, China, 8–10 November 2014; pp. 445–449.
22. Do, N.A.; Dias, D.; Oreste, P.; Djeran-Maigre, I. Three-dimensional numerical simulation for mechanized tunnelling in soft ground: The influence of the joint pattern. *Acta Geotech.* **2014**, *9*, 673–694. [CrossRef]
23. Guo, P.P.; Gong, X.N.; Wang, Y.X. Displacement and force analyses of braced structure of deep excavation considering unsymmetrical surcharge effect. *Comput. Geotech.* **2019**, *113*, 103102. [CrossRef]
24. Li, S.C.; Li, S.C.; Xu, B.S.; Wang, H.P.; Ding, W.T. Study on determination method for minimum rock cover of subsea tunnel. *J. Rock Mech. Eng.* **2007**, *26*, 2289–2295.
25. Zhang, D.M.; Ma, L.X.; Zhang, J.; Hicher, P.Y.; Juang, C.H. Ground and tunnel responses induced by partial leakage in saturated clay with anisotropic permeability. *Eng. Geol.* **2015**, *189*, 104–115. [CrossRef]
26. Zhang, D.M.; Liu, Y.; Huang, H.W. Leakage-induced settlement of ground and shield tunnel in soft clay. *J. Tongji Univ.* **2013**, *41*, 1185–1190+1212.
27. Zhang, P.; Chen, R.P.; Wu, H.N.; Liu, Y. Ground settlement induced by tunneling crossing interface of water-bearing mixed ground: A lesson from Changsha, China. *Tunn. Undergr. Space Technol.* **2020**, *96*, 103224. [CrossRef]
28. Tang, Z.L.; Yao, W.; Zhang, J.C.; Xu, Q.J.; Xia, K.W. Experimental evaluation of PMMA simulated tunnel stability under dynamic disturbance using digital image correlation. *Tunn. Undergr. Space Technol.* **2019**, *92*, 103039. [CrossRef]
29. Yu, H.; Chen, J.; Bobet, A.; Yuan, Y. Damage observation and assessment of the Longxi tunnel during the Wenchuan earthquake. *Tunn. Undergr. Space Technol.* **2016**, *54*, 102–116. [CrossRef]
30. Hu, B.; Li, X.Q.; Huang, D. Safety risk analysis and protective control of existing pipelines affected by deep pit excavation in metro construction. *Model. Simul. Eng.* **2019**, *2019*, 3643808. [CrossRef]
31. Ahmed, M.; Iskander, M. Analysis of tunneling-induced ground movements using transparent soil models. *J. Geotech. Geoenviron. Eng.* **2010**, *137*, 525–535. [CrossRef]
32. Vu, M.N.; Broere, W.; Bosch, J. Effects of cover depth on ground movements induced by shallow tunnelling. *Tunn. Undergr. Space Technol.* **2015**, *50*, 499–506. [CrossRef]
33. Guo, P.P.; Gong, X.N.; Wang, Y.X.; Lin, H.; Zhao, Y.L. Minimum cover depth estimation for underwater shield tunnels. *Tunn. Undergr. Space Technol.* **2021**, *115*, 104027. [CrossRef]
34. Bai, Y.; Wu, Z.; Huang, T.; Peng, D.P. A dynamic modeling approach to predict water inflow during karst tunnel excavation. *Water* **2022**, *14*, 2380. [CrossRef]
35. Hu, R.; Liu, Q.; Xing, Y.X. Case study of heat transfer during artificial ground freezing with groundwater flow. *Water* **2018**, *10*, 1322. [CrossRef]
36. Wang, W.X.; Faybishenko, B.; Jiang, T.; Dong, J.Y.; Li, Y. Seepage characteristics of a single ascending relief well dewatering an overlying aquifer. *Water* **2020**, *12*, 919. [CrossRef]

37. Pan, Y.H.; Qi, J.R.; Zhang, J.F.; Peng, Y.X.; Chen, C.; Ma, H.N.; Ye, C. A comparative study on steady-state water inflow into a circular underwater tunnel with an excavation damage zone. *Water* **2022**, *14*, 3154. [CrossRef]
38. Niu, F.Y.; Cai, Y.C.; Liao, H.J.; Li, J.G.; Tang, K.J.; Wang, Q.; Wang, Z.C.; Liu, D.D.; Liu, T.; Liu, C.; et al. Unfavorable geology and mitigation measures for water inrush hazard during subsea tunnel construction: A global review. *Water* **2022**, *14*, 1592. [CrossRef]
39. Zhang, Y.; Zhang, D.L.; Fang, Q.; Xiong, L.J.; Yu, L.; Zhou, M.Z. Analytical solutions of non-Darcy seepage of grouted subsea tunnels. *Tunn. Undergr. Space Technol.* **2020**, *96*, 103182. [CrossRef]
40. Fahimifar, A.; Zareifard, M.R. A new elasto-plastic solution for analysis of underwater tunnels considering strain-dependent permeability. *Struct. Infrastruct. Eng.* **2014**, *10*, 1432–1450. [CrossRef]
41. Fang, Q.; Song, H.R.; Zhang, D.L. Complex variable analysis for stress distribution of an underwater tunnel in an elastic half plane. *Int. J. Numer. Anal. Meth. Geomech.* **2015**, *39*, 1821–1835. [CrossRef]
42. Mi, B.; Xiang, Y.Y. Model experiment and calculation analysis of excavation-seepage stability for shallow shield tunneling in sandy ground. *Rock Soil Mech.* **2020**, *41*, 837–848.
43. Yang, X.L.; Huang, F. Stability analysis of shallow tunnels subjected to seepage with strength reduction theory. *J. Cent. South Univ. Technol.* **2009**, *16*, 1001–1005. [CrossRef]
44. Chen, X.W.; Lei, P.; Yang, X.Y. Seepage and stability analysis of dike based on Geo-Studio. *J. Chang. Univ. Sci. Technol.* **2015**, *12*, 64–68.
45. Chen, H.R.; Zhang, G.J.; Li, Y.; Liu, L.L.; Meng, Y.Q.; Tian, D.Z. Study on the influence of river-crossing pipeline to the permeability safety of embankment. *Yellow River* **2017**, *39*, 34–37+41.
46. Duan, X.B.; Xie, L.F. Unsteady seepage test under condition of rapid drawdown. *J. Yangtze River Sci. Res. Inst.* **2009**, *26*, 7–12.

water

MDPI

Correction

Correction: Liu et al. Failure Characteristics of the Water-Resisting Coal Pillar under Stress-Seepage Coupling and Determination of Reasonable Coal Pillar Width. *Water* 2023, *15*, 1002

Quanhui Liu [1,2], Yuanbo Xue [2], Dan Ma [2,3,*] and Qiang Li [2]

1 Kekegai Coal Mine Shaanxi Yanchang Petr Yulin Coal Chem Co., Yulin 719000, China
2 School of Mines, China University of Mining and Technology, Xuzhou 221116, China
3 MOE Key Laboratory of Deep Coal Resource Mining, China University of Mining and Technology, Xuzhou 221116, China
* Correspondence: dan.ma@cumt.edu.cn; Tel.: +86-176-2650-0518

In the original publication [1], there was a mistake in Figure 4 as published. The geographical location label "Mongolia" should be changed to "Inner Mongolia" in Figure 4. The corrected Figure 4 appears below. The authors apologize for any inconvenience caused and state that the scientific conclusions are unaffected. This correction was approved by the Academic Editor. The original publication has also been updated.

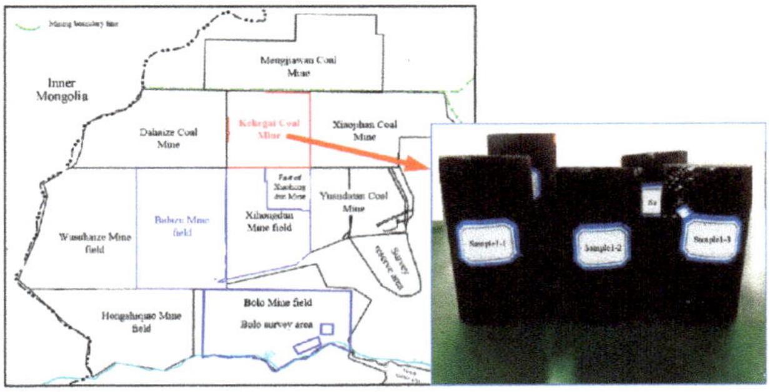

Figure 4. The location of sampling and coal samples.

Reference

1. Liu, Q.; Xue, Y.; Ma, D.; Li, Q. Failure characteristics of the water-resisting coal pillar under stress-seepage coupling and determination of reasonable coal pillar width. *Water* **2023**, *15*, 1002. [CrossRef]

Citation: Liu, Q.; Xue, Y.; Ma, D.; Li, Q. Correction: Liu et al. Failure Characteristics of the Water-Resisting Coal Pillar under Stress-Seepage Coupling and Determination of Reasonable Coal Pillar Width. *Water* 2023, *15*, 1002. *Water* **2023**, *15*, 1804. https://doi.org/10.3390/w15101804

Received: 17 April 2023
Accepted: 20 April 2023
Published: 9 May 2023

water

MDPI

Article

Detection of Landfill Leachate Leakage Based on ERT and OCTEM

Yulong Lu [1,2], Jialuo Tao [1], Chuanghua Cao [3,*], Hanlin Liu [1], Yang Liu [1] and Zhengbin Ge [4]

1 School of Earth Sciences and Spatial Information Engineering, Hunan University of Science and Technology, Xiangtan 411201, China; 1010106@hnust.edu.cn (Yulong Lu)
2 Institute of Geological Survey of Hunan Province, Changsha 410083, China
3 Hunan Geological Disaster Monitoring, Early Warning and Emergency Rescue Engineering Technology Research Center, Changsha 410004, China
4 Hunan Remote Sensing Geological Survey and Monitoring Institute, Changsha 410015, China; 1901050120@mail.hnust.edu.cn
* Correspondence: 21010103009@mail.hnust.edu.cn

Abstract: Leakage in the impervious layer of a domestic waste landfill seriously pollutes the soil and groundwater. Therefore, it is necessary to carry out rapid nondestructive leakage location detection. In this research, the electrical resistivity tomography (ERT) method and the opposing-coils transient electromagnetic method (OCTEM) were used to detect the leakage location. The inversion sections of both methods showed a clear low–middle–high resistivity spectrum in the longitudinal direction that could be used to speculate the distribution pattern of the upper waste body layer, the bottom impermeable layer, and the lower limestone layer. The leakage area was identified in Zone B of the landfill on the basis of inversion results and drilling verification results. The results indicate that OCTEM and ERT were both sensitive to leakage detection. However, OCTEM had higher longitudinal resolution and more refined inversion results, resulting in more effective delineation for the location of the damage and leakage of the impervious landfill layer, thereby providing a new technical basis for landfill leakage detection.

Keywords: electrical resistivity tomography method; opposing-coil transient electromagnetic method; landfill; leakage area

Citation: Lu, Y.; Tao, J.; Cao, C.; Liu, H.; Liu, Y.; Ge, Z. Detection of Landfill Leachate Leakage Based on ERT and OCTEM. *Water* **2023**, *15*, 1778. https://doi.org/10.3390/w15091778

Academic Editors: Lucila Candela and Giorgio Mannina

Received: 17 March 2023
Revised: 19 April 2023
Accepted: 28 April 2023
Published: 5 May 2023

1. Introduction

Toxic and harmful waste gases, slag, and liquids are produced form mining, mineral processing, smelting, chemical production, and other industrial activities. If not handled properly, these harmful substances can cause serious pollution to the air, soil, and water [1–6]. In recent years, with the rapid development of the social economy, the problem of domestic waste disposal has become increasingly prominent [7]. Due to problems such as excessive investment and air pollution in new waste disposal methods such as waste incineration, sanitary landfills are still an important domestic waste disposal method [8]. The structure of a landfill consists of a solid waste layer, a geotextile layer, a pebble layer, an impermeable layer, and a soil layer. High-density polyethylene (HDPE) is a popular impermeable material. It has stable chemical properties, strong corrosion resistance, and low permeability to water and air [9–11]. As the last line of defense to prevent the leakage of landfill leachates, HDPE is widely used in landfills. HDPE could be used as impermeable material by cutting off the leakage channel of the earth dam, and HDPE's large tensile strength and elongation can withstand water pressure and adapt to the deformation of the dam. HDPE also has good resistance to bacteria and chemical attacks, including acid, alkali, salt erosion. HDPE is always combined with other impermeable material, such as geosynthetic clay liners [12–14]. However, in actual application, due to artificial or natural factors, HDPE film breakage occurs [15–20]. Landfill leachates are rich in iron, copper, zinc, other heavy metals, and other types of pollutants [21–25]. Once the leachate leaks, it seriously pollutes the soil and groundwater, causing serious damage to local residents [26–31]. The ability to detect the location of seepage points in a timely and accurate manner is greatly important to carry out effective disposal [6–8].

Leakage detection systems are usually preburied for the construction of landfills, and traditional methods are electrical and acoustic [20–22]. Electrical methods lay detection electrodes or conductive gridlike fibers under the impermeable layer, and determine the leakage location according to the electromotive force distortion or short circuit of conductive fibers. However, they have a low safety factor, and conductive fibers are prone to misalignment [32–36]. Acoustic methods lay multiple vibration sensors under the impermeable layer that are used to locate and receive the instantaneous damage signals of the HDPE film. However, clutter interference leads to distorted wave velocity and susceptible signals that indicate the low accuracy and easy misdetections of this method. Therefore, landfills are often required to conduct noninvasive leak detection. Traditional leakage location detection methods, such as alternating electrical current detection, affect the accuracy of terminal detection results because of the electromagnetic crosstalk between transmission lines [37–43]. The disadvantage of the dipole method in the direct electric current method detection system is that the current significantly attenuates with an increase in measurement distance, that is, the covering the thickness and conductivity of the waste layer affects the potential distribution. The thicker the garbage layer is, the weaker the surface potential distribution, and the lower the detection sensitivity are, and, thereby, the greater the error is in determining the location and number of holes [32–36]. The potential distribution of the current field in the electrode grid method is directly related to the layout of the detection electrodes under the film [44]. The decreasing diffusion effect of the field interferes with the detection electrodes near the leakage point, which may easily cause misjudgments in the detection results. In view of these encountered leak-detection problems, we performed a detection test on a landfill site using the transient opposing-coil electromagnetic method (OCTEM) on the basis of the electrical resistivity tomography method (ETR). The comprehensive anomaly of the two methods could effectively determine the location of the leakage point to aid in landfill leakage detection.

The purpose of this paper is, therefore, to apply two geophysical methods, namely, the opposing-coils transient electromagnetic method and the electrical resistivity tomography method, to detect the leakage point in the landfill and provide useful information for predicting leakages and delineating the possible leakage zone for future explorative work. The OCTEM and ETR methods were both used in the detection of landfill leakage points, and their effectiveness and resolution are compared. Lastly, this study provides useful methods for the detection of leakage points in landfills, and establishes a basis for future exploration and antiseepage work.

2. Materials and Methods

The research area was the domestic waste landfill in Chenzhou city, Guiyang county, Hunan province, China, where a population of 900,000 individuals lives. The landfill site is generally high in the east and low in the west, and surrounded by hills (Figure 1). The landfill consists of Zones A and B. Zone A is located on the northeastern side of the landfill site, which belongs to the upper part of the river valley and has a relatively high altitude (255 m). Zone B is located on the southern and western sides of the landfill site, which belong to the lower part of the river valley and have relatively low elevation (247 m). The actual elevation of the landfill site is 230.82–263.06 m, and the maximal elevation difference is 32.24 m. The research area has a subtropical humid monsoon climate, which is characterized by warm temperatures, four distinct seasons, abundant heat, concentrated rain, changeable spring temperatures, frequent droughts in summer and autumn, a short cold period, and a long hot period. Annual rainfall is from 1992.7 to 1075.1 mm, generally about 1500 mm, and the average annual rainfall is 1363 mm. Annual rainfall is mainly concentrated in April to July, accounting for more than 60% of annual rainfall. The dry season is from September to January. The average annual temperature is 19.5 °C, the highest temperature is 41 °C, and the lowest temperature is 2 °C. Frost weather in the research area lasts for an average of 14.7 days. The first frost often occurs from late November to early December, and the last frost occurs in mid-to-late February of the following year. The annual frost-free period is 277 days.

Figure 1. Location and survey line layout of the research area (from Google Earth).

2.1. Geological Background

The research area is located in the mid–southern section of the Qinhang junction zone, the southern end of the north–south Leiyang-Linwu tectonic belt, the middle part of the Nanling tectonic belt, and the northeastern Yanling-Chenzhou-Lanshan basement fault passing through this area. The oldest outcropping strata in the region are the Sinian system, the newest is Quaternary, and Silurian is missing. Tectonic activity in this area successively experienced Caledonian, Indosinian, and Yanshanian tectonic activities. The fold structure in this area is mainly overturned anticline and syncline composed of Devonian–Permian strata. The fold structure in this area is a complex north–south fold belt. The fault structure in the area is relatively developed, which is mainly near the north–south compression–torsional fault, followed by nearly EW tension-torsional fault, and the NW tension-torsional fault is the least developed.

The main strata exposed in the study are as follows. The Quaternary alluvial-proluvial layer located in the valley zone in the area, which is gray black, gray, yellow, and hard plastic. It is mainly composed of clay particles containing a small amount of gravel and debris, and mainly composed of limestone and sandstone with general toughness, medium dry strength, and thickness of 3.0–9.2 m. It contains plant roots within 0.4 m of its surface layer. The Quaternary slope residual layer is mainly distributed in the hillside, and is grayish yellow, yellowish brown, slightly wet, and plastic to hard plastic. Its composition is mainly silty clay and it contains strong weathering granite. The particles of the gravel are about 1~3 cm, and the content is about 10~25%. The Devonian Xikuangshan Formation is the outcropped bedrock in the area, and is grayish black cyan, and gray, partially contains carbonaceous limestone, the upper part is strongly weathered, and the fracture is relatively developed, with broken cores with thickness of 0.5–5.8 m. The lower part of the rock is relatively complete, and karst is locally developed. This layer is the main aquifer in the area with good permeability and water abundance. The lower carboniferous Yanguan stage limestone is mainly distributed in the northwest and southeast of the landfill. It is moderately weathered, containing carbonaceous with an aphanitic structure and calcite mesh filling. The landfill site is located in the core of the small syncline structure, and the fault structure is not developed.

On the basis of the geological conditions of the landfill site, sand, gravel, and especially clay minerals provide a better base liner to prevent possible leakage to the surrounding environment [5]. The sand and gravel layer provide the drainage channel for the landfill leachate collection [38]. Furthermore, the geological structure of the landfill location is not developed, and the syncline structure is helpful in containing the leachates [15].

2.2. Basic Physical Structure of the Domestic Waste Landfill

Generally, the physical layered structure of a landfill mainly consists of a top capping system, a bottom liner system, and a lower bedrock layer. The waste landfill in the research area is currently in use and has not been closed; the capping system on the top has not yet been completed (Figure 2). The capping system of the landfill in the research area is only covered with an impermeable layer consisting of a geomembrane and a small amount of clay on the surface to prevent surface water infiltration as much as possible; a layer of landfill domestic waste ranging from a few to a dozen meters' thickness is below the impermeable layer. The bottom liner system is an impermeable layer located at the bottom and around the landfill consisting of an absorption and a filtration layer, a leachate collection layer, HDPE, and a clay layer from top to the bottom. It collects and isolates the leachate from waste, and prevents it from leaking downward. The lower bedrock layer is mainly composed of clay layer and bedrock, with 10–20 m clay thickness; the bedrock is the limestone of the Devonian Xikuangshan Formation and the lower Carboniferous Yanguan Stage. The stratum dips to the NW with a small dip angle.

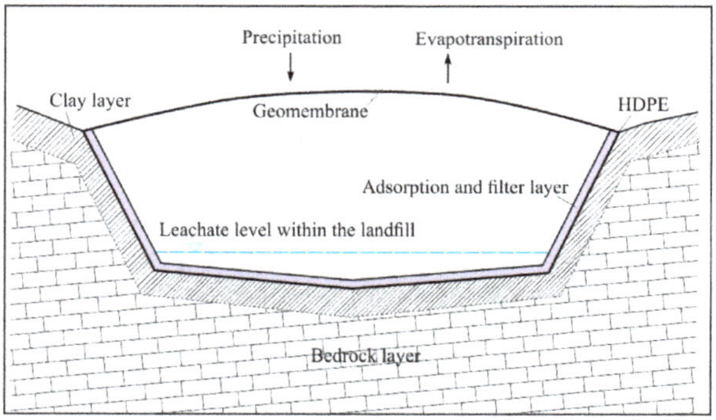

Figure 2. Structural diagram of the landfill.

2.3. Geophysical Properties

The composition of domestic waste is complex and varies from region to region. Generally, the composition of waste in cottage areas is mainly ash, followed by food, paper, and plastic. Residents living in areas with buildings mainly produce domestic waste. China's urban domestic waste has the basic characteristics of a mixed composition, high moisture, and a low calorific value. Garbage bags contain ash, plastic, and paper. The composition of domestic waste sites in the research area is extremely complex, with the waste composition of both building and cottage areas, and moisture content of above 50%.

According to previous physical-property data [45,46], combined with the results of field experiments in the research area, waste bodies are mostly domestic waste, which is easily corroded to produce leachate. The leachate has high conductivity and low-resistance with resistivity values of 1~100 Ω-m. The impermeable system is mainly composed of materials nonwoven geotextiles, a gravel layer, HDPE, and a clay layer with resistivity values of about 200 Ω-m. The lower part of the antiseepage layer limestone, showing high resistance characteristics, has resistivity generally in the range of 300–2500 Ω-m. The electrical difference between the layers of the landfill in the research area is obvious, and prerequisites for electrical exploration are available.

3. Experimental Methods

3.1. Electrical Resistivity Tomography Method

Electrical resistivity tomography (ERT) is a noninvasive geophysical method that measures the spatial distribution of the resistivity of a geological body. A total of 18 electrical resistivity survey lines with a direction of 140° and distance of 10 m were set up in the landfill, totaling 288 detection points. The ERT instrument model used in this study was SyscalProSwitch-48, IRIS Instruments. The SyscalPro was equipped with a 10-channel data receiver connected to 4 electrodes through cables (Figure 3). It works by supplying power to two powered electrodes, A and B, and measuring the potential difference between two potential electrodes, M and N, to perform resistivity measurements [34]. The use of four electrodes for measurements minimizes the effect of contact resistance at the interface between the soil pore water and the electrodes [46,47]. One resistivity data were obtained for each four-electrode combination measurement, and the location and depth of measurement points were determined via the distance between the powered electrodes and the type of the electrode device.

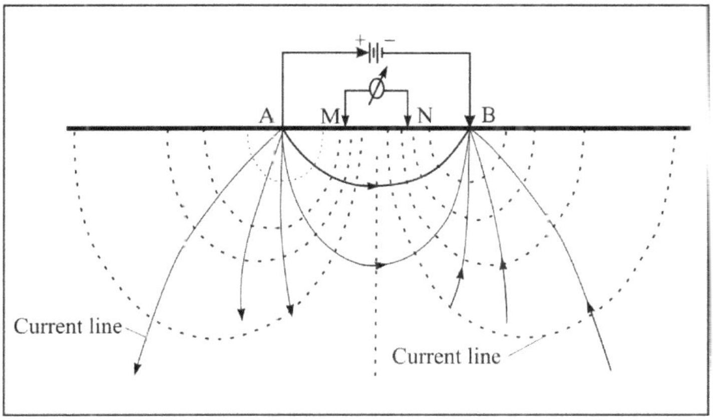

Figure 3. Working principle of electrical resistivity tomography.

This system was characterized by large storage capacity, accurate and fast measurements, easy operation, and stable and reliable performance, which meet the relevant specifications and design requirements. Resistivity measurements along a measuring line with different combinations of four electrodes were performed to obtain the two-dimensional resistivity distribution of the geological body. Resistivity measured with the electrical resistivity tomography method is called apparent resistivity ρ, expressed as follows:

$$\rho = 2\pi\alpha\frac{\Delta V}{I} \quad (1)$$

where I is the input current between the power supply electrodes A and B, ΔV is the potential difference between potential electrodes M and N, α is the electrode spacing, and the unit of apparent resistivity ρ is Ohm·m.

3.2. Opposing-Coil Transient Electromagnetic Method

The opposing-coil transient electromagnetic method uses an underground return source to send a primary-step electromagnetic field to the subsurface to detect the electrical characteristics of the subsurface medium by detecting the response of the secondary induction field generated by turning off the excitation source [48,49]. Unlike the conventional transient electromagnetic method, the opposing-coil transient electromagnetic method is sent using a dual-coil source in which Current I of the same magnitude and opposite direction is simultaneously supplied to the dual-coil source and measured after the current is turned off. The magnetic lines of the force at the receiving coil were horizontal and had a vertical magnetic field of 0 before and after shutdown, with a magnetic flux of 0 in the plane of that position and a vertical magnetic field in other planes. The formation of a zero-flux surface can eliminate the effect of primary field shutdown at any moment; by measuring at the zero-flux position, it can eliminate the blind spot and detect shallow underground spaces. Figure 4 shows the placement of the center-return device of the opposing-coil transient electromagnetic method.

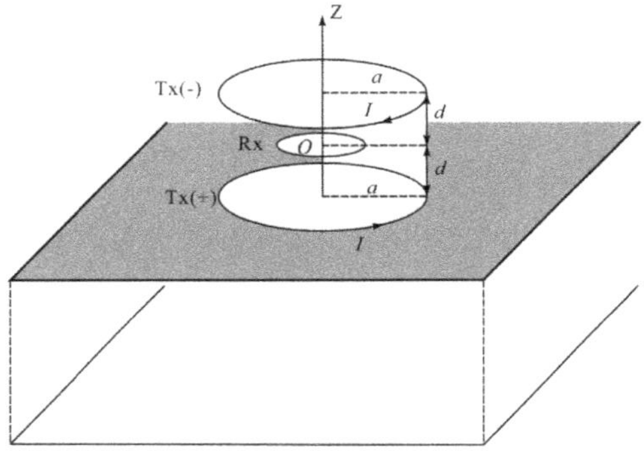

Figure 4. Schematic of the OCTEM device.

Field source analysis was performed for the oppposing-coil transient electromagnetic method, and a column coordinate system with the center of the horizontal current loop as the origin was established (unit vectors $u\rho$, $u\theta$, and uz). We used the vector superposition principle to obtain the equation for the primary field of equivalent inverse flux transient electromagnetic method 2 [50]:

$$B_{primary}(\rho, z) = [B_\rho(\rho, z-d) + B_\rho(\rho, z+d)]u_\rho + [B_\rho(\rho, z-d) - B_\rho(\rho, z+d) \quad (2)$$

where d is the distance between the transmitting coil and the receiving coil, and B is the magnetic induction intensity. The secondary opposing-coil transient electromagnetic method field can be calculated using the superposition principle, which is calculated by referring to the algorithm of Coggon [51–53], starting from the Maxwell spin equation shown as Equation (3):

$$\Delta \times H = \varepsilon\frac{\partial E}{\partial t} + \sigma E + J_e \quad (3)$$

Consider the following boundary conditions:

$$n \times \left(H^1 - H^2\right)|_\Gamma = 0 \tag{4}$$

$$J_e = P_e \omega(t) \delta(x) \delta(y) \delta(z) \tag{5}$$

$$\omega(t) = \begin{cases} 0, & (t < 0) \\ 1, & (t \geq 0) \end{cases} \tag{6}$$

where H is the magnetic field strength; E is the electric field strength; J_e is the current density; ω is the dielectric constant; t is time; $\delta(x)$ is the pulse function; P_e is the magnetic moment of the electric or magnetic field; $\omega(t)$ is the source current waveform.

The transient electromagnetic response of an arbitrary geoelectric model could be obtained by solving the magnetic field using the finite element method, of which the implementation is found in the literature [19].

4. Results and Discussion

4.1. Electrical Resistivity Tomography Method

After preprocessing the original data, the resistivity inversion section was obtained by using RES2DINV software. The inversion section was inferred and interpreted according to the electrical characteristics. The inversion section showed uniformly changed resistivity and an obviously layered law. The shallow part of the inversion section was the low-resistivity domestic waste layer. The middle part was the medium-resistivity impervious layer and the Quaternary sedimentary strata. The lower part was the high-resistivity limestone of the Xikuangshan Formation of the Devonian System. In the longitudinal direction, the resistivity of the inversion section showed an obvious low–medium–high three-layer electrical structure, which was in good agreement with the hierarchical structure of the physical properties of the landfill site (Figure 5).

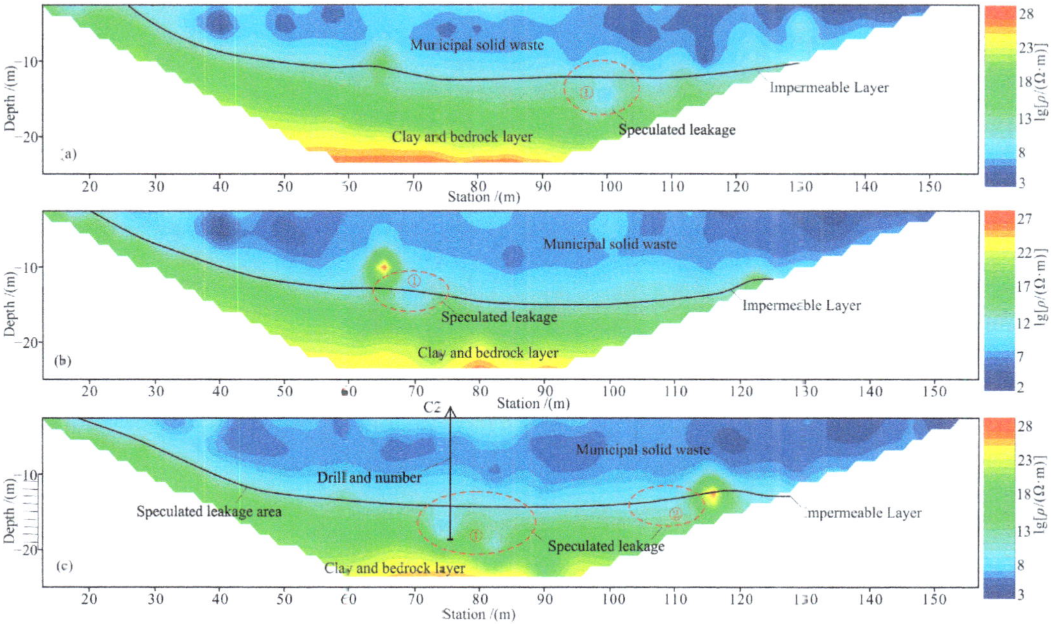

Figure 5. Inversion results and interpretation of ERT. (**a**) G100 line; (**b**) G110 line; (**c**) G120 line.

The resistivity of the domestic waste leachate was very low, so a low-resistivity anomaly could be observed in the seepage area, and a funnel-shaped low-resistivity anomaly appeared at the bottom or lower part of the inversion section impervious layer. In this research, obvious ground resistivity anomalies were observed in the depth of resistivity inversion sections of the G100, G110 and G120 lines (Figure 5). For example, there was an obvious low-resistivity anomaly at 100 m depth of the G100 line with a depth of 8–15 m and sectional width of about 10 m (Figure 5a). There was an obvious low-resistivity anomaly at 70 m depth of the G110 line with a depth of 10–15 m and a sectional width of about 10 m (Figure 5b). There were two low-resistivity anomalies at 80 and 110 m depth of the G120 line with a depth of 12–22 m. The sectional widths were about 20 and 10 m, respectively (Figure 5c).

4.2. Opposing-Coil Transient Electromagnetic Method

On the basis of the electrical resistivity tomography method, we performed research on the opposing-coils transient electromagnetic method. Its survey line coincided with the survey line of the resistivity method, and the survey points were distributed in the range of 40–115 m. The following data were acquired in fixed-point measurement mode: measurement point distance of 10 m, the direction of 140°, emission frequency of 6.25 Hz, emission current of 10 A, off time of 50 μs, a superposition period of 200 times, 306 measurement points, and a total of 18 cross-sections. After preprocessing the original data, HPTEM Data Process software was used to invert the resistivity inversion sections, which were inferred and interpreted according to the electrical characteristics. The inversion section showed uniformly changed resistivity and an obviously layered law. The shallow part of the inversion section was the low-resistivity domestic garbage layer. The middle part was the medium-resistivity impervious layer and the Quaternary sedimentary strata. The lower part was the high-resistivity limestone of the Xikuangshan Formation of the Devonian System. In the longitudinal direction, the resistivity of the inversion section showed an obvious low–medium–high three-layer electrical structure, which was in good agreement with the hierarchical structure of the physical properties of the landfill site (Figure 6).

Compared with the electrical resistivity tomography method, the opposing-coil transient electromagnetic method was more sensitive for retrieving the longitudinal resistivity of the inversion section. Six inversion sections showed low-resistivity anomalies, namely, S90, S110, S120, S130, and S140. There was an obvious low-resistivity anomaly at 95 m depth of the S90 line with a depth of 10–20 m and a sectional width of about 10 m (Figure 6a). There was an obvious low-resistivity anomaly at 100 m depth of the S100 line with a depth of 10–15 m and a sectional width of about 10 m (Figure 6b). There were two obvious low-resistivity anomalies at 85 and 100 m depth of the S110 line with a depth of about 10–20 m and a sectional width of about 10 m (Figure 6c). There were three low-resistivity anomalies at 50, 70, and 115 m depth of the S120 line with a depth of 10–18 m. The sectional width of the low-resistivity anomaly at the 50 m measuring point was about 15 m, and the section width of the other two low-resistivity anomalies (70 and 120 m) was about 10 m (Figure 6d). There were two obvious low-resistivity anomalies in the deep part of the S130 line located at 50 and 75 m with a depth of 10–18 m and sectional widths of about 15 and 10 m, respectively (Figure 6e). There was an obvious low-resistivity anomaly at 90 m depth of the S140 line with a depth of 18–24 m and width of about 10 m (Figure 6f). These low-resistivity anomalies may have been caused by the seepage of a low-resistance leachate if the damage of the impervious layer had occurred; therefore, these low-resistance anomalies may be leakage areas.

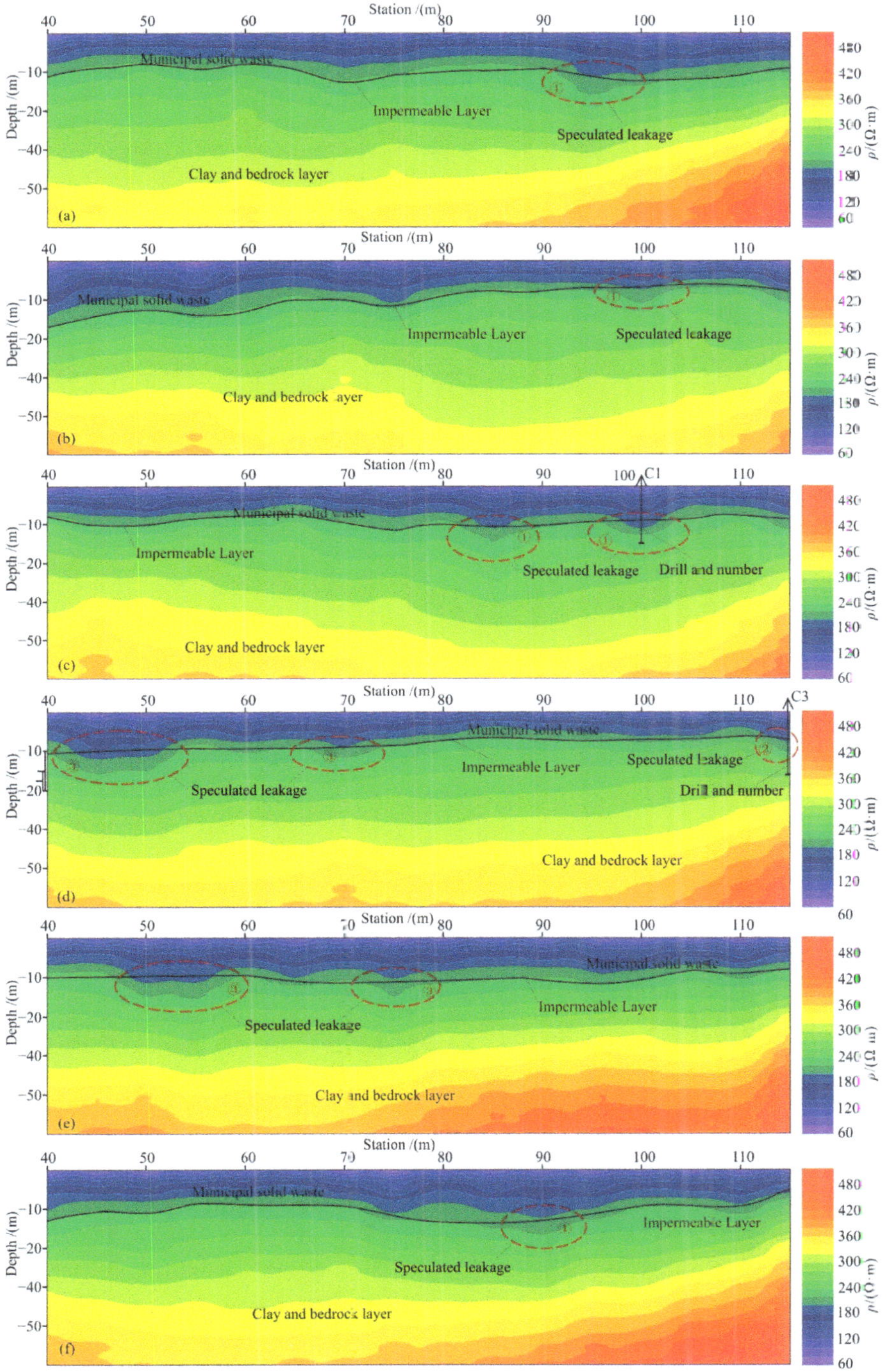

Figure 6. Inversion result and interpretation of ERT. (**a**) S90 line; (**b**) S100 line; (**c**) S110 line; (**d**) S120 line; (**e**) S130 line; (**f**) S140 line.

4.3. Distribution Characteristics of the Leakage Area and Borehole Verification

As shown in Figure 7, drilling was used to verify the low-resistivity anomaly area. C1, implemented at 100 m of the G110 line, revealed that the thickness of the domestic waste layer was 10.91 m. The depth of the gravel and clay layers of the antiseepage system was 10.91–14.10 m with a thickness of about 3.19 m. The deep part showed obvious leachate infiltration and was mainly dark brown silty clay (Figure 6c). C2 and C3 were drilled at 75 and 115 m of the G120 line. The thickness of the domestic waste layer exposed by C2 was 10.10 m. The gravel and clay layers of the antiseepage system were at 10.10–13.21 m with a thickness of 3.11 m. The deep part showed obvious infiltration and was dark brown silty clay (Figure 6c). The thickness of the domestic waste layer exposed by C3 was 12.23 m. The gravel and clay layers of the antiseepage system were at 12.23–15.16 m, with a thickness of about 2.96 m. The deep part showed obvious infiltration and was dark brown silty clay (Figure 6d).

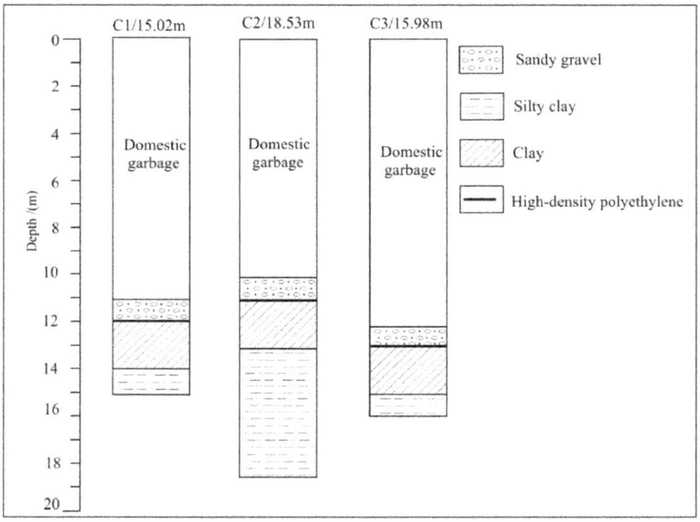

Figure 7. Geological borehole column.

The low-resistivity anomalies delineated with the electrical resistivity tomography method and opposing-coil transient electromagnetic method were consistent each other. The longitudinal resistivity of the opposing-coil transient electromagnetic method was more sensitive, and the low-resistivity anomalies were relatively more detailed. According to the comprehensive low-resistivity anomaly obtained with the two methods and combined with the results of drilling verification, three leachate leakage areas were delineated in Zone B of the landfill (Figure 8). ① The leakage area is located in the middle of Zone B between Lines 100 and 150 with a scope of about 260 m^2 and a depth of about 10–15 m; ② the leakage area is located near the drop well in the middle of Zone B between Lines 110 and 140 with a scope of about 50 m^2 and a depth of about 10~15 m; ③ the leakage area is located at the northwestern side of Leakage Area ① between Lines 120 and 140 with a scope of about 120 m^2 and a depth of about 10~15 m.

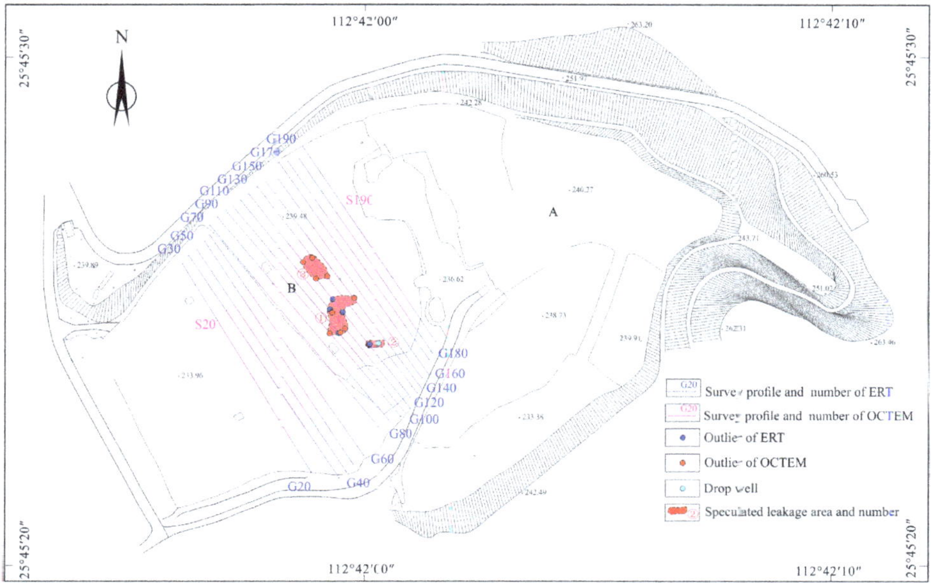

Figure 8. Distribution map of the speculated leakage area with a damaged impervious layer.

5. Conclusions

(1) The inversion sections of the electrical resistivity tomography method and opposing-coil transient electromagnetic method showed a clear low–middle–high resistivity spectrum in the longitudinal direction corresponding to an upper landfill layer, impermeable bottom layer, and the lower part of the argillaceous sandstone. The anomalous area of the pollution plume was observed at the bottom of the leakage area, and the gradually decreased concentration reflected the characteristics of pollution diffusion.

(2) On the basis of the characteristics of the low-resistance anomalies of the inversion section found with the electrical resistivity tomography method and opposing-coil transient electromagnetic method, three low-resistance anomalous areas were delineated in Zone B of the landfill that were identified as impermeable layer seepage areas after being verified using drilling.

(3) The opposing-coil transient electromagnetic method is more sensitive to abnormal changes in resistivity, and is accurate and effective for landfill leakage detection. OCTEM has broad application prospects in landfill leakage detection.

(4) Future studies could focus on the exposure of subsurface information obtained with this study by using the ETR and OCTEM methods. Various exploration methods such as trail trenches and drilling could be used to determine whether the zone was accurately detected or not. Anti-seepage measures could also be taken to prevent possible leakage in the landfill.

Author Contributions: Writing—original draft, Y.L. (Yulong Lu); Writing—review & editing, Y.L. (Yang Liu); Project administration, C.C.; Methodology, J.T.; Software, Z.G. Data curation, H.L. All authors have read and agreed to the published version of the manuscript.

Funding: This research was supported by the Natural Science Foundation of Hunan Province (grant no. 2022JJ30244), the Research Project of Teaching Reform of Hunan Province (grant no. HNJG-2022-0790), the Research Foundation of the Department of Natural Resources of Hunan Province (grants no. 2022-25 and 2020-04), the Key Team Project of Science and Technology Innovation in Xiangtan City (grant no. ZY-CXTD20221003), and the Key Technology and Application of Groundwater Detection in the Arid Area of Hunan Province (grant no. 2023-24).

Data Availability Statement: No new data were created or analyzed in this study. Data sharing is not applicable to this article.

Conflicts of Interest: The authors declare no conflict of interest.

References

1. Huangpu, H.H.; Li, H.Y. Analysis of the factors affecting the production of municipal solid waste. *Sci. Technol. Manag.* **2018**, *20*, 44–49.
2. Xiang, R.; Lei, G.Y.; Xu, Y.; Zhou, Q.; Liu, Y.Q.; Dong, L.; Liu, J.C.; Huang, Q.F. Aging Behaviors of HDPE geomembrane in Landfill environment and its impact on pollution risk of surrounding groundwater. *Res. Environ. Sci.* **2020**, *33*, 978–986.
3. Wang, Q. What causes landfill leakage? *Environ. Econ.* **2015**, *21*, 27.
4. Yuan, W.X.; Chen, S.P.; Tai, J.; Song, L.J. Present situation, problems and development countermeasures of landfill in China. *Environ. Sanit. Eng.* **2016**, *24*, 8–11.
5. Liu, Y.; Bouazza, A.; Gates, W.P.; Rowe, R.K. Hydraulic performance of geosynthetic clay liners to sulfuric acid solutions. *Geotext. Geomembr.* **2015**, *43*, 14–23. [CrossRef]
6. Wang, L.Q.; Li, X.M.; Zhu, F.H. Current situation of municipal solid wastes disposal and development proposals in China. *Environ. Pollut. Control.* **2015**, *37*, 106–109.
7. Li, Y.C.; Xu, Z.; Ma, H.; Hursthouse, A.S. Removal of Manganese (II) from acid mine wastewater: A review of the challenges and opportunities with special emphasis on Mn-Oxidizing bacteria and microalgae. *Water* **2019**, *11*, 2493. [CrossRef]
8. Meng, D.; Li, J.; Liu, T.; Liu, Y.; Yan, M.; Hu, J.; Li, H.; Liu, X.; Liang, Y.; Liu, H.; et al. Effects of redox potential on soil cadmium solubility: Insight into microbial community. *J. Environ. Sci.* **2019**, *75*, 224–232. [CrossRef]
9. Deng, R.J.; Jin, C.S.; Ren, B.Z.; Hou, B.L.; Hursthouse, A.S. The potential for the treatment of antimony-containing wastewater by iron-based adsorbents. *Water* **2017**, *9*, 794. [CrossRef]
10. Wang, Z.; Liao, L.; Hursthouse, A.; Song, N.; Ren, B.Z. Sepiolite-based adsorbents for the removal of potentially toxic elements from water: A strategic review for the case of environmental contamination in Hunan, China. *Int. J. Environ. Res. Public Health* **2018**, *15*, 1653. [CrossRef]
11. Zhao, Y.L.; Zhang, C.S.; Wang, Y.X.; Lin, H. Shear-related roughness classification and strength model of natural rock joint based on fuzzy comprehensive evaluation. *Int. J. Rock Mech. Min. Sci.* **2021**, *137*, 104550. [CrossRef]
12. Xie, Q.; Ren, B.Z.; Hursthouse, A.S.; Shi, X.Y. Effects of mining activities on the distribution, controlling factors, and sources of metals in soils from the Xikuangshan south mine, Hunan Province. *Integr. Environ. Assess. Manag.* **2022**, *18*, 748–756. [CrossRef]
13. Feng, H.; Liu, F.; Luo, P.; Xie, G.; Xiao, R.; Hu, W.; Peng, J.; Wu, J. Performance of integrated ecological treatment system for decentralized rural wastewater and significance of plant harvest management. *Ecol. Eng.* **2018**, *124*, 69–76.
14. Zhao, Y.L.; Zhang, L.Y.; Liao, J.; Wang, W.J.; Liu, Q.; Tang, L.M. Experimental study of fracture toughness and subcritical crack growth of three rocks under different environments. *Int. J. Geomech.* **2020**, *20*, 04020128. [CrossRef]
15. Liu, Y.; He, B.; Xie, J.; Lu, Y.; Zhang, L. Compatibility of geosynthetic clay liners at different temperatures. *J. Environ. Prot. Ecol.* **2021**, *22*, 2295–2306.
16. Pandey, L.M.S.; Shukla, S.K. An insight into waste management in Australia with a focus on landfill technology and liner leak detection. *J. Clean. Prod.* **2019**, *225*, 1147–1154. [CrossRef]
17. Tang, L.; Tang, X.W.; Liu, Y.; Qu, S.X. Prediction of pore size characteristics of woven slit-film geotextiles subjected to unequal biaxial tensile strains. *Geotext. Geomembr.* **2020**, *48*, 724–734. [CrossRef]
18. Zhang, J.; Deng, R.J.; Ren, B.Z.; Hou, B.L.; Hursthouse, A. Preparation of a novel Fe_3O_4/HCO composite adsorbent and the mechanism for the removal of antimony (III) from aqueous solution. *Sci. Rep.* **2019**, *9*, 13021. [CrossRef]
19. Liu, Y.; Hao, Y.; Lu, Y. Improved design of risk assessment model for PPP project under the development of marine architecture. *J. Coast. Res.* **2018**, *83*, 74–80. [CrossRef]
20. Zhu, G.; Wang, C.; Dong, X. Fluorescence excitation-emission matrix spectroscopy analysis of landfill leachate DOM in coagulation-flocculation process. *Environ. Technol.* **2017**, *9*, 1489–1497. [CrossRef]
21. Tian, K.; Benson, C.H.; Yang, Y.M.; Tinjum, J.M. Radiation dose and antioxidant depletion in a HDPE geomembrane. *Geotext. Geomembr.* **2018**, *46*, 426–435. [CrossRef]
22. Li, L.; Yu, X.; Wang, T.J.; Tan, Y.G. The in fluence analysis of Jinkou landfill on the ground water quality in Wuhan City. *Environ. Pollut. Control.* **2016**, *38*, 7–12.
23. Li, Y.; Hu, X.; Ren, B.Z. Treatment of antimony mine drainage: Challenges and opportunities with special emphasis on mineral adsorption and sulfate reducing bacteria. *Water Sci. Technol.* **2016**, *73*, 2039–2051. [CrossRef] [PubMed]
24. Luo, X.; Ren, B.Z.; Hursthouse, A.S.; Thacker, J.R.M.; Wang, Z.H. Soil from an abandoned manganese mining area (Hunan, China): Significance of health risk from potentially toxic element pollution and its spatial context. *Int. J. Environ. Res. Public Health* **2020**, *17*, 6554. [CrossRef]
25. Zhou, Y.; Ren, B.Z.; Hursthouse, A.S.; Zhou, S. Antimony ore tailings: Heavy metals, chemical speciation, and leaching characteristics. *Pol. J. Environ. Stud.* **2019**, *28*, 485–495. [CrossRef]
26. Liu, Y.; Gates, W.P.; Bouazza, A. Acid induced degradation of the bentonite component used in geosynthetic clay liners. *Geotext. Geomembr.* **2013**, *36*, 71–80. [CrossRef]

27. He, Z.; Ren, B.Z.; Hursthouse, A.; Wang, Z. Efficient removal of Cd(II) using SiO_2-$Mg(OH)_2$ nanocomposites derived from sepiolite. *Int. J. Environ. Res. Public Health* **2020**, *17*, 2223. [CrossRef]
28. Zhang, Y.; Ren, B.Z.; Hursthouse, A.; Deng, R.J.; Hou, B.L. Leaching and releasing characteristics and regularities of Sb and As from antimony mining waste rocks. *Pol. J. Environ. Stud.* **2019**, *28*, 4017–4025. [CrossRef]
29. Zhao, Y.L.; Luo, S.L.; Wang, Y.X.; Wang, W.J.; Zhang, L.Y.; Wan, W. Numerical analysis of karst water inrush and a criterion for establishing the width of water-resistant rock pillars. *Mine Water Environ.* **2017**, *36*, 508–519. [CrossRef]
30. Yuan, Q.; Zhu, G. A review on metal organic frameworks (MOFs) modified membrane for remediation of water pollution. *Environ. Eng. Res.* **2021**, *26*, 190435. [CrossRef]
31. Deng, R.J.; Shao, R.; Ren, B.Z.; Hou, B.L.; Tang, Z.E.; Hursthouse, A. Adsorption of antimony(III) onto Fe(III)-treated humus sludge adsorbent: Behavior and mechanism Insights. *Pol. J. Environ. Stud.* **2019**, *28*, 577–586. [CrossRef]
32. Liu, S.; Fan, S.G.; Sun, Z.H.; Zhang, X.X. Analysis of metal ion content in different treatment stages of landfill leachate. *Appl. Chem. Ind.* **2018**, *47*, 1304–1307.
33. Xie, J.; Liu, Y.; Lu, Y.; Zhang, L. Application of the high-density resistivity method in detecting a mined-out area of a quarry in Xiangtan City, Hunan Province. *Front. Environ. Sci.* **2022**, *10*, 1068956.
34. Lu, Y.L.; Yang, T.C.; Abdollah, T.T.; Liu, Y. Fast recognition on shallow groundwater and anomaly analysis using frequency selection sounding method. *Water* **2023**, *15*, 96. [CrossRef]
35. Lu, Y.L.; Cao, C.H.; Liu, Y.Q.; Liu, Y. Study on application of comprehensive geophysical prospecting method in urban geological survey-Taking concealed bedrock detection as an example in Dingcheng district, Changde City, Hunan province, China. *Appl. Sci.* **2023**, *13*, 417. [CrossRef]
36. Zhao, Y.L.; Liu, Q.; Zhang, C.; Liao, J.P.; Lin, H.; Wang, Y. Coupled seepage-damage effect in fractured rock masses: Model development and a case study. *Int. J. Rock Mech. Min. Sci.* **2021**, *144*, 104822. [CrossRef]
37. Zhang, Y.; Ren, B.Z.; Hursthouse, A.; Deng, R.J.; Hou, B.L. Study on the migration rules of Sb in antimony ore soil based on HYDRUS-1D. *Pol. J. Environ. Stud.* **2018**, *28*, 965–972. [CrossRef]
38. Liu, Y.; Gates, W.P.; Bouazza, A. Impact of acid leachates on microtexture of bentonites used in geosynthetic clay liners. *Geosynth. Int.* **2019**, *26*, 136–145. [CrossRef]
39. Hou, B.L.; Liu, X.; Li, Z.; Ren, B.Z.; Kuang, Y. Heterogeneous fenton oxidation of butyl xanthate catalyzed by iron-loaded sewage sludge. *Fresenius Environ. Bull.* **2022**, *31*, 4125–4131.
40. Zhao, Y.L.; Zhang, L.; Wang, W.; Tang, J.; Lin, H.; Wan, W. Transient pulse test and morphological analysis of single rock fractures. *Int. J. Rock Mech. Min. Sci.* **2017**, *91*, 139–154. [CrossRef]
41. Hou, B.L.; Li, Z.; Deng, R.J.; Ren, B.Z. Advanced treatment of coal chemical industry wastewater by expansive flow biological aerated filter. *Fresenius Environ. Bull.* **2017**, *26*, 4517–4521.
42. Shi, X.; Ren, B.Z.; Hursthouse, A. Source identification and groundwater health risk assessment of PTEs in the stormwater runoff in an abandoned mining area. *Environ. Geochem. Health* **2022**, *44*, 3555–3570. [CrossRef] [PubMed]
43. Yang, X.; Liu, Y.; Yang, C. Research on the slurry for long-distance large-diameter pipe jacking in expansive soil. *Adv. Civ. Eng.* **2018**, *9*, 47–54. [CrossRef]
44. Calamita, G.; Brocca, L.; Perrone, A.; Piscitelli, S.; Lapenna, V.; Melone, F.; Moramarco, T. Electrical resistivity and TDR methods for soil moisture estimation in central Italy test-sites. *J. Hydrol.* **2012**, *454*, 101–112. [CrossRef]
45. Xie, Q.; Ren, B.Z. Pollution and risk assessment of heavy metals in rivers in the antimony capital of Xikuangshan. *Sci. Rep.* **2022**, *12*, 14393. [CrossRef]
46. Cheng, Y.X.; Liu, H.S.; Zhao, Z.Y. Investigation of urban landfill contamination using geophysical methods. *Chin. J. Eng. Geophys.* **2004**, *1*, 26–30.
47. Wang, L.; Long, X.; Wang, T.T.; Xi, Z.Z.; Chen, X.F.; Zhong, M.F.; Dong, Z.Q. Application of the opposing-coils transient electromagnetic method in detection of urban shallow cavities. *Geophys. Geochem. Explor.* **2022**, *46*, 1289–1295.
48. Yang, T.C.; Gao, Q.S.; Li, H.; Fu, G.H.; Hussain, Y. New insights into the anomaly genesis of the frequency selection method: Supported by numerical modeling and case studies. *Pure Appl. Geophys.* **2023**, *180*, 969–982. [CrossRef]
49. Xi, Z.Z.; Long, X.; Zhou, S.; Huang, L.; Song, G.; Hou, H.T.; Wang, L. Opposing coils transient electro-magnetic method for shallow subsurface detection. *Chin. J. Geophys.* **2016**, *59*, 3428–3435.
50. Coggon, J.H. Electromagnetic and electrical modeling by the finite element method. *Geophysics* **1970**, *36*, 132–153. [CrossRef]
51. Leng, J.H.; Lu, Y.L.; Li, X.Q.; Zhao, X.Y.; Liu, Y. Metallogenetic Potential of the Ziyunshan Pluton in Central Hunan, South China: Insights from Element Geochemistry of Granites. *Minerals* **2023**, *13*, 144. [CrossRef]
52. Lu, Y.L.; Li, X.Q.; Liu, Y.; Leng, J.H. The Establishment of Ore-Controlling Fracture System of Baoginshan Gold Mine Based on Fracture-Tectonic Analysis. *Mob. Inf. Syst.* **2021**, *2021*, 5887680. [CrossRef]
53. Key, K. MARE2DEM: A 2-D inversion code for controlled-source electromagnetic and magnetotelluric data. *Geophys. J. Int.* **2016**, *207*, 571–588. [CrossRef]

water

MDPI

Article

Application and Automatic Monitoring and Analysis of Hybrid Support Structure in Ultra-DEEP Foundation Pit Engineering in the Lanzhou Area under Complex Environmental Conditions

Jian Wu [1], Shuaihua Ye [2,*], Zhiquan Wang [1] and Dong Yang [1]

1 Gansu Civil Engineering Research Institute Co., Ltd., Lanzhou 730020, China
2 School of Civil Engineering, Lanzhou University of Technology, Lanzhou 730050, China
* Correspondence: yesh@lut.edu.cn

Abstract: This paper takes the deep foundation pit project of Lanzhou Hospital of Traditional Chinese Medicine as the background. The design and construction of the foundation pit is relatively difficult due to the complex environment around the pit, the dense surrounding buildings, the complex underground soil layer and the influence of groundwater on the pit. In order to detect problems in the construction process, the pit was monitored in real time through an automated monitoring system for the whole process of excavation and backfilling of the pit. The analysis of the actual monitoring data shows that: (i) the support scheme of bored pile + prestressed anchor cable support combined with concrete corner bracing can meet the design of this type of foundation pit without causing disturbance to the surrounding buildings; (ii) combined with the actual case of the influence of groundwater on the excavation process of the foundation pit, it proves that the real-time measurement by the robot can timely detect the safety hazards caused by external factors during the construction process of the foundation pit. The project is a very important one for deep pits and complex pits. This project provides a good reference case for deep foundation pits and foundation pit projects in complex environments.

Keywords: ultra-deep foundation pits; hybrid support; automated monitoring; complex environments

check for updates

Citation: Wu, J.; Ye, S.; Wang, Z.; Yang, D. Application and Automatic Monitoring and Analysis of Hybrid Support Structure in Ultra-DEEP Foundation Pit Engineering in the Lanzhou Area under Complex Environmental Conditions. *Water* **2023**, *15*, 1335. https://doi.org/10.3390/w15071335

Academic Editors: Andreas Angelakis and Bommanna Krishnappan

Received: 8 February 2023
Revised: 20 March 2023
Accepted: 24 March 2023
Published: 28 March 2023

1. Introduction

With the continuous development of urbanization in the western region, the foundation pit projects in the west of loess area have gradually increased. Because the topography and geomorphology region of the west of China are relatively complex, the available land resources are relatively small, and the requirements for the foundation pit project are becoming more and more stringent. In addition, because the safety performance requirements of deep foundation pits are relatively high, and the impact on the surrounding existing buildings is rather large, it makes it a complex problem in geotechnical engineering research. To have a more specific understanding of the deep foundation pit project, scholars have created comprehensive monitoring of the foundation pit project to ensure safe progress of the foundation pit project.

Dmochowski, G. et al. [1] analyse various risk factors for the implementation of infill buildings and the revival of historic buildings using only facades. In addition, examples of modern solutions for fixing existing building walls are presented, as well as methods for monitoring vertical deformations using a hydrostatic level (HLC) system. Zhandos Y. Orazalin, et al. [2] analytically highlighted the effects of three-dimensional excavation and support geometry on wall and ground movements by simulating the basement of the Stata Center building on the MIT campus. L. Sebastian Bryson, et al. [3] proposed a relative stiffness ratio theory and analysed several case studies using 3D finite element analysis. The new relative stiffness ratio performed well in predicting the bending stiffness of the braced system and the lateral deformation caused by the actual excavation of the braced system. Luca Masini et al. [4] using the Rome Metro Line C as a case study, and through extensive instrumental testing during construction, as well as field observations, concluded that a robust retaining system and strict control of the construction sequence were key to minimising the impact of deep excavation. T. SCHWAMB et al. [5] numerical simulations of a circular shaft in London were carried out and compared with pre-monitoring data in order to understand the performance of the shaft during excavation and to draw conclusions for the future design of similar structures. L. Sebastian Bryson et al. [6] presented a case study, using the results of a three-dimensional (3D) finite element (FE) analysis of the study, which involved the observed response of a building in the vicinity of a deep foundation pit. Panpan Guo, Xiaonan Gong et al. [7] report on a rare deep braced approach in the context of deep foundation excavation in Suzhou, China, which is facing severe constraints from underground facilities or structures, and provide a comprehensive analysis of the performance of this deep excavation, which provides a good case study for practical engineering. Panpan Guoa, Xiaonan Gong et al. [8] With the aim of revealing the displacement and force behaviour of deep foundation support under asymmetric overload loading, the displacement and force behaviour of diaphragm walls, AM piles and ground were analysed using the finite difference method through a metro interchange station in Hefei, China, as the engineering background. The validity of the finite element method was verified by comparison with the field monitoring results and the elastic foundation beam method.

Zhou Yong et al. [9] took the pile bracing support structure of the deep foundation pit at Dongfanghong Square Station of Lanzhou Metro Line 1 as the engineering background and analysed the displacement variation law of the enclosure structure, surrounding buildings and ground settlement based on the field monitoring results and numerical calculation. Ye Shuaihua et al. [10] relied on the foundation pit project of a station in Lanzhou Metro to monitor the horizontal and vertical displacement of pile top, surface settlement, steel support axial force, and groundwater level during the foundation pit construction, and systematically analyzed the monitoring data. Mitew-Czajewska, M. [11] describes four excavation scenarios carried out in the immediate vicinity of an existing metro structure, with a special description and analysis of the experimental data and a discussion of the displacements, leading to conclusions about the effects of deep excavation on the structure in the vicinity of the excavation. In addition, the 2D finite element modelling of the selected example is described in detail and the displacement results of the tunnel are compared with the measured results. Jarosław Rybak et al. [12] based on previous experience, outline typical damage occurring in the construction of supporting walls (sheet pile driving, diaphragm wall driving and formation) and in tunnelling. Other damage is caused by unloading of the soil (caused by the excavation phase) and horizontal loads during prestressing of columns or ground anchors. Selected case studies of sheet pile wall installations are given for typical failures caused by unplanned excavations and their effects on adjacent structures. Miłosz Just, presents [13] the results of deep foundation pit analysis using the finite element method and the effects of underground diaphragm wall construction on surrounding objects. Paolo Castaldo et al. [14] Using the design of a new underground station in Naples (Italy) as a research context, the structural safety of existing buildings in the vicinity of deep excavations is assessed by calculating the probability of exceedance for different damage criteria in a simplified probabilistic approach based on univariate or multivariate probabilistic analysis, using the results of a numerical model for boundary value problems. Ryszard Dachowski et al. [15] propose the application of the Preference Ranking Organisation Method for Enriched Evaluation (PROMETHEE) II when selecting the best solution and ranking the selected foundation underpinning methods. the analysis of the selected foundation support methods is based on a complex decision problem that includes aspects of the three pillars of sustainability and it is extended to include technical and construction criteria. Maria Mrowczynska et al. [16] By presenting a method of measuring and processing the results, it has been made possible to determine the vertical displacement of the measurement and control network points inside and outside the building, where variations in the form of scratches and cracks in the external and internal walls are noted.

Ye Shuaihua et al. [17] relied on the deep and large foundation pit project in a complex environment in Lanzhou City, and analyzed the deformation characteristics of the foundation pit and the influence of foundation pit excavation on adjacent buildings during the excavation process according to the monitoring data of the soil, supporting structure, adjacent buildings around the foundation pit and the numerical simulation of foundation pit excavation, To provide a reference for the design of deep and large foundation pit projects in complex environments. Yin Lijie, Li Yujie, Zhu Yanpeng [18] and others proposed a diaphragm wall with internal bracing support system for deep foundation pits in red sandstone using the Lanzhou Metro Yan Yuan Road Station as an example. Numerical simulations of the excavation process were carried out using Midas GTS finite element software, and the simulated data and monitoring data were compared and analysed. The results of the study can provide experience reference for the design of similar red sandstone foundation pit support structures encountered in the subsequent construction of Lanzhou Metro. Ye Shuaihua et al. [19] took a deep foundation pit project of China Mobile Gansu Company as an example, and analysed the deformation characteristics of three types of support structures, namely composite soil nail walls with soil nails plus prestressed anchors, soil nail walls and row piles plus prestressed anchors, respectively, based on the monitoring results, which showed the practicality of such support effects in similar areas.

This paper is based on the deep foundation pit project of Lanzhou Hospital of Traditional Chinese Medicine; because the surrounding environment of the foundation pit is relatively complex, it is necessary to fully consider the influence of surrounding buildings, complex strata, and groundwater on the deep foundation pit project. In order to ensure safety during the construction of foundation pits and to gain experience in working with deep foundation pits in complex environments. By automatically monitoring the vertical and horizontal displacement, deep horizontal displacement, settlement deformation of foundation pit, internal force of bolts, settlement of surrounding buildings and surrounding pipelines, and groundwater level of supporting structures and analyzing the monitoring data, this paper provides a good case for similar foundation pit projects in loess areas, and also provides a reference for the design and construction of deep and large foundation pit projects in complex environments.

2. Project Overview

2.1. Site Overview

The Northwest Regional Medical Center Project of Gansu Provincial Hospital of Traditional Chinese Medicine is located in Qilihe District, Lanzhou City, and the project is close to urban roads. The ground elevation of the site is 1528.88~1528.35 m, considering the thickness of 0.1 m cushion + 0.1 m waterproof layer, the depth of the podium pit of the medical part on the north side is 12.0~12.5 m, the depth of the foundation pit of the main building of the medical interest on the north side is 13.1 m. The depth of the foundation pit of the three-dimensional garage on the south side is 22.3 m. The surrounding environment of the foundation pit is complex and there are many buildings. Part: (1) a medical area, (2) an underground garage, the surrounding environment as shown in Figure 1.

The medical area is located on the north side of the underground garage; the foundation pit is in the shape of an "L", the west and north sides are all D Block Inpatient Building, pile raft foundation, buried about 10.0 m, the west side is about 15 m from the outdoor basement wall, and the north side is about 4.7 m from the outdoor basement wall. The east side of the site is spacious, the closest buildings to the foundation pit are residential buildings 6 and 7, which are about 13 m~18 m away from the outer wall of the underground garage, and the pile foundation is buried at a depth of about 11 m.

Figure 1. Map of the surrounding environment of the foundation pit.

The underground garage area has a rectangular plan and a depth of 22.3 m. On the west side, there is a 6-story frame house, an independent foundation, buried about 2.0 m deep. It is about 12.6 m from the outer wall of the underground garage. The south side is 8.4 m away from the outer wall of the underground garage for Guazhou Road, with heavy traffic. On the east side, the podium of the Medical Technology Complex Building (Building A) of the Provincial Hospital of Traditional Chinese Medicine is adjacent to the foundation pit, and the nearest distance is about 7.7 m from the outer wall of the underground garage. 2 basement floors with independent foundation. The main building of the Medical Technology Complex Building (Building A) is a pile raft foundation, buried 12 m deep.

2.2. Hydrogeological Conditions

The maximum excavation depth of the foundation pit of this project is about 22.3 m, and the lithology of the site strata within the proposed site is composed of the fourth series of new artificial fill and alluvium, which is divided into 6 main engineering geological layers and 4 engineering sublayers according to its engineering geological characteristics, which are now described as follows: ① Miscellaneous fill (Q_4^{ml}): variegated, slightly wet, loose, uneven soil quality, the surface layer of 0.10~0.20 m is mostly concrete ground, the overall backfill is uneven, irregular, the backfill contains silt, bricks, concrete blocks, white ash, gravel, Coal ash and dark impurities, etc. ② Plain fill (Q_4^{ml}): yellow-brown, slightly wet, slightly dense, mainly backfill silt, containing a small amount of gravel, construction waste, etc. ③ Loess-like silt (Q_4^{al+pl}): black-brown, slightly wet-wet, slightly dense, extremely uneven soil, the composition is mainly fine sand, local powder and clay particles are interlayered, hand rubbing has a sticky feeling, smell. ③-1 Gravelly sand (Q_4^{al+pl}): distributed in the form of sandwich or lens body, blue-gray, slightly wet, slightly dense, uneven particle size, particle size greater than 2 mm particle content accounts for about 30.0~45.0%. ④ Peaty silty fine san (Q_4^{al+pl}): black-brown, slightly wet-wet, slightly dense, extremely uneven soil, the composition is mainly silty sand, local powder and clay particles are interlayered, hand rubbing has a sticky feeling, smelly smell. ⑤ Fine sand (Q_4^{al+pl}): blue-gray, wet ~ saturated, slightly dense, locally sandwiched with a thin layer of silt. ⑤-1 Round gravel (Q_4^{al+pl}): distributed in the form of sandwich or lens body, blue-gray, wet ~ saturated, medium dense. The general particle size is about 2~20 mm, accounting for more than 50% of the total amount; the maximum visible particle size is more than 30 mm, sub-circular, good roundness, general sorting, large sand content, most of the medium coarse sand filling, local sand filling. ⑥ Cobble (Q_4^{al+pl}): blue-gray, wet ~ saturated, medium dense. See the boulder; the maximum particle size seen in the borehole is more than 300 mm. It is sub-circular, the roundness is good, the sorting is general, and most of the medium coarse sand filling, is local sand filling. There is a thin layer of fine sand or a thin layer of silt, and the whole field is continuously distributed. ⑥-1 fine sand (Q_4^{al+pl}): distributed in the form of sandwich or lens body, blue-gray, wet ~ saturated, slightly dense, the core is discrete, sandwiched with individual small gravel. ⑥-2 Round gravel (Q_4^{al+pl}): distributed in the form of sandwich or lens body, blue-gray, wet ~ saturated, medium dense, general particle size in 2~20 mm, the maximum visible particle size is more than 40 mm, sub-circular, general sorting, large sand content. The thickness and physical and mechanical parameters of each soil layer in the foundation pit are shown in Table 1.

Table 1. Basic parameters of each rock and soil layer.

Soil Layer Sequence	Layer Name	Soil Thickness/m	γ/(kN/m^3)	c/kPa	ϕ/(°)
①	Miscellaneous fill (Q_4^{ml})	1.00~2.20	16.0	5.0	18.0
②	Plain fill (Q_4^{ml})	0.50~7.50	17.0	18.0	19.0
③	Loess-like silt (Q_4^{al+pl})	2.70~10.10	17.0	19.0	25.0
③-1	Gravelly sand (Q_4^{al+pl})	1.9	21.0	3.0	27.0
④	Peaty silty fine san (Q_4^{al+pl})	0.70–2.10	20.0	11.0	20.0
⑤	Fine sand (Q_4^{al+pl})	0.40~1.70	20.0	2.0	22.0
⑤-1	Round gravel (Q_4^{al+pl})	0.50~1.70	21.0	3.0	27.0
⑥	Cobble (Q_4^{al+pl})	10.30~18.30	22.0	3	38
⑥-1	Fine sand (Q_4^{al+pl})	0.70	20.0	2.0	22.0
⑥-2	Round gravel (Q_4^{al+pl})	0.50~5.10	21.0	3.0	27.0

Note: (Definitions: γ is soil weight, c is soil cohesion force, and ϕ is the angle of internal friction).

A layer of groundwater was exposed at the depth of exploration, and the groundwater type was Quaternary loose rock pore diving. It is mainly found in the pores of the pebble layer, and the amount of water is relatively abundant. During the survey, the groundwater level was buried at a depth of 10.20~12.20 m, and the elevation was 1516.84~1517.85 m. The water inflow of a single well is 700~2000 m^3/d. The permeability coefficient of the pebble layer is 30~50 m/d. The primary recharge sources are the lateral replenishment of the Yellow River in the west and north of the site, followed by atmospheric precipitation, lateral replenishment of the Huangyugou south of the site, irrigation infiltration, and infiltration of residents' production and domestic water, which is discharged from the west, south to northeast of the site, and the groundwater variation is 1.00~1.50 m.

3. Pit Support Design Scheme and Construction

3.1. Selection of Supporting Structure

The excavation depth of the foundation pit project is considerable, the soil layer is relatively complex, the influence of the groundwater table is profound, the surrounding buildings are dense, and the deformation requirements are strict. Comprehensively considering the safety, economic rationality, and construction convenience of the project, the foundation pit is divided into nine sections, and two design schemes are adopted: bored pile + prestressed anchor cable foundation pit support and cantilever pile + concrete diagonal brace foundation pit support.

A~B, B~C, D~E section, C~D section, E~F section, F~G section, G~H~I section, I~J, J~K section, K~L section and L~A section adopt bored pile + prestressed anchor foundation pit support scheme, and area A~E adopts cantilever pile + concrete diagonal brace foundation pit support form. A typical profile is shown in Figure 2. According to the geological survey report, the groundwater level around the foundation pit is buried at a depth of about 10 m. The precipitation scheme of the tube well outside the hole + catchment well in the hole is adopted.

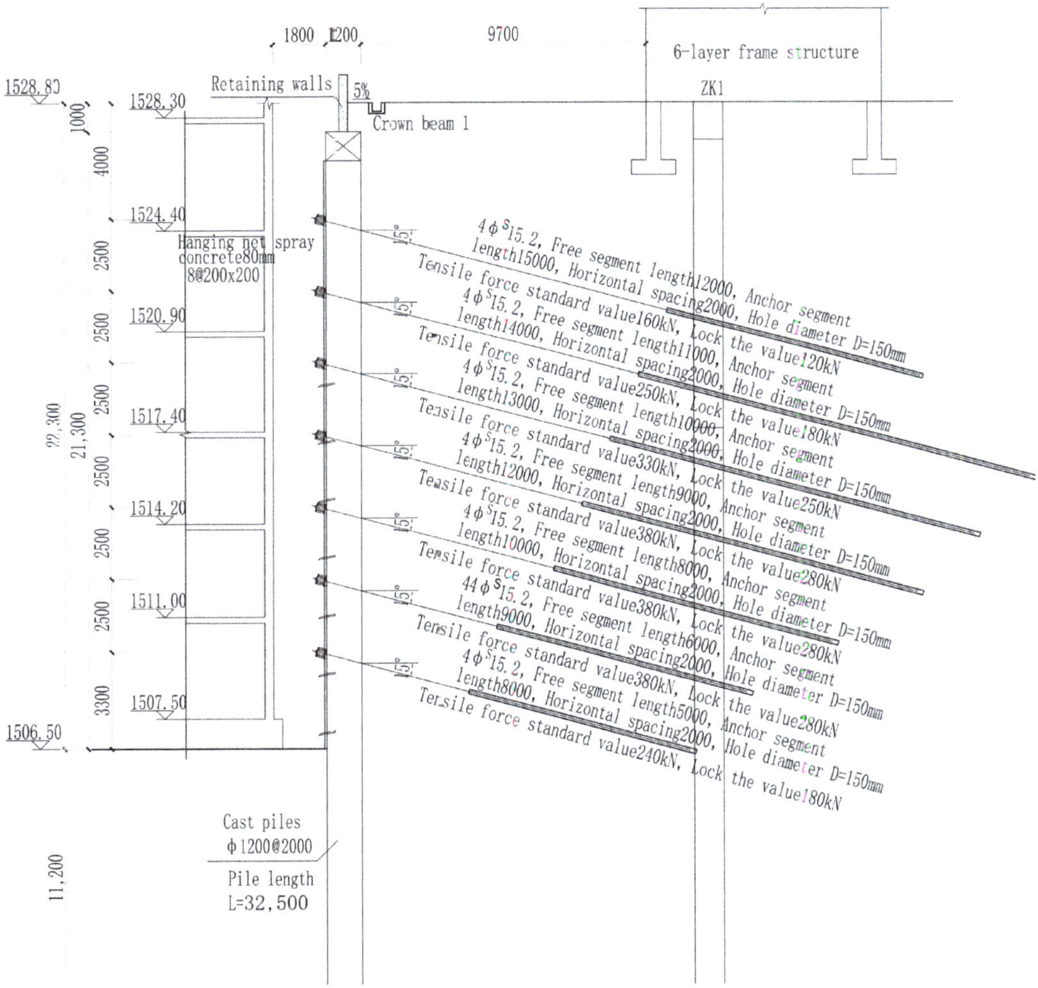

(**a**) Pile anchor support structure

Figure 2. *Cont.*

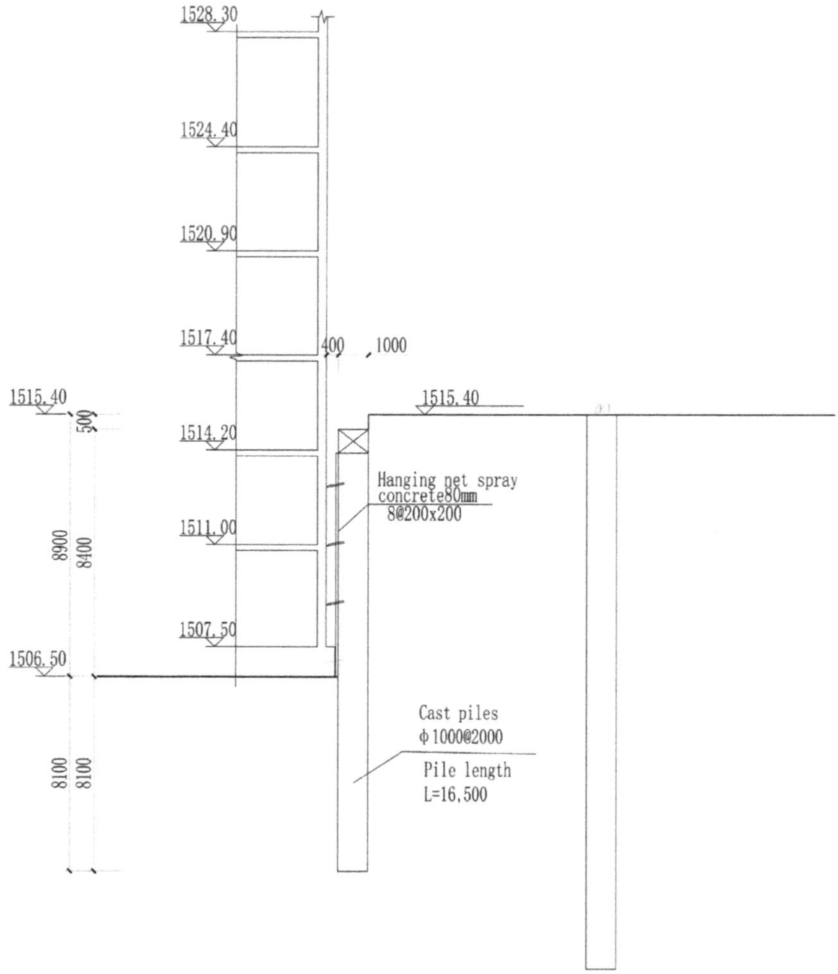

(**b**) Cantilever pile + concrete diagonal brace support structure

Figure 2. Profile view of a typical support structure.

3.2. Reinforced Concrete Support Piles

Support pile is a commonly used support method in foundation pit engineering, which mainly bears the active earth pressure of the foundation pit and combines with the crown beam to form a stable support structure [20,21]. The plane size of the foundation pit is large, and the working space is sufficient, so a reinforced concrete support pile + prestressed anchor cable is used to form an effective support scheme.

To ensure the construction quality, the primary reinforcement of the supporting pile is preferably connected by straight threaded sleeves, and lap welding connections can also be used; The relationship of longitudinal steel bar of the crown beam can be connected by lap welding. The welding adopts double-sided welding; the length of the help strip is not less than 5 d, the weld height is not less than 0.3 d, and the weld width is not less than 0.8 d. When welding, the joints should be staggered, the common area of the same section should not exceed 50%, and the interval arrangement; And must first do the welding test, qualified before formal welding. The main steel bar combinations are shown in Table 2.

Table 2. Reinforcement table of support piles by section.

Section	D (Diameter/mm)	① Longitudinal Rebar	Stiffeners	Spiral Ribs
M-M	1200	22C28	2C16@2000	C12@150
N-N	1200	23C28	2C16@2000	C12@150
O-O	1000	22C25	1C16@2000	C12@150
P-P	1000	22C25	1C16@2000	C10@150

3.3. Prestressed Anchor Cables

In this foundation pit project, the combination of prestressed anchor cable and supporting pile produces a more effective supporting effect for the foundation pit project. In this foundation pit project, the anchor cable adopts a 15.2 mm diameter steel strand, and the hole diameter is 150 mm. When applying to prestress to the anchor cable, it is necessary to wait for the strength of the grouting consolidation body of the anchor section to reach 20 MPa or reach more than 80% of the design strength and for the supporting pile gets more than 80% of the design strength, and the anchor cable can be tensioned and locked, and the five-stage tension is carried out according to the locking value, and the specific construction process is shown in Figure 3.

Figure 3. Drilling construction of anchor cable.

3.4. Crown Beam, Angle Brace

Considering the safety, economy, and stability of the foundation pit, the A–E sections are connected by an angle brace, and Figure 4 is the construction drawing of the corner brace site. The primary function of the crown beam is to communicate between stressed members and structure's integrity, and the node connection is shown in Figure 5.

Figure 4. On-site construction of corner braces.

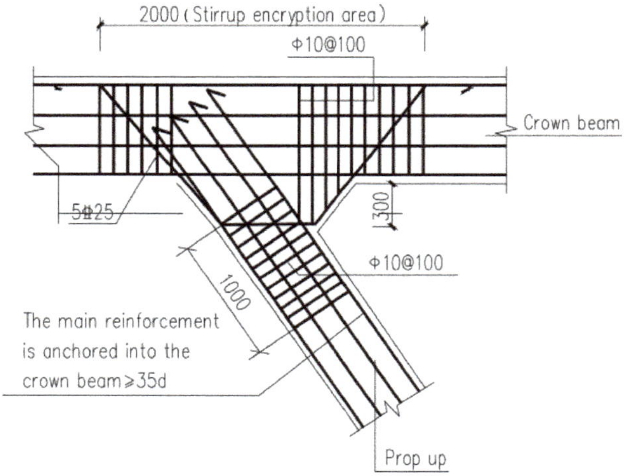

Figure 5. Construction drawings of angular braces/diagonal braces and crown joints.

The crown beam and angle brace of the foundation pit project are made of reinforced concrete material, and the bottom elevation position of the support beam is first excavated during its construction; the foundation is rammed, the bottom mold of the support beam is installed, and then the steel bar is installed on the bottom mold, and when the steel bar is installed, the support beam and the waist beam node should be connected as a whole, and the concrete is poured as a whole. For concrete braces, arching is used during construction, and the arching height is L/400.

3.5. Foundation Pit Precipitation

The garage on the south side of the foundation pit of Gansu Provincial Hospital of Traditional Chinese Medicine is about 22.3 m deep, so the garage foundation pit needs to be designed for precipitation; after the rain of the foundation pit on the south side, the medical area on the north side does not need rain, so no precipitation wells are set up, and the medical room on the north side only sets up blind ditches and catchment wells in the space of the fertilizer trough, and Figures 6 and 7 are the effect drawings of the construction process and the completion of construction. The tube well precipitation + open drainage scheme in the foundation pit of this project is adopted, and the catchment wells need to be laid at a distance of 30 m around the bottom of the hole. The precipitation cycle of the foundation pit is determined according to the construction progress, and the precipitation well near the post-pouring belt is considered according to the specific situation after the excavation of the foundation. The well is sealed after the post-pouring belt is poured, and waterproof measures such as waterproof rings should be set up. After the precipitation outside the pit, the lower part is backfilled with graded sand and gravel to prevent the formation of groundwater channels and cause the foundation of surrounding buildings to collapse.

Figure 6. Borehole downpipe-buried.

Figure 7. Tubewell lowering diagram.

4. On-Site Monitoring and Data Analysis

Considering the density of surrounding buildings and their low water level, the horizontal displacement of deep soil, water level change, horizontal removal of supporting piles, vertical displacement of supporting banks, settlement of surrounding buildings, settlement of surrounding pipelines, the axial force of anchor cables, and monitoring of cracks at the top and around the supporting structure were monitored in this foundation pit project. To ensure the accuracy of monitoring, the foundation pit monitoring adopts automatic monitoring instruments and quantitative equipment monitoring (see Figure 8 for the work schematic diagram), the manual patrol inspection is qualitatively combined, and the selected measuring robot meets the accuracy requirements of the polar coordinate method for the total station, and the distance between the on-site work base station and the monitoring point is controlled within 150 m. Set up a data automatic collection base station at the site; the monitoring cycle of the foundation pit project starts on 1 September 2020, a period of 25 months of monitoring; the whole monitoring process adopts automatic reading, uploading, recording, alarm, the layout plan of the foundation pit monitoring point is shown in Figure 9, and the number of monitoring points is detailed in Table 3.

Table 3. Summary of monitoring points.

Monitor Content	Number of Monitoring Points
Settlement and horizontal displacement monitoring of supporting structures	22 points
Support anchor cable internal force monitoring	7 sets of sections
Deep horizontal displacement monitoring	6 wells
Groundwater level monitoring	3 points
Settlement monitoring of buildings around foundation pits	58 points
Perimeter pipeline displacement monitoring	18 points
Crack monitoring	Laid out according to the actual situation

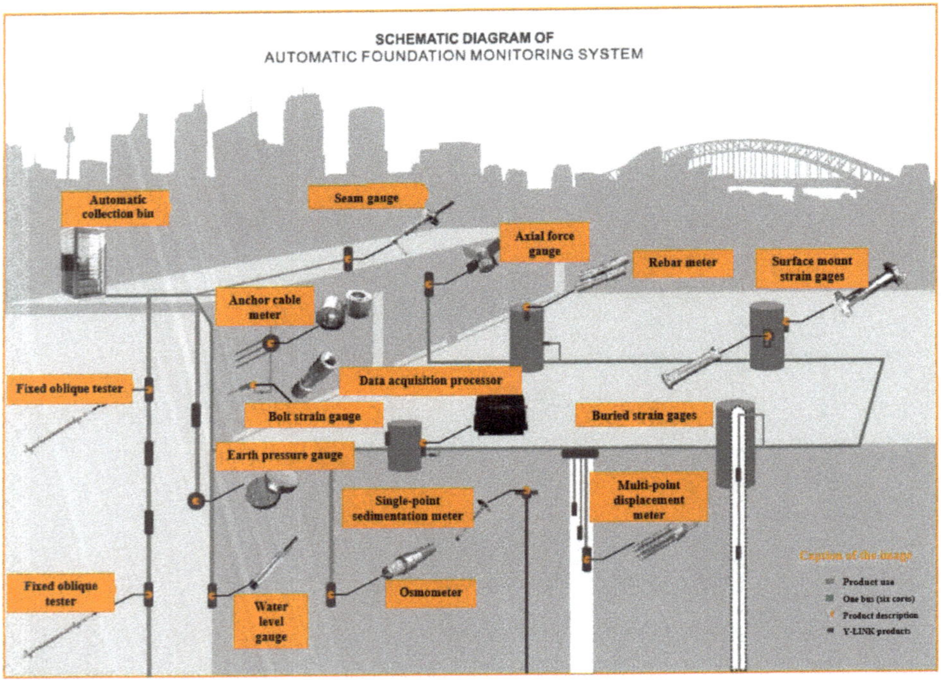

Figure 8. Schematic diagram of foundation pit online monitoring system.

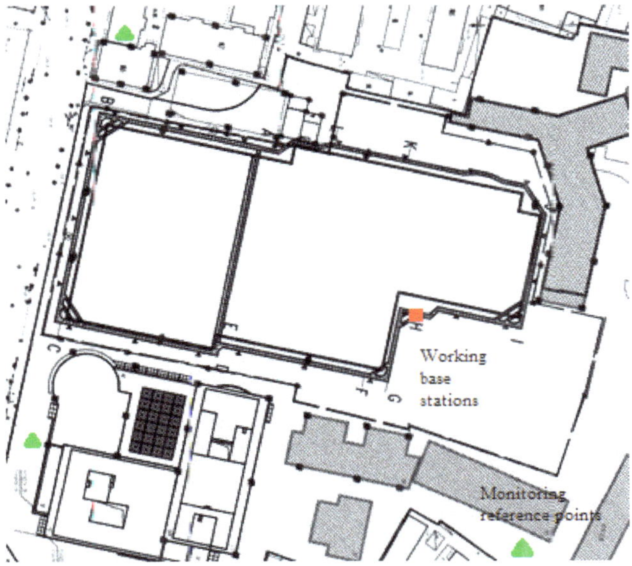

Figure 9. Layout plan of monitoring points of foundation pit.

The data of each monitoring point can be obtained by measuring the robot monitoring foundation pit, and the data processing center has a variety of functional requirements according to its object orientation. Its available composition structure is shown in Figure 10. The data of each monitoring point can be summarized to plot the change curve of the monitoring point quickly.

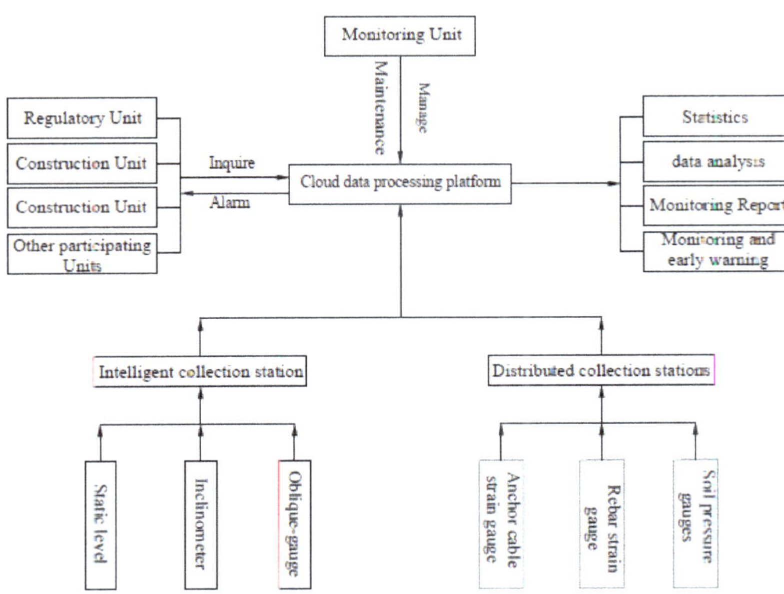

Figure 10. Function structure diagram of the automatic monitoring system.

4.1. Settlement and Horizontal Displacement Monitoring of Supporting Structures

The settlement and horizontal displacement monitoring of the supporting structure were monitored by the Tianbao S9 measurement robot. Due to the significant disturbance and more changes in the excavation process of the foundation pit, the measurement robot can monitor the foundation pit in an all-around and full-time real-time manner to ensure a timely and effective discovery, and the robot details are shown in Figure 11.

Figure 11. Ten Pao S9 measuring robot.

The measuring robot is based on the principle of establishing a planar control network within the pit, establishing a corresponding planar coordinate system and setting up control points in undisturbed areas around the pit. The measuring robot will set up the corresponding stations and monitor the monitoring points in the pit one by one, obtaining distances and angles, back-calculating the corresponding coordinates X,Y,H as the initial measurement values and comparing them with the data taken at a later stage to see if there are any changes in the data and whether the pit has settled or not.

When the horizontal displacement of the supporting structure is carried out, a total of 22 displacement monitoring points are arranged, and the XY value measured by the robot is further converted into horizontal displacement. Combined with the monitoring data analysis, it can be seen that the main distribution range of the cumulative change of horizontal displacement is 0.000 mm~38.54 mm, the cumulative value of the early warning value of horizontal displacement is 25 mm, and the change rate is 2 mm/d, so it can be concluded that the incremental change obtained by the flat displacement monitoring point of the supporting structure during the entire monitoring period exceeds the range of the allowable early warning value of the specification.

Through data observation, the horizontal displacement of the supporting structure mainly occurred in the excavation stage of the foundation pit, during which the cumulative value of the horizontal displacement of each monitoring point was 14.52~19.83 mm, and its deformation accounted for about 60% of the entire monitoring cycle. After the foundation pit is excavated to the foundation, a certain amount of horizontal displacement is continuously generated in adjusting the internal force of the supporting structure system. After the cumulative deformation value of each monitoring point exceeds the early warning range, the staff of the project department encrypts the monitoring point and strengthens the on-site inspection frequency; according to the monitoring data and inspection results, the side wall of the foundation pit is not abnormal in the subsequent construction until the foundation pit is backfilled, the deformation rate of each monitoring point is less than 2 mm/d, and the monitored deformation data converges until the horizontal displacement of the supporting structure is stable (see Figure 12 for the specific change of monitoring point).

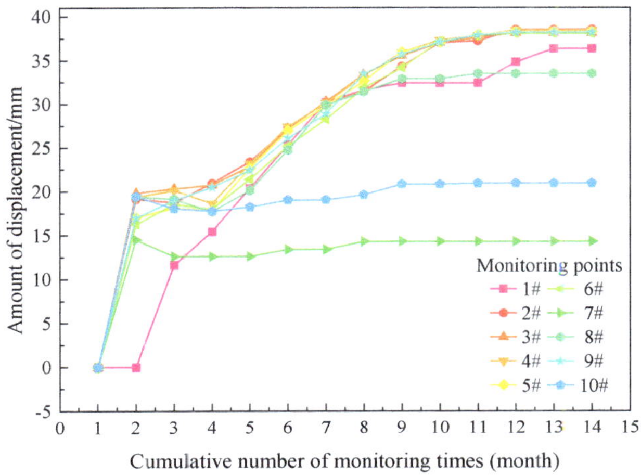

(**a**) 1#–10# deformation diagram of monitoring points

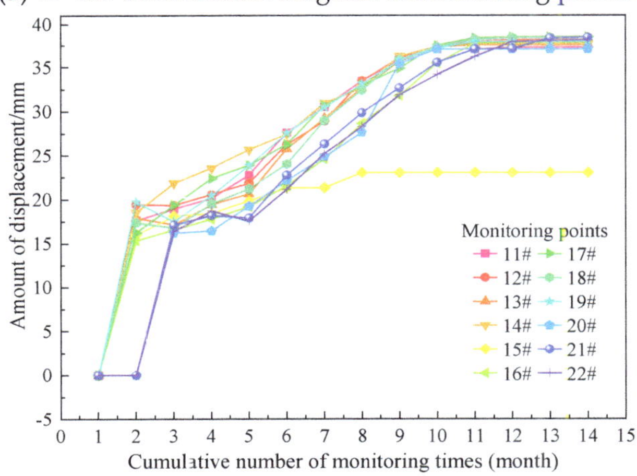

(**b**) 11#–22# deformation diagram of monitoring points

Figure 12. Horizontal displacement monitoring results of supporting structures.

Figure 13 shows the change curve of the vertica displacement monitoring point of the supporting structure. A total of 22 vertical displacement monitoring points at the top of the foundation pit support structure were arranged. The vertical displacement measured by the static level sensor was combined with the H value measured by the measuring robot, which was converted into vertical displacement for comparative analysis.

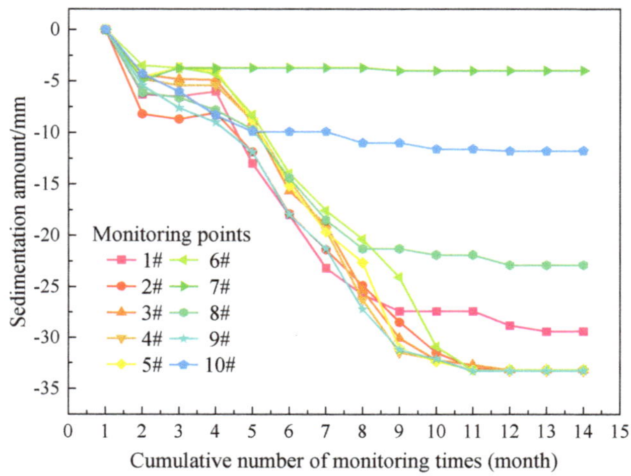

(**a**) 1#–10# deformation diagram of monitoring points

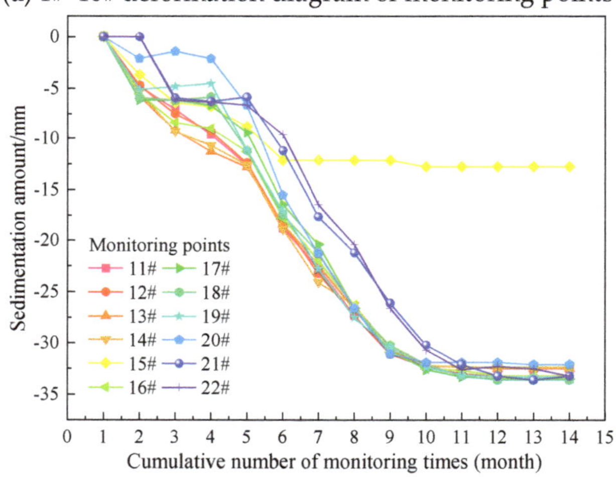

(**b**) 11#–22# deformation diagram of monitoring points

Figure 13. Monitoring results of vertical displacement of supporting structures.

By measuring the value obtained by the robot for analysis, the cumulative value of settlement displacement of each monitoring point is 4.00 mm~33.600 mm; according to the requirements of the specification, the vertical displacement early warning value is a cumulative value of 15 mm, and the change rate is 2 mm/d, which can be judged that the incremental change of some monitoring points of all monitoring points in the foundation pit during the entire settlement displacement monitoring period exceeds the allowable warning value range of the specification. After the cumulative deformation value exceeds the scope of the early warning value, the on-site monitoring point is encrypted monitoring, and the on-site inspection is strengthened; the side wall of the foundation pit has not been continuously deformed, and the surrounding cracks have not continued to develop, and the monitoring points with large vertical deformation after analysis are all at the edge road and passage, there are edge loads and vehicle vibrations, and the construction unit removes the edge load, eliminating the potential safety hazards around the foundation pit and the foundation pit.

4.2. Horizontal Displacement of the Soil of the Foundation Pit

When monitoring the horizontal displacement of the soil mass of the foundation pit, a total of 6 flat displacement monitoring points of the bottomless foundation pit were arranged (three groups of horizontal displacement of deep soil and three groups of horizontal displacement of the deep pile), and the sinusoidal function of the inclination angle on the standard spacing L of the upper and lower guide wheels of the inclinometer was converted into horizontal displacement.

As shown in Figures 14 and 15, when the sensitive axis of the accelerometer is on a horizontal plane, the projection of the vector g on the sensitive axis is zero and the output of the accelerometer is zero. When there is a tilt angle θ between the sensitive axis of the accelerometer and the horizontal plane, the accelerometer outputs a voltage signal.

$$A+ = K_0 + K_{gSin} \tag{1}$$

where: K_0 is the accelerometer bias value; K is the accelerometer sensitivity, g is the acceleration of gravity.

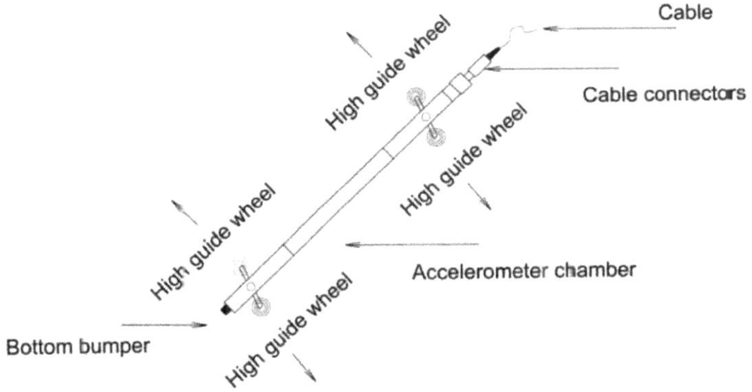

Figure 14. Oblique-gauge.

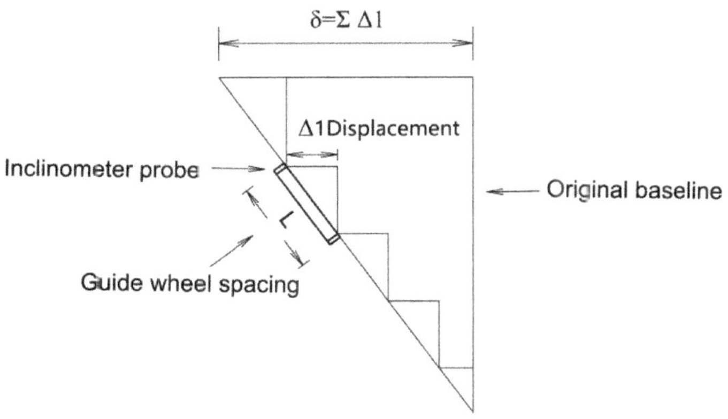

Figure 15. How inclinometers work.

To eliminate the effect of K_0, the probe can be turned 180° and a second measurement can be taken at the point where the text following an equation need not be a new paragraph. Please punctuate equations as regular text.

$$A- = K_0 - K_{gSin} \quad (2)$$

(1) − (2) Eliminate the bias value K_0 to obtain the difference; (1) + (2) cancels out the projection of the accelerometer's inclination, leaving only two times the bias value, called the 'difference'.

When these incremental horizontal deviations are added up and plotted from the bottom of the borehole, the result is a curve of the change in horizontal offset between the initial observation and any subsequent observation, representing the deformation of the soil during this observation, i.e., the horizontal displacement.

By observing the monitoring data, it can be seen that the cumulative change of deep horizontal displacement is mainly distributed in 0.000 mm~11.876 mm, and the flat displacement changes occur at all monitoring points. According to the existing specifications, the early warning value of the deep horizontal displacement of the foundation pit is 0.3% h~0.4% h, and h is the excavation depth of the foundation pit. The rate of change is 2 mm/d~3 mm/d. Through the monitoring data, it can be judged that the amount of change, individual change rate, and an individual cumulative change of all monitoring points of the foundation pit during the monitoring period are beyond the range of the allowable warning value of the specification. Still alarm value is not reached, which does not affect the safety of the foundation pit.

In the process of on-site monitoring, according to the apparent findings of the monitoring data, the deep horizontal displacement has no influence on the external environment during the monitoring process, the on-site data changes are more sensitive, and the data provided are more timely and accurate; Moreover, the deep horizontal displacement of the three groups of piles is relatively stable, and there is no huge fluctuation, but the change of the flat displacement data of the three groups of soil is slightly more significant than that of the horizontal displacement of the pile body. Still, they are all within the safe and controllable range of the foundation pit. The deep horizontal removal of the pile and the horizontal displacement of the soil are shown in Figure 16.

According to the automatic early warning system of the monitoring system, it was found that on 17 December 2020, the water pipeline of the pump room on the west side of the site foundation pit ruptured, resulting in a change in the horizontal displacement data at this location (Figures 17 and 18). Through the timely warning of the monitoring system, the relevant units immediately take corresponding treatment measures to eliminate the risk of the foundation pit in time.

4.3. Vertical Displacement Monitoring of the Static Level

This time, a total of 22 foundation pit static level displacement monitoring points were arranged, of which Figure 19 is the static curve of the foundation pit, and it can be seen from data observation: 1. The cumulative change of vertical displacement of static level in all monitoring points is mainly distributed in 0.000 mm~−32.297 mm. The existing specifications stipulate that the early warning value is 0.4% h~0.6% h, and h is the excavation depth of the foundation pit; The rate of change is 2 mm/d~3 mm/d. It can be judged that the amount of change, individual change rate, and an individual cumulative change of all monitoring points of the foundation pit during the whole monitoring period are beyond the allowable alarm value range of the specification. After data analysis according to the site situation, it was judged that the cumulative change exceeded the capacity, and corresponding measures were taken on the site in time, and the data change was mainly concentrated in the early excavation process, and the data change in the later excavation process was stable, which did not affect the safety of the foundation pit, and the foundation pit as a whole was in a steady state.

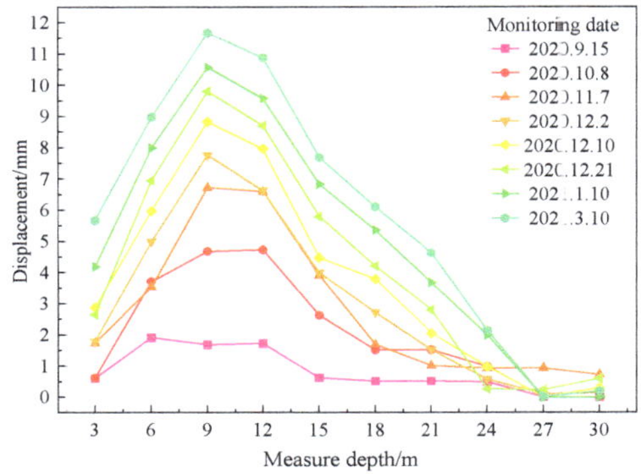

(**a**) Curve diagram of the east gate of the deep horizontal displacement of the Chinese Hospital pile

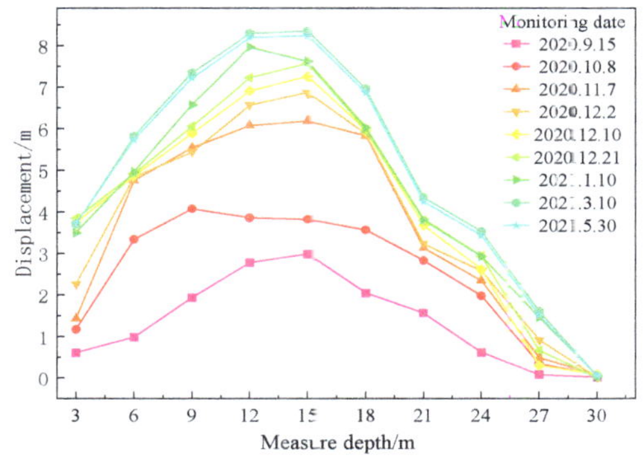

(**b**) East side profile of deep horizontal displacement of piles in Chinese Hospital

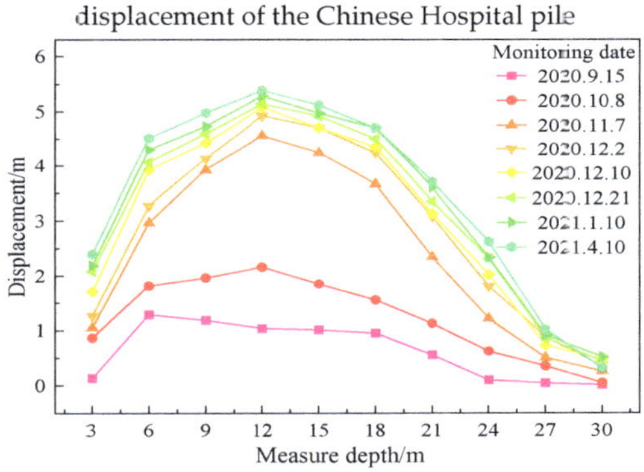

(**c**) West side curve of deep horizontal displacement of piles in Chinese Hospital

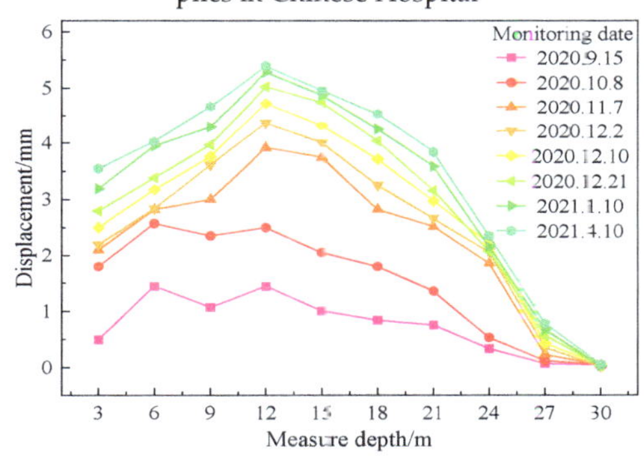

(**d**) Curve diagram of the deep horizontal displacement of the soil on the south side of the Chinese Hospital 1#

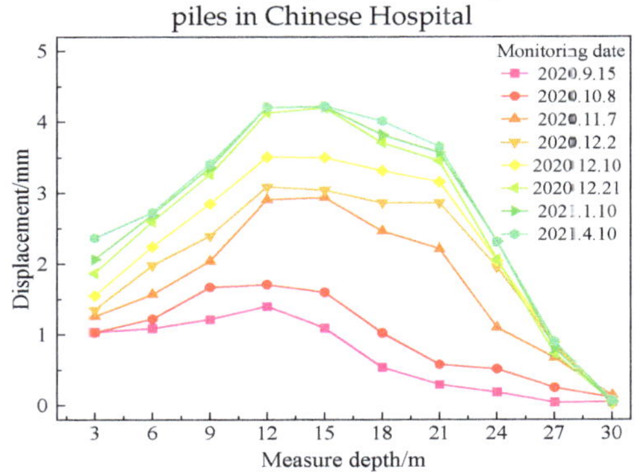

(**e**) Deep horizontal displacement of the soil of the Chinese Hospital 2# curve on the south side

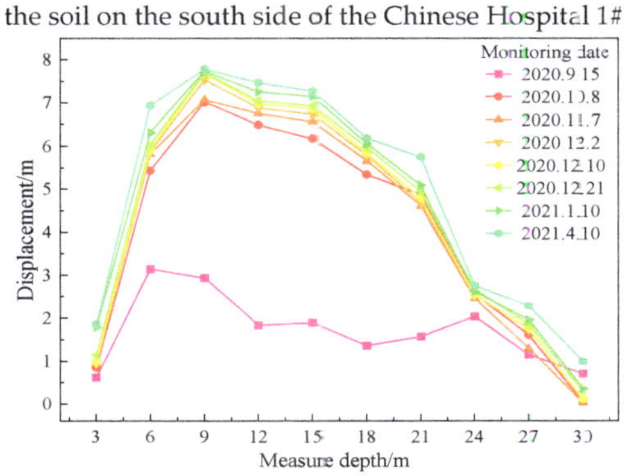

(**f**) Deep horizontal displacement of the soil of the Chinese Hospital 3# curve on the south side

Figure 16. Deep horizontal displacement plot.

Figure 17. Ruptured pipe on site.

Figure 18. Pit site conditions.

After the pit raft has been poured, the elevation of the datum points monitored by the static level around the pit is in a stable state and the elevation measured by the static level is compared with the real-time monitoring data from the measurement robot in the direction H of each monitoring point. The monitoring data obtained by two different monitoring methods can see that the two sets of monitoring data change are the same, which has been verified with each other, indicating that there are indeed changes in the excavation process of the foundation pit. Still, the amount of change is within the controllable range, which does not affect the safe use of the foundation pit.

4.4. Perimeter Pipeline Monitoring

Figure 20 shows the pipeline monitoring map around the foundation pit, and a total of 18 pipeline monitoring points around the foundation pit are arranged in this foundation pit. The elevation of the measurement point can be calculated by measuring the height of the reference point.

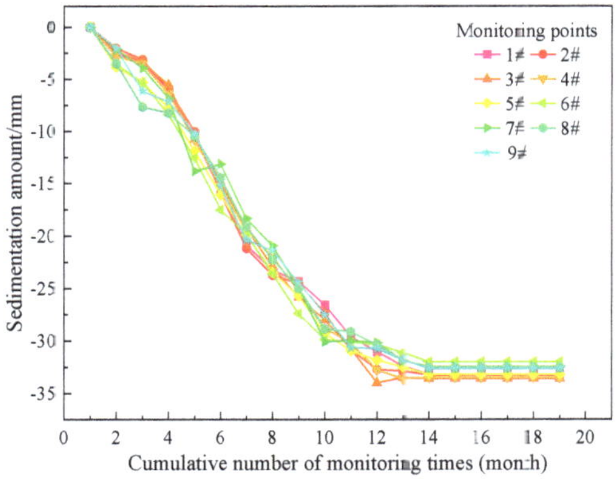

(**a**) Static curves of the south-eastern side of the Chinese Hospital

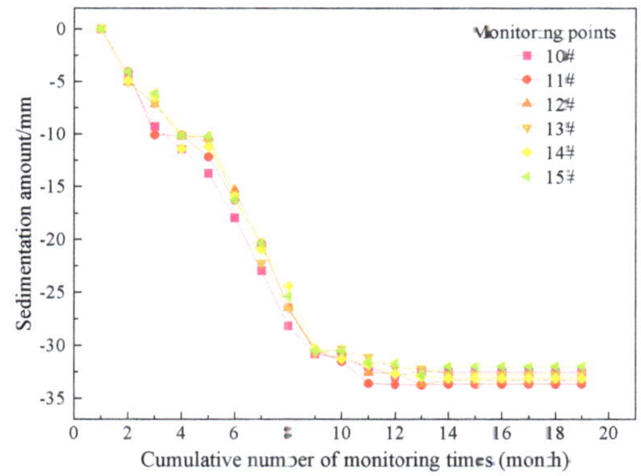

(**b**) Static curves on the west side of the Chinese Hospital

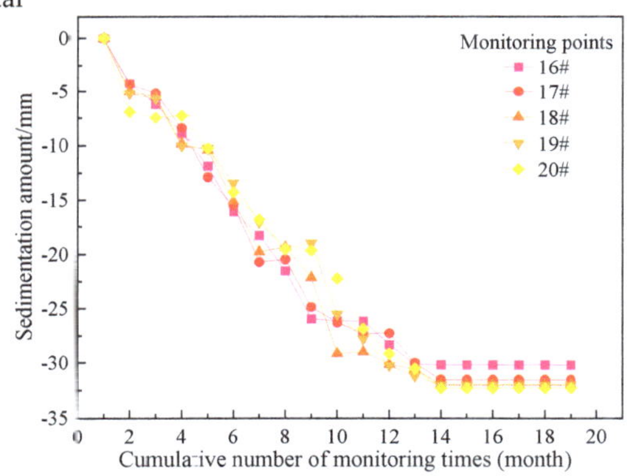

(**c**) Hydrostatic curve of the north side of the Chinese Hospital

Figure 19. Static graph.

Through the observation data, the cumulative changes of pipeline monitoring around the foundation pit were mainly distributed in 0.000 mm~−1.140 mm, and vertical displacement changes occurred in all monitoring points. According to the existing specifications, the surrounding pipeline monitoring alarm value can be known: the cumulative settlement is 10 mm, and the settlement rate is continuously greater than 1 mm/d. It can be judged that the amount of change, individual change rate, and an individual cumulative change of all monitoring points of the foundation pit during the whole monitoring period have not exceeded the range of the allowable alarm value of the specification; After data analysis according to the site situation, it was judged that the cumulative data change was mainly concentrated in the early excavation process, and the data change was stable during the later excavation process, which did not affect the safety of the pipelines around the foundation pit. The channels around the foundation pit were in a stable state as a whole.

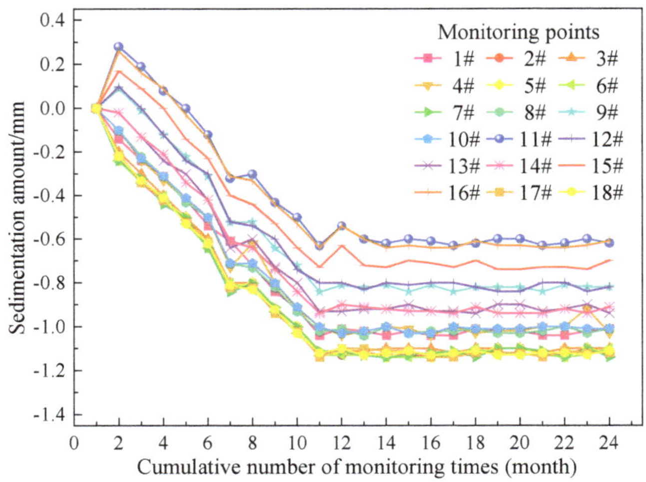

Figure 20. Perimeter pipeline monitoring diagram.

After the pouring of the foundation pit raft slab, the pipeline monitoring data around the foundation pit is in a stable and safe state, and the construction in the foundation pit at that time does not affect the safety of the pipeline around the foundation pit, through the means of manual patrol and the monitoring data comparison, the surrounding channel does not have displacement and damage, although there is a slight vibration caused by a tiny settlement, the settlement change is within the controllable range, does not affect the safe use of the pipeline around the foundation pit.

4.5. Internal Force Monitoring of Anchor Cables

Figure 21 shows the value of the internal force of the anchor cable in the pit. For the anchor cable in the pit, the frequency of each monitoring point is measured by a vibrating string strain gauge (the vibrating string gauge has several vibrating string strain gauges distributed on the force measuring cylinder. The electromagnetic coil excites the vibrating chord and measures its vibration frequency. The frequency signal is transmitted via cable to the reading device, where the strain variable causing the deformation of the stressed steel cylinder can be measured and the load value felt by the anchor cable gauge can be calculated by substituting the calibration coefficient.) The load values are calculated and 7 sets of internal force monitoring points are arranged for the anchor cable internal force monitoring in the pit.

Through the observation and monitoring data, the cumulative change of internal force of the anchor cable is mainly distributed in 0.00 kN~25.77 kN; according to the existing specifications, it can be seen that the early warning value of the internal force detection value of anchor cable in the foundation pit is the cumulative value (60%~80%), the maximum value f2, and f2 is the load design value. Rate of change (70%~80%)f2. It can be judged that the amount of change, change rate, and a cumulative change of all monitoring points of the foundation pit during the entire internal force monitoring period does not exceed the allowable alarm value range of the specification.

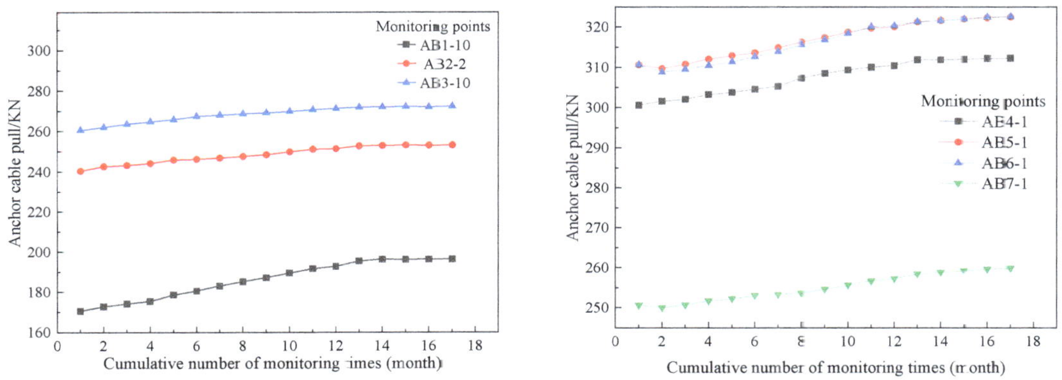

(**a**) Group 1–3 anchor cable internal force monitoring curve (**b**) Group 4–7 anchor cable internal force monitoring curve

Figure 21. Anchor cable internal force monitoring curve.

In the early monitoring process, it was found that the locking value of all anchor cables did not reach the design locking value, and there would be an inevitable loss of internal force during the pressure relief process after closing, which was preliminarily judged to be the loss of internal power generated at the lock and clamp. Therefore, in the later stage of monitoring, the tensile force value is increased to offset the internal force loss caused in the last stage in the process of pressure relief, so that the final locking value can reach the designed locking value.

4.6. Water Level Monitoring

Figure 22 is the analysis chart of the water level monitoring of the foundation pit, and a total of 3 groups of monitoring points are arranged during the water level monitoring of the foundation pit. Advanced isolated diffused silicon sensitive elements or ceramic capacitive pressure sensing sensors are used for monitoring. The static pressure is converted into an electrical signal using a sensor. Then temperature compensation and linear correction are carried out to convert it into a standard electrical signal, based on the principle that the water pressure is proportional to the height of the water.

Taking point 0 as the precipitation reference point, through observing the monitoring data, the cumulative change of the water level gauge is mainly distributed in 0.00 mm~0.386 mm; according to the existing specifications, it can be known that the incremental change early warning value of water level is 1000~2000 mm, the change rate is 500 mm/d, all monitoring points have water level changes, and the amount of change, change rate and a cumulative change of all monitoring points in the foundation pit during the entire water level monitoring period individually exceed the range of allowable alarm values of the specification.

As can be seen from Figure 20, the water level at point 3 of water level monitoring has a significant change, with a maximum variation of 3.718 m. The most significant groundwater increase occurred on 29 March 2021, 1.749 m higher than the designed scheduled water level, due to damage to the pumping pump in the local precipitation well, and the pumping volume was less than the recharge, resulting in the site being flooded with groundwater. In the later stage, the construction party strengthened the pumping capacity of the pump, reduced the water level to 0.5 m below the design foundation elevation, and the water level changed steadily.

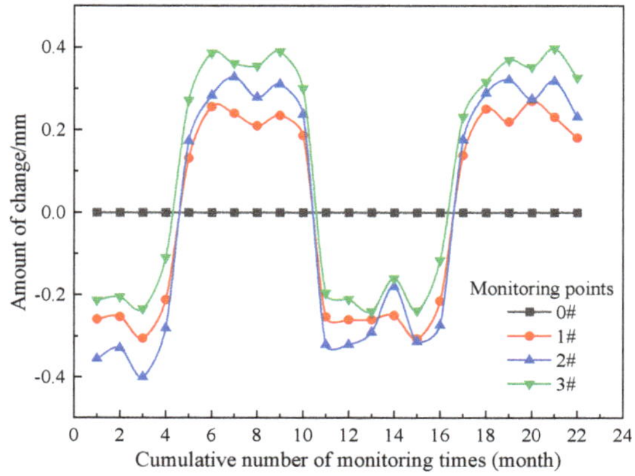

Figure 22. Curve of water level change in foundation pit.

4.7. Vertical Displacement of Surrounding Buildings

There are many complex buildings around this foundation pit, and it is necessary to monitor them in this foundation pit project. The foundation pit project used static level sensors to watch the surrounding buildings, and 58 monitoring points were deployed. The subsidence observation uses the observation point and the reference point to form a closed level route for observation, the reference point is selected as the starting point, counter-clockwise direction measurement and control, the route will be the road, the surrounding buildings and structures, underground pipelines monitoring points in an orderly manner as one, the final starting point closed as a circular route.

By observing the settlement data of surrounding buildings, it can be seen that the settlement generated by every single building is relatively uniform, and its difference value and cumulative value do not exceed the specification. Combined with the results of the wall crack survey of surrounding buildings, the excavation of the foundation pit did not have a significant impact on the surrounding buildings, and the whole was in a stable state.

For monitoring different buildings, the monitoring results are shown in Table 4, and the maximum cumulative settlement is displayed for different measurement results of the same building; see Figure 23 for details. Due to the large area of the family hospital on the west side of the foundation pit, has a significant impact on the foundation pit, so it is divided into 6 groups of different location measurement points for monitoring. The inpatient building on the north side of the foundation pit is monitored in two groups of varying location measurement points.

Real time monitoring is more responsive to realism than the simulation of the pit using finite elements in [6]. However it is necessary to carry out finite element analysis in the design of the pit.

Table 4. Cumulative settlement by building.

Monitor Content	Number of Monitoring Points	Cumulative Sedimentation (mm)
Family building on the east side of the foundation pit	6 points	9.71–9.83
Bank office building and MRI room on the west side of the foundation pit	6 points	9.32–9.84

Table 4. *Cont.*

Monitor Content	Number of Monitoring Points	Cumulative Sedimentation (mm)
Family building on the west side of the foundation pit	5 points	9.31–9.82
Family building on the west side of the foundation pit	5 points	9.62–9.82
Family building on the west side of the foundation pit	6 points	9.53–9.92
Family building on the west side of the foundation pit	6 points	9.62–9.84
Family building on the west side of the foundation pit	8 points	9.87–9.93
Family building on the west side of the foundation pit	6 points	9.03–9.14
Inpatient building on the north side of the foundation pit	4 points	8.82–9.16
Inpatient building on the north side of the foundation pit	6 points	9.30–9.73

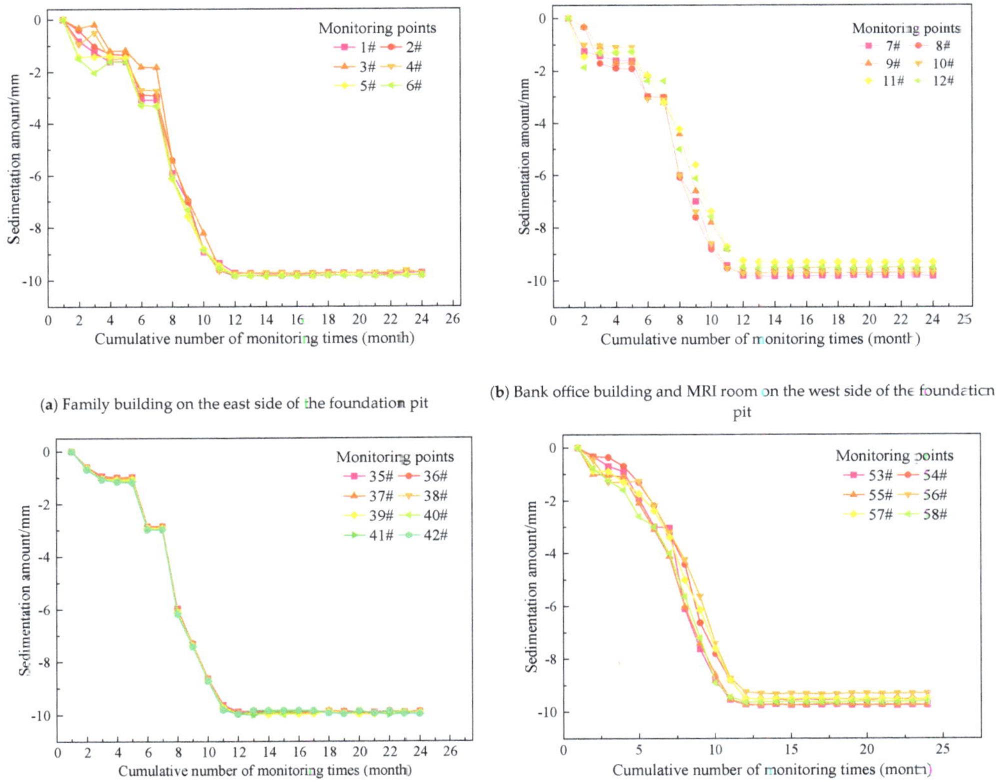

(**a**) Family building on the east side of the foundation pit

(**b**) Bank office building and MRI room on the west side of the foundation pit

(**c**) Family building on the west side of the foundation pit

(**d**) Inpatient building on the north side of the foundation pit

Figure 23. Curve of settlement of surrounding buildings.

5. Conclusions

A new automated monitoring system was adopted for the foundation pit support project of the Lanzhou City Chinese Hospital in Gansu Province, enabling real-time collection of monitoring data, automatic calculation and transmission to a data processing platform via radio signals. In order to ensure the safety of the pit and at the same time to prevent the safety of the surrounding buildings and pipelines from being affected by the construction of the pit. Data processing was carried out for vertical and horizontal displacements of the support structure, deep horizontal displacements, vertical displacements of the static level, surrounding pipelines, internal forces of anchor cables, groundwater level and settlement monitoring results of surrounding buildings respectively. Based on the analysis of the data obtained, the following summary can be obtained:

(1) Although the vertical and horizontal displacements of the support structure exceeded the specified warning values, the pit side walls were not continuously deformed and the surrounding cracks were not continuously developed through the later encrypted monitoring and on-site inspection by the site personnel, and the pit could be safely carried out. Therefore, it is suggested that the construction process of the foundation pit needs to focus on monitoring the displacement of the supporting structure to ensure the safety of the foundation pit.

(2) The deep horizontal displacement monitoring point of the foundation pit partially exceeds the early warning value, but does not reach the alarm value, which does not affect the safety of the foundation pit and the safety of the foundation pit, and the impact of the water pipeline rupture on the horizontal displacement of the foundation pit is found in time through the early warning of the online monitoring system. It can be proved that the real-time monitoring system can timely discover the influence of external unfavourable factors on the foundation pit.

(3) All monitoring points have been vertically displaced, and some displacement values exceed the early warning value. However, the foundation pit is still in a stable state, which does not affect the safety of the foundation pit.

(4) Through the internal force monitoring of the prestressed anchor cable, it can be obtained that the prestress loss will occur after the anchor cable is locked. The tensile force value of the anchor cable monitoring can be increased to ensure that the locking value can reach the design lock value.

(5) By comparing the precipitation reference points, it is found that the individual monitoring data exceeded the warning range during the water level monitoring period, which was due to the damage of the pumping pump in the local precipitation well. The pumping pump collection and drainage capacity was strengthened in time in the later stage to stabilize the water level change.

(6) During the excavation of the foundation pit, the surrounding pipelines, and surrounding buildings have obvious impacts, and the settlement occurs within the controllable range to ensure that the whole is in a stable state.

Author Contributions: Conceptualization, J.W. and S.Y.; methodology, J.W.; validation, J.W. and S.Y.; formal analysis, J.W.; investigation, Z.W.; resources, S.Y.; data curation, D.Y.; writing—original draft preparation, S.Y.; writing—review and editing, S.Y.; supervision, J.W.; project administration, S.Y.; funding acquisition, S.Y. All authors have read and agreed to the published version of the manuscript.

Funding: This research was funded by National Natural Science Foundation of China (Grant No. 52168050).

Data Availability Statement: Due to legal issues and trade secrets, the monitoring data in this article cannot be provided free of charge. Nevertheless, all concepts and data are explained in the submitted studies, and some studies are available upon request.

Conflicts of Interest: Declare The authors declare no conflict of interest.

References

1. Dmochowski, G.; Szolomicki, J. Technical and structural problems related to the interaction between a deep excavation and adjacent existing buildings. *Appl. Sci.* **2021**, *11*, 481. [CrossRef]
2. Zhandos, Y.; Orazalin, M.; Andrew, J.; Whittle, M.; Matthew, B.; Olsen, M. Three-Dimensional Analyses of Excavation Support System for the Stata Center Basement on the MIT Campus. *J. Geotech. Geoenviron. Eng.* **2015**, *141*, 05015001. [CrossRef]
3. Sebastian Bryson, M.; David, G.; Zapata-Medina, S. Method for Estimating System Stiffness for Excavation Support Walls. *J. Geotech. Geoenviron. Eng.* **2012**, *138*, 1104–1115. [CrossRef]
4. Masini, L.; Gaudio, D.; Rampello, S.; Romani, E. Observed Performance of a Deep Excavation in the Historical Center of Rome. *J. Geotech. Geoenviron. Eng.* **2021**, *147*, 05020015. [CrossRef]
5. Schwamb, T.; Soga, K. Numerical modelling of a deep circular excavation at Abbey Mills in London. *Géotechnique* **2015**, *65*, 604–619. [CrossRef]
6. Sebastian Bryson, L.; Michael, J.; Kothe mer, M. Cracking in Walls of a Building Adjacent to a Deep Excavation. *J. Perform. Constr. Facil.* **2011**, *25*, 491–503. [CrossRef]
7. Guo, P.; Gong, X.; Wang, Y.; Lin, H.; Zhao, Y. Analysis of observed performance of a deep excavation straddled by shallowly buried pressurized pipelines and underneath traversed by planned tunnels. *Tunn. Undergr. Space Technol.* **2023**, *132*, 104946. [CrossRef]
8. Guoa, P.; Gong, X.; Wang, Y. Displacement and force analyses of braced structure of deep excavation considering unsymmetrical surcharge effect. *Comput. Geotech.* **2019**, *113*, 103102. [CrossRef]
9. Zhou, Y.; Wang, H.; Zhu, Y. Construction Mechanics Behavior Analysis of Pile—Strut Supporting Structureof a Subway Deep Foundation Pit. *J. Railw. Eng. Soc.* **2019**, *1*, 86–92.
10. Ye, S.; Ding, S.; Gong, X.; Gao, S.; Chen, C. Monitoring and numerical simulation of deep foundation pit of a subway station in Lanzhou. *Chin. J. Geotech. Eng.* **2018**, *40* (Suppl. 1), 177–182.
11. Mitew-Czajewska, M. A study of displacements of structures in the vicinity of deep excavation. *Arch. Civ. Mech. Eng.* **2019**, *19*, 547–556. [CrossRef]
12. Rybak, J.; Ivannikov, A.; Kulikova, E.; Zyrek, T. Deep excavation in urban areas—Defects of surrounding buildings at various stages of construction. *MATEC Web Conf.* **2018**, *146*, 02012. [CrossRef]
13. Just, M. Analysis of a Deep Excavation in Diaphragm Walls, on Surrounding Buildings. *Arch. Inst. Civil Eng.* **2018**, 86–104. [CrossRef]
14. Castaldo, P.; Calvello, M.; Palazzo, B. Structural safety of existing buildings near deep excavations. *Int. J. Struct. Eng.* **2014**, *5*, 163–187. [CrossRef]
15. Dachowski, R.; Gałek, K. Selection of the Best Method for Underpinning Foundations Using the Promethee II Method. *Sustainability* **2020**, *12*, 5373. [CrossRef]
16. Mrowczynska, M.; Grochowska, E.; Gibowski, S. Monitoring V ertical Displacements of an Engineeriong Object with Masonry Walls. *J. Civ. Eng. Environ. Arch.* **2018**, *65*, 53–62.
17. Ye, S.; Li, D. Monitoring and simulation analysis of deep and large foundation pit excavation in complex environment. *CHINA Civ. Eng. J.* **2019**, *52*, 117–126.
18. Yin, L.; Li, Y.; Zhu, Y.; Huo, B.; Yang, Z.; Duan, Q. Monitoring and numerical simulation of support for foundation pit at Yanyuan Road Station of Lanzhou Metro. *Chin. J. Geotech. Eng.* **2021**, *43* (Suppl. 1), 111–116.
19. Ye, S.; Chen, C.; Kou, W.; Zhu, Y. Displacement monitoring and deformation characteristics of a deep foundation pit in Lanzhou. *Chin. J. Geotech. Eng.* **2014**, *36* (Suppl. 1), 440–445.
20. Liu, L.; Mo, H.; Cao, H. Analysis and Application of Circular Section Prestressed Reinforced Concrete Cantilever Retaining Piles. *J. Build. Struct.* **2005**, *35*, 42–44.
21. Wang, T.; Zhou, L.; Sun, H. Simplified calculation of asymmetrical reinforcement of circular section reinforced concrete support piles. *J. China Foreign Highw.* **2014**, *34*, 206–209.

water

MDPI

Article

Mining Leachates Effect on the Hydraulic Performance of Geosynthetic Clay Liners under Different Temperatures

Yang Liu [1,2,*], Xinxin Li [1], Yuanzhuo Tu [1] and Yulong Lu [1,2]

1 School of Earth Sciences and Spatial Information Engineering, Hunan University of Science and Technology, Xiangtan 411201, China
2 Hunan Geological Disaster Monitoring, Early Warning and Emergency Rescue Engineering Technology Research Center, Changsha 410004, China
* Correspondence: 1020143@hnust.edu.cn

Abstract: Geosynthetic clay liners (GCLs) are often used as anti-seepage systems in landfills and at the bottom of tailing ponds. The anti-seepage performance of GCL will change under different temperatures. In this study, bentonite was mixed with test solutions at different temperatures to measure the basic performance indexes of bentonite components and analyze the permeability. The composition and microstructure of bentonite at different temperatures were analyzed by X-ray diffraction, X-ray fluorescence spectrum, and SEM, and the change rule of permeability property with the mine leachates at different temperatures was understood by combining the macro-measured parameters with the microscopic analysis results. The research results indicate that the fluid loss of two bentonites increased with the increasing temperature due to the inhibition of ion exchange between bentonite and mixture by the increased temperature. The swelling index of the bentonite increased at high temperatures. The microstructure analysis showed the increase of the pore size attributed to high temperature, and the uneven distribution of the pore size resulted in the increase of the intrinsic permeability. The study would provide the reference for the application of GCL in mining.

Keywords: mining leachates; hydraulic performance; geosynthetic clay liner; different temperatures

check for updates

Citation: Liu, Y.; Li, X.; Tu, Y.; Lu, Y. Mining Leachates Effect on the Hydraulic Performance of Geosynthetic Clay Liners under Different Temperatures. *Water* **2023**, *15*, 1132. https://doi.org/10.3390/w15061132

Academic Editor: Bommanna Krishnappan

Received: 23 February 2023
Revised: 10 March 2023
Accepted: 12 March 2023
Published: 15 March 2023

1. Introduction

There are various and large nonferrous metal mines in Hunan province. During the mining process, large amounts of tailings and mineral waste residue pollute the groundwater and soil. Agentia, such as foaming agent, inhibitor, conditioning agent, and flocculant, would be added in a beneficiation process [1–6]. These mining leachates are stored in the tailing impoundment; therefore, heavy metal pollutants and chemical agents are in the tailings. The mining leachates flowing into the groundwater system and soil affect crop growth and the people's health [7–12]; therefore, the tailing impoundment should establish an anti-seepage system to prevent seepage of the leachates.

The geosynthetic clay liner (GCL) is an industrial product that has very low permeability and greater freeze–thaw cycle resistance [13–16]. Due to its easy transportation and low cost, GCL has been widely used in the anti-seepage systems of railways, roads, etc., and the cover or bottom layer for the tailings and landfills [17–20]. The low permeability of GCL is due mainly to the swelling of its bentonite component. All the tests in the study are conducted primarily on the bentonite component.

Temperature has great influence on the impermeability of GCL, and the reaction between the metallic cations in the mining leachates and the acidic or alkaline solution release heat; thus, the increased temperature may affect the hydraulic properties of GCL [20–23].

The bentonite component of GCL absorbs a lot of water from the soil at the time landfill construction, and the ambient temperature around GCL in landfills can reach 55 °C or higher, which greatly reduces the amount of water absorbed by the bentonite component [24–27]. However, at room temperature, the water content of bentonite reaches more than 100%. Therefore, the high temperature generated in the natural environment will have a negative impact on the hydraulic performance of GCL [28–30]. The decrease of water absorption increases the external pressure of bentonite, which results in cracking and seriously reduces the anti-seepage performance of GCL. Lin et al. [31] discussed the engineering performance of GCL under the temperature effect, and summarized Southen and Rowe's assessment of the cracking possibility of GCL's bentonite component under the thermal effect by using the SUMMIT model. Arden et al. [32] indicated that the increased temperature exerts a negative impact on the permeability of GCL by reducing the amount of water absorbed and leading to cracking and subsequent degradation of the hydraulic performance of the GCL [33–38]. Southen and Rowe indicated that the cracking of the bentonite component is greatly influenced by the initial water content of the underlying soil and the temperature gradient. Lower water content and higher temperature gradient result in a greater possibility of cracking [39–45]. The cracking of the bentonite component at high temperature may be one of the main factors affecting the permeability of GCL. Therefore, this study analyzes the cracking mechanism of the bentonite component and the change of GCL permeability through a series of tests on the basic properties of GCL at high temperatures.

Hanson et al. [46] studied the influence of temperature and humidity on the shear strength of double-layer nonwoven GCL and high-density polyethylene mesh geomembranes (T-GM). In direct shear tests at the interface, pressures of 10, 20, and 30 kPa were applied to the upper covered GCL, and pressures of 100, 200, and 300 kPa were applied to the bottom GCL. The temperature environments were set at 20 °C and 40 °C. It was found that the influence of temperature on the shear strength was greater than that resulting from the moisture content, and the shear strength varied with temperature by up to 54%. For the bottom liner, the shear strength was affected more by moisture content than by temperature. The degree of variation with moisture content was 43%. Jiang [47] believed that the components in landfill leachate are complex, and GCL's bentonite component has different absorption degrees of heavy metal ions at different temperatures.

Many studies on the hydraulic performance of GCL have already been conducted. He [48] found that the permeability increased with the increasing temperature, and the k increased 5.5× when the temperature increased from 20 °C to 70 °C. Shao et al. [49] studied the liquid limit and swell index of three types of clay under 5 °C–50 °C. They found that the liquid limit and swell index were increased with the increase of the temperature, which was attributed to the water absorbed amount and the absorbed types. Research by Ishimori [50] indicated that the free swell index values were higher at 60 °C than those at 20 °C in NaCl solutions. Ye et al. [51] obtained similar results by conducting the swell index and hydraulic conductivity tests on Gaomianzi bentonite under different temperatures.

In this study, parameters such as fluid loss and free swell index are determined to clarify the changing rule of the hydraulic performance of GCL under different temperatures. Based on the previous research, this study conducts the tests under room temperature (25 °C) and at 50 °C, 60 °C, and 70 °C. The composition and micro-structure of bentonite is studied by using micro-observation, including XRD, XRF, and SEM. The hydraulic performance of the GCL with different mining leachates are evaluated in this study.

2. Materials and Methods

2.1. Materials

2.1.1. Geosynthetic Clay Liners

The geosynthetic clay liners used in this study are both granular and those with nonwoven geotextile in the bottom and woven geotextile at the top. The bentonite components were peeled out, and the constituent of two bentonites are shown in Table 1.

Table 1. Basic constituents of bentonite.

		Bentonite A	Bentonite B
Mineral Constituents/%	Montmorillonite	78.1	74.1
	Quartz	7.6	5.3
	Feldspar	9.6	9.4
	Calcite	4.7	6.1
	Mica	—	5.1
Element Constituents/%	O	47.5	45.4
	Na	2.9	8.01
	Mg	1.59	1.05
	Al	8.67	8.18
	Si	29.37	28.61
	K	1.07	1.07
	Ca	3.48	2.85
	Others	5.42	4.38

2.1.2. Test Solutions

The three test solutions used in this study are deionized water and mining leachates from Xikuangshan and Baojinshan mines, which are both in Loudi City, Hunan Province (Figure 1). The main mineral resources of Xikuangshan are antimony, and mineral processing agents include mainly lead nitrate, xanthate, sulfur nitrogen, and foaming agent. Baojinshan is a typical gold mine; xanthate, black medicine, copper sulfate, and other agents are adding in the process of mineral processing. The leachates constituents are determined and shown in Tables 2 and 3.

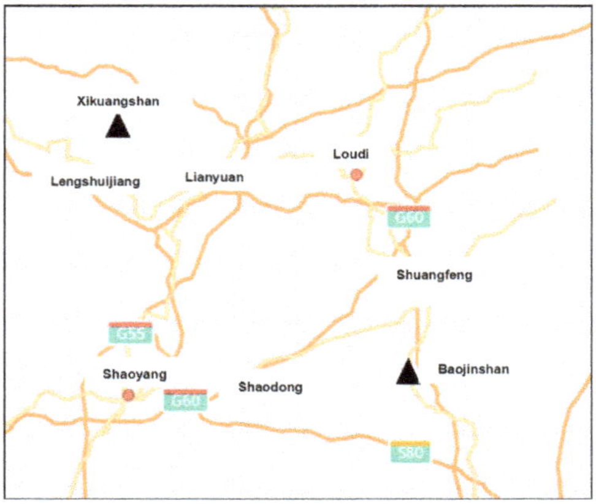

Figure 1. Location map of the study area.

Table 2. Element content table of Xikuangshan ore leachate.

Element	Al	Ca	K	Mg	Na	Pb	S	Si	Sb	Sr	W
	mg/L	mg/L	mg/L	mg/L	mg/L	mg/L	mg/L	mg/L	mg/L	mg/L	mg/L
Content	0.03	37.1	19.9	2	480	0.02	192	9	0.02	0.07	1.8

Table 3. Element content table of Baojinshan ore leachate.

Element	Al	Ca	K	Mg	Li	Na	S	Si	Sb	Sr
	mg/L	mg/L	mg/L	mg/L	mg/L	mg/L	mg/L	mg/L	mg/L	mg/L
Content	0.26	20.9	22.0	3.9	0.124	441	316	8.4	0.023	0.208

As shown in the tables, in the leachates of Xikuangshan tailings and Baojinshan tailings, the sodium contents are higher and the calcium contents are relatively lower, and the different compositions and concentrations of the two types of leachates have different influences on the impermeability of GCL.

2.2. Conducted Tests

2.2.1. Fluid Loss Test

A modified fluid loss test was conducted by initially combining the American Petroleum Institute (API) methods (API Spec 13A, 13B) with that of the American Society of Testing and Materials (ASTM standard D5891). The main differences in our approach from the ASTM method were that solutions were collected after 7.5 min, 30 min, 60 min, and 90 min, the thickness of the filter cakes was measured using a Vernier caliper, and the wet and dry masses of the filter cake were measured. In order to clarify the influence of the temperature, the suspensions were placed in an incubator and heated in a water bath for 24 h to the test temperature (50 °C, 60 °C, and 70 °C). The temperature-controlled devices were used to prevent the heat loss during the tests. The detailed procedures are as follows: The bentonite is ground to 100% through 200 mesh sieve; 22.5 g bentonite is mixed with 350 mL leachates using blender; after stirring for 5 ± 0.5 min, the mixing cup is removed from the mixer, and the bentonite adhering to the cup wall is scraped off; and the prepared suspensions are placed in a water bath incubator and heated for 24 h with the temperature setting at 50 °C, 60 °C, and 70 °C. Heating plates are used throughout the subsequent test processes to prevent heat loss. At 7.5 min, remove all adhesive liquid from the beaker and the drain, immediately place the clean, dry 100 mL cylinder under the drain line, and collect the liquids at 30.0 min, 60.0 min, and 90.0 min. The collected liquid volume, the bentonite cake thickness, and the weight of the cake are recorded.

2.2.2. Free Swell Index Test

Free swell index tests were performed according to ASTM D5890, except that de-ionized water was replaced with mining leachates. The detailed procedures are as follows: (1) The bentonite was dried in the oven at 105 °C, ground to powdered bentonite 100%, and passed through a 200 mesh sieve. (2) Weigh 2.00 ± 0.01 g dry powdered bentonite on a weighing paper. (3) Add 90 mL de-ionized water to a clean 100 mL measuring cylinder. (4) Take up no more than 0.1 g of powdered bentonite from the weighing paper and carefully sprinkle it over the entire water surface in the cylinder within approximately 30 s. Let the powdered bentonite moisten and settle to the bottom of the cylinder. Repeat Step 4 until the entire 2.00 g sample has been added. (5) Rinse all the attached particles from the side of the cylinder into the cylinder and pour the test liquid to 100 mL level. (7) After standing for 16 h, the observed volume of hydrated bentonite was recorded. All the free swell index tests were conducted under the setting temperature, and the temperature-controlled wrap was used during the tests to ensure that the test was performed at the setting temperature.

2.2.3. X-ray Observation

X-ray diffraction (XRD) and X-ray fluorescence spectrum (XRF) were performed for the bentonite powder in order to obtain quantitative analysis, and samples of the bentonites after fluid loss test were sent to Changsha Institute of Mining and Metallurgy for quantitative mineralogical analysis. The typical mineralogical methods used were as follows: the amount of 1.5 g of oven-dried material was milled (<10 micron). The samples were then mixed thoroughly before lightly back-pressing into stainless steel sample holders for XRD analysis. XRD patterns were collected on a PANalytical X' Pert Pro Multi-purpose diffractometer using Fe-filtered Co K radiation (λ = 1.78897 Å), variable divergence slit, 1 °C anti-scatter slit, and fast X' Celerator Si strip detector. The diffraction patterns were recorded in 0.017° 2θ steps with a 0.5 s counting time per step and logged to data files for analysis.

2.2.4. Scanning Electron Microscope

Scanning electron microscope images were obtained on a Hitachi SU3500 SEM equipment at Hunan University of Science and Technology. Powdered specimens of bentonites (unreacted or reacted with mining leachates at different temperatures) were air-dried to constant weight and then were affixed onto a stainless steel sub with carbon tape. The prepared specimens were kept in a desiccator before testing to avoid moisture absorption. Platinum coating was conducted on the specimens to improve image quality. The detailed procedures are as follows: (1) Bentonite sample preparation: the filter cake obtained after the fluid loss test is put into the oven at 105 °C for drying, and the dry cake is taken out for future use. (2) Black conductive tape with a length smaller than the sample stage was cut off and pasted on the sample stage. The conductive tape was smoothed with tweezers to make the adhesive tape adhere more firmly. After smoothing, the protective film on the other side of the tape was torn off with tweezers. The sample was taken out and fixed on the conductive adhesive surface with tweezers and slightly pressed to make it stick well to the conductive adhesive. After the sample was placed, a WD-40 powerful dust removal tank was used to dust the sample surface. (3) The sample stage was placed into the holder in the electron microscope. Since multiple samples were placed onto the stage, the numbers were marked in a certain sequence, and photos were taken to prevent sample confusion during observation. (4) The representative characteristic areas of each sample were selected, and the bentonite micro-structure was observed in accordance with the magnifications of 200×, 500×, 1000×, 2000×, and 5000×.

3. Results and Discussion

3.1. Fluid Loss Test

The fluid loss changes of bentonite under different test solutions at various temperatures are shown in Figure 2. As shown, the fluid loss of both bentonites increased with the gradual rise of temperature. This could be attributed to the inhibition of the ion exchange between bentonite and the mining leachates since ionic adsorption of bentonite is an exothermic reaction. The increased temperature resulted in the reduction of the number of absorbed water molecular layers, which indicates the degradation of the anti-seepage performance of GCL. At 70 °C, the leachate volume of bentonite B with Xikuangshan mine tailings reached 196.6 mL after 90 min. This indicates that the high temperature may even lead to the failure of the anti-seepage system and greatly degrade the anti-seepage performance. Moreover, the fluid loss values of bentonite in tailing leachate during the increase from room temperature to 50 °C was greater than those from 60 °C to 70 °C. This indicates that the temperature gradient also exerts an influence on the permeability of bentonite, and the increase of the temperature gradient resulted in the increase in the amplitude of change for the permeability.

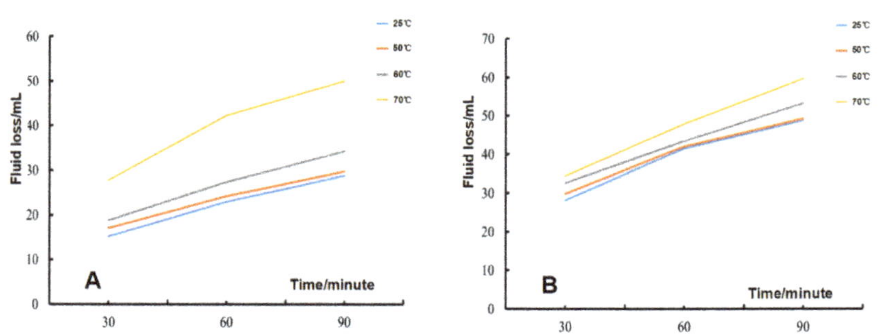

(a)

Figure 2. *Cont.*

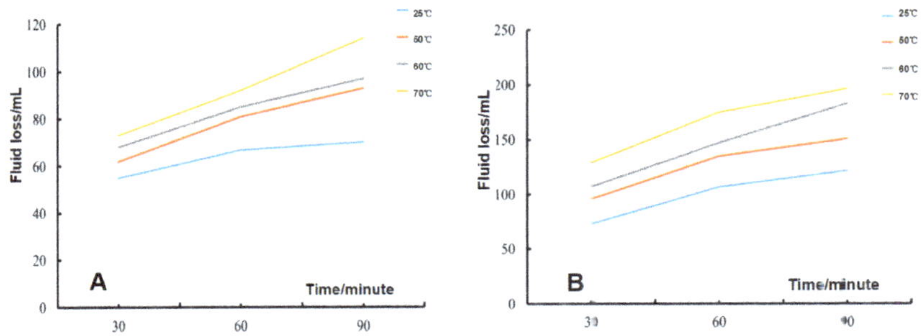

(**b**)

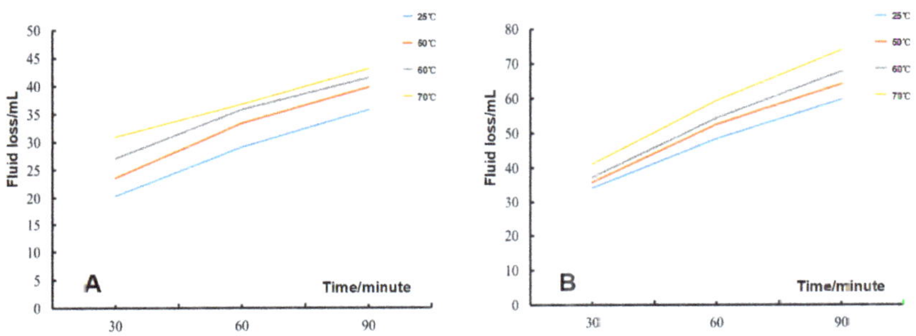

(**c**)

Figure 2. Fluid loss of bentonite with different test solutions at various temperatures ((**a**–**c**) are bentonite with deionized water, Xikuangshan, and Baojinshan, respectively; A, B refer to bentonite A and bentonite B, respectively).

3.2. Free Swell Index Test

The results of free swell index tests under different temperatures with various mining leachates are shown in Figure 3. Generally, the free swell index of bentonite increases slightly with the increase of temperature. For both bentonite A and B, the swell index values are greatest in Baojinshan mining leachates. The species and content of cations in the leachate of Baojinshan tailing mine make the ion exchange reaction of bentonite more adequate than other leachates, and the high temperature improves the ion transport and the bentonite dispersion.

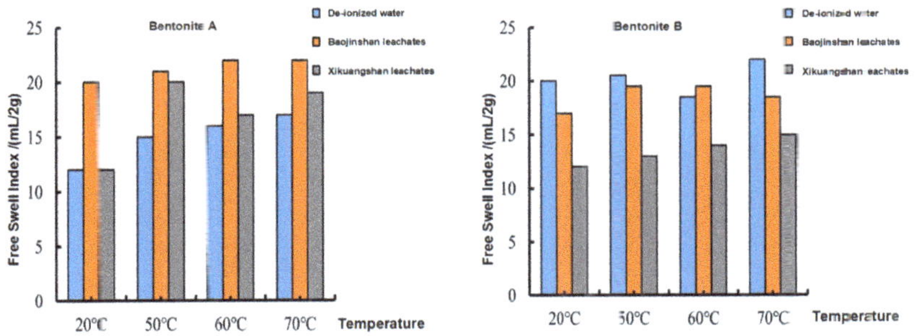

Figure 3. Free swell index of bentonite A and B at different temperatures.

3.3. X-ray Diffraction and X-ray Fluorescence Spectrum

MDI Jade 6.5 was applied to analyze the XRD spectrum; as shown in Figure 4, the diffraction angles of each peak of bentonite basically remain unchanged at different temperatures, which indicates that the increased temperature did not result in the change of the bentonite mineral composition. XRF shows an increased content of sodium but a decreased content of calcium in the leachates at high temperatures (as see in Table 4). The increased temperature speeds up the ion exchange reaction between the bentonite and the mining leachates. The loss of sodium in the bentonite structure resulted in fewer absorbed water layers microscopically and, therefore, increased the bentonite permeability.

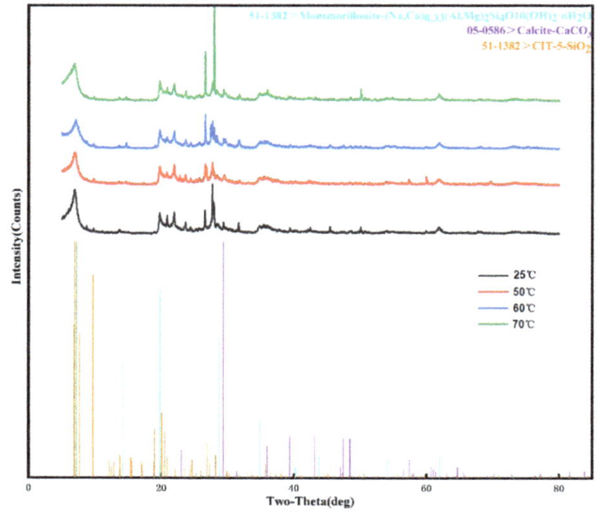

(**a**)

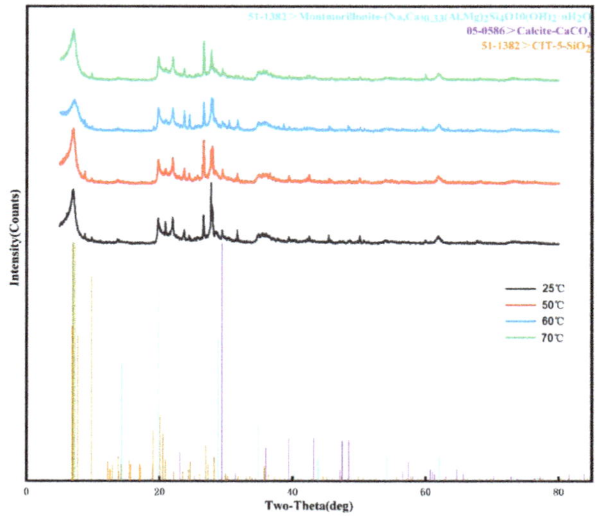

(**b**)

Figure 4. XRD spectrum of bentonite at different temperatures ((**a**) bentonite B+Xikuangshan mining leachates and (**b**) bentonite B+Baojinshan mining leachates).

Table 4. X-ray fluorescence element content of Bentonite in Baojinshan leachate.

Sample	Element							
	O	Na	Mg	Al	Si	P	S	Cl
A	47.200	2.650	1.490	8.952	28.050	0.114	0.055	1.227
A + 50 °C	46.100	2.800	1.500	7.999	27.230	0.089	1.479	1.290
A + 70 °C	47.500	3.510	1.580	8.163	27.490	0.099	0.715	1.200
	K	Ca	Ti	Cr	Mn	Fe	Ni	Cu
A	2.060	4.151	0.445	0.005	0.064	3.409	-	0.007
A + 50 °C	1.050	3.838	0.435	-	0.087	3.103	-	-
A + 70 °C	0.958	3.709	0.447	-	0.077	3.156	-	0.003
	Zn	Ga	Rb	Sr	Zr	Nb	Ba	Others
A	0.010	-	0.005	0.011	0.009	-	0.073	0.010
A + 50 °C	0.009	-	0.004	0.027	0.016	-	0.065	2.154
A + 70 °C	0.009	-	0.004	0.029	0.014	-	0.069	1.962
	O	Na	Mg	Al	Si	P	S	Cl
B	42.200	5.460	1.350	8.877	28.960	0.087	0.398	0.886
B + 50 °C	47.900	3.380	1.500	8.240	28.260	0.101	0.551	1.280
B + 70 °C	47.400	3.830	1.480	8.079	30.150	0.091	0.362	1.240
	K	Ca	Ti	Cr	Mn	Fe	Ni	Cu
B	1.060	3.151	0.445	0.005	0.064	3.409	-	0.007
B + 50 °C	1.239	2.918	0.454	-	0.074	3.147	-	-
B + 70 °C	1.100	2.890	0.436	-	0.073	3.131	-	-
	Zn	Ga	Rb	Sr	Zr	Nb	Ba	Others
B	0.010	-	0.005	0.011	0.009	-	0.073	0.013
B + 50 °C	0.009	-	0.004	0.028	0.012	-	0.075	0.390
B + 70 °C	0.009	-	0.004	0.028	0.012	-	0.074	0.021

3.4. Scanning Electron Microscope

As shown by the SEM images in Figure 5, the bentonite surfaces were coarser and denser at high temperature; this is because the rise of temperature resulted in the bentonite crack, thereby making the particles look uneven and rough [52–56]. According to the microscopic observation on bentonite SEM images under high temperature environment, it can be found that the porosity of particles tended to increase, while the surface fractal dimension slightly increased with the increase of temperature, indicating that the surface structure and morphology of bentonite changed greatly at high temperatures. According to the microscopic data of bentonite in a high temperature environment, it can be found that the porosity of particles tended to increase with the increase of temperature. The increased porosity resulted in higher permeability. This may be because the ionic exchange reaction occurred after bentonite was mixed with the mining leachates, which is an exothermic reaction. The increased temperature inhibited the reaction of ion exchange to some extent. The cationic adsorption capacity of bentonite was weakened, the mutual attraction between each particles was affected, and the distance of double layer decreased. All these changes may degrade the hydraulic performance, which is consistent with the results of fluid loss. The crack could be observed at high temperatures, and the space between particles increased; therefore, the free swelling of bentonite increased as well, which is consistent with the free swell index test results obtained in this study.

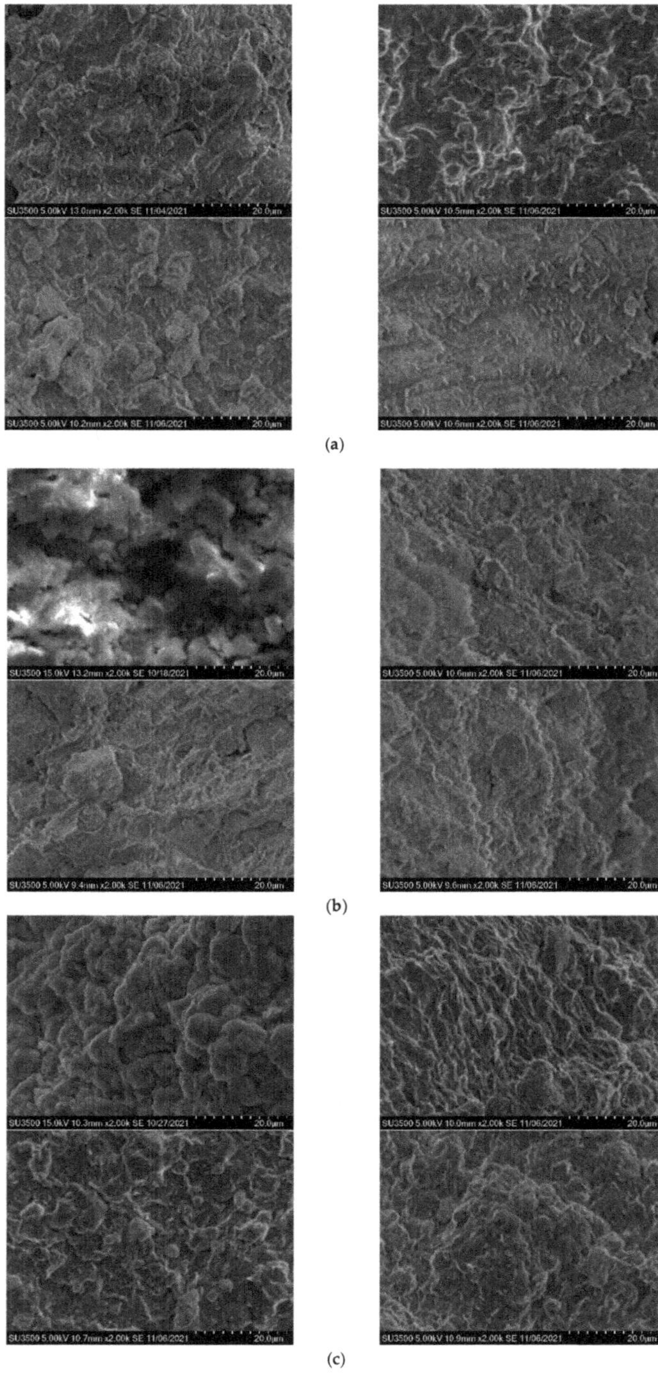

Figure 5. SEM images of bentonite A with test solutions at various temperature. ((**a–c**) are Bentonite A with deionized water, Baojinshan mining leachates, and Xikuangshan mining leachates, respectively, at room temperature, 50 °C, 60 °C, and 70 °C).

3.5. Hydraulic Conductivity Analysis

One of the important indexes to test the impermeability of bentonite is its permeability coefficient, which is estimated by Equations (1)–(5) through the bentonite fluid loss tests under different temperatures.

As reported by Rushton et al. (2000) [57], materials under a constant applied stress condition (i.e., 690 kPa in this study) should yield a filter cake of roughly constant density. Therefore, the filter cake formed during the 30 min test is expected to be incompressible with a constant void ratio (cake density) within the time frame. As such, the filter cake volume is expected to linearly increase for each unit volume of suspension filtered, and the constant of proportionality, β, can then be used to give an equation for cake thickness at any time (Chung and Daniel, 2008) [58]:

$$L = \frac{\beta V}{A} \tag{1}$$

where L is the thickness of the filter cake (m), V is the filter cake volume (m^3), and A is the filter cake area (m^2).

Chung and Daniel (2008) calculated the β value with the method of a mass balance on the solid and liquid in the filter system:

$$\beta = \frac{C_m \rho_w}{(1 - C_m)(1 - n_c)\rho_s - C_m n_c \rho_w} \tag{2}$$

where C_n is the mass fraction of solids in the suspension, ρ_w is liquid density, n_c is the porosity of the filter cake, and ρ_s is the particle density of solid.

The porosity (n_c) of the filter cake can be calculated as:

$$e = \frac{\rho_s}{\rho_d}, \qquad n_c = \frac{1+e}{e} \tag{3}$$

where e is the void ratio of filter cake, ρ_s is particle density of clay (assumed to be 2.7 g/cm^3 for the bentonites used in this study), and ρ_d is the dry density of clay calculated from the ratio of dry mass and volume of filter cake.

The filtration model derived by Rushton et al. (2000) from Darcy's equation is:

$$\frac{dV_f}{dt} = \frac{k_c}{\gamma_w}\frac{P_0}{L}A \tag{4}$$

where V_f is the filtrate volume (m^3), t is the filtration time (s), k_c is the hydraulic conductivity of filter cake (m/s), γ_w is the unit weight of liquids (kN/m^3), P_0 is the applied stress (kPa), and L and A are defined as above.

Therefore, Chung and Daniel (2008) put forward an equation to calculate the hydraulic conductivity (k_c) of the filter cake as:

$$k_c = \frac{\beta\gamma_\omega}{2A^2}\frac{{V_f}^2}{P_0 t} \tag{5}$$

To assess the validity of Equation (5), Chung and Daniel plotted the data of $P_0 t/V_f$ against V_f to check for linearity. In the present study, the viscosity (η) changed with the temperature; therefore, the equation is shown as:

$$k_c = \frac{\beta\gamma_\omega}{2A^2}\frac{{V_f}^2}{P_0 t\eta} \tag{6}$$

The parameters including filtrate volume, cake thickness, and cake quality were recorded at different times. The permeability coefficients of bentonite at various temperatures are calculated according to Equation (5) as shown in Table 5.

Table 5. Calculation value of permeability coefficient of bentonite at high temperature.

Bentonite	A				B			
Temperature(°C)	0	50	60	70	0	50	60	70
De-ionized water	6.96×10^{-9}	8.12×10^{-9}	1.22×10^{-8}	3.24×10^{-8}	2.04×10^{-8}	4.63×10^{-8}	6.11×10^{-8}	8.56×10^{-8}
Baojinshan leachates	2.62×10^{-8}	4.46×10^{-8}	7.15×10^{-8}	9.35×10^{-8}	4.37×10^{-8}	5.12×10^{-8}	7.26×10^{-8}	9.44×10^{-8}
Xikuangshan leachates	9.18×10^{-9}	6.04×10^{-7}	8.81×10^{-7}	2.12×10^{-6}	4.14×10^{-7}	4.78×10^{-7}	6.88×10^{-7}	8.66×10^{-7}

As shown in Table 5, the permeability coefficient of bentonite gradually increases with the rise of temperature. The increase of temperature inhibited the ion exchange reaction of bentonite, resulting in the weakening of the cationic adsorption capacity of bentonite in the test solutions. The particles dispersed and cracked, which was consistent with the microscopic observation results. With the increase of permeability coefficient, the seepage resistance of bentonite decreased, resulting in a gradual increase in the volume of filtered liquid during the fluid loss test.

Katsumi et al. [59] showed that there is a strong correlation between bentonite permeability coefficient (k) and free swell index (SI) in different test solutions (Figure 6a). The correlation between the SI values of bentonite and calculated k_c at different temperatures in this study is shown in Figure 6b. As can be seen from the figure, the calculated k values are consistent with those values from Katsumi et al. under different temperatures. The calculated k_c is slightly greater than the k from Katsumi, which may be because the samples used in the Katsumi test are GCLs with the test pressure at 20–35.4 kPa, while the samples in the fluid loss tests in this study are bentonite with the filtration pressure at 690 kPa; thus, the calculated k value is greater than the measured k value from Katsumi et al.

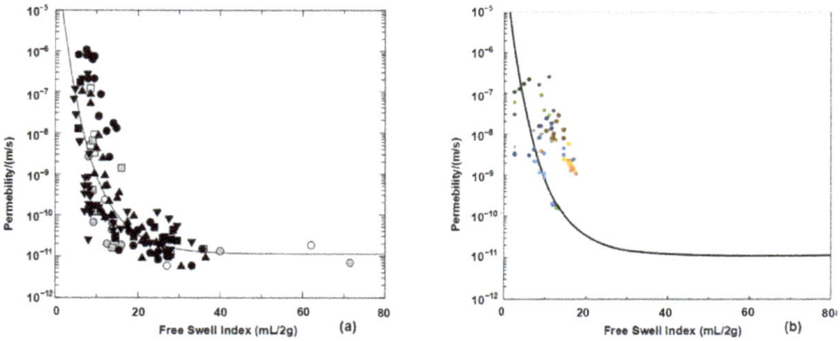

Figure 6. Relationship between the free swell index and permeability ((**a**) from Katusmi and (**b**) from the data in this study).

The changes of bentonite permeability are complicated at different temperatures; thus, the bentonite intrinsic permeability was studied by Equation (7) in order to further investigate the change mechanism of bentonite permeability under different temperatures. Generally, the permeability coefficient can be expressed as:

$$k = K\frac{\rho g}{\mu} = K\frac{g}{\eta} \quad (7)$$

where k is permeability coefficient, m/s; K is the intrinsic permeability, m^2; ρ is fluid density, kg/m^3; μ is fluid viscosity, Pa·s; η is the dynamic viscosity coefficient, m^2/s; and g is the acceleration of gravity, 9.81 m/s^2. The density of fluids at different temperatures was determined by reference [32].

By calculating the natural permeability, the relationship between the natural permeability at different temperatures was obtained (see Figure 7). Intrinsic permeability is a parameter used to characterize the fluid conductivity of soil itself, and the values are related to particle size, interface physical and chemical properties, pore characteristics, etc., but have no relationship with the properties of the testing leachates. It can be seen from Figure 6 that the intrinsic permeability is not a constant, and it changes with the change of temperatures. When the temperature gradually increases, the intrinsic permeability increases with the increase of temperature. This may also be attributed to the bentonite cracks with the increase of temperature, which results in the uneven distribution of the pore size, finally leading to the increase of the intrinsic permeability.

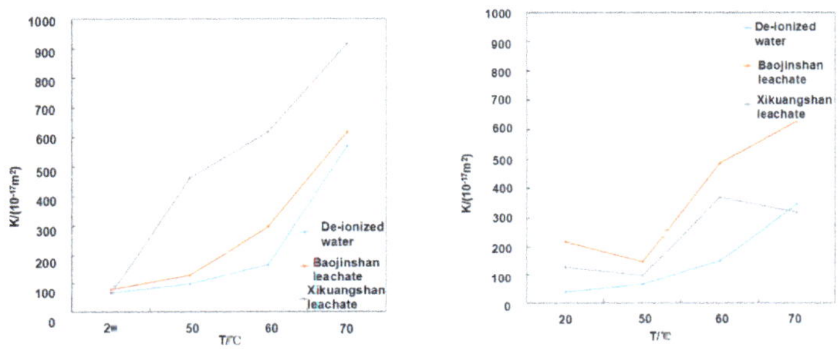

Figure 7. Intrinsic permeability of two bentonites ((**a**) for bentonite A and (**b**) for bentonite B) at different temperatures

Through the calculation of the permeability coefficient and the study on the intrinsic permeability, it is found that the permeability coefficient and the intrinsic permeability have the same change trend under different temperatures. Both of them gradually increase as the temperature rises. This indicates that the permeability coefficient and the intrinsic permeability have superimposed effects on the impermeability of bentonite, rather than being in conflict with each other.

4. Conclusions

In this study, three types of mine tailing leachate were used by setting different temperature environments. The change rule of bentonite permeability was analyzed by measuring the basic properties of bentonite components. X-ray diffraction, X-ray fluorescence spectroscopy, and SEM were used to analyze the composition of bentonite at different temperatures and observe the micro-structure changes. The fluid loss of both bentonite gradually increased with the increase of temperature because the increase in temperature inhibited the ion exchange between bentonite and the filtrate. In the free swell index test, the cracking of bentonite resulting from the high temperature also provided more swelling space. X-ray diffraction and fluorescence spectrum analysis of bentonite found that the main composition of bentonite did not change with the change of temperature. XRF showed an increased content of sodium but a decreased content of calcium in the leachates at high temperatures. The increased temperatures sped up the ion exchange reaction between the bentonite and the mining leachates, which influenced the bentonite swelling and the hydraulic performance of GCL. SEM images showed that the surface of bentonite became denser and coarser with the increase of temperature during to the cracking.

The permeability coefficient and intrinsic permeability of bentonite increased gradually with the increase of temperature because the increase of temperature inhibited the ion exchange reaction, leading to the particle dispersion and cracking and thus resulting in the increase. Through the calculation of the permeability coefficient and the study on the intrinsic permeability, it was found that the influence of the permeability coefficient and the intrinsic permeability on the impervious performance of bentonite was concurrent.

Author Contributions: Software, X.L.; Writing—original draft, Y.T.; Writing—review & editing, Y.L. (Yang Liu); Project administration, Y.L. (Yang Liu); Methodology, Y.L. (Yulong Lu). All authors have read and agreed to the published version of the manuscript.

Funding: This work was supported by the Natural Science Foundation of Hunan Province (Grant No. 2022JJ30244) and the Research Project of Teaching Reform of Hunan Province (Grant No. HNJG-2022-0790).

Data Availability Statement: No new data were created or analyzed in this study. Data sharing is not applicable to this article.

Conflicts of Interest: The authors declare no conflict of interest.

References

1. Komnitsas, K.; Kontopoulos, A.; Lazar, I. Risk assessment and proposed remedial actions in coastal tailings disposal sites in Romania. *Miner. Eng.* **1998**, *11*, 1179–1190. [CrossRef]
2. Li, Y.C.; Xu, Z.; Ma, H.; Hursthouse, A.S. Removal of Manganese(II) from Acid Mine Wastewater: A Review of the Challenges and Opportunities with Special Emphasis on Mn-Oxidizing Bacteria and Microalgae. *Water* **2019**, *11*, 2493. [CrossRef]
3. Meng, D.; Li, J.; Liu, T.; Liu, Y.; Yan, M.; Hu, J.; Li, H.; Liu, X.; Liang, Y.; Liu, H.; et al. Effects of redox potential on soil cadmium solubility:Insight into microbial community. *J. Environ. Sci.* **2019**, *75*, 224–232. [CrossRef]
4. Deng, R.J.; Jin, C.S.; Ren, B.Z.; Hou, B.L.; Hursthouse, A.S. The Potential for the Treatment of Antimony-Containing Wastewater by Iron-Based Adsorbents. *Water* **2017**, *9*, 794. [CrossRef]
5. Wang, Z.; Liao, L.; Hursthouse, A.; Song, N.; Ren, B. Sepiolite-based adsorbents for the removal of potentially toxic elements from water: A strategic review for the case of environmental contamination in Hunan, China. *Int. J. Environ. Res. Public Health* **2018**, *15*, 1653. [CrossRef]
6. Liu, Y.; Bouazza, A.; Gates, W.P.; Rowe, R.K. Hydraulic performance of geosynthetic clay liners to sulfuric acid solutions. *Geotext. Geomembr.* **2015**, *43*, 14–23. [CrossRef]
7. Rowe, R.K.; Mukunoki, T.; Bathurst, R.J. Hydraulic conductivity to Jet-A1 of GCLs after up to 100 freeze–thaw cycles. *Géotechnique* **2008**, *58*, 503–511. [CrossRef]
8. Zhao, Y.L.; Zhang, C.S.; Wang, Y.X.; Lin, H. Shear-related roughness classification and strength model of natural rock joint based on fuzzy comprehensive evaluation. *Int. J. Rock Mech. Min. Sci.* **2021**, *137*, 104550. [CrossRef]
9. Xie, Q.; Ren, B.Z.; Hursthouse, A.S.; Shi, X.Y. Effects of mining activities on the distribution, controlling factors, and sources of metals in soils from the Xikuangshan South Mine, Hunan Province. *Integr. Environ. Assess. Manag.* **2022**, *18*, 748–756. [CrossRef]
10. Liu, Y.; He, B.; Xie, J.; Lu, Y.; Zhang, L. Compatibility of geosynthetic clay liners at different temperatures. *J. Environ. Prot. Ecol.* **2021**, *22*, 2295–2306.
11. Feng, H.; Liu, F.; Luo, P.; Xie, G.; Xiao, R.; Hu, W.; Peng, J.; Wu, J. Performance of integrated ecological treatment system for decentralized rural wastewater and significance of plant harvest management. *Ecol. Eng.* **2018**, *124*, 69–76.
12. Zhao, Y.L.; Zhang, L.Y.; Liao, J.; Wang, W.J.; Liu, Q.; Tang, L.M. Experimental study of fracture toughness and subcritical crack growth of three rocks under different environments. *Int. J. Geomech.* **2020**, *20*, 04020128. [CrossRef]
13. Zhao, Y.L.; Tang, J.Z.; Chen, Y.; Zhang, L.Y.; Wang, W.J.; Liao, J.P. Hydromechanical coupling tests for mechanical and permeability characteristics of fractured limestone in complete stress-strain process. *Environ. Earth Sci.* **2017**, *76*, 24. [CrossRef]
14. Tang, L.; Tang, X.W.; Liu, Y.; Qu, S.X. Prediction of pore size characteristics of woven slit-film geotextiles subjected to unequal biaxial tensile strains. *Geotext. Geomembr.* **2020**, *48*, 724–734. [CrossRef]
15. Wang, T.; Huang, C.; Du, G.; Liu, Y.; Xie, J.; Li, H. Geochronology, geochemistry and zircon Hf-isotopes of the early Mesoproterozoic Yaopengzi dolerite in SW Yangtze block (Sichuan, SW China): Implications for the Columbia supercontinent breakup. *Geosci. J.* **2019**, *23*, 557–573. [CrossRef]
16. Zhang, J.; Deng, R.; Ren, B.; Hou, B.; Hursthouse, A. Preparation of a novel Fe_3O_4/HCO composite adsorbent and the mechanism for the removal of antimony (III) from aqueous solution. *Sci. Rep.* **2019**, *9*, 13021. [CrossRef] [PubMed]
17. Tang, Z.; Deng, R.; Zhang, J.; Ren, B.; Hursthouse, A. Regional distribution characteristics and ecological risk assessment of heavy metal pollution of different land use in an antimony mining area-Xikuangshan, China. *Hum. Ecol. Risk Assess.* **2020**, *26*, 1779–1794. [CrossRef]
18. Liu, Y.; Hao, Y.; Lu, Y. Improved design of risk assessment model for PPP project under the development of marine architecture. *J. Coast. Res.* **2018**, *83*, 74–80. [CrossRef]
19. Zhu, G.; Wang, C.; Dong, X. Fluorescence excitation-emission matrix spectroscopy analysis of landfill leachate DOM in coagulation-flocculation process. *Environ. Technol.* **2017**, *9*, 1489–1497. [CrossRef]
20. Zhou, L.; Liu, X.; Li, J.; Liao, J. Robust AVO inversion for the fluid factor and shear modulus. *Geophysics* **2021**, *86*, 471–483. [CrossRef]
21. Liu, Y.; Gates, W.P.; Bouazza, A. Acid induced degradation of the bentonite component used in geosynthetic clay liners. *Geotext. Geomembr.* **2013**, *36*, 71–80. [CrossRef]
22. Li, Y.; Hu, X.; Ren, B. Treatment of antimony mine drainage: Challenges and opportunities with special emphasis on mineral adsorption and sulfate reducing bacteria. *Water Sci. Technol.* **2016**, *73*, 2039–2051. [CrossRef] [PubMed]

23. Luo, X.; Ren, B.; Hursthouse, A.S.; Thacker, R.M.; Wang, Z. Soil from an Abandoned Manganese Mining Area (Hunan, China): Significance of Health Risk from Potentially Toxic Element Pollution and Its Spatial Context. *Int. J. Environ. Res. Public Health* **2020**, *17*, 6554. [CrossRef]
24. Zhou, Y.; Ren, B.; Hursthouse, A.S.; Zhou, S. Antimony Ore Tailings: Heavy Metals, Chemical Speciation, and Leaching Characteristics. *Pol. J. Environ. Stud.* **2019**, *28*, 485–495. [CrossRef]
25. Liu, Y.; Gates, W.P.; Bouazza, A. Impact of acid leachates on microtexture of bentonites used in geosynthetic clay liners. *Geosynth. Int.* **2019**, *26*, 136–145. [CrossRef]
26. Lu, Y.; Cao, C.; Liu, Y.; Liu, Y. Study on Application of Comprehensive Geophysical Prospecting Method in Urban Geological Survey—Taking Concealed Bedrock Detection as an Example in Dingcheng District, Changde City, Hunan Province, China. *Appl. Sci.* **2023**, *13*, 417. [CrossRef]
27. Song, N.; Hursthouse, A.; McLellan, I.; Wang, Z. Treatment of environmental contamination using sepiolite: Current approaches and future potential. *Environ. Geochem. Health* **2021**, *43*, 2679–2697. [CrossRef]
28. He, Z.; Ren, B.; Hursthouse, A.; Wang, Z. Efficient removal of Cd(II) using SiO_2-$Mg(OH)_2$ nanocomposites derived from sepiolite. *Int. J. Environ. Res. Public Health* **2020**, *17*, 2223. [CrossRef]
29. Ling, J.; Dai, S.; Zhou, Y.; Chen, Q.; Zhang, Y.; Li, K. Three-Dimensional DC Anisotropic Resistivity Modeling Using a Method in the Mixed Space-Wavenumber Domain. *Pure Appl. Geophys.* **2022**, *179*, 2183–2200. [CrossRef]
30. Song, K.; Ren, X.; Mohanmod, A.; Liu, J.; Wang, F. Research on drinking groundwater source safety management based on numerical simulation. *Sci. Rep.* **2020**, *10*, 15481. [CrossRef] [PubMed]
31. Lin, B.; Zhang, C.; Feng, Y. Research progress on temperature effect of impervious clay liners in sanitary landfills. In Proceedings of the Second National Geotechnical and Engineering, Taipei, Taiwan, 28 October–1 November 2006; Chinese Society of Rock Mechanics and Engineering: Wuhan, China, 2006; Volume 2, p. 8.
32. Arden, B.; Mohammad, C. Effect of temperature on hydration of geosynthetic clay liners in landfills. *Waste Manag. Res.* **2013**, *22*, 58–71.
33. Zhao, Y.L.; Luo, S.L.; Wang, Y.X.; Wang, W.J.; Zhang, L.Y.; Wan, W. Numerical analysis of karst water inrush and a criterion for establishing the width of water-resistant rock pillars. *Mine Water Environ.* **2017**, *36*, 508–519. [CrossRef]
34. Yuan, Q.; Zhu, G. A review on metal organic frameworks (MOFs) modified membrane for remediation of water pollution. *Environ. Eng. Res.* **2021**, *26*, 190435.1–190435.2. [CrossRef]
35. Deng, R.J.; Shao, R.; Ren, B.Z.; Hou, B.L.; Tang, Z.E.; Hursthouse, A. Adsorption of Antimony(III) onto Fe(III)-Treated Humus Sludge Adsorbent: Behavior and Mechanism Insights. *Pol. J. Environ. Stud.* **2019**, *28*, 577–586. [CrossRef] [PubMed]
36. Yang, X.; Liu, Y.; Yang, C. Research on the slurry for long-distance large-diameter pipe jacking in expansive soil. *Adv. Civ. Eng.* **2018**, *9*, 47–54. [CrossRef]
37. Zhao, Y.L.; Liu, Q.; Zhang, C.; Liao, J.; Lin, H.; Wang, Y. Coupled seepage-damage effect in fractured rock masses: Model development and a case study. *Int. J. Rock Mech. Min. Sci.* **2021**, *144*, 104822. [CrossRef]
38. Zhang, Y.; Ren, B.; Hursthouse, A.; Deng, R.; Hou, B. Study on the migration rules of Sb in antimony ore soil based on HYDRUS-1D. *Pol. J. Environ. Stud.* **2018**, *28*, 965–972. [CrossRef]
39. Zhang, Y.; Ren, B.; Hursthouse, A.; Deng, R.; Hou, B. An Improved SWAT for predicting manganese pollution load at the soil-water interface in a manganese mine area. *Pol. J. Environ. Stud.* **2018**, *27*, 2357–2365. [CrossRef]
40. Xie, J.; Liu, Y.; Lu, Y.; Zhang, L. Application of the high-density resistivity method in detecting a mined-out area of a quarry in Xiangtan City, Hunan Province. *Front. Environ. Sci.* **2022**, *10*, 1068956.
41. Hou, B.L.; Liu, X.; Li, Z.; Ren, B.Z.; Kuang, Y. Heterogeneous fenton oxidation of butyl xanthate catalyzed by iron-loaded sewage sludge. *Fresenius Environ. Bull.* **2022**, *31*, 4125–4131.
42. Zhao, Y.L.; Zhang, L.; Wang, W.; Tang, J.; Lin, H.; Wan, W. Transient pulse test and morphological analysis of single rock fractures. *Int. J. Rock Mech. Min. Sci.* **2017**, *91*, 139–154. [CrossRef]
43. Hou, B.; Li, Z.; Deng, R.; Ren, B. Advanced treatment of coal chemical industry wastewater by expansive flow biological aerated filter. *Fresenius Environ. Bull.* **2017**, *26*, 4517–4521.
44. Shi, X.; Ren, B.; Hursthouse, A. Source identification and groundwater health risk assessment of PTEs in the stormwater runoff in an abandoned mining area. *Environ. Geochem. Health* **2022**, *44*, 3555–3570. [CrossRef] [PubMed]
45. Zhang, Y.; Huang, F. Indicative significance of the magnetic susceptibility of substrate sludge to heavy metal pollution of urban lakes. *Sci. Asia* **2021**, *47*, 374. [CrossRef]
46. Hanson, J.L.; Chrysovergis, T.S.; Yesiller, N. Temperature and moisture effects on GCL and textured geomembrane interface shear strength. *Geotextile Geomembr.* **2015**, *22*, 110–124. [CrossRef]
47. Jiang, J. *Experimental Study on Application Characteristics of GCL under Complex Conditions*; Wenzhou University: Wenzhou, China, 2017.
48. He, J.: Hu, X.; Yan, X.; Wan, J. Temperature effect test on clay permeability. *Adv. Sci. Technol. Water Resour.* **2017**, *37*, 55–60.
49. Shao, Y.; Shi, B.; Liu, C.; Gu, K.; Tang, C. Study on temperature effect of hydrophysical properties of clay. *J. Geotech. Eng.* **2011**, *33*, 1576–1582.
50. Ishimori, H.; Katsumi, T. Temperature effects on the swelling capacity and barrier performance of geosynthetic clay liners permeated with sodium chloride solutions. *Geotext. Geomembr.* **2012**, *33*, 25–33. [CrossRef]
51. Ye, W.M.; Wan, M.; Chen, B.; Chen, Y.; Cui, J. Temperature effects on the unsaturated permeability of the densely compacted GMZ01 bentonite under confined conditions. *Eng. Geol.* **2012**, *126*, 1–7. [CrossRef]

52. Tang, C.; Shi, B. Swelling and shrinkage deformation characteristics of expansive soil during the dry-wet cycle. *Chin. J. Geotech. Eng.* **2011**, *33*, 1376–1384.
53. Xie, Q.; Ren, B. Pollution and risk assessment of heavy metals in rivers in the antimony capital of Xikuangshan. *Sci. Rep.* **2022**, *12*, 14393. [CrossRef] [PubMed]
54. Zhou, S.; Hursthouse, A. The impact of physical properties on the leaching of potentially toxic elements from antimony ore processing wastes. *Int. J. Environ. Res. Public Health* **2019**, *16*, 2355. [CrossRef] [PubMed]
55. Song, K.; Yang, G.; Wang, F.; Liu, J.; Liu, D. Application of Geophysical and Hydrogeochemical Methods to the Protection of Drinking Groundwater in Karst Regions. *Int. J. Environ. Res. Public Health* **2020**, *17*, 3627. [CrossRef] [PubMed]
56. Lu, Y.; Yang, T.; Abdollah, T.; Liu, Y. Fast Recognition on Shallow Groundwater and Anomaly Analysis Using Frequency Selection Sounding Method. *Water* **2023**, *15*, 96.
57. Rushton, A.; Ward, A.S.; Holdich, R.G. *Solid-Liquid Filtration and Separation Technology*, 2nd ed.; Wiley-VCH Verlag GmbH: Weinbeim, Germany, 2000.
58. Chung, J.; Daniel, D.E. Modified fluid loss test as an improved measure of hydraulic conductivity for Bentonite. *Geotech. Test. J.* **2008**, *31*, 243–251.
59. Katsumi, T.; Ishimori, H.; Onikata, M.; Fukagawa, R. Long-term barrier performance of modified bentonite materials against sodium and calcium permeant solutions. *Geotext. Geomembr.* **2008**, *26*, 14–30. [CrossRef]

water

MDPI

Article

Failure Characteristics of the Water-Resisting Coal Pillar under Stress-Seepage Coupling and Determination of Reasonable Coal Pillar Width

Quanhui Liu [1,2], Yuanbo Xue [2], Dan Ma [2,3,*] and Qiang Li [2]

1 Kekegai Coal Mine Shaanxi Yanchang Petr Yulin Coal Chem Co., Yulin 719000, China
2 School of Mines, China University of Mining and Technology, Xuzhou 221116, China
3 MOE Key Laboratory of Deep Coal Resource Mining, China University of Mining and Technology, Xuzhou 221116, China
* Correspondence: dan.ma@cumt.edu.cn; Tel.: +86-176-2650-0518

Abstract: Groundwater inrush hazard has always been a great threat to the construction of vertical shafts in coal mines. Generally, the failure of the water-resisting coal pillar under coupled stress-seepage conditions around the vertical shaft is the main reason for the generation of the water inrush channel. In order to understand the mechanical behaviors of the water-resisting coal pillar, the strength of typical coal affected by the size and water content was investigated, and the stress sensitivity of permeability was investigated by a stress-seepage coupling test. The stress field and deformation of the water-resisting coal pillar were investigated by numerical simulation, the stability of the water-resisting coal pillars with different widths was evaluated, and the reasonable width of the coal pillars under coupled stress-seepage condition was determined. Results show that the water content and coal pillar width have a great influence on the mechanical characteristics of coal samples. Under the conditions of lower water content and larger coal sample width, the coal sample presents higher strength, smaller axial deformation, smaller permeability and porosity, and weak sensitivity to stress. The simulation results show that the boundary of the main roadway at the end of the coal pillar is dominated by tensile stress, and fractures can significantly contribute to the destruction of coal pillars. With the increase in the width of the water-resisting coal pillar, the internal damage variable, maximum tensile stress, porosity, and average water flow velocity of the coal pillar decrease, which reduces the risk of water inrush and improves the safety of the water-resisting coal pillar. An evaluation model of the reasonable width of the water-resisting coal pillar under the stress-seepage coupling was proposed, and the model was verified by the shear slip law and experimental results. This study provides theoretical and experimental guidance for the risk management of groundwater inrush disaster during the construction of vertical shafts in coal mines.

Keywords: groundwater inrush risk; stress-seepage coupling; water-resisting coal pillar; reasonable coal pillar width

Citation: Liu, Q.; Xue, Y.; Ma, D.; Li, Q. Failure Characteristics of the Water-Resisting Coal Pillar under Stress-Seepage Coupling and Determination of Reasonable Coal Pillar Width. *Water* **2023**, *15*, 1002. https://doi.org/10.3390/w15051002

Academic Editors: Yixian Wang, Panpan Guo, Hang Lin and Yanlin Zhao

Received: 9 February 2023
Revised: 26 February 2023
Accepted: 27 February 2023
Published: 6 March 2023
Corrected: 9 May 2023

1. Introduction

The complex hydrogeological conditions of mines have seriously affected the safe production activities of mines in recent years, and the prevention and control of water inrush in mines has received wide attention [1,2]. To realize the rapid recovery of mining economic costs, the shaft construction period should be reduced as far as possible. Conventional shaft construction methods include the freezing method, drilling method, grouting method, and open caisson method. The air reverse circulation drilling method has the advantages of high efficiency, energy conservation, environmental protection, and a wide application range [3]. However, high-pressure mud will be produced during this shaft construction method, which is prone to induce water inrush accidents. Therefore, the application of the air reverse circulation drilling method in the complex geological environment is limited.

To prevent water inrush in the process of air reverse circulation drilling, it is necessary to study the evolution law of hydraulic characteristics of the coal pillar under complex hydrogeological conditions in this drilling process.

A large number of studies on the prevention and control of water inrush under complex geological conditions have been conducted. Zhang et al. [4] proposed that formation pressure, aquifer water pressure, mining scale, and geological structure are the main factors controlling water inrush. Ma et al. [5,6] reviewed the water-inrush seepage mechanism from three aspects: stress-seepage coupling mechanism, flow transition mechanism and rock erosion mechanism. The evolution law of stress field and seepage field during water inrush is fully studied by numerical method and experimental analysis. Ganji et al. [7] evaluated the structural stability and safety of heterogeneous rock dam under loads such as dead weight and hydrostatic pressure using overstress and sliding safety indexes. Guo et al. [8] analyzed the water inrush mechanism qualitatively and quantitatively. Through acoustic emission (AE) technology and scanning electron microscope (SEM), Chen et al. [9] analyzed the mechanical and micro-fracture behavior of water-bearing coal samples. It was concluded that the coal pillar should be prevented from being subjected to inclined load and water erosion to avoid water inrush accidents. Hosseinzadeh et al. [10] proposed that seepage flow is a type of load widely distributed in the bedrock, which has an important effect on the water inrush of the dam. Chen et al. [11] believed that retaining a water-resisting coal pillar is an effective measure to prevent water inrush, and water pressure and seepage rate of overlying strata are key factors to predict potential water inrush. Ma et al. [12] drew the stress distribution of the floor under different coal pillar widths and derived the stress distribution data of the floor. Numerical simulation method is used to compare the stress distribution of roadway floor and the deformation characteristics of surrounding rock under different coal pillar widths. Finally, they determine the width and optimize the size of the coal pillar. Hu et al. [13] summarized that retaining a water-resisting coal pillar is the most effective water-saving method for coal seam mining, and proposed the conditions for determining the critical width of the water-resisting coal pillar. Ma et al. [14] proposed that when mining roadway excavation encountered fault fracture zone, permeability and porosity increased with time, and the granular structure of broken surrounding rock was easy to form water channel, which may cause water gusher hazard. The above research shows that retaining a water-resisting pillar is the most convenient and economic measure to prevent water inrush from coal seams.

The width of the water-resisting coal pillar is an important parameter of the coal pillar. It is very important to determine a reasonable width of coal pillar to ensure mine safety and save coal resources. Chen et al. [15] obtained the empirical formula and the safe width of the coal pillar in the section based on simulation analysis and mathematical measurement. Wang et al. [16] derived the evaluation formula of the width of the protective coal pillar and analyzed the failure form of the main roof. It is concluded that the breaking position of the main roof has a great influence on the width of the protective coal pillar. The critical width of the protective coal pillar in coal seam stope is determined to ensure the stability of the surrounding rock. Chen et al. [11] proposed that it is necessary to understand the influence of water seepage and loading rate on the mechanical properties of coal rock samples and the development of fracture network, so as to determine the width of the water-resisting coal pillar. Jaiswal et al. [17] conducted field monitoring and numerical simulation for the whole process of coal mass strain softening during coal pillar mining. The monitoring results show that the yield of coal pillar begins almost 2/3 of the peak strength. On this basis, the statistical expressions of the pillar strength and the estimation after failure are obtained, and the reasonable width range of the pillar is given. Dong et al. [18] established the mechanical model of the main roof rock beam, deduced the limit span and limit deflection of the rock beam fracture, and verified the reliability of the thickness of the water-resisting coal pillar. Through theoretical analysis and simulation analysis, Hu et al. [13] provided a theoretical basis for the rationality of the width of the water-resisting coal pillar. It was reported that the maximum width of the water-resisting coal pillar is the critical width, and the coal pillar was divided as mine pressure affected zone, effective waterproof zone, and water level affected zone with respect of water resistance. Das et al. [19] concluded that there is no suitable calculation formula to estimate the strength of inclined coal pillars considering many factors at present. The strength of inclined coal pillars with different widths is estimated by numerical simulation. Based on the hydraulic model of floor water inrush, Shi et al. [20] put forward the calculation formula of the fault water-resisting coal pillar in the stope considering the influencing factors such as confined water pressure and coal seam dip angle. Li et al. [21] determined the safe width of the water-resisting coal pillar in combination with the mechanical expression of the confined water pressure and the width of the water-resisting coal pillar under the limit equilibrium state. Wang et al. [22] classified the structure of the water-resisting coal pillar into the plastic zone affected by rock pressure, the core zone with effective waterproof and the water pressure damage zone, and measured the physical shape distribution of the water-resisting coal pillar using ultrasonic detection technology. This study provides a theoretical basis for determining the width of the water-resisting coal pillar. Wang et al. [23] determined the safe width of the water-resisting coal pillar by combining the physical model test results with the elastic-plastic limit equilibrium theory, and concluded that the coal seam before mining can be divided into fault-affected zone, elastic zone, and plastic yield fracture zone according to the stress and displacement characteristics. Based on field experience and numerical model studies, Mathey et al. [24] proposed another South African squat coal pillar strength formula. Combined with numerical simulation research, Yao et al. [25] monitored coal samples with different water contents and strain rates through a non-destructive immersion test, which is significant for the design of the water-resisting coal pillar.

The above research has deeply explored the determination method of coal pillar width. However, the evolution law of the stress-seepage coupling characteristics of water inrush from rocks under the action of the high-confined water has not been considered. Bukowski [26] analyzed the water inrush event in Poland and comprehensively analyzed the wellbore water disaster in combination with influencing factors (such as the intensity of water inrush, and the content of suspended substances in the water flowing into the wellbore). Ma et al. [27] obtained the temporal and spatial distribution of hydraulic characteristics through numerical simulation. The rock particles near the fluid outlet first fluidized and continuously migrated outward, resulting in the increase of the porosity and permeability of fault rock. Through establishing the hydrogeological model of water inrush, Huo et al. [28] analyzed the relationship between hydraulic gradient, water inrush and relevant factors, and summarized the general law of water inrush disasters in aquifers. Kerimov et al. [29] developed a numerical work flow to study the effect of non-frictional, convex and particle size polydispersity on the stress-seepage coupling of porosity, permeability, and elastic bulk modulus in granular porous media. Ma et al. [30] used the coupling equation of stress and seepage to model the seepage law. It was concluded that under the support pressure and unloading of the coal mining face, the dense seepage pathways can be formed by the fracture connection, thus causing groundwater inrush. Chen et al. [31] proposed that the evolution of fracture in rock mass causes the mining-induced fracture to evolve along the direction of principal stress and connect with each other, resulting in the formation of water inrush pathways. Liang et al. [32] analyzed the change rule of stress, displacement and water pressure during tunnel excavation, providing corresponding observation indicators for the water inrush process. Ma et al. [33] proposed that in the process of water inrush, the nonlinear characteristics of rocks are gradually changed due to the erosion of flowing water, and the particle loss is the main reason for the decline of nonlinear hydraulic characteristics. Li et al. [34] proposed that the lower the compressive stress level, the stronger the compressibility of broken rock mass, the stronger the stress sensitivity of small granular broken gangue, and the more likely the water inrush accident occurs. Yin et al. [35] put forward that mining activities disturb the original stress state, which leads to redistribution of stress, rock failure, and the formation of a water-conducting failure zone around the mining panel. Ma et al. [36,37] proposed the uniaxial compressive strength and elastic modulus of coal samples gradually deteriorate during the dry-wet cycle. According to the cyclic failure of the samples caused by wet and dry cycles, the deterioration process of the mechanical properties of the samples was divided into initial deterioration stage and re-deterioration stage, which verified the periodic failure of the cemented structure. Behrenbruch et al. [38] introduced a model to classify the characteristics of geological sedimentary environments and associated pore structures. In "reverse modeling" applications, pore structure parameters, porosity, and permeability are estimated for specific geological conditions. High confined water pressure is an important factor in mine water inrush, and also the basic power of coal seam water inrush [39]. Wu et al. [40,41] elaborated the influence mechanism of aggregate size distribution on rock sample and analyzed the contribution of aggregate size distribution to the seepage performance of rock sample. When the typical fractured coal body is subject to high seepage pressure, water inrush disasters are more likely to occur. However, the above research mainly focuses on the seepage and mechanical characteristics of brittle rock materials such as sandstone and granite, and less attention is paid to the seepage mechanical characteristics of coal rock.

In fact, the mechanical properties of coal rock are different from the above rock materials. In particular, the mechanical properties of coal rock will be changed greatly after it has been softened by water. Therefore, the stress-seepage coupling characteristics of coal rock are extremely complex, and an effective method to evaluate the stability of the water-resisting coal pillar is urgently required. In addition, the calculation results of the current calculation criteria for the width of the water-resisting coal pillar are conservative and have a large deviation from the actual results, resulting in a waste of coal resources. To analyze the stress-seepage coupling characteristics of the water-resisting coal pillar, the stress-seepage coupling tests of coal samples with different water contents and inner diameters were carried out, and the stress-seepage coupling characteristics of coal samples were obtained in this study. Then, the numerical simulation of water inrush in the shaft under different conditions was carried out, and the spatial distribution characteristics of stress and deformation of the water-resisting coal pillar were obtained. Finally, the simulation results of water inrush from water-resisting coal pillar with different widths were discussed, the evolution laws of the damage variable, maximum tensile stress, porosity, and average velocity of the coal pillar were obtained, and the calculation method of reasonable reserved width of the water-resisting coal pillar under compressed air reverse circulation drilling method was proposed. This study provides a theoretical and experimental reference for the prevention and control of coal seam water inrush.

2. Engineering Background

Kekegai Coal Mine is located in Yuyang District, Yulin City, Shaanxi Province. In this coal mine, the compressed air reverse circulation drilling method is innovatively adopted, which effectively improves the excavation speed of the shaft. However, in this method, mud at the depth of the wellbore will form high pore pressure, exposing the coal around the deep wall to high water pressure. As shown in Figure 1, during the shaft excavation, the internal fractures of the coal and rock mass around the shaft continue to expand and the permeability increases significantly under the influence of mining stress and high water pressure. As a result, the mud will enter the inclined shaft through the seam fractures.

To explore the change rule of mud pressure with the shaft driving depth, the relationship curve between mud pressure and shaft depth is drawn in combination with the data of Kekegai Coal Mine, as shown in Figure 2.

During the excavation of the air inlet and return shaft in Kekegai Coal Mine, the mud pressure P shows a linear increasing trend with the increase in the well depth. Through field measurement, the linear fitting equation of mud pressure under different well depths is $P = 0.0113H + 0.056$, and the fitting goodness R^2 is 0.9523. The coefficient of the first order term reflects the variation in mud pressure with the well depth. The constant term reflects the formation properties of the well.

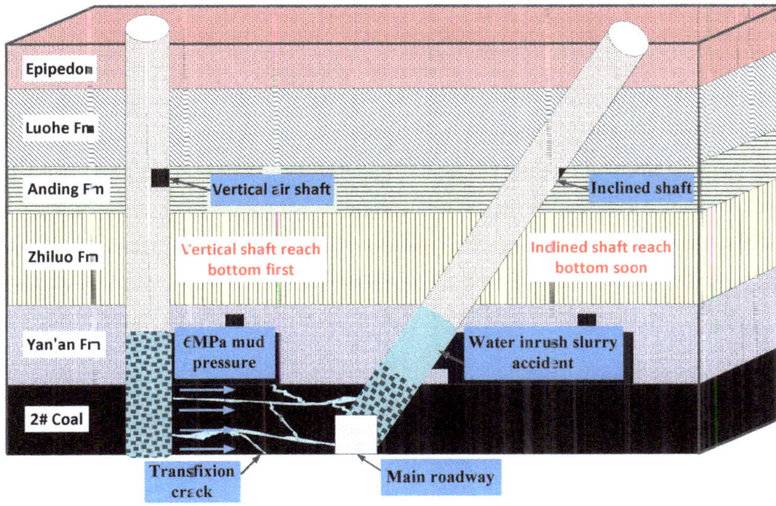

Figure 1. Principle of water inrush from water-resisting coal pillar in Kekegai coal mine.

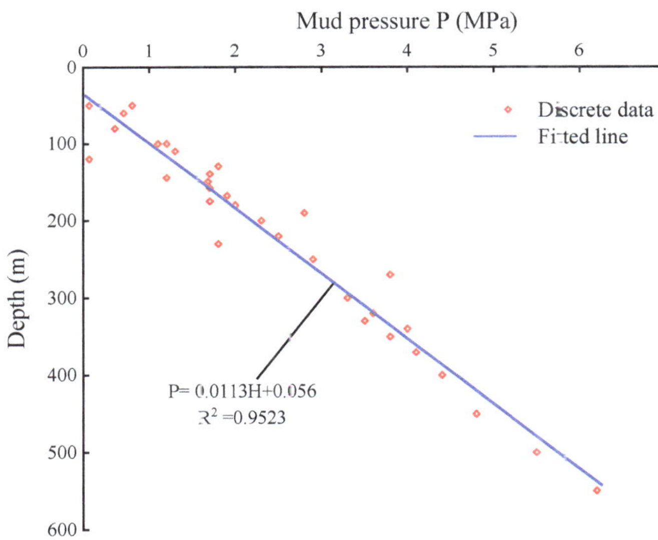

Figure 2. Relationship between pressure and well depth.

As shown in Figure 3, with the continuous increase in the excavation depth in shaft, the stress of rock mass around the shaft wall and at the outer edge keeps rising. In particular, the stress concentration is greatest at the outer edge of the shaft bottom. At the same time, the mud pressure used in the shaft excavation also has a certain loading effect on the shaft wall. Under the joint action of stress and mud pressure, a complex fracture network is formed in the surrounding rock of the shaft, leading to the fracture evolution of the rock mass.

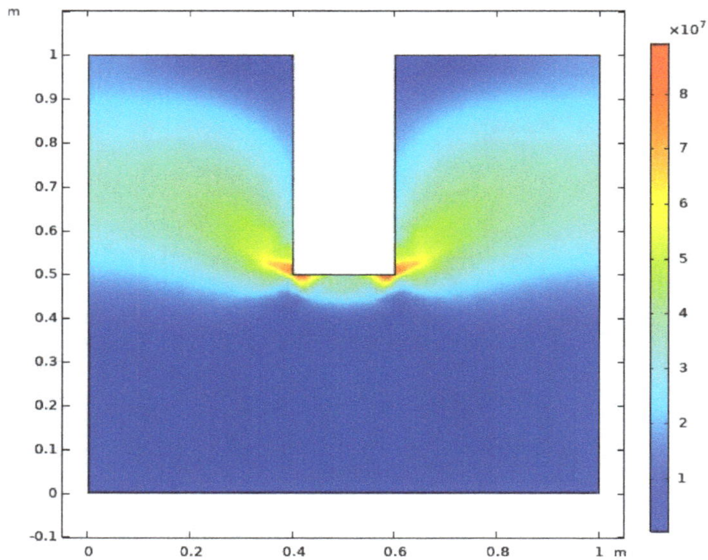

Figure 3. Distribution of water pressure varying with well depth.

3. Stress-Seepage Coupling Test on Coal Rock

3.1. Selection and Preparation of Coal Samples

The coal rock sample was taken from the surrounding rock around the shaft of Coal Seam 2 in Kekegai Coal Mine, Yulin City, Shaanxi Province (see Figure 4). The coal rock density was 1.5 kg/m^3, and the saturated water content was 14%. To explore the mechanical and seepage evolution characteristics of the coal pillar, hollow cylinder coal samples were made, as shown in Figure 5. The sample was processed into a coal core with a diameter of 70 mm and a length of 140 mm by the drilling sampler and rock cutter. Coal samples were put into the air circulation oven to dry for 12 h, and then the natural immersion method was used to obtain coal samples with different moisture contents.

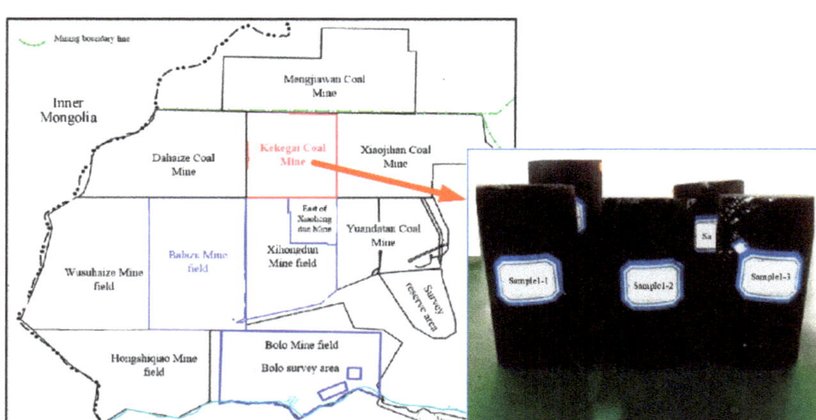

Figure 4. The location of sampling and coal samples.

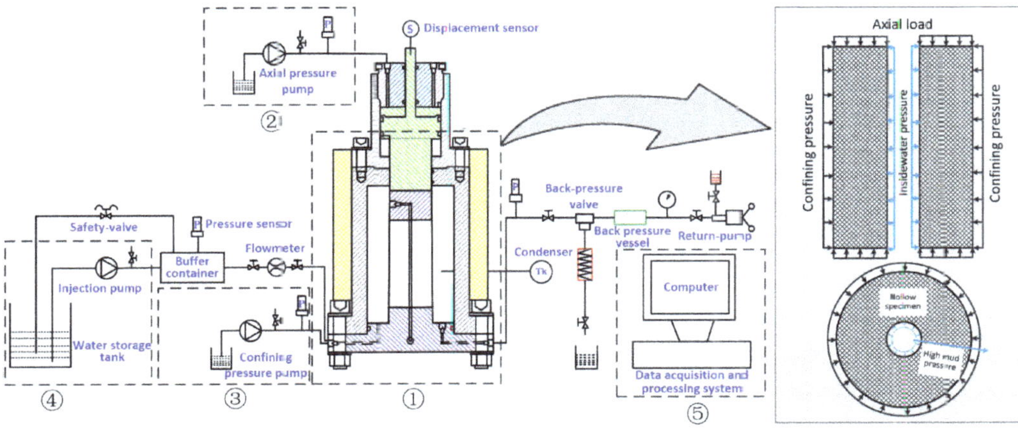

①—stress–seepage coupling instrument,
②—axial pressure control system,
③—confining pressure control system,
④—water pressure control system,
⑤—automatic data acquisition system.

Figure 5. Triaxial stress-seepage coupling testing system.

3.2. Experimental Methods, Apparatus and Scheme

To obtain the evolution law of the stress-seepage mechanical characteristics of coal rock under the action of high-pressure mud inside the shaft, the stress-seepage coupling test method for the surrounding rock of the shaft was proposed, and the stress-seepage coupling environment of the surrounding rock of the deep shaft was restored. As shown in Figure 5, axial pressure and confining pressure were applied to the hollow cylindrical coal sample, and water pressure was applied to the inner wall of the coal sample, so that the radial seepage of water was realized in the sample.

To achieve the above test methods, a thermal-hydraulic-mechanical (THM) triaxial coupling test system was independently developed, as shown in Figure 5. At the beginning of the test, the sample was placed in the stress-seepage coupling instrument, and the axial pressure and confining pressure control systems simultaneously imposed loads on the sample. The water pressure control system provided different water pressures to the inner wall of the coal sample. The automatic data acquisition system was used to monitor the seepage and mechanical characteristics of coal samples in real time.

Table 1 shows the experimental scheme of the sample. The inner diameter and water content of coal samples were taken as variables, and the change rule of seepage mechanical characteristics of different coal samples under the water pressure of 6 MPa and axial load of 10 MPa was studied. The inner diameter dimensions were classified into five groups of 15 mm, 20 mm, 25 mm, 30 mm and 35 mm, corresponding to tests A, B, C, D, and E. For each group of tests, samples with three different water content (0%, 6.28%, and 13.65%) were designed.

Table 1. Experimental scheme of stress-seepage coupling in the coal sample.

Group Number	Inner Size (mm)	Water Content (%)	Axial Compressive Pressure (MPa)	Confining Pressure (Mpa)
A1	15	0	10	6
A2	15	6.28	10	6
A3	15	13.65	10	6
B1	20	0	10	6
B2	20	6.28	10	6
B3	20	13.65	10	6
C1	25	0	10	6
C2	25	6.28	10	6
C3	25	13.65	10	6
D1	30	0	10	6
D2	30	6.28	10	6
D3	30	13.65	10	6
E1	35	0	10	6
E2	35	6.28	10	6
E3	35	13.65	10	6

3.3. Experimental Results

Figure 6 shows the axial strain evolution curve of coal samples with different water contents under the stress-seepage coupling (taking samples in Group C as an example). The three curves successively experience the initial deformation stage, stable deformation stage, and accelerated deformation stage. Comparing the three curves, it can be found that with the increase in water content, the axial strain of the sample increases faster, and the accelerated deformation stage is reached earlier. This shows that with the increase in water content, the lower the internal structural strength of the coal sample, the greater the deformation of the coal sample under the stress-seepage coupling, and the shorter time required for the structural failure.

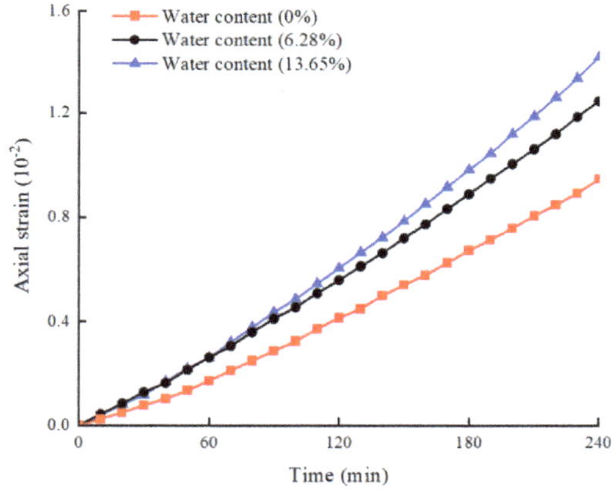

Figure 6. Axial strain evolution of samples with different water contents under stress-seepage coupling (taking samples in Group C as an example).

Figure 7a shows the change rule of the average flow velocity of coal samples with different water contents under the stress-seepage coupling (taking samples in Group C as an example). With the increase in time, the average flow rate inside the coal sample increases continuously. At the initial stage of the test, the lower the water content of the coal sample, the greater the average flow rate. However, with the increase in seepage time, the average flow velocity in the sample with high water content gradually exceeds that in the sample with low water content. In the sample with large water content, the increase in flow rate is greater, and the main increase in flow rate occurs after 120 min.

This is because the structural strength of the sample with higher water content is lower. Under the stress-seepage coupling effect, the fracture expands quickly. As a result, the average velocity increases rapidly in the middle and late periods and finally exceeds the coal samples with a low water content. Figure 7b shows the relationship between porosity and water content of coal samples under the stress-seepage coupling. It is observed that the coal sample with higher water content has a more significant porosity growth, which indicates that the cracks in the coal sample have a more significant expansion under the stress-seepage coupling effect. After 180 min, the porosity of coal samples with different water contents begins to decrease. This is due to the continuous compaction of coal samples under the loading, and the pores are gradually closed.

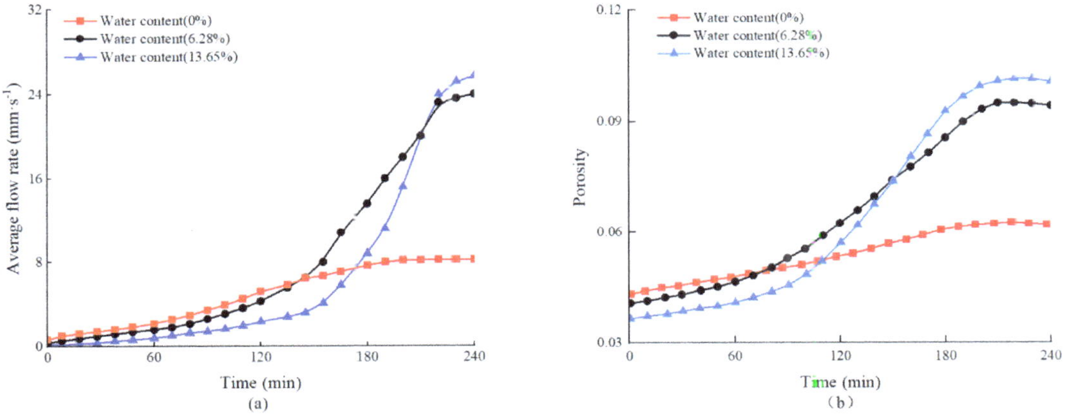

Figure 7. Evolution of hydraulic characteristics of samples with different water contents under stress-seepage coupling (taking samples in Group C as an example): (**a**) Average flow rate under different water contents; (**b**) Porosity under different water contents.

Figure 8 shows the axial-strain evolution curve of coal samples with different inner diameters under the stress-seepage coupling (taking samples in Group C as an example). By comparing the three curves, it can be found that for the sample with a large inner diameter, the strain of the sample increases significantly under the stress-seepage coupling effect, and the axial strain of the coal body reaches 0.0158 at 240 min. For the specimen with an inner diameter of 35 mm, the accelerated deformation stage is reached after 130 min. However, for the sample with the smallest inner diameter (15 mm), the accelerated deformation stage is not reached until 170 min later, and the axial strain after the test is only 0.0053. The reason for this phenomenon is that due to the longer seepage path and smaller water pressure gradient, the sample with a larger inner diameter has stronger resistance to the stress-seepage coupling effect.

Figure 9a shows the change rule of the average velocity of coal sample with an inner diameter under the stress-seepage coupling (taking samples in Group C as an example). It can be seen that the larger the inner diameter of the coal sample, the larger the peak value of the average velocity. The reason is that with the increase in the inner diameter of the coal sample, the mechanical strength of the coal sample decreases continuously, the internal fractures of the coal and rock mass are well developed, and the seepage resistance gradually decreases. Besides, in the sample with a larger inner diameter, the fluid seepage path is shorter, the hydraulic pressure gradient is larger, and the flow velocity is faster. At the same time, it can be observed that the flow rate increases faster in coal samples with a larger inner diameter. This shows that the narrower the coal pillar is, the faster the water inrush occurs. Figure 9b shows the evolution law of the average porosity of coal samples with different inner diameters under the stress-seepage coupling. It can be observed that the larger the inner diameter of the coal sample, the faster the growth rate of porosity, and the greater the growth range. This is because in the process of stress-seepage coupling, the larger the inner diameter of coal sample, the lower the structural strength of the coal body, and the easier to form through cracks in its interior, resulting in a substantial increase in porosity. At the same time, the porosity decreases slowly in the late period of stress-seepage coupling (t > 200 min). This is because after the formation of through cracks, the rock mass continues to undergo compression deformation under the influence of stress loading, and the porosity decreases after reaching the peak.

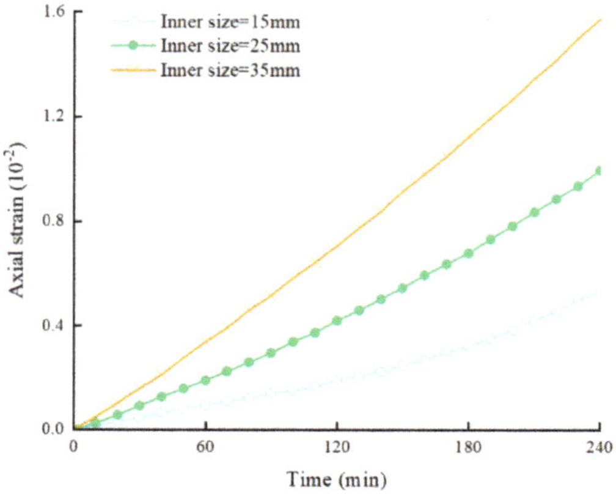

Figure 8. Axial strain evolution of samples with different inner diameters under stress-seepage coupling (Group C; water content: 6.28%).

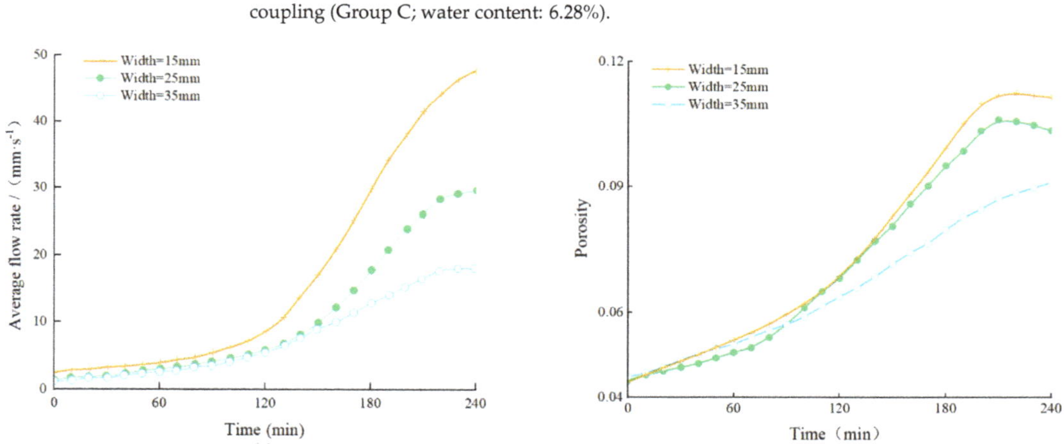

Figure 9. Evolution of hydraulic characteristics of samples with different inner diameters under stress-seepage coupling (Group C; water content: 6.28%): (**a**) Average flow rate under different water contents; (**b**) Porosity under different water contents.

4. Numerical Simulation of Stress-Seepage Coupling in the Coal Pillar

4.1. Simulation Schemes

The water-resisting coal pillar is located between the bottom transportation roadway and the shaft, which is an important structure to prevent the high-pressure mud from flowing into the bottom roadway. The mechanical and seepage characteristics of the water-resisting coal pillar are affected by many factors. To study the water-resisting capacity of the water-resisting coal pillar under different conditions, numerical simulation on coal rock under different water pressures and rock integrity was carried out in this section.

Three simulation schemes were designed in the study. Scheme I: the mechanical characteristics of the intact water-resisting coal pillar under the action of mining stress were simulated; Scheme II: the evolution of the mechanical and hydraulic characteristics of the intact water-resisting coal pillar under the stress-seepage coupling effect was simulated; Scheme III: the evolution of the mechanical and hydraulic characteristics of the water-resisting coal pillar with a large number of natural joints and fissures was simulated under the stress-seepage coupling effect. By comparing the simulation results of the above schemes, the appropriate width of the water-resisting coal pillar can be determined.

Taking the water-blocking coal pillars of the main and auxiliary inclined shafts of Kekegai Coal mine as the simulation object, the relevant numerical calculation model is established according to the engineering geology and hydrogeological conditions on site, as shown in Figure 10. The overall geometric size of the model is 11.4 × 20 × 4.2 m^3. Stress and displacement boundary conditions: the front, back and bottom boundaries of the model are fixed constraint normal displacement. According to the dead weight of overlying strata, a vertically downward uniform load of 5 MPa is applied to the top boundary of the model. In simulation schemes II and III, the water pressure of 6 MPa was set to the shaft wall. Both the applied water pressure boundary and the excavation roadway boundary are free boundaries, as shown in Figures 10 and 11.

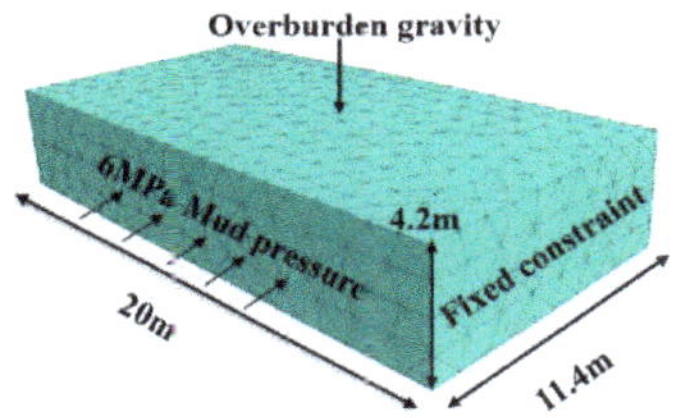

Figure 10. Schematic diagram of a water-resisting coal pillar model.

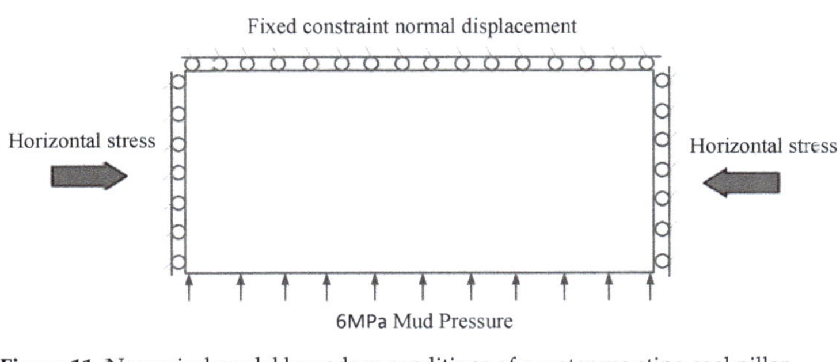

Figure 11. Numerical model boundary conditions of a water-resisting coal pillar.

4.2. Simulation Results

(1) Mechanical distribution characteristics of the intact water-resisting coal pillar under mining stress

Figures 12 and 13 show the stress and displacement distribution of the intact water-resisting coal pillar under mining stress. It can be concluded that the effective stress of the intact water-resisting coal pillar shows a significant non-uniform distribution. The coal pillar is subjected to compressive stress at the shaft wall and tensile stress at the boundary of the main roadway. From the shaft wall to the inside of the coal pillar, the compressive stress gradually decreases and decreases to zero at a certain position in the middle of the coal pillar. Under the interaction of high-water pressure and mining stress, there is an area with increased compressive stress at the shaft wall and an area with increased tensile stress at the main roadway wall. The maximum compressive stress on the coal pillar is 3 MPa, and the tensile stress is 0.338 MPa. Through the test, the average compressive strength of the intact water-resisting coal pillar is 21.52 MPa, and the tensile strength is 1.27 MPa. Therefore, the failure of the water-resisting coal pillar will not occur under this condition. As shown in Figure 12, the displacement of coal pillar also presents non-uniform spatial distribution. The displacement gradually increases from the roadway wall to the shaft wall. For the shaft wall and roadway wall, the displacement at the center is greater than that at both ends.

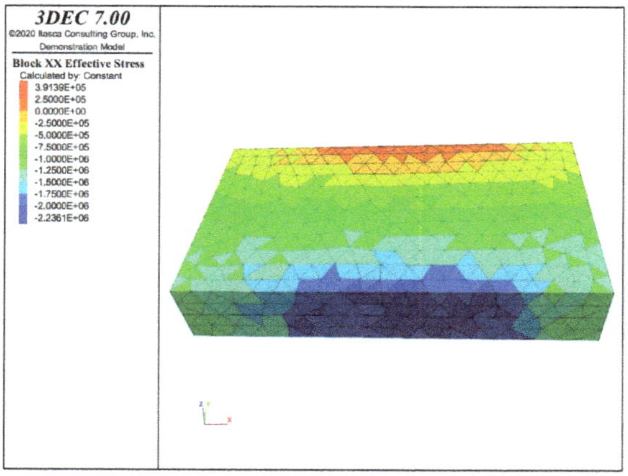

Figure 12. Effective stress distribution under the intact water-resisting coal pillar. (Unit: Pa).

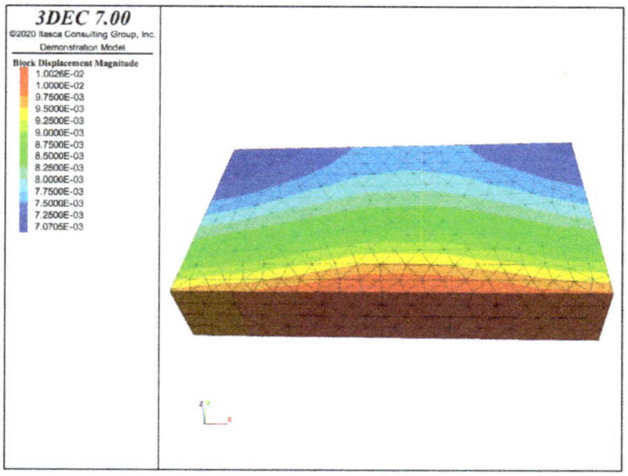

Figure 13. Displacement distribution under the intact water-resisting coal pillar. (Unit: m).

(2) Distribution of mechanical characteristics of intact water-resisting coal pillar under the stress-seepage coupling

Figures 14 and 15 depict the stress and displacement distribution of the intact water-resisting coal pillar under the stress-seepage coupling. It can be observed that the distribution mode of effective stress of the water-resisting coal pillar under the stress-seepage coupling is similar to that of the water-resisting coal pillar only under the influence of mining stress. Under the coupling effect of stress and seepage, there is an area with increased compressive stress at the shaft wall and an area with increased tensile stress at the main roadway wall. The maximum compressive stress of the coal pillar is 4.12 MPa, and the maximum tensile stress of the coal pillar is 0.378 MPa. However, the average saturated compressive strength and saturated tensile strength of the intact water-resisting coal pillar obtained through the test are 14 MPa and 0.6 MPa. Therefore, the failure of the water-resisting coal pillar will not occur under this condition. Comparing the simulation results of Scheme I and Scheme II, the maximum tensile stress value in Scheme II is much larger than that in Scheme I, and the tensile stress area of the coal pillar in Scheme II is 1.36 times of that in Scheme I. As shown in Figure 14, the displacement of the coal pillar presents non-uniform spatial distribution, and the displacement gradually increases from the roadway wall to the shaft wall. For the shaft wall and roadway wall, the center displacement is greater than that in the two ends. Through the above comparison, it can be concluded that the stress-seepage coupling effect has a significant impact on the stability of the intact water-resisting coal pillar.

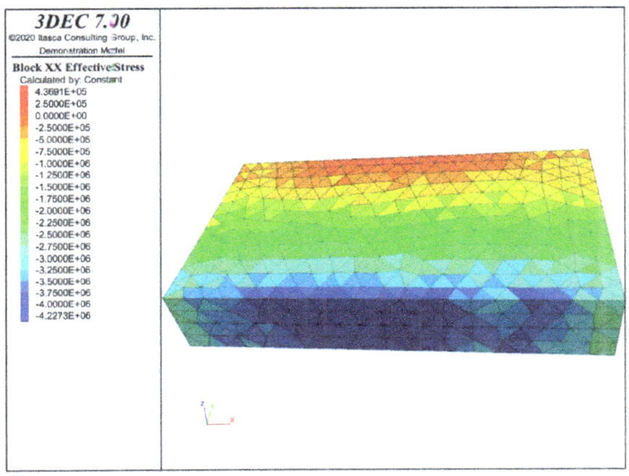

Figure 14. Effective stress of the water-resisting coal pillar with natural joints under stress-seepage coupling effect. (Unit: Pa).

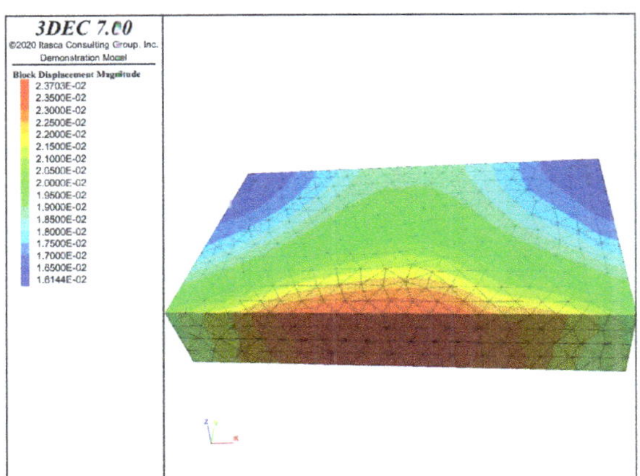

Figure 15. Displacement distribution of the water-resisting coal pillar under the stress-seepage coupling effect. (Unit: m).

(3) Distribution of mechanical characteristics of the water-resisting coal pillar with fractures under the stress-seepage coupling

Figures 16 and 17 show the stress and displacement distribution of the water-resisting coal pillar with natural fissures under the stress-seepage coupling effect. The maximum compressive stress and the tensile stress of the coal pillar are 4.23 MPa and 1.072 MPa, respectively. The test results show that the average compressive strength and tensile strength of the water-resisting coal pillar with a large number of natural joint fissures under the stress-seepage coupling action are 14 MPa and 0.6 MPa, respectively. Therefore, under this condition, the wall of the coal pillar roadway is unstable under tension, and the failure of the water-resisting coal pillar occurs. As presented in Figure 16, the displacement inside the coal pillar is significantly different. The closer to the shaft wall, the greater the displacement. Besides, the displacement at both ends of the coal pillar is smaller than that at the center, regardless of whether it is the inner wall of the shaft or the inner wall of the roadway. The above results show that under the stress-seepage coupling effect, the bearing capacity of the water-resisting coal pillar with nature fissures is poor, and the natural fissures can reduce the mechanical properties of the rock. These fissures continue to expand under the stress-seepage coupling effect, and eventually develop into through fractures, leading to the failure of the coal pillar and water inrush accidents.

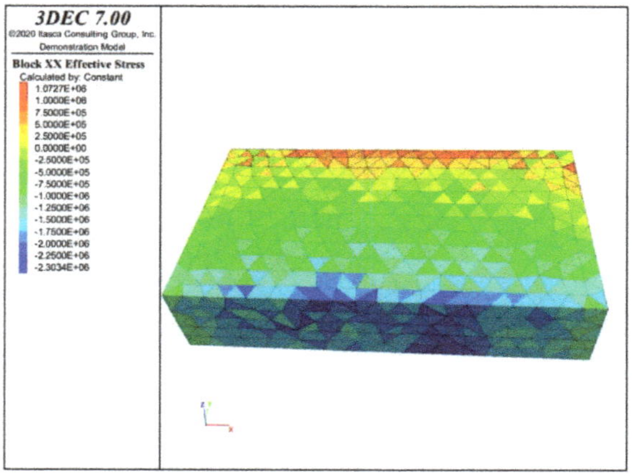

Figure 16. Effective stress distribution of the water-resisting coal pillar with joints under stress-seepage coupling effect. (Unit: Pa).

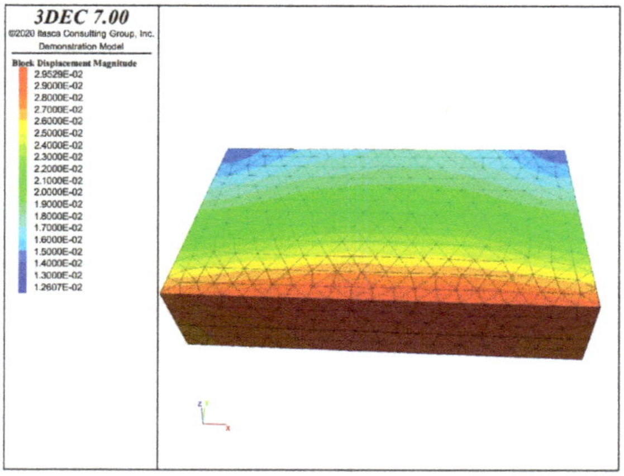

Figure 17. Displacement distribution of the water-resisting coal pillar with joints under the stress-seepage coupling effect. (Unit: m).

To sum up, the effective stress and displacement of the water-resisting coal pillar show non-uniform spatial distribution under the three schemes. The coal pillar is subjected to compressive stress at the shaft wall and tensile stress at the boundary of the main roadway. There is a stress increase area at the shaft wall and a tensile stress increase area at the roadway wall. The water-resisting coal pillar with a large number of natural joint fissures will be damaged under the stress-seepage coupling, while the coal pillar will not be damaged under other conditions. Considering both the stress-seepage coupling effect and the natural fracture of the coal pillar, the stability of the water-resisting coal pillar in the Kekegai coal mine is poor and the coal pillar is prone to failure.

5. Determination of the Reserved Width of the Water-Resisting Coal Pillar under the Stress-Seepage Coupling

According to the numerical simulation results in the previous section, when there are stress-seepage coupling effect and the natural fracture of the water-resisting coal pillar, the wall area of the water-resisting pillar roadway will be subject to tensile failure, which will lead to water inrush accidents. Therefore, the existing coal pillar width cannot meet the needs of the project. To solve this problem, the damage coefficient is introduced in this section to evaluate the stability of the water-resisting coal pillars with different widths. The evaluation results are combined with the test and numerical simulation results to determine the reserved width of the water-resisting coal pillars under the stress-seepage coupling effect.

In the detailed rules for water prevention and control of coal mines, the following provisions are reserved for the width of the water-resisting coal pillars:

$$L = 0.5KM\sqrt{\frac{3\text{p}}{K_\text{p}}} \tag{1}$$

where K is the safety factor, ranging from 2 and 4. According to the above equation, the width L of the water-resisting pillar should be between 15.82–31.62 m. According to the above interval, 3DEC numerical simulation of the coal pillar with different widths is carried out, and the effective stress and displacement distribution results of coal pillars with joints and fissures under different coal pillar widths are calculated, as shown in Figures 18–22.

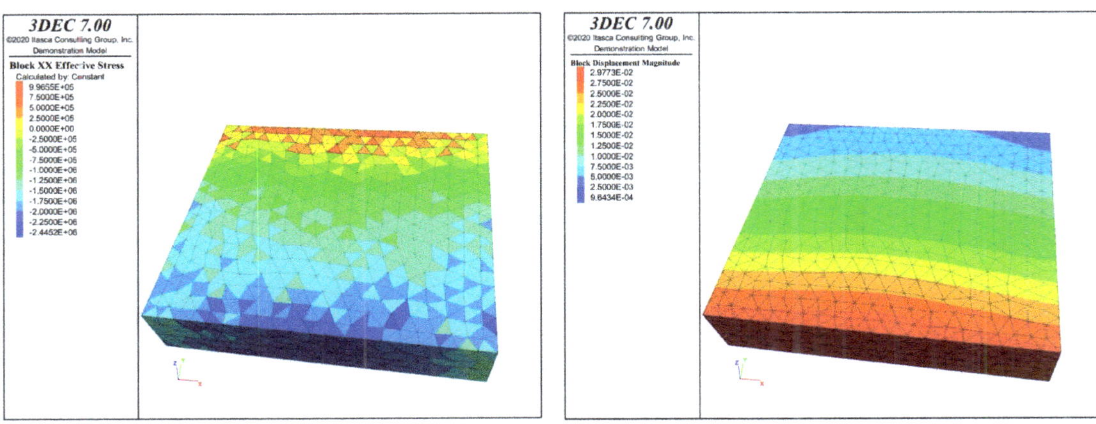

(**a**) (**b**)

Figure 18. Stress and displacement distribution of the water-resisting coal pillar with a width of 15 m containing natural fissures under the stress-seepage coupling effect: (**a**) Stress (Unit: Pa); (**b**) Displacement (Unit: m).

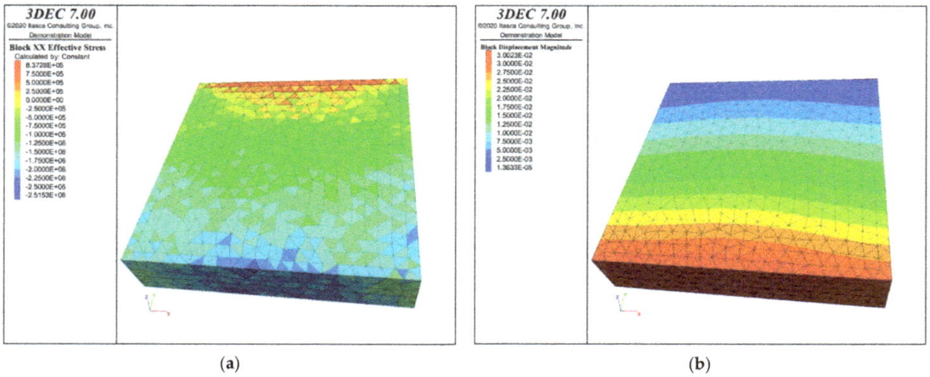

Figure 19. Stress and displacement distribution of the water-resisting coal pillar with a width of 20 m containing natural fissures under the stress-seepage coupling effect: (**a**) Stress (Unit: Pa); (**b**) Displacement (Unit: m).

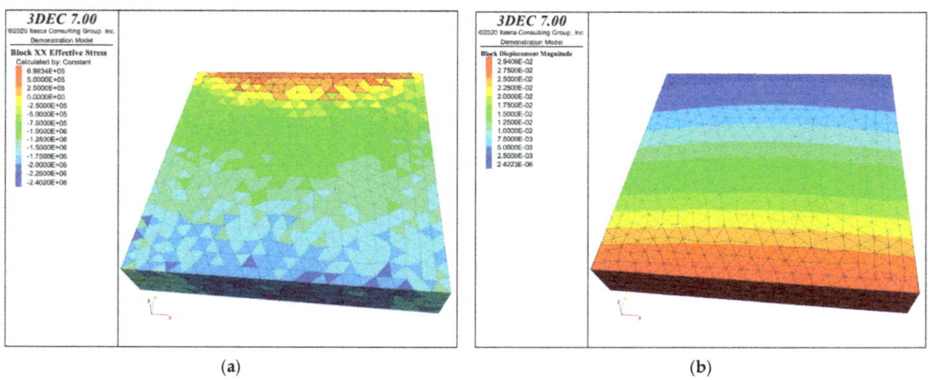

Figure 20. Stress and displacement distribution of the water-resisting coal pillar with a width of 25 m containing natural fissures under the stress-seepage coupling effect: (**a**) Stress (Unit: Pa); (**b**) Displacement (Unit: m).

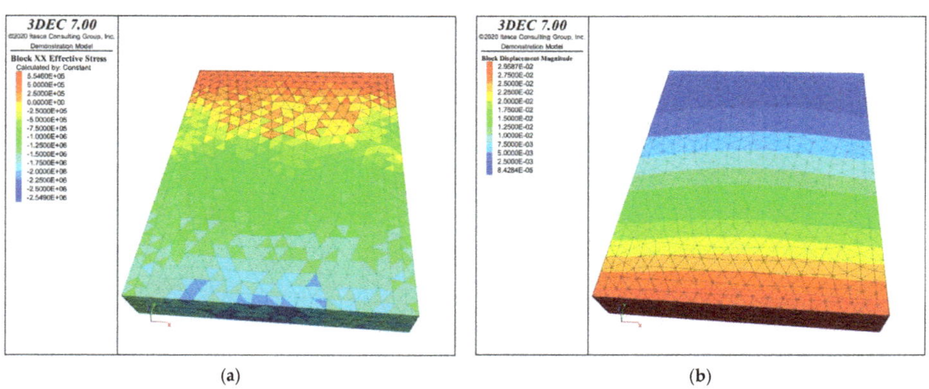

Figure 21. Stress and displacement distribution of the water-resisting coal pillar with a width of 30 m containing natural fissures under the stress-seepage coupling effect: (**a**) Stress (Unit: Pa); (**b**) Displacement (Unit: m).

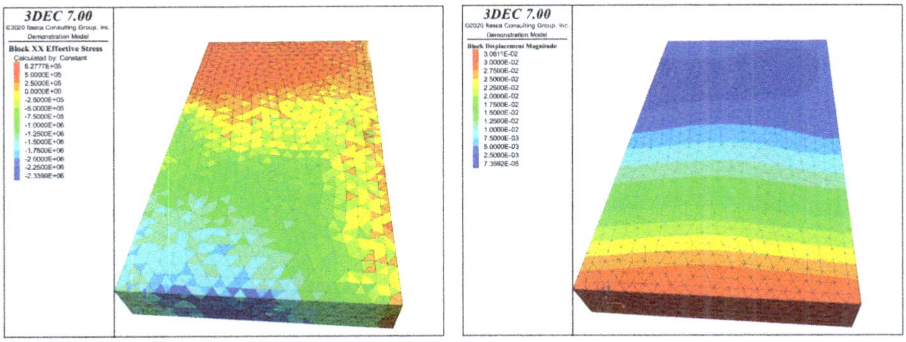

(a) (b)

Figure 22. Stress and displacement distribution of the water-resisting coal pillar with a width of 15–35 m containing natural fissures under the stress-seepage coupling effect: (**a**) Stress (Unit: Pa); (**b**) Displacement (Unit: m).

It can be seen from Figures 18–22 that after the water-resisting coal pillar is reserved, the stress of the coal pillar shows a significant non-uniform distribution due to the influence of the overburden movement and evolution height. To further analyze the stability characteristics of the coal pillar in the above results, the damage variable is introduced to describe the fracture development. The damage variable can be used to reflect the internal defects of the coal body, and the essence of the sliding instability of the coal pillar on the macro level is the result of the damage accumulation of internal micro units. The damage variable D of the water-resisting coal pillar can be calculated by the following equation:

$$D = \frac{S_D}{S_O} \tag{2}$$

where S_D is the number of the failure nodes; S_O is the total number of nodes. When $D = 0$, the coal pillar is not affected by its internal stress distribution; When D is closer to 1, it indicates that the failure and instability risk of the coal pillar is higher.

By using the node statistics function in 3DEC software, the number of nodes with damage and failure (that is, the tensile stress exceeds the tensile strength of coal and rock) is counted in the above stress-displacement distribution diagram, and the damage variables of the water-resisting coal pillars with different widths are calculated. Figure 23 shows the specific calculation results.

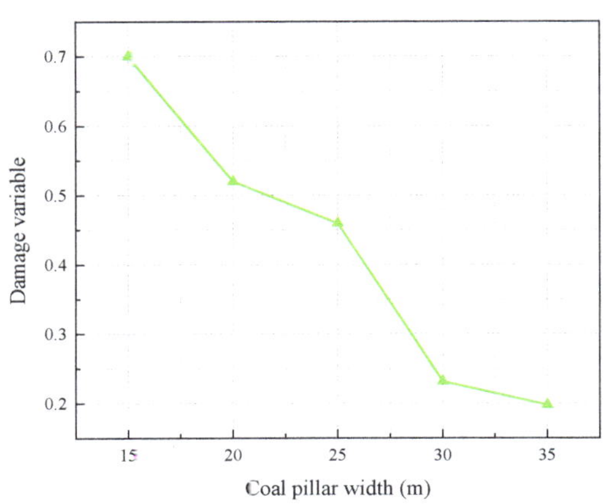

Figure 23. Damage variables of coal pillars within different widths.

Figure 23 shows the calculated results of the damage variable of coal pillars with different widths. It can be seen that the damage variable of the water-resisting coal pillar decreases with the increase in coal pillar width. The results of the previous research [42,43] have shown that when the damage variable of the water-resisting coal pillar is 0.25 and below, the coal pillar is in a safe state. Under this condition, the number of cracks in the coal body under the stress-seepage coupling effect is less. According to this point of view, it is determined that the reasonable width interval for the water-resisting coal pillar is 28.3–31.62 m. This value will be further analyzed from the following three aspects: maximum tensile stress, porosity, and flow velocity.

Figure 24 shows the results of the maximum tensile stress of coal pillars with different widths. With the increase in the width of the water-resisting coal pillar, the maximum tensile stress inside the coal pillar decreases as a whole. However, in the width interval of 30–35 m, the tensile stress increases rapidly. When the width of the water-resisting coal pillar is 28.3–31.62 m, the maximum tensile stress of the coal pillar is 0.58 MPa, which is less than its saturated tensile strength of 0.6 MPa. This means that the coal pillar will not be unstable or damaged.

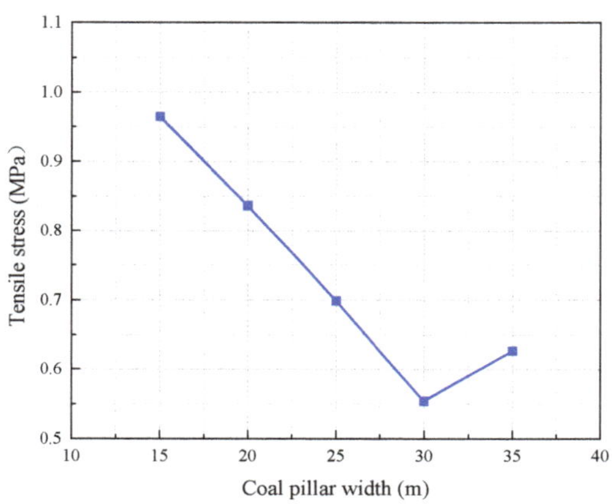

Figure 24. Maximum tensile stress of coal pillars within different widths.

Figure 25 shows the porosity results of coal pillars with different widths. It can be seen that the porosity of the water-resisting coal pillar decreases gradually with the increase in coal pillar width. It indicates that with the increase in coal pillar width, the water-resisting coal pillar is safer and the risk of water inrush is smaller. When the width of the water-resisting coal pillar is 28.3–31.62 m, the porosity in the coal pillar is less than 0.0975, and the coal pillar has a high-water resistance.

Figure 26 depicts the average flow velocity values of the water-resisting coal pillars with different widths. It can be seen that the average velocity inside the water-resisting coal pillar decreases gradually with the increase in coal pillar width. This shows that the wider the coal pillar, the stronger the water-resisting capacity of the water-resisting coal pillar, and the lower the risk of water inrush. When the width of the water-resisting coal pillar is 28.3–31.62 m, the average velocity in the coal pillar is about 25 mm/s. At this time, the risk of water inrush from the coal pillar is extremely low.

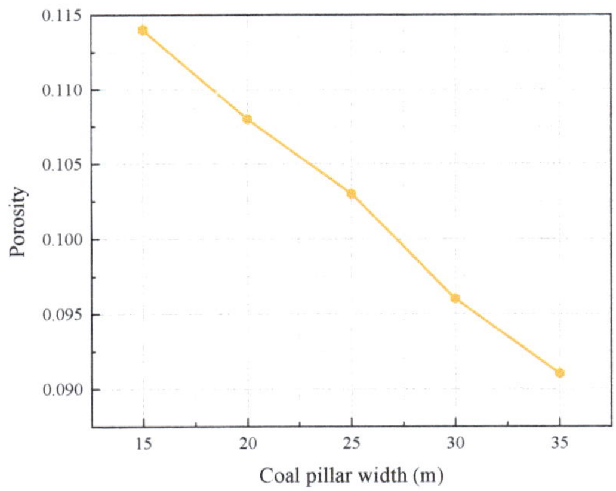

Figure 25. Porosity of coal pillars within different widths.

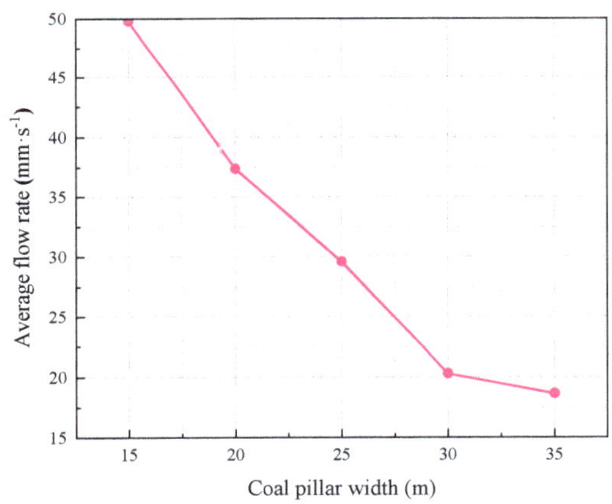

Figure 26. Average flow velocity of coal pillars within different widths.

From the above analysis, it is obtained that the width of the water-resisting pillar should be set between 28.3–31.62 m. To effectively save coal resources, the design width of the water-resisting coal pillar should be reduced as much as possible. However, considering the influence of mud on the stress-seepage coupling effect of the internal cracks of the water-resisting coal pillar, the width of the coal pillar needs to be appropriately increased to ensure safety. The safety factor of the water-resisting coal pillar is set as 1.03, and the reasonable value of the width of the water-resisting coal pillar can be obtained as 30.57 m.

The water-resisting coal pillar with the determined width of 30.57 m is verified by the shear slip law of the coal body. The shear slip equation of coal pillars can be described as follows:

$$CFS = \tau - \mu(\sigma - \rho) - C \tag{3}$$

where τ is the shear stress along the fault plane; μ is the friction coefficient in the fault plane; σ is the normal stress on fault plane (the value of the compression stress is positive, while the value of the tensile stress is negative); ρ is the tensile stress generated by the interstices in the crust (the value of the compression stress is positive, while the value of the tensile stress is negative); C is the cohesive stress.

Through the calculation results, it is concluded that under the stress-seepage coupling action, no shear slip under Coulomb fracture stress will occur for the water-resisting coal pillar of this width, which verifies the reliability of the calculation method proposed in this study.

6. Conclusions

To analyze the failure characteristics of the water-resisting coal pillar under the stress-seepage coupling effect, a series of stress-seepage coupling tests of the surrounding rock of the shaft were carried out, and the evolution law of the stress-seepage coupling characteristics of coal rock under the action of high-pressure mud was analyzed. The numerical simulation of water inrush in the shaft under different conditions was carried out to study the water-resisting ability of the water-resisting coal pillar under different conditions; The damage coefficient was introduced to evaluate the stability of the water-resisting coal pillars with different widths, and the reserved width of the water-resisting coal pillars under the coupling effect of stress and seepage was determined. The main conclusions are as follows:

The test results show that the stress-seepage coupling characteristics of coal samples are greatly affected by water content and coal pillar width. Under the stress-seepage coupling effect, the peak strength of coal samples with higher water content is lower, the failure process of the coal pillar is more complex, and the axial strain, average velocity and porosity increase more significantly. The larger the inner diameter of the coal sample, the lower the peak strength, and the greater and faster the axial strain, average velocity, and porosity increase. This shows that the higher the water content and the narrower the width of the coal pillar, the higher the risk of coal pillar damage and water inrush.

Through numerical simulation, it is found that the effective stress and displacement of the water-resisting coal pillar under the three schemes show non-uniform distribution. The coal pillar is subjected to compressive stress at the shaft wall and tensile stress at the boundary of the main roadway. The tensile stress area of the coal pillar in Scheme II is significantly larger than that in Scheme I, which indicates that the stress-seepage coupling effect has a significant impact on the stability of the intact water-resisting coal pillar. In addition, the intact water-resisting coal pillar under mining stress and the intact water-resisting coal pillar under the stress-seepage coupling action are free from coal pillar failure. However, the failure of the water-resisting coal pillar with a large number of natural joints and fissures occurs under the stress-seepage coupling action.

The numerical simulation of the water-resisting coal pillar with different reserved widths has a certain guiding significance for engineering practice. The simulation results show that with the increase in the width of the water-resisting coal pillar, the internal damage variable, maximum tensile stress, porosity, and average flow velocity of the coal pillar decrease, the water-resisting coal pillar is safer and the risk of water inrush decreases. According to the simulation results of damage variable, maximum tensile stress, porosity, and flow velocity, the width of the water-resisting coal pillar under the stress-seepage coupling are determined. Finally, this result is validated by the shear slip law of the coal body, which verifies the reliability of the proposed calculation method of coal pillar width in this study.

Author Contributions: Conceptualization, Funding acquisition, Project administration, Resources, D.M.; Data curation, Formal analysis, Investigation, Visualization, Writing—original draft, Q.L. (Quanhui Liu); Methodology, Software, Writing—review and editing, Y.X.; Supervision, Validation, Q.L. (Qiang Li). All authors have read and agreed to the published version of the manuscript.

Funding: This study was supported by the National Science Fund for Excellent Young Researchers of China (52122404) and the National Natural Science Foundation of China (41977238).

Data Availability Statement: Data is contained within the article.

Conflicts of Interest: The authors declare no conflict of interest.

References

1. Wen, Z.; Xia, Y.P.; Ji, Y.G.; Liu, Y.M.; Xiong, Z.M.; Lu, H. Study on risk control of water inrush in tunnel construction period considering uncertainty. *J. Civ. Eng. Manag.* **2019**, *25*, 757–772. [CrossRef]
2. Jose, M.-M.F.; Fernando, D.-R.; Jesus, G.-Z.; Wenceslao, M.-R.; Manuel, L.-C.; Lourdes, G.-C. Identification of leakage and potential areas for internal erosion combining ERT and IP techniques at the Negratin Dam left abutment (Granada, southern Spain). *Eng. Geol.* **2018**, *240*, 74–80.
3. Bo, K.; Wang, M.S.; Zhao, Z.Q. Numerical simulation on bottom hole flow fields of reverse circulation bit. *Appl. Mech. Mater.* **2013**, *256–259*, 2826–2830. [CrossRef]
4. Zhang, J.C. Investigations of water inrushes from aquifers under coal seams. *Int. J. Rock Mech. Min. Sci.* **2005**, *42*, 350–360. [CrossRef]
5. Ma, D.; Duan, H.Y.; Zhang, J.X.; Bai, H.B. A state-of-the-art review on rock seepage mechanism of water inrush disaster in coal mines. *Int. J. Coal Sci. Technol.* **2022**, *9*, 50. [CrossRef]
6. Li, C.; He, S.F.; Hou, W.T.; Ma, D. Experimental study on expansion and cracking properties of static cracking agents in different assembly states. *Int. J. Min. Sci. Technol.* **2022**, *32*, 1259–1272. [CrossRef]
7. Ganji, H.T.; Alembagheri, M. Stability of monolithic gravity dam located on heterogeneous rock foundation. *Arab. J. Sci. Eng.* **2018**, *43*, 1777–1793. [CrossRef]
8. Guo, X.; Chai, J.R.; Qin, Y.; Xu, Z.G.; Fan, Y.N.; Zhang, X.W. Mechanism and treatment technology of three water inrush events in the Jiaoxi River Tunnel in Shaanxi, China. *J. Perform. Constr. Facil.* **2019**, *33*, 04018098. [CrossRef]
9. Chen, L.; Zhang, D.S.; Yao, N.; Wang, L.; Fan, G.W.; Wang, X.F.; Zhang, W. Coupling influence of inclination angle and moisture content on mechanical properties and microcrack fracture of coal specimens. *Lithosphere* **2022**, *2021*, 6226445. [CrossRef]
10. Hosseinzadeh, A.; Nobarinasab, M.; Soroush, A.; Lotfi, V. Coupled stress-seepage analysis of Karun III concrete arch dam. *Proc. Inst. Civ. Eng. Geotech. Eng.* **2013**, *166*, 483–501. [CrossRef]
11. Chen, L.W.; Feng, X.Q.; Xie, W.P.; Zeng, W.; Zheng, Z.Y. Using a fluid-solid coupled numerical simulation to determine a suitable size for barrier pillars when mining shallow coal seams beneath an unconsolidated, confined aquifer. *Mine Water Environ.* **2017**, *36*, 67–77. [CrossRef]
12. Ma, Q.; Zhang, Y.D.; Gao, L.S.; Li, Z.X.; Song, G.Y.; Zheng, Y. The optimization of coal pillars on return air sides and the reasonable horizon layout of roadway groups in highly gassy mines. *Sustainability* **2022**, *14*, 9417. [CrossRef]
13. Hu, M.L.; Zhao, W.L.; Lu, Z.; Ren, J.X.; Miao, Y.P. Research on the reasonable width of the waterproof coal pillar during the mining of a shallow coal seam located close to a reservoir. *Adv. Civ. Eng.* **2019**, *2019*, 3532784. [CrossRef]
14. Ma, D.; Duan, H.Y.; Zhang, J.X. Solid grain migration on hydraulic properties of fault rocks in underground mining tunnel: Radial seepage experiments and verification of permeability prediction. *Tunn. Undergr. Space Technol.* **2022**, *126*, 104525. [CrossRef]
15. Chen, S.H.; Zhang, Z.H. Determination of coal pillar width and support parameters in deep coal mines—A case study. *J. Test. Eval.* **2019**, *47*, 3160–3173. [CrossRef]
16. Wang, B.N.; Dang, F.N.; Gu, S.C.; Huang, R.B.; Miao, Y.P.; Chao, W. Method for determining the width of protective coal pillar in the pre-driven longwall recovery room considering main roof failure form. *Int. J. Rock Mech. Min. Sci.* **2020**, *130*, 104340. [CrossRef]
17. Jaiswal, A.; Shrivastva, B.K. Numerical simulation of coal pillar strength. *Int. J. Rock Mech. Min. Sci.* **2009**, *46*, 779–788. [CrossRef]
18. Dong, Y.; Huang, Y.C.; Du, J.F.; Zhao, F. Study on overburden stability and development height of water flowing fractured zone in roadway mining with cemented backfill. *Shock Vib.* **2021**, *2021*, 6661168. [CrossRef]
19. Das, A.J.; Mandal, P.K.; Paul, P.S.; Sinha, R.K.; Tewari, S. Assessment of the strength of inclined coal pillars through numerical modelling based on the ubiquitous joint model. *Rock Mech. Rock Eng.* **2019**, *52*, 3691–3717. [CrossRef]
20. Shi, L.Q.; Wang, Y.; Qiu, M.; Han, L.; Zhao, Y.P. Research on the required width of a fault waterproof coal pillar based on underground pressure control theory. *Arab. J. Geosci.* **2019**, *12*, 480. [CrossRef]
21. Li, A.; Ji, B.N.; Ma, Q.; Liu, C.Y.; Wang, F.; Ma, L.; Mu, P.F.; Mou, L.; Yang, Y.X.; Ding, X.S. Design of longwall coal pillar for the prevention of water inrush from the seam floor with through fault. *Geofluids* **2021**, *2021*, 5536235. [CrossRef]
22. Wang, R.; Bai, J.B.; Yan, S.; Pan, G.Q.; Zhang, D.; Zhu, Q.C. Structure partition and reasonable width determination of waterproof coal pillar in strip mining. *Lithosphere* **2021**, *2021*, 3339797. [CrossRef]
23. Wang, Y.Q.; Wang, X.; Zhang, J.S.; Yang, E.S.; Zhu, W.J.; Wang, Z.P. Similar experimental study on retaining waterproof coal pillar in composite strata mining. *Sci Rep.* **2022**, *12*, 1366. [CrossRef] [PubMed]
24. Mathey, M.; van der Merwe, J.N. Critique of the South African squat coal pillar strength formula. *J. S. Afr. Inst. Min. Metall.* **2016**, *116*, 291–299. [CrossRef]
25. Yao, Q.L.; Yu, L.Q.; Chen, N.; Wang, M.N.; Xu, Q. Experimental study on damage and failure of coal-pillar dams in coal mine underground reservoir under dynamic load. *Geofluids.* **2021**, *2021*, 5623650. [CrossRef]
26. Bukowski, P. Water hazard assessment in active shafts in Upper Silesian Coal Basin Mines. *Mine Water Environ.* **2011**, *30*, 302–311. [CrossRef]
27. Ma, D.; Duan, H.Y.; Zhang, J.X.; Liu, X.W.; Li, Z.H. Numerical simulation of water–silt inrush hazard of fault rock: A three-phase flow mode. *Rock Mech. Rock Eng.* **2022**, *55*, 5163–5182. [CrossRef]

28. Huo, Q.; Li, Z.; Ye, Q.Q. Research on the mechanism of water inrush disaster in the construction of tunnel based on the aquifer inrush model. In Proceedings of the 2nd International Conference on Civil Engineering (ICCEHB 2011), Shijiazhuang, China, 16–18 December 2011.
29. Kerimov, A.; Mavko, G.; Mukerji, T.; Dvorkin, J.; Al Ibrahim, M.A. The influence of convex particles' irregular shape and varying size on porosity, permeability, and elastic bulk modulus of granular porous media: Insights from numerical simulations. *J. Geophys. Res. Solid Earth* **2018**, *123*, 10563–10582. [CrossRef]
30. Ma, D.; Miao, X.X.; Bai, H.B.; Huang, J.H.; Pu, H.; Wu, Y.; Zhang, G.M.; Li, J.W. Effect of mining on shear sidewall groundwater inrush hazard caused by seepage instability of the penetrated karst collapse pillar. *Nat. Hazards* **2016**, *82*, 73–93. [CrossRef]
31. Chen, J.T.; Zhao, J.H.; Zhang, S.C.; Zhang, Y.; Yang, F.; Li, M. An experimental and analytical research on the evolution of mining cracks in deep floor rock mass. *Pure Appl. Geophys.* **2020**, *177*, 5325–5348. [CrossRef]
32. Liang, D.X.; Jiang, Z.Q.; Zhu, S.Y.; Sun, Q.; Qian, Z.W. Experimental research on water inrush in tunnel construction. *Nat. Hazards* **2016**, *81*, 467–480. [CrossRef]
33. Ma, D.; Duan, H.Y.; Li, X.B.; Li, Z.H.; Zhou, Z.L.; Li, T.B. Effects of seepage-induced erosion on nonlinear hydraulic properties of broken red sandstones. *Tunn. Undergr. Space Technol.* **2019**, *91*, 102993. [CrossRef]
34. Li, Q.; Ma, D.; Zhang, Y.D.; Liu, Y.; Ma, Y.J. Insights into controlling factors of pore structure and hydraulic properties of broken rock mass in a geothermal reservoir. *Lithosphere* **2022**, *2021*, 3887832. [CrossRef]
35. Yin, S.X.; Zhang, J.C.; Liu, D.M. A study of mine water inrushes by measurements of in situ stress and rock failures. *Nat. Hazards* **2015**, *79*, 1961–1979. [CrossRef]
36. Ma, D.; Kong, S.B.; Li, Z.H.; Zhang, Q.; Wang, Z.H.; Zhou, Z.L. Effect of wetting-drying cycle on hydraulic and mechanical properties of cemented paste backfill of the recycled solid wastes. *Chemosphere* **2021**, *282*, 131163. [CrossRef]
37. Ma, D.; Zhang, J.X.; Duan, H.Y.; Huang, Y.L.; Li, M.; Sun, Q.; Zhou, N. Reutilization of gangue wastes in underground backfilling mining: Overburden aquifer protection. *Chemosphere* **2021**, *264*, 128400. [CrossRef] [PubMed]
38. Behrenbruch, P.; Biniwale, S. Characterisation of clastic depositional environments and rock pore structures using the Carman-Kozeny equation: Australian sedimentary basins. *J. Pet. Sci. Eng.* **2005**, *47*, 175–196. [CrossRef]
39. Li, C.Y.; Cui, C.Y.; Lei, G.R.; Zuo, J.P.; Yu, X.; He, T.; Li, X.S.; Du, W.S. Mechanism of confining pressure unloading and seepage induced tensile fracture of rock mass in Deep mining. *J. China Coal Soc.* **2022**, *47*, 3069–3082.
40. Wu, J.Y.; Jing, H.W.; Gao, Y.; Meng, Q.B.; Yin, Q.; Du, Y. Effects of carbon nanotube dosage and aggregate size distribution on mechanical property and microstructure of cemented rockfill. *Cem. Concr. Compos.* **2022**, *127*, 104408. [CrossRef]
41. Wu, J.Y.; Wong, H.S.; Yin, Q.; Ma, D. Effects of aggregate strength and mass fraction on mesoscopic fracture characteristics of cemented rockfill from gangue as recycled aggregate. *Compos. Struct.* **2023**, *311*, 116851. [CrossRef]
42. Zhao, K.; Huang, Z.; Yu, B. Damage characterization of red sandstones using uniaxial compression experiments. *RSC Adv.* **2018**, *8*, 40267–40278. [CrossRef] [PubMed]
43. Zhang, J.H.; Chen, H.K.; Wang, H.; Zhou, Z. Experimental study on damage evolution characteristics of rock-like material. *Arab. J. Sci. Eng.* **2019**, *44*, 8503–8513.

water

Article

Impact of the Boreholes on the Surrounding Ground

Sudip Shakya [1], Koki Nakao [1], Shuichi Kuwahara [2] and Shinya Inazumi [3,*]

[1] Graduate School of Engineering and Science, Shibaura Institute of Technology, Tokyo 135-8548, Japan
[2] Japan Association for Pulling-Out Existing Piles, Tokyo 152-0004, Japan
[3] College of Engineering, Shibaura Institute of Technology, Tokyo 135-8548, Japan
* Correspondence: inazumi@shibaura-it.ac.jp; Tel.: +81-3-5859-8360

Abstract: The infrastructures that were constructed decades ago do not meet the present structural benchmark, and they need to be demolished. In order to reclaim these lands, the existing pile foundations must be removed; otherwise, the land will lose its value. Since the piles are pulled out, vacant spaces are created in the ground. This causes the surrounding ground to experience settlement, jeopardizing its stability. The degree of influence depends upon the number of boreholes, the saturated condition of the ground, the time period of the vacant condition, the presence of loading, etc. It is important to understand the scope of the probable settlement under various situations. This study focused on determining the amount of displacement and its range for three different saturated soil types under loaded and unloaded conditions using the finite element method (FEM) analysis. It was observed that stiff ground underwent maximum deformation, while soft ground experienced the maximum influence from external factors. Moreover, the presence of loading not only increased the displacement amount and range, but it also caused a change in the location of the maximum displacement.

Keywords: land reclaim; pile removal; finite element method; ground settlement; loading condition

check for updates

Citation: Shakya, S.; Nakao, K.; Kuwahara, S.; Inazumi, S. Impact of the Boreholes on the Surrounding Ground. *Water* **2023**, *15*, 188. https://doi.org/10.3390/w15010188

Academic Editors: Panpan Guo, Yixian Wang, Hang Lin and Yanlin Zhao

Received: 15 October 2022
Revised: 5 November 2022
Accepted: 24 November 2022
Published: 2 January 2023

1. Introduction

The industrialization period brought rapid economic growth, with lots of infrastructures being constructed all over the world. Those infrastructures have now become obsolete due to the constant updating of building standards, and many have surpassed their lifespan [1]. Thus, they need to be demolished in the near future [2,3]. The chances of these infrastructures being supported by pile foundations are high; thus, the removal of the existing piles is compulsory if the land is to be resold [4–7]. The pile removal process should be carried out carefully so that errors, such as crushing the piles or leaving portions of them in the ground, can be avoided [5–10]. The pile tip chucking method [11] is a recently developed pile pulling method in which the entire pile is wrapped by chucking and efficiently pulled upward, even in difficult cases of mid-breaks or defective joints. Meanwhile, if the boreholes that remain after the existing piles are pulled out are left hollow, the collapse and deformation of the surrounding ground may occur [12]. This causes the concern of ground instability in the vicinity of the boreholes even during the borehole excavation operation. The loading exerted by heavy machinery and the vibrations produced during the operation contribute to the deformation of the surrounding ground. Moreover, the material deposited in the surrounding ground also contributes to the ground deformation. If these boreholes are left vacant, the ground may undergo further deformation. However, filling the excavation holes with treated soil, such as sand, fluidized soil, cement milk, or the equivalent, after removing the existing piles will help reduce the adverse effects [13–15]. Unfortunately, despite these measures, the chances of ground deformation cannot be completely prevented because the treated soil will require 7 to 28 days to fully gain strength, and if there are other important infrastructures nearby, the deformation of the ground may then have adverse effects on their structural integrity. Therefore, it is very important to have clear knowledge of the deformation behavior of the ground.

Wheeler et al. [16], Wheeler [17], Alonso et al. [18], and Fredlund and Morgenstern [19,20] showed that the degree of saturation has a significant influence on the stress-strain behavior of unsaturated soil and that the soil deformation undergoes a continuous elasto-plastic change process. In this study, the authors attempted to quantify the amount of ground deformation for different degrees of saturation using an elasto-plastic model, i.e., the modified Cam-Clay model, with and without the influence of loading, by a finite element method analysis [21–24] so that the adverse effect of ground settlement and instability during pile removal could be tackled.

2. Analysis

2.1. Model and Method

Figure 1a,b shows the analytical models, 50 × 50 × 25 m, for the single and double boreholes. Both models are composed of two different layers showing the typical ground that needs pile foundation for constructing infrastructures. The upper 18 m is a ground layer composed of clay, and the lower 7 m is a bearing layer composed of sand. The properties of the bearing layer remain constant in all the simulations. Meanwhile, the ground layer is assumed to comprise three different clays with different parameters for each simulation. The boreholes are 2 m in diameter, 20 m in depth, and 4 m apart from each other.

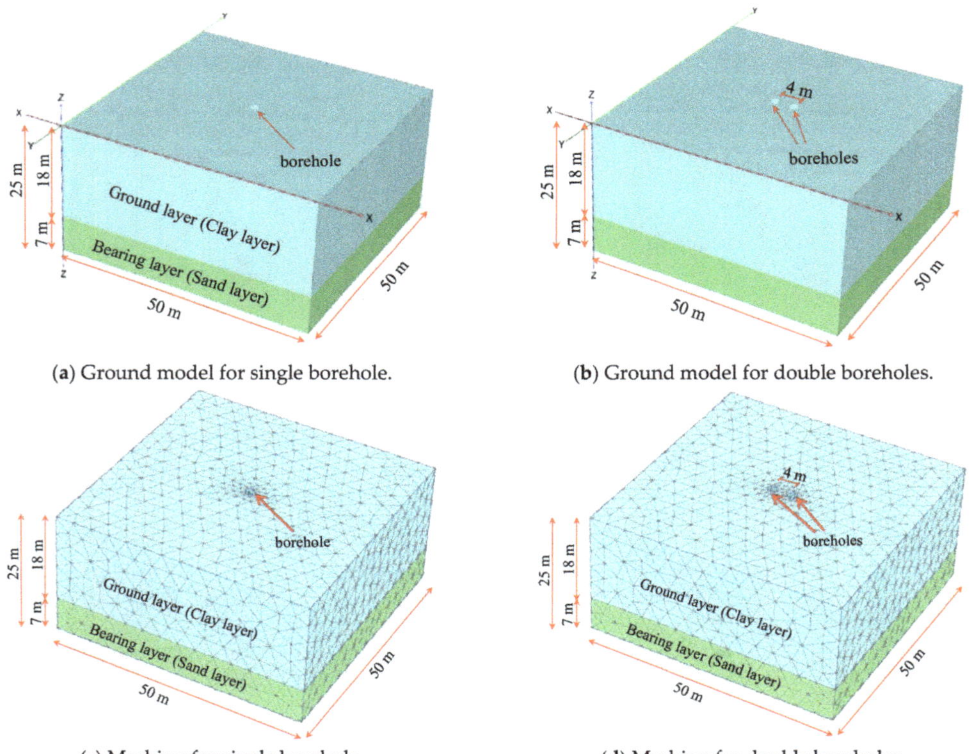

(**a**) Ground model for single borehole. (**b**) Ground model for double boreholes. (**c**) Meshing for single borehole. (**d**) Meshing for double boreholes.

Figure 1. Analytical ground models for single and double boreholes.

Figure 1c,d shows the meshing of the analytical models. Each analytical model was subjected to a finite element method (FEM) analysis using PLAXIS 3D software after the model was divided into a certain number of triangular meshes during the simulations. The mesh size was set to be of smaller size and more concentrated around the boreholes in comparison to other areas so that the simulation accuracy could be increased. The literature review revealed that the soil deformation is the continuous elasto-plastic change process, and the soil showed the complex behavior as a three-dimensional analysis object. Thus, for clay, the elastoplastic model [25] was selected because of its elasticity and plasticity properties. An advanced geotechnical analysis that uses the finite element method is often based on an isotropic elastoplastic model, e.g., the modified Cam-Clay model for clay [26,27]. The modified Cam-Clay model [28–30] is an elasto-plastic strain-hardening model that is based on the critical state theory. Additionally, this model produces the more accurate shear behavior of soil, and it requires only three main parameters, i.e., compression index (λ), swelling index (κ), and the critical stress ration (M), which can be easily determined. Thus, the modified Cam-Clay model was applied to the ground layer, whereas the Mohr-Coulomb model was applied to the bearing layer and the filler material. The drainage condition was set as top-bottom drainage and left-right drainage to increase the accuracy of the simulation.

2.2. Parameters

Table 1 shows the input parameters for the clays of different hardness, while Table 2 shows the input parameters for the filler material and bearing layer. The values for the irrelevant parameters in this simulation are not given. The values for the filler material and bearing layer parameters in Table 2 were taken from Kawahara et al. [11]. The modified Cam-Clay model was applied to the ground layer; Xu et al. [31] reported the applicability of this model to consolidated soil and over-consolidated soft clays. In this study, all three clays were considered to have an over-consolidation ratio (OCR) of 1, representing a normal consolidation state. Ikegami et al. [32] and Murakami et al. [33] analyzed the model using actual data measured in the field for the parameter values or set the values by referring to the measured N-values. In this study, however, the values for the parameters in the simulations were derived from the relations of one parameter with another. Kawamura et al. [34] suggested relationships for calculating the values of the compaction index and swelling index from the natural water contents for the simulations. Meanwhile, the methods used to determine other parameters are explained below.

Table 1. Input parameters for the clays of different hardness.

w_n	N-Value	γ_{unsat} (kN/m^3)	γ_{sat} (kN/m^3)	ν	λ	κ	M	K_0	OCR	e_0	k_p (m/d)	Ground Classification
30%	8.88	16	17	0.277	0.107	0.012	1.555	0.383	1	0.817	2.23×10^{-2}	Stiff
40%	6.99	15	16	0.275	0.164	0.018	1.562	0.380	1	1.089	4.34×10^{-3}	Medium
80%	2.68	13	14	0.274	0.389	0.04	1.569	0.378	1	2.177	8.47×10^{5}	Soft

Table 2. Input parameters for the filler material and bearing layer.

Material	γ_{unsat} (kN/m^3)	γ_{sat} (kN/m^3)	q_u (kN/m^3)	E (kN/m^3)	ν (-)	Φ (°)	C (kN/m^2)	k_p (m/d)	N-Value
Filler material	14	15	100	136,223	0.48	26	50	8.64×10^{-5}	-
Bearing layer	20	21	-	1.4×10^{5}	0.3	-	0	0.864	50

Uniaxial compressive strength (q_u): The uniaxial compressive strength of clay with a certain water content was calculated with Equation (1), while the N-value of the clay was calculated with Equation (2):

$$q_u = 255.5e^{-0.024w_n} \quad (1)$$

$$q_u = 14N \quad (2)$$

where q_u is the uniaxial compressive strength, N is the N-value of the clay, w_n is the natural water content, and e is the base of the natural logarithm.

Unit weight volume (γ): The consistency of each clay type was confirmed by the calculated value for N and was categorized as stiff, medium, or soft for 30, 40, or 80% water contents, respectively. After that, the unit weight of each clay was determined based on its consistency from the Road Earthworks-Embankment Guidelines 2022 [35].

Compression index (C_c): The compression index was calculated from the natural water content (w_n), as shown in Equation (3), while the compression index (λ) used in the modified Cam-Clay model was calculated with Equation (4):

$$C_c = 0.013\ (w_n - 11) \tag{3}$$

$$\lambda = 0.434C_c \tag{4}$$

where C_c is the compression index, and λ is the compression index used in the modified Cam-Clay model.

Swelling index (C_s): The swelling index was calculated from the natural water content (w_n), shown in Equation (5), while the swelling index (κ) used in the modified Cam-Clay model was calculated with Equation (6).

$$C_s = 0.0013\ (w_n - 8.8) \tag{5}$$

$$\kappa = 0.434C_s \tag{6}$$

where C_s is the swelling index, and κ is the swelling index used in the modified Cam-Clay model.

Critical stress ratio (M): The irreversible ratio (Λ) was calculated from the compressive index (λ) and the swelling index (κ), shown in Equation (7), while the critical stress ratio (M) was calculated from the irreversible ratio (Λ), shown in Equation (8).

$$\Lambda = 1 - \frac{\kappa}{\lambda} \tag{7}$$

$$M = \Lambda * 1.75 \tag{8}$$

where Λ is the irreversible ratio and M is the critical stress ratio.

Coefficient of earth pressure at rest (K_0): Using Equation (9), sin Φ was calculated from the critical stress ratio, and the coefficient of static earth pressure was calculated from sin Φ, as shown in Equation (10):

$$M = \frac{6 \sin \Phi}{3 - \sin \Phi} \tag{9}$$

$$K_0 = 1 - \sin \Phi \tag{10}$$

where Φ is the internal friction angle, and K_0 is the coefficient of static earth pressure.

Poisson's ratio (ν): Poisson's ratio was calculated with Equation (11):

$$\nu = \frac{K_0}{1 - K_0} \tag{11}$$

where ν is Poisson's ratio, which is calculated from the static earth pressure coefficient.

Void ratio (e_0): As shown in Equation (12), the void ratio was calculated from the compression index:

$$C_c = 0.478\ (e_0 - 0.3) \tag{12}$$

where e_0 is the void ratio.

Hydraulic conductivity (k_p): The hydraulic conductivity was calculated as shown in Equation (13):

$$k_p = 6322.7{w_n}^{-5.68} \tag{13}$$

where k_p is the hydraulic conductivity, which is calculated from the natural water content.

3. Results and Discussion

3.1. Influence of Boreholes on Surrounding Ground under Unloaded Condition

3.1.1. In Case of Single Boreholes Left Vacant

Figure 2 shows the top views of the displacement contour for single boreholes left vacant. It is seen that the surrounding area of influence has a circular shape. The maximum displacement vectors for single boreholes without loading for stiff, medium, and soft grounds were observed at (25, 26, 0), (25, 26, 0), and (25.903, 25.429, −0.710), respectively. Figure 3 shows a comparison of the rates of change in the maximum displacement vectors of single boreholes. Soft ground underwent the least displacement, whereas stiff and medium grounds underwent approximately equal displacement at the end of the simulation. The initial increment in displacement had a higher value for stiffer ground and a lower value for the stabilization time for the softer ground. The total displacement values at the end of the 29-day simulation for stiff, medium, and soft grounds were 35, 34.3, and 13.6 mm, respectively. Additionally, the time periods required for the stabilization were 8, 15, and 1.25 days, respectively. The reason for the stiff clay having the maximum deformation might be in the optimum water content of the soil. It is established fact that the maximum soil deformation occurs at the optimum water content under loading, and as confirmed from the result, the chances of the stiff soil having a 30% natural water content, which would be closer to the optimum water content than 40% and 80%, is high.

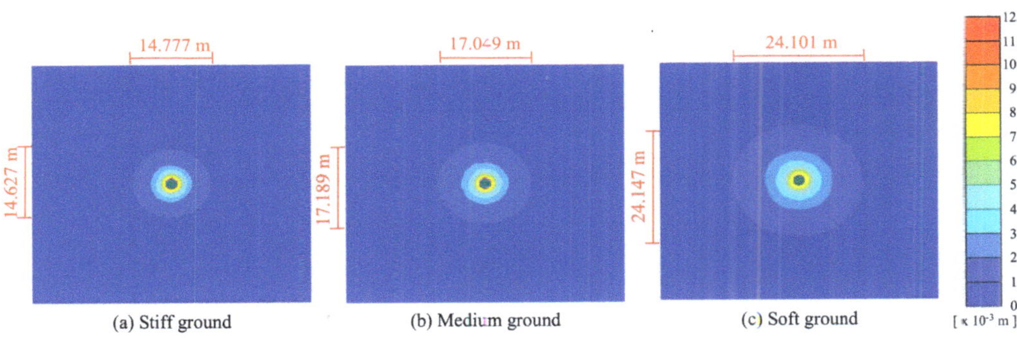

Figure 2. Top views of displacement contour for single boreholes left vacant.

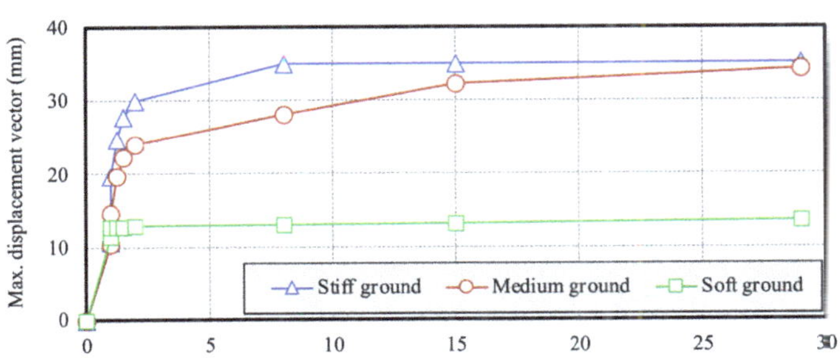

Figure 3. Rates of change in maximum displacement vectors for single boreholes left vacant.

3.1.2. In Case of Double Boreholes Left Vacant

Figure 4 shows the top views of the displacement contour for double boreholes left vacant. The influential area is elliptical in shape, with the highest degree of influence around the vicinity of the borehole. The maximum displacement vector was observed at the coordinates of (26.866, 25.5, 0), (26.866, 25.5, 0), and (23.966, 25.259, −0.428) for stiff, medium, and soft grounds, respectively. Figure 5 shows a comparison of the rates of change in the maximum displacement vectors of double boreholes. Figures 6 and 7 show a comparison of the total stress and pore water pressure of the double boreholes, respectively. The displacement and stabilization patterns, when compared to those of the single boreholes left vacant, were found to be similar, except for medium ground, which continued to deform until the end of the simulation period, and the final displacement value became greater than that of the stiff ground. Considering the result for the single borehole deformation, it can be concluded that the 40% water content also might not be farther from the optimum water content than 30%, but the presence of the extra borehole might have triggered an additional factor in deformation. The total displacement values for stiff, medium, and soft grounds were observed to be 46.4, 49, and 17.8 mm, respectively. The displacement value was higher compared to the single borehole, suggesting that the increase in the number of boreholes contributed to a higher deformation.

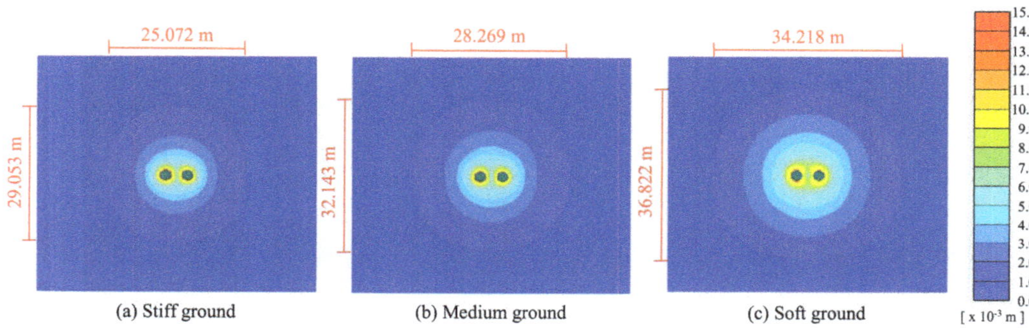

Figure 4. Top views of displacement contour for double boreholes left vacant.

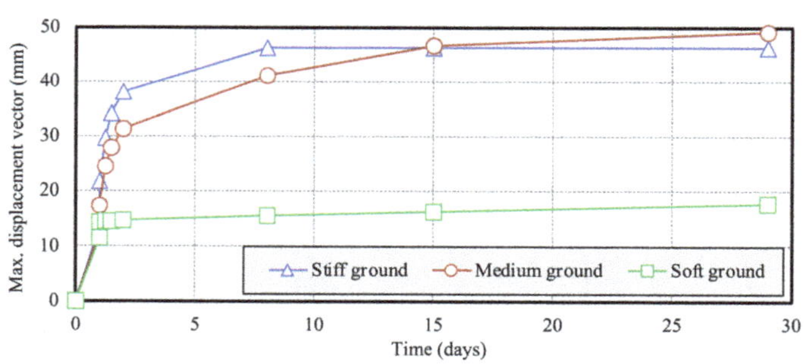

Figure 5. Rates of change in maximum displacement vectors for double boreholes left vacant.

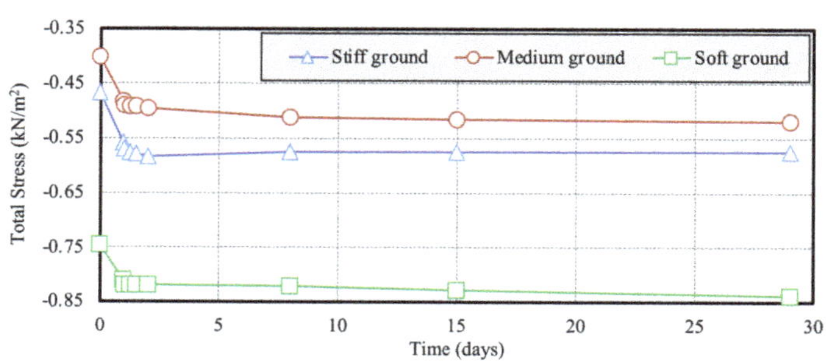

Figure 6. Rates of change in total stress for double boreholes left vacant.

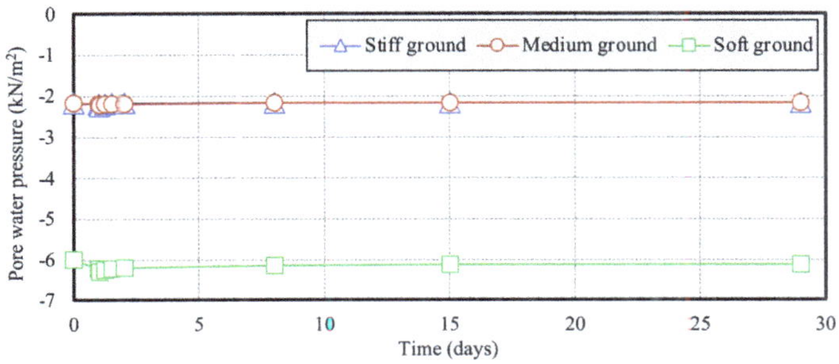

Figure 7. Rates of change in pore water pressure for double boreholes left vacant.

Soft ground had lower values for the initial total stress and pore water pressure compared to stiff and medium grounds. All the ground types experienced a reduction in value during the simulation. The total stress amounts for stiff and medium grounds started stabilizing after an 8-day period, but they continued to decrease for soft ground. Meanwhile, the pore water pressure initially decreased rapidly, but it continued to increase after 1 day until it reached a stabilized value. This recovery period was the shortest for stiff ground and the longest for soft ground.

3.1.3. Comparison of Cases of Vacant and Filled Double Boreholes

From the analysis of the vacant and filled conditions, it was seen that the rate of the surrounding ground deformation immediately after the borehole excavation was rapid and of the maximum value. In this study, a comparison was made of the amount of sinking of the surrounding ground and the pore water pressure for vacant and filled conditions. Figure 8 shows a comparison of the frontal cross sectional, i.e., the XZ plane total stress contour, in the cases of double boreholes left vacant and after being filled with the filler material. After the boreholes were filled, the stress concentrating around the boreholes decreased, and the stress at the bottom of the boreholes dissipated, resulting in an increase in stress.

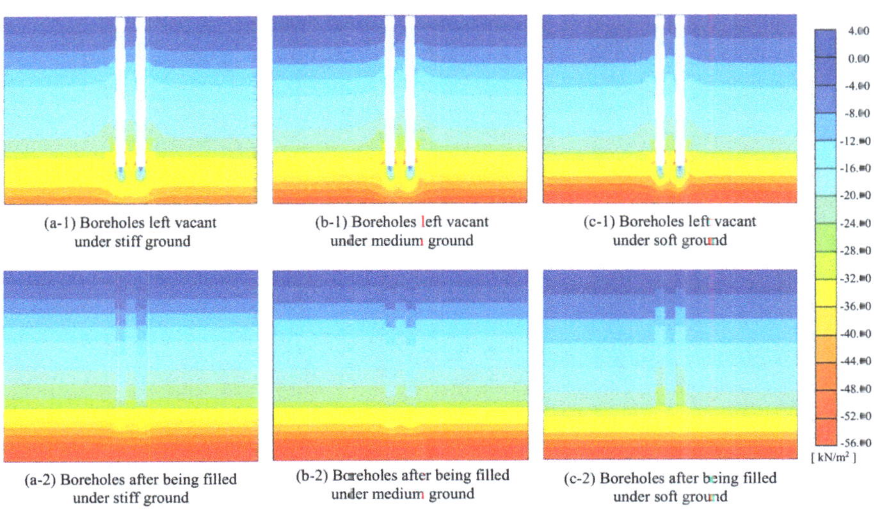

Figure 8. Comparison of frontal cross sectional total stress contour for vacant and filled cases of double boreholes.

Figure 9 shows a comparison of the amounts of total sinking of the different ground types for the vacant and filled double boreholes. Figure 10 shows a comparison of the pore water pressure of stiff, medium, and soft grounds in succession for the vacant and filled double boreholes. When the boreholes were filled with the filler material, stiff and medium grounds were seen to be uplifted by a small amount instead of experiencing settlement. The probable reason might be the upward stress caused by the action of filling the boreholes. As for soft ground, the amount of settlement was greatly reduced by the remaining concentrated stress. Overall, the settlement of the ground was largely avoided in the case of the filled boreholes. The pore water pressure recovered in all the ground types after the boreholes were filled. It was higher in stiff ground, with a recovery value of 0.06 kN/m^2, and it was same for medium and soft ground, with a recovery value of 0.04 kN/m^2.

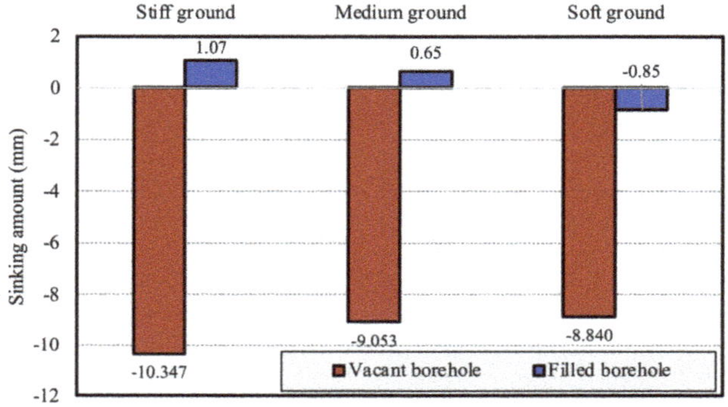

Figure 9. Comparison of sinking amounts for vacant and filled cases of double boreholes.

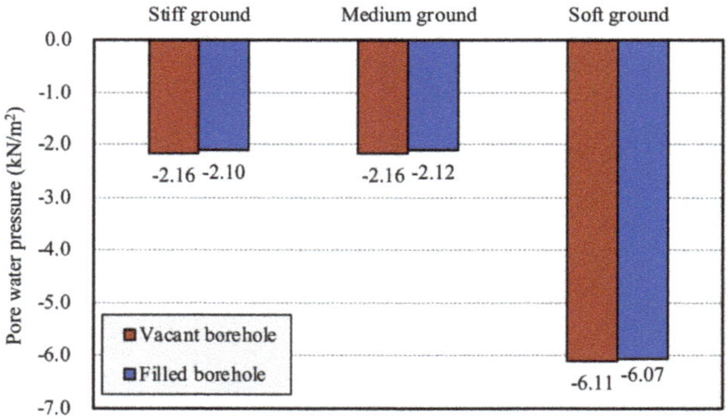

Figure 10. Comparison of porewater pressure for vacant and filled cases of double boreholes.

3.2. *Influence of Loading on Ground Surrounding Boreholes*

Figure 11 shows analytical models with loading for single and double boreholes at positions 1 and 2 in succession. Two steel paving plates, each 1.5 × 3.0 m in size and 22 mm thick, were placed over the area of 3 × 3 m, which acted as the dead load of 0.5 kN/m^2 magnitude. The load was set by considering that the load value would not become too large and affect the accuracy of the analysis results.

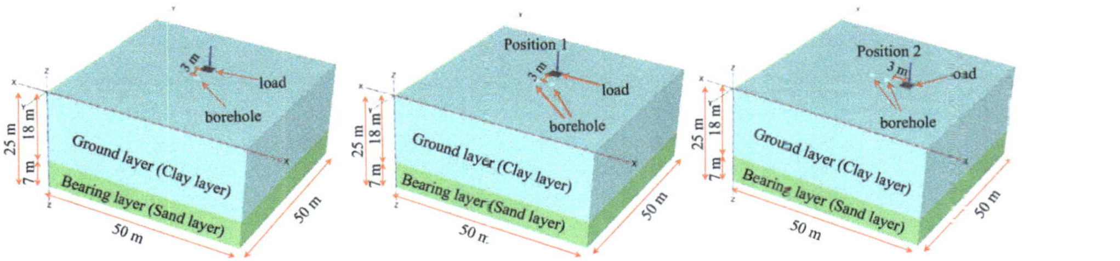

Figure 11. Analytical ground models for single and double boreholes under the loading condition.

3.2.1. In Case of Single Boreholes Left Vacant

Figure 12 shows the top views of the displacement contour for single boreholes with loading. Figure 13 shows a comparison of the rates of change in maximum displacement vectors for single boreholes with loading. The maximum displacement vectors for single boreholes without loading were observed at (25, 26, 0), (25, 26, 0), and (25.903, 25.429, −0.710) for stiff, medium, and soft grounds, respectively. When the load was applied, the maximum displacement vector coordinates shifted to (25, 26, 0), (25, 26, 0), and (25, 26, 0), respectively. The coordinates of the maximum displacement vectors remained unchanged for stiff and medium grounds, but they shifted toward the positive Y axis, in the direction of the loading location, for soft ground. This implies that only soft ground is likely to deform at different places depending upon the location of loading, which makes it more vulnerable than other ground types. The displacement and stabilization patterns were found to be the same as those of the unloaded case for all ground types. The displacement value increased by approximately 4 mm for the loading case in stiff and medium grounds, but it was only approximately 1 mm for soft ground. The overall deformation amount increased due to loading, and the value was higher for stiffer ground. Meanwhile, the deformation amount and the coordinates of the maximum deformation changed for soft ground, with the deformation inclining in the loading location.

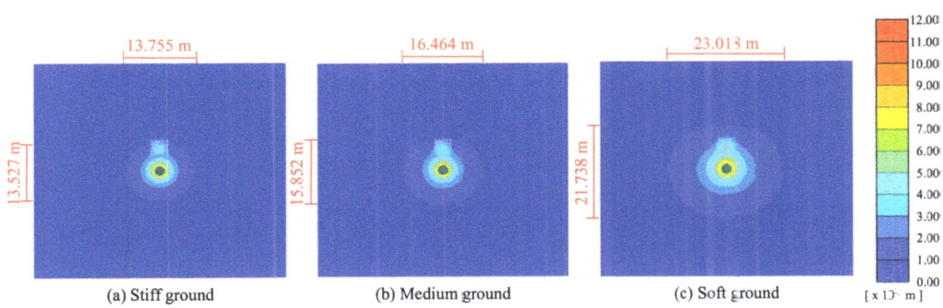

Figure 12. Top views of the displacement contour for single boreholes under the loading condition.

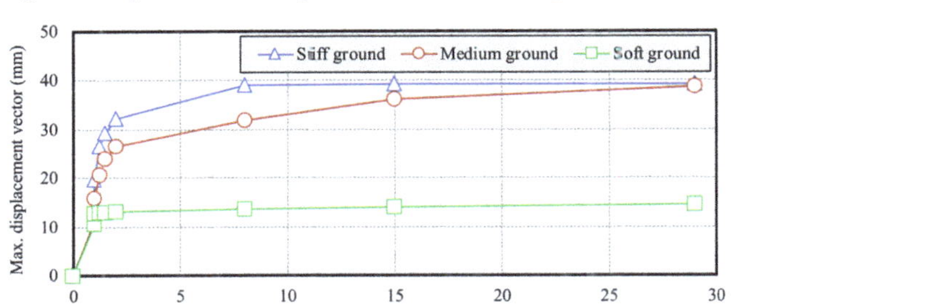

Figure 13. Rates of change in maximum displacement vectors for single boreholes under the loading condition.

3.2.2. In Case of Two Boreholes Left Vacant and Loading at Different Position

Figure 14 shows the top views of the displacement contour for the double borehole models, with loading at locations 1 and 2 in succession for grounds of different stiffness levels. Figure 15 shows a comparison of the rates of change in the maximum displacement vectors for double boreholes, with loading at position 1 or 2 in succession. The maximum displacement vectors for the double boreholes in a non-loaded condition were observed at the coordinates of (26.866, 25.5, 0), (26.866, 25.5, 0), and (23.966, 25.259, −0.428) for stiff, medium, and soft grounds, respectively. When loading was applied at position 1, the maximum displacement vector coordinates shifted to (23, 26, 0), (23, 26, 0), and (26.231, 25.639, −0.356), respectively, leaning toward the positive Y-axis. Additionally, when the loading was applied at position 2, the maximum displacement vector coordinates shifted to (26.134, 24.5, 0), (26.134, 24.5, 0), and (26.034, 24.741, −0.428), respectively, leaning toward the positive X axis. Thus, it can be concluded that the maximum displacement tends to occur in the direction of the load application.

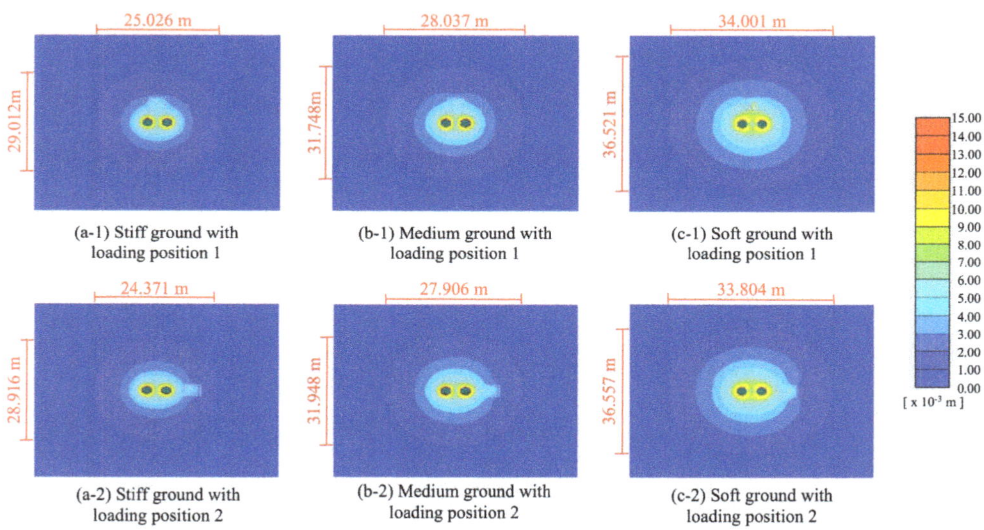

Figure 14. Top views of displacement contour for double boreholes under the loading condition.

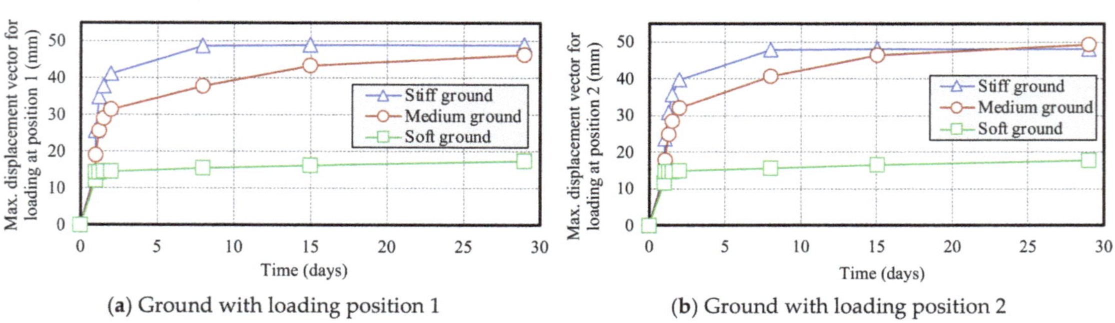

Figure 15. Rates of change in maximum displacement vectors for double boreholes under the loading condition.

However, there were no distinct changes in the intensity of the maximum displacement vector for either non-loading or loading at the different locations. The probable reason for this might lie in the assumption made during the simulation to reduce the calculation load.

Firstly, the load was not heavy enough to cause the maximum impact on the surrounding ground, as it was only 0.5 kN/m^2 spanning over 3 × 3 m. Next, the ground model contained only two boreholes; considering that the model is 50 × 50 m, the number of boreholes was too low to impart the maximum impact. For stiff ground, the maximum displacement vector value increased by approximately 2 mm in the case of loading at either position. For soft ground, the changes were negligible in both cases. Meanwhile, for medium ground, the value decreased by approximately 3 mm in the case of loading at position 1, but it remained unchanged in the case of loading at position 2. Although the overall influence of loading at either position remained negligible, the displacement of the ground between the load and the borehole increased, and the range of influence shifted toward the direction of the loading.

4. Conclusions

In this study, single and double boreholes were studied for cases with and without loading. The main focus was placed on the amount of maximum displacement experienced in each case. Additionally, a comparison was made for double boreholes in the cases of the boreholes being left vacant and being filled with filler material, and the parameters, such as the sinking amount and pore water pressure, were examined. From this study, the following conclusions can be drawn.

(1) The amount of ground displacement was seen to depend upon the stiffness of the ground. The maximum initial displacement was observed for stiff ground in all cases, except the case of double boreholes, in which the final stabilized displacement value was slightly higher or equal to that of the stiff ground.
(2) Soft ground was found to be relatively more unstable than stiff and medium grounds, as the location of the maximum deformation of this ground was different than that of the other grounds.
(3) The increase in the amount of displacement was observed to be larger for the case of an increased number of boreholes than that due to the loading.
(4) The surrounding ground remained settled if the boreholes were left vacant, but this settlement was prevented if the holes were immediately filled with appropriate filling material. Moreover, the pore water pressure recovery was higher for the filled condition.
(5) The presence of external loading not only contributed to an increase in the amount of displacement, but it also affected the location of the maximum displacement. It was observed that the inclination tended to occur in the direction of loading, indicating susceptibility to external loading.
(6) The influence on the horizontal range and maximum displacement vector of the surrounding ground was lower in location 2. In other words, borehole-related work conducted with machinery located in the existing pile alignment resulted in less influence on the ground.

This study was conducted for a maximum of two boreholes and was focused only on the deformation behavior. It did not consider the effect on the borehole wall during the progress of the pile removal. Future research should focus on improving the practicality of the conclusions by developing models based on the actual conditions of the site.

Author Contributions: Conceptualization, S.I. and S.K.; methodology, S.I. and K.N., software, S.I.; validation, S.S., K.N. and S.I.; formal analysis, S.S.; investigation, S.K. and K.N.; resources, S.I.; data curation, S.S. and K.N.; writing—original draft preparation, S.S.; writing—review and editing, S.I.; visualization, S.I.; supervision, S.I.; project administration, S.I.; funding acquisition, S.I. and S.K. All authors have read and agreed to the published version of the manuscript.

Funding: This research received no external funding.

Institutional Review Board Statement: Not applicable.

Informed Consent Statement: Not applicable.

Data Availability Statement: Not applicable.

Conflicts of Interest: The authors declare no conflict of interest.

References

1. Yoshioka, S.; Kawasaki, H. Japan's high-growth postwar period (the role of economic plans). *ESRI Res. Note* **2016**, *27*, 1–82.
2. Inazumi, S.; Tanaka, S.; Komaki, T.; Kuwahara, S. Effect of insertion of casing by rotation on existing piles in removal of existing pile. *Geotech. Res.* **2021**, *8*, 25–37. [CrossRef]
3. Ministry of Land, Infrastructure, Transport and Tourism (MLIT). *Public Building Construction Standard Specification (Building Work)*, 2013th ed.; Ministry of Land, Infrastructure, Transport and Tourism: Tokyo, Japan, 2016.
4. Inazumi, S.; Kuwahara, S.; Jotisankasa, A.; Chaiprakaikeow, S. Construction method for pulling out existing piles and influence of pulling-out holes on the surrounding ground. *Geotech. Geol. Eng.* **2020**, *38*, 6107–6123. [CrossRef]
5. Kuwahara, S.; Inazumi, S.; Jotisankasa, A.; Chaiprakaikeow, S. Influence of the condition of pullout holes on the surrounding ground. *Int. J. Geo Eng.* **2020**, *11*, 10. [CrossRef]
6. Inazumi, S.; Hashimoto, R.; Shinsaka, T.; Nontananandh, S.; Chaiprakaikeow, S. Applicability of additives for ground improvement utilizing fine powder of waste glass. *Materials* **2021**, *14*, 5169. [CrossRef] [PubMed]
7. Inazumi, S.; Kuwahara, S.; Jotisankasa, A.; Chaiprakaikeow, S. Improvement mechanism of sodium carbonate on traditional composite filler. *Ground Improv.* **2021**, *174*, 132–139. [CrossRef]
8. Inazumi, S.; Namikawa, T.; Kuwahara, S.; Hamaguchi, S. Influence of pulling out existing piles on the surrounding ground. *Int. J. GEOMATE Geotech. Constr. Mater. Environ.* **2017**, *13*, 16–21. [CrossRef]
9. Inazumi, S.; Kuwahara, S.; Jotisankasa, A.; Chaiprakaikeow, S. MPS-CAE simulation on dynamic interaction between steel casing and existing pile when pulling out existing piles. *Int. J. Geomate Geotech. Constr. Mater. Environ.* **2020**, *18*, 68–73. [CrossRef]
10. Inazumi, S.; Kuwahara, S.; Nontananandh, S.; Jotisankasa, A.; Chaiprakaikeow, S. Numerical analysis for ground subsidence caused by extraction holes of removed piles. *Appl. Sci.* **2022**, *12*, 5481. [CrossRef]
11. Kawahara, T.; Kuwahara, S.; Inazumi, S.; Eguchi, T. Evaluation of the influence of filling material of existing pile extraction holes on the surrounding ground. In Proceedings of the JSCE 73rd Annual Conference, Sapporo, Japan, 29–30 August 2018; pp. 255–256.
12. Nontananandh, S.; Kuwahara, S.; Shishido, K.; Inazumi, S. Influence of perforated soils on installation of new piles. *Appl. Sci.* **2022**, *12*, 7712. [CrossRef]
13. Inazumi, S.; Kaneko, M.; Tomoda, Y.; Shigematsu, Y.; Shishido, K. Evaluation of flow-ability on fluidization treated soils based on flow analysis by MPS method. *Int. J. GEOMATE Geotech. Constr. Mater. Environ.* **2017**, *12*, 53–58. [CrossRef]
14. Inazumi, S.; Kaneko, M.; Shigematsu, Y.; Shishido, K. Fluidity evaluation of fluidisation treated soils based on the moving particle semi-implicit method. *Int. J. Geotech. Eng.* **2018**, *12*, 325–336. [CrossRef]
15. Inazumi, S.; Shigematsu, Y.; Nakao, K.; Shishido, K. 3-D particle flow analysis for fluidization treated soils. *Am. J. Civ. Environ. Eng.* **2018**, *3*, 59–67.
16. Wheeler, S.J.; Sharma, R.J.; Buisson, M.S.R. Coupling of hydraulic hysteresis and stress-strain behaviour in unsaturated soils. *Geotechnique* **2003**, *53*, 41–54. [CrossRef]
17. Wheeler, S.J. Inclusion of specific water volume within an elasto-plastic model for unsaturated soil. *Can. Geotech. J.* **1996**, *33*, 42–57. [CrossRef]
18. Alonso, E.E.; Gens, A.; Hight, D.W. Special Problem Soils. In Proceedings of the 9th European Conference on Soil Mechanics and Foundation Engineering, Dublin, Ireland, 31 August–3 September 1987; Volume 3, pp. 1087–1146.
19. Fredlund, D.G.; Morgenstern, N.R. Stress state variables for unsaturated soils. *J. Geotech. Eng. Div.* **1977**, *103*, 447–466. [CrossRef]
20. Fredlund, D.G.; Morgenstern, N.R.; Widger, R.A. The shear strength of unsaturated soils. *Can. Geotech. J.* **1978**, *15*, 313–321. [CrossRef]
21. Pradhan, K.K.; Chakraverty, S. Chapter Four-Finite Element Method. *Comput. Struct. Mech. Stat. Dyn. Behav.* **2019**, 25–28. [CrossRef]
22. Okamoto, K. Chapter 6—Finite Element Method. In *Fundamentals of Optical Waveguides*, 3rd ed.; Academic Press: Cambridge, MA, USA, 2022; pp. 271–338. [CrossRef]
23. Ramirez, W.F. Chapter 8—Solution of Partial Differential Equations. In *Computational Methods in Process Simulation*, 2nd ed.; Elsevier: Amsterdam, The Netherlands, 1997; pp. 353–430. [CrossRef]
24. Rapp, B.E. Finite Element Method. In *Microfluidics: Modelling, Mechanics and Mathematics*; Elsevier: Amsterdam, The Netherlands, 2017.
25. Ghasemzadeh, H.; Soujoudi, M.H.; Ghoreishian Amiri, S.A.; Karami, M.H. Elastoplastic model for hydro-mechanical behaviour of unsaturated soils. *Soils Found.* **2017**, *57*, 371–383. [CrossRef]
26. Roscoe, K.H.; Burland, J.B. *On the Generalised Stress-Strain Behaviour of Wet Clay*; Cambridge University Press: Cambridge, UK, 1968; pp. 535–609.
27. Liu, K.; Chen, S.L.; Voyiadjis, G.Z. Integration of anisotropic modified Cam Clay model in finite element analysis: Formulation, validation, and application. *Comput. Geotech.* **2019**, *116*, 103198. [CrossRef]

28. Borja, R.I. Cam-Clay plasticity, Part II: Implicit integration of constitutive equation based on a nonlinear elastic stress predictor. *Comput. Methods Appl. Mech. Eng.* **1991**, *88*, 225–240. [CrossRef]
29. Borja, R.I.; Lee, S.R. Cam-clay plasticity, Part 1: Implicit integration of elasto-plastic constitutive relations. *Comput. Methods Appl. Mech. Eng.* **1990**, *78*, 49–72. [CrossRef]
30. Perić, D. Analytical solutions for a three-invariant Cam clay model subjected to drained loading histories. *Int. J. Numer. Anal. Methods Geomech.* **2006**, *30*, 363–387. [CrossRef]
31. Xu, S.; Xu, G.; Cheng, Y. Review of Cam-Clay model development of soil. *Acad. J. Yangtze River Acad. Sci. People's Repub. China* **2007**, *24*, 3.
32. Ikegami, S.; Mizuno, K.; Kiyama, N.; Kinoshita, H.; Tsuchida, T. Evaluation and case analysis of lateral deformation prediction of soft ground. *Geotech. Eng. Constr.* **2007**, *25*, 133–144.
33. Murakami, T.; Niihara, Y.; Yamada, T.; Ohno, S.; Noguchi, T.; Miyata, M. Deformation prediction of large-scale seawall structures by elasto-viscoplastic FEM analysis. *J. Jpn. Soc. Civil. Eng. C* **2012**, *68*, 224–238.
34. Kawamura, S.; Ekisaka, K.; Aso, K. Consideration on soil characteristic values in ground in Yamaguchi prefecture. *J. Chugoku Branch Jpn. Geotech. Soc. Soils Constr.* **2017**, *35*, 185–190.
35. Japan Road Association. *Road Earthworks-Embankment Guidelines*, 14th ed.; Japan Road Association: Tokyo, Japan, 2022; p. 101.

water

MDPI

Article

Fast Recognition on Shallow Groundwater and Anomaly Analysis Using Frequency Selection Sounding Method

Lu Yulong [1,*], Yang Tianchun [1,*], Abdollah Taheri Tizro [2] and Liu Yang [1]

[1] School of Earth Sciences and Spatial Information Engineering, Hunan University of Science and Technology, Xiangtan 411201, China
[2] Department of Water Engineering, College of Agriculture, Bu-Ali Sina University, Hamedan 65174, Iran
* Correspondence: 1010106@hnust.edu.cn (L.Y.); 1020076@hnust.edu.cn (Y.T.); Tel.: +86-130-1715-5739 (L.Y.)

Abstract: The validity of the frequency selection method (FSM) in shallow (<150 m) groundwater exploration was illustrated by practical applications, and the relationship between potential electrode spacing *MN* and groundwater depth in FSM sounding method was analyzed and preliminary theoretical research was carried out by a simple geologic-geophysical model of sphere. Firstly, under the combined action of horizontal alternating electric field and alternating magnetic field, a simplified geophysical model of low resistivity conductive sphere in homogeneous half space was established, and the forward calculation was performed on the FSM sounding curve. Then, the water yield of 131 wells in the application of FSM in the Rural Drinking Water Safety Project of 12th Five-Year Plan in Guangxi Province was counted. In addition, detailed tabular statistical analysis was carried out on the drilling results of 98 drilling wells, and the relationship between potential electrode spacing *MN* at abnormal sounding curve and actual drilling water depth was compared and studied. Theoretical analysis and practical application show that FSM has obvious effectiveness in shallow groundwater exploration, and it is an effective method to determine shallow groundwater well locations in the future. The cause of FSM anomaly is the comprehensive effect of the natural 3D alternating electromagnetic signal underground. At the same time, the practical statistics show that there is 1:1 approximation between the size of potential electrode spacing *MN* at the anomaly curve of the frequency selection method and the actual drilling water depth, which verifies the correctness of the theoretical simulation results. FSM could be widely used in the shallow groundwater exploration in the future, and it is an effective, non-destructive, fast, and low-cost geophysical method.

Keywords: geophysics; groundwater exploration; forward modeling; frequency selection method (FSM); sounding; electromagnetic field; stray current

Citation: Yulong, L.; Tianchun, Y.; Tizro, A.T.; Yang, L. Fast Recognition on Shallow Groundwater and Anomaly Analysis Using Frequency Selection Sounding Method. *Water* **2023**, *15*, 96. https://doi.org/10.3390/w15010096

Academic Editors: Panpan Guo, Hang Lin, Yanlin Zhao and Yixian Wang

Received: 13 December 2022
Revised: 16 December 2022
Accepted: 18 December 2022
Published: 28 December 2022

1. Introduction

The frequency selection method of telluric electricity field is simply called the frequency selection method (FSM), which is a passive source electromagnetic exploration method. It is based on the difference of electrical conductivity between underground rocks (ores) and rocks, using magnetotelluric field as the source of the working field. By measuring and studying the variation law of the horizontal component of electric field produced by several natural alternating electromagnetic fields with different frequencies on the ground, the electrical properties of underground geoelectric sections are studied. This method was proposed by Chinese scholars [1]. Since the early 1980s, Chinese scholars have put forward the stray current method (or the audio geoelectric potential method), the acoustic geoelectric field method, the low-frequency geoelectric field method, the frequency selection method of natural electric field, the telluric electrical-field frequency selecting method, the audio frequency telluric electric field method, the natural alternating electric field method, the interfering electric field method, the underground magneto fluid detection method, and so on [1–9]. However, the principle and characteristics of the instruments developed or used by the above-mentioned methods are essentially similar to one another. Therefore, the authors hereby call them FSM in this paper. Up to now, scholars have paid much attention to the practical application of this method, especially in the development of instruments and the application of this method in the exploration of groundwater resources and groundwater disasters [10–17], but there is little theoretical research on this method.

FSM is similar to the audio magnetotelluric method (AMT), but it only measures the horizontal electric component of the ground and does not measure the magnetic field component. The frequency selection of this method is mostly at 15 Hz~1.5 kHz, and the potential electrodes spacing of *MN* is usually 20 m or 10 m when profiling method is used. Therefore, it is an efficient, cost-effective tool for groundwater exploration, and more adaptable than other geophysical prospecting methods in mountainous areas and towns with dense building facilities.

At the same time, FSM is similar to the telluric current prospecting and telluric-telluric profiling (TTP) method, which only measures the electric field component [18–22]. However, telluric current prospecting is suited primarily to sedimentary structures associated with lateral changes in electrical resistivity that are frequently encountered in petroleum and geothermal exploration. The TTP method is an effective and economical geophysical method for detecting the fluctuation of high resistivity bedrock strata, with its frequency 0.01~0.1 Hz and potential electrodes spacing 1~n km. The TTP method is derived from the magnetotellurics (MT) method, and in general, a fixed reference point is chosen in application, and it is often used in conjunction with MT method to calculate the false apparent resistivity, and the work frequency is also different with FSM [18].

According to the previous research results, the source of natural field method is very complex and there are three viewpoints for the field source problem of FSM. The first is that its field source is exactly the same as MT, mainly from the outside of the earth, which is also the view of most scholars. The second is the electromagnetic waves radiated upward from the mantle asthenosphere and the geomagnetic field below the Earth's interior [2]. The third is that the field source is the industrial stray current field [1].

In recent years, the author has carried out a preliminary study on the causes of profile anomalies in FSM through practical applications and theoretical research [23–30]. The author considers that it is unnecessary to pay special attention to whether the field source of FSM is from the outside of the earth, the inside of the earth or the industrial stray current field. Past successful examples indicate a shallow prospecting depth of FSM in groundwater exploration (generally < 200 m), and anthropogenic electromagnetic noise, such as high-voltage lines and substation equipment, will be avoided as far as possible in practice. According to the controlled source audio-frequency magnetotellurics (CSAMT), it can be approximately considered as the far zone when the separation between receiver and transmitter is greater than three times the skin depth (depth of penetration, investigation depth) [31–35]. The anthropogenic noise has generally been avoided in FSM applications, and the alternating electromagnetic field caused by surface anthropogenic noise can be considered as a far field source. Therefore, the author considers that the field source of FSM can be considered as the comprehensive effect of aforementioned three sources and other electromagnetic field sources that are not recognized. Because all these sources can be considered as far field sources, the primary field source can be simplified to the interaction of horizontal electric field and horizontal magnetic field components in Cartesian coordinate system. Based on this, the author has carried out forward simulation research in the past and obtained the anomalous curve of FSM profiles similar to the measured results [28]. Groundwater is considered as the preferred source of water for meeting domestic, industrial and agricultural requirements, its exploration, rational development and utilization are the issues of great concern to all of us [36–45]. Since the 1980s, FSM has been gradually widely used and developed in groundwater resources exploration, and instruments have also been gradually improved and automated. For example, the TC series frequency selector manufactured by Hunan Puqi Geological Exploration Equipment Research Institute of Changsha, Hunan, China has now been exported to more than 77 countries in the word and has attained satisfactory results, the technology of this method has been widely recognized by scholars; and the project "Research and Application of Key Technologies for Groundwater Exploration Based on FSM" jointly undertaken by Hunan University of Science and Technology and Hunan Puqi Geological Exploration Equipment Research Institute was awarded the second prize of Hunan Science and Technology Progress Award in 2018. However, a few scholars in China still have doubts about this method, and some even consider it a pseudoscience.

In this paper, the author takes the application of FSM in 12th Five-Year Plan "the rural drinking water safety project" in Guangxi Province as an example and demonstrates the effectiveness of FSM in shallow groundwater exploration from the practical results of 131 boreholes and theoretical research. At the same time, the objective of this paper is to highlight that the anomaly of FSM sounding is related to the potential electrodes spacing. This is in contradiction with the current theory of the magnetotelluric method. It is well known that depth of investigation of the MT method or AMT method is not related to the receiver electric field dipole size. Rather, it is dependent on frequency and resistivity, but not on geometry. This shows that FSM is worthy of further study.

2. Description of the Study Area

2.1. Physical Environment

Guangxi Province is located in the southern of China. The area encompasses a total area of about 23.76×10^4 km^2, and it extends between latitude 20°54′ and 26°24′ N and longitude 104°28′ and 112°04′ E, as shown in Figure 1. Guangxi is located in the southeastern edge of the Yunnan-Guizhou plateau, the second step of China's terrain. It is located in the western part of the hills of Guangxi and Guangdong provinces and faces the Beibu Gulf in the south. The terrain in this province is high in northwest and low in southeast, inclining from northwest to southeast, and the whole area is geomorphology of mountainous and hilly basins. Because of the compression of the Pacific plate and the Indian Ocean plate, the mountains are mostly arced, and the highest peak in the area is 2141 m above mean sea level (a.m.s.l.). It belongs to the subtropical monsoon climate area, with the average yearly temperature of 17.5~23.5 °C and the average annual rainfall of 841.2~3387.5 mm throughout the province, and the rainy season is concentrated in May~September.

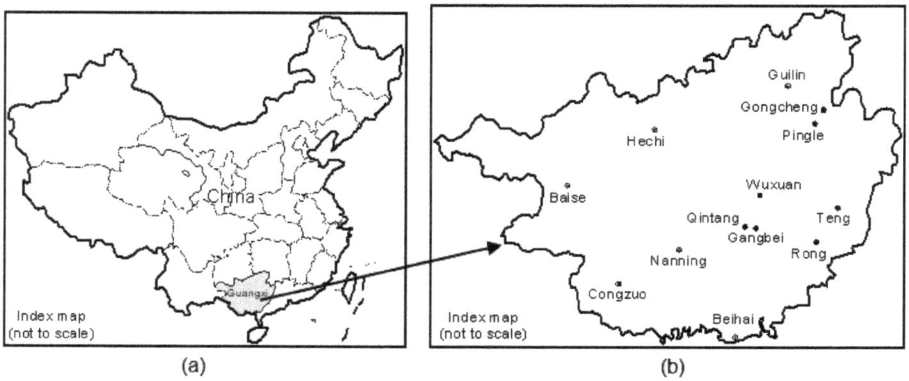

Figure 1. Map showing the location of FSM application in Guangxi study area. (**a**) The location of the study province in China. (**b**) The enlarged map of the study province.

2.2. Geological Environment

Geotectonically, the Guangxi area belongs to the South-China Platform, which was strongly activated in Mesozoic era. The pre-Sinian shallow metamorphic rocks constitute the basement of the platform. According to the difference of geological development history in this area, it can be divided into three units. The northern area is the southern end of the Jiangnan paleocontinent exposed by large basement rock series. The eastern area is mainly the Lower Carboniferous sedimentary caprock, which belongs to a part of the Xiangxi Depression. However, in the vast area, it belongs to Guangxi depression area of the Middle Carboniferous to Triassic sedimentary caprock distribution.

The strata in Guangxi are quite well developed, with Paleozoic sediments as the main deposits, and their depositional development is controlled by Cathaysian tectonic system. The pre-Devonian rock series is composed of clastic sandy shale, schist, phyllite, etc., which distribute in the paleocontinental uplift area and its marginal zone. From Devonian to Triassic, the thick carbonate rocks were mainly accumulated and most widely exposed because of Caledonian movement. Devonian strata can be divided into two groups according to their lithological characteristics, the lower part is quartz sandstone and shale conglomerate, the upper part is mainly gray-white pure limestone, sandy shale and siliceous layer are sandwiched in the middle. The Carboniferous-Permian strata are mainly distributed in the south of the Jiangnan paleocontinent, and the lithofacies differentiation is remarkable, especially in the Permian strata. The bottom of Lower Carboniferous is grey and grey-black layered limestone, and the upper part is coal-measure with varying thickness and drastic changes. The Middle and Upper Carboniferous is mainly composed of gray and gray-white thick-bedded limestone with stable lithofacies. The Lower Permian is black and dark grey limestone, while the upper-Permian is not widely exposed, and its lithofacies changes at the bottom are similar to those described above. The western part of Guangxi accumulated huge thick clastic sandy shale facies during the Triassic period, and there is a thin limestone layer at the bottom. Tertiary strata are developed in eroded valleys, with red calcareous conglomerate and sandstone beds in the lower part and lacustrine sandy shale beds in the upper part. Quaternary strata are not widely distributed in this area, but mainly alluvium, usually with a thickness of only tens of meters.

In addition to aforementioned sedimentary rocks, there are also pre-Sinian granites, Caledonian gneissic granites and Yanshanian granites in the northeastern and southern of Guangxi, of which Yanshanian granites are dominant.

The tectonics of Guangxi is mainly controlled by the mountain-shaped tectonics, the EW tectonics and the Neocathaysian tectonic system. Groundwater in this area can be divided into karstic water and fissure and pore water in sandy shale. Generally speaking, the underground runoff is very good. The former circulates through limestone cave fissures and underground rivers, and it belongs to bicarbonate water. The latter circulates in the fissures and pores of sand shale, mainly in sulfate bicarbonate water. Groundwater is recharged by atmospheric precipitation and discharged by rivers.

Guangxi Province is recognized as one of the most typical and main karst areas in the world, and the area of karst area is about 9.58×10^4 km^2, accounting for 40.9% of the total area. The underground fractured karst cave system causes the surface water to leak rapidly and massively into the underground, resulting in water shortage on the ground, which makes it very difficult to use water in Guangxi karst area. The specific location of the application of FSM is located in seven counties of central and eastern Guangxi, such as Wuxuan County, Gongcheng County, Pingle County, Teng County, Rong County, Qingtang District, and Gangbei District, as shown in Figure 1. The regional geology of the area is shown in Figure 2.

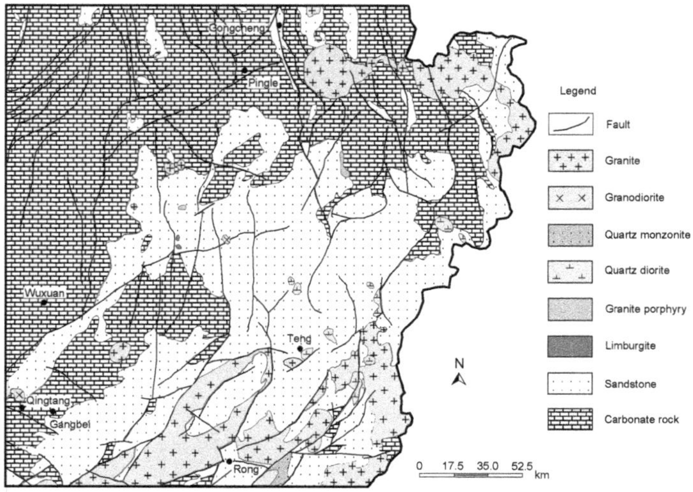

Figure 2. Regional geological map showing the location of FSM in Guangxi study area.

3. Methodology

The approach adopted is similar to conventional resistivity profile exploration. The profile detection device of FSM generally adopts parallel movement mode in field measurement, i.e., the potential electrodes M and N move along the survey line with a fixed electrode spacing, and the midpoint O of the measuring electrode spacing *MN* is a recording point, as shown in Figure 3a. The field electrode spacing *MN* is usually 10 m or 20 m. The magnitudes of horizontal electric field component ΔV on the surface produced by several different fixed frequency natural electromagnetic fields will be measured at each measuring point. The station spacing is usually 5 m, 10 m, or 20 m, while the station spacing of 1 m or 2 m is used to determine the borehole location. Figure 4a is the profiling detection results of FSM on a survey line in Chetian village, Limu town, Gongcheng County, Guangxi Province, with operating frequencies of 25 Hz, 67 Hz, and 170 Hz. The horizontal axis represents the measuring point, and the vertical axis represents the potential difference in Figure 4a. Because groundwater is generally a relatively good conductor, the place where the relative low potential appears in the profile curve of FSM is the abnormal point, i.e., points of 6, 10, and 13 in Figure 4a are the abnormal points, which can be considered and recommended for the borehole location of groundwater.

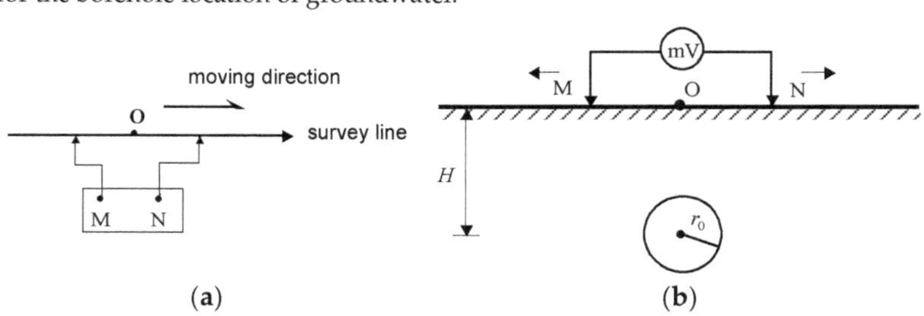

Figure 3. (**a**) A schematic diagram of a profile detection device and (**b**) a sounding device of FSM.

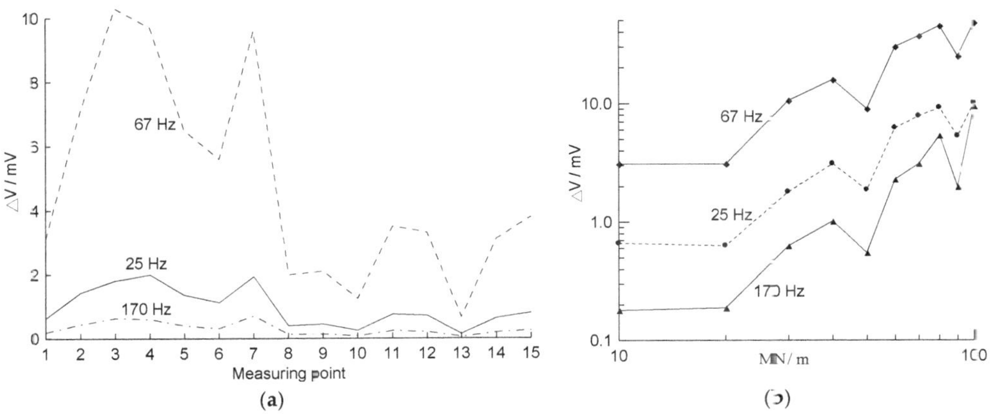

Figure 4. (**a**) Profile survey results and (**b**) sounding results of FSM in Chetian village.

The sounding configuration of FSM is similar to the Schlumberger configuration of the conventional vertical electrical sounding method. The measured point lies at O, as shown in Figure 3b, and potential electrodes M and N move outwards synchronously respectively. The increment of electrode spacing is usually 5 m or 10 m, and 1 m in a particular case. The exploration depth increases with the increase of potential electrode spacing *MN* (Figure 3b) [46–48]. Thus, the variation of potential difference with depth (the potential electrodes spacing *MN*) at a certain sounding point can be obtained, and the burial depth of groundwater can be judged according to the sudden variations of potential difference curves with the potential electrodes spacing *MN*. Figure 4b shows the sounding curves of FSM at point 13 in Figure 4a. The horizontal axis and the vertical axis represent the potential electrodes spacing and the potential difference respectively. The increment of electrode spacing *MN* is 10 m in Figure 4b. The potential difference ΔV is generally increase with the gradual enlargement of electrode spacing *MN*. The obvious inflection point of the sounding curve represents the anomaly, i.e., there may be conductivity anomalous bodies, such as groundwater at the depths reflected by the electrode spacing *MN*. For example, the anomaly appears near *MN* equal to 50 m and 90 m in Figure 4b.

4. Results and Discussion

4.1. Results

From 2013 to 2015, the 273 geology prospecting team undertook the work of groundwater geophysical exploration and drilling work according to the 12th Five-Year Plan of the rural drinking safety project in Wuxuan County, Gongcheng County, Pingle County, Teng County, Rong County, Qingtang District and Gangbei District of Guangxi Province. FSM was used in preliminary geophysical prospecting work, and then 131 boreholes with an average depth of 107.5 m were carried out according to the exploration results of FSM. The minimum drilling depth is 75.2 m, i.e., borehole PL-12 in Tangnao Hamlet, Pengtian Village, Zhangjia Town of Pingle County; the maximum drilling depth is 142.8 m, i.e., borehole WX32-1 in Boyao Village, Ertang Town of Wuxuan County.

Among 131 boreholes them, there are 17 boreholes with water yield less than 1 m^3/h, it accounts for 13.0% of the total number of boreholes; 29 boreholes with water yield (1 m^3/h, 5 m^3/h) accounted for 22.1%; 85 boreholes with water yield more than 5 m^3/h and the rate is approximately 64.9%. On the basis of profile survey results of FSM, FSM sounding is carried out at the abnormal points, and then the final drilling position and depth are determined according to the sounding results. Abnormal sounding results and drilling results of some typical borehole wells are shown in Table 1.

Table 1. The detail of some typical borehole wells.

No.	Borehole Location				Drilling Depth (m)	Spacing *MN* of Sounding Anomaly (m)		Depth to Water of Borehole (m)		Ratio of *MN*/Aquifer Depth
	County	Town	Village	Hamlet		Range	Mean	Depth Range	Mean	
1			Huama	Huameng	82.0	60~70	65	46.1~52.1	49.1	1.324
2		Tongwan	Ancun	Lukuan	113.6	20~30	25	20.0~20.5	20.3	1.235
3					135.2	60~80	70	68.0	68.0	1.029
4			Guzuo	Lufeng	105.4	80~100	90	50.3~85.3	67.8	1.327
5			Fudan	Team 1–10	117.4	80~90	85	70.4~79.0	74.7	1.138
6				Wanlong	110.2	40~50, 70~90	45	52.0~55.0	53.5	0.841
7			Xinxue	Fangxue	114.5	60~70, 80~90	65	57.6~68.0, 68.0~79.6	62.8	1.035
8		Luxin		Xiaoxue	112.0	20~30, 70~80	25	30.8~33.6, 71.20~78.0	32.2	0.776
9			Diyou	Diyou	109.2	70~100	85	70.0~80.0	75.0	1.133
10				Gubei	111.6	50~60	55	46.7~48.0	47.4	1.162
11				Gunan	128.6	50~60	55	74.0~85.0	79.5	0.692
12			Guhang	Xiaoxue	95.2	60~70	65	69.5~71.3	70.4	0.923
13			Gantang	Gantang	100.6	34~48	41	32.8~40.0	36.4	1.126
14		Siling	Guzhang	Taocun	120.7	60~70, 80~90	65	62.2~66.6, 73.6~75.8, 101.8~103.2	64.4	1.009
15			Silao	Gutie	108.7	70~80	75	69.2~90.8	80.0	0.938
16					98.8	30~40	35	39.2~49.2, 71.2~85.5	44.2	0.792
17			Shuicun	Pingdong	82.4	50~60, 70~80	55	35.4~56.1, 70.0~72.0	45.8	1.202
18			Dalin	Guzhai	81.0	40~50	45	40.9~45.6	43.3	1.040
19			Boyao	Boyao	142.8	60~80	70	56.5~65.0, 105.0~115.0	60.8	1.152
20		Ertang		Jieshou	127.3	30~40, 80~90	35	43.3~55.9, 92.0~96.7	49.6	0.706
21			Ludang	Yiqiao	140.9	60~70	65	62.0~75.0	68.5	0.949
22				Ludang	128.5	40~50, 80~90	45	45.4~65.6, 102.5~107.1	55.5	0.811
23			Hema	Wuhe	83.0	70~80	75	64.5~67.6	66.05	1.136
24				Lingding	112.5	70~80	75	70.0~75.0	72.5	1.034
25	Wuxuan		Mocun	Mocun	93.1	50~60, 80~90	55	68.0~69.0, 72.6~74.0	68.5	0.803
26				Mumian	88.6	40~50, 60~70	45	39.8~45.5	42.65	1.055
27		Dongxiang	Liyun	Ludong	93.0	30~40, 50~60	35	40.5~44.6, 70.0~77.2	42.55	0.823
28			Luoqiao	Nasha	116.9	80~90	85	86.1~86.4	86.25	0.986
29				Luoqiao	107.9	50~70	60	54.3~61.0, 78.6~93.5	57.65	1.041
30			Fengyan	Xiafengyan	138.8	40~50, 80~90	45	44.5~54.0	49.25	0.914
31			Dengsi	Longpu	122	20~30, 40~50, 80~90	25	33.4~33.9, 33.9~44.1,75.5~87.0	33.65	0.743
32			Laishan	Daren	130.4	70~80	75	54.5~76.8	65.65	1.142
33					128.8	60~70, 90~100, 110~120	65	64.5~78.5, 90.5~115.7	71.5	0.909
34		Jinji	Shixiang	Cunweihui	114.5	50~60, 70~90	55	48.8~56.2, 96.4~104.0	52.5	1.048
35			Daping	Bashou	120.8	20~30, 70~80	25	22.9~25.3	24.1	1.037
36					99.8	40~60, 70~80	50	39.3~62.0	50.65	0.987
37			Xianglong	Lubo	110.0	40~50, 70~80	45	32.7~39.5	36.1	1.247
38					110.5	40~60	50	39.2~70.7	54.95	0.910
39		Tongling	Xinlong	Pinglong	133.5	20~30, 50~70	25	24.3~25.4, 64.3~74.8	24.85	1.006
40					110.5	50~60, 70~80	55	53.9~77.4	65.65	0.838
41			Daxiang	Jiuxu	116.2	60~70	65	60.0~71.0	65.5	0.992

Table 1. *Cont.*

No.	Borehole Location				Drilling Depth (m)	Spacing *MN* of Sounding Anomaly (m)		Depth to Water of Borehole (m)		Ratio of *MN*/Aquifer Depth
	County	Town	Village	Hamlet		Range	Mean	Depth Range	Mean	
42		Wuxuan	Yacun	Yacun	97.04	80~90	85	54.2~64.8	59.5	1.429
43				Matou	106.8	70~80	75	69.2~87.0	78.1	0.960
44			Dalu	Xinbeihan	104.7	60~70	65	47.0~70.2	58.6	1.109
45					113.3	50~60	55	28.8~35.1, 41.4~49.1	45.25	1.215
46					109.5	60~70, 80~90	65	51.0~62.3, 80.5~87.6	56.65	1.147
47			Qingshui	Qingshui	84.4	40~50	45	54.5~55.0, 55.0~64.7	54.75	0.822
48		Sanli	Sanjiang	Shiziling	90.0	40~50	45	50.8~64.0	57.4	0.784
49			Guli	Guli	110.2	70~80	75	61.9~76.0	68.95	1.088
50			Wangcun	Wangcun	104.5	50~60	55	48.2~68.2	58.2	0.945
51			Wuxing	Longtou	109.5	50~70	60	66.1~87.9	77.0	0.779
52					111.7	50~60	55	46.2~52.5	49.35	1.114
53				Tianliao	102.3	50~70	60	60.0~80.4	70.2	0.855
54			Wufu	Jiacun 6–7	138.6	60~70	65	75.0~85.0	80.0	0.813
55				Xingcun	120.5	80~90	85	70.9~99.5	85.2	0.998
56				Jiacun 8–10	119.8	70~80	75	70.8~94.4	82.6	0.908
57					112.9	60~70, 80~90	65	60.0~70.0	65	1.0
58			Changle	Laicun	80.7	40~50	45	46.0~54.4	50.2	0.896
59	Pingle	Pingle	Taolin	Jiajian	110.3	40~60	50	53.74~57.4	55.57	0.900
60			Longwo	Xiaoxue	135	80~100	90	82.2~102.0	92.1	0.977
61		Zhangjia	Rongjin	Zongxue	100.3	30~50	40	43.4~47.5	45.45	0.880
62			Shuishan	Tianliaochong	130.1	60~70, 100~110	65	76.3~98.6	87.45	0.743
63			Pengtian	Tangnao	75.2	40~50, 80~90	45	50.7~75.2	62.95	0.715
64		Ertang	Hongjiang-kou	Laocun	120.1	50~60	55	50.0~60.0	55	1.0
65					120.1	80~90	85	100.0~105.0	102.5	0.829
66					103.0	70~110	90	90.0~100.0	95	0.947
67			Majia	Xianghuayan	111.0	50~60, 80~90	55	50.5~57.0, 83.2~84.0	53.75	1.023
68		Qiaoting	Qiaoting	Xinglong	96.1	60~70	65	75.5~82.4	78.95	0.823
69		Yuantou	Jinhua	Nanshe	83.2	70~80	75	44.7~64.0	54.35	1.380
70		Shazi	Anquan	Yangtijing	120.2	70~80	75	59.0~87.0	73.0	1.027
71			Xiezhong	Geshuitang	102.2	40~60	50	54.0~62.0	58.0	0.862
72				Xiaziling	112.2	80~100	90	83.6~100.0	91.8	0.980
73	Gong-cheng	Pingan	Beixi	Beixi	90	40~50, 60~70	45	30.7~39.8, 65.8~90.0	35.25	1.277
74			Jutang	Niulutou	82.2	30~40, 60~70	35	47.3~52.1	49.7	0.704
75		Gongcheng	Menlou	Menlou	95.1	60~70, 80~90	65	60.0~70.0, 83.0~89.0	65.0	1.0
76					105.5	50~60, 80~90	55	38.4~40.5, 82.8~84.0	39.45	1.394
77			Jiangbei	Jiangbei	82.2	40~50	45	49.4~61.2	55.3	0.814
78		Longhu	Shizi	Laohulei	81.3	60~80	70	48.5~55.45	51.98	1.347
79		Jiahui	Baiyang	Panlong	86.1	40~60, 80~90	50	60.5~67.45	63.98	0.782
80		Lianhua	Dongzhai	Baima	83.2	40~50, 60~70	45	54.0~65.0	60.0	0.750
81			Dushi	Liaolou	125.9	30~40	35	36.1~41.8	38.95	0.899
82			Hushan	Liangshan	81.2	40~50	45	52.4~55.0	53.7	0.838
83			Lianhua	Team 10	86.5	40~50, 60~70	45	47.5~50.0	48.75	0.923
84			Jiantou	Shetang	82.5	40~60	50	55.5~58.0	56.75	0.881
85		Xiling	Xiasong	Mojiaping	80.3	40~50	45	40.9~54.5	47.7	0.943
86					112.0	70~90	80	101.5~112.0	106.75	0.749
87			Huwei	Yangliu	126.6	60~70	65	75.6~93.3	84.45	0.770
88					109.6	50~60, 70~90	55	60.95~69.88, 96.25~102.13	65.42	0.841

Table 1. *Cont.*

No.	Borehole Location				Drilling Depth (m)	Spacing *MN* of Sounding Anomaly (m)		Depth to Water of Borehole (m)		Ratio of *MN*/Aquifer Depth
	County	Town	Village	Hamlet		Range	Mean	Depth Range	Mean	
89		Limu	Liuling	Liuling	108.1	60~70	65	90.5~99.5	95.0	0.684
90			Wufu	Miaoleikou	114.4	60~70	65	66.6~73.6, 98.1~105.7	70.1	0.927
91			Changjia	Changjia	119.7	40~60, 80~90	50	53.9~56.3	55.1	0.907
92			Shangguan	Futang	110.5	30~40, 60~80	35	33.9~35.5	34.7	1.009
93			Chetian	Chetian	107.3	50~70, 90~100	60	52.5~64.0	58.25	1.030
94				Zhoujia	83.5	40~50	45	49.0	49.0	0.918
95			Liangxi	Qingshuigou	119.6	40~50, 70~80, 90~100	45	56.5~59.45	57.98	0.776
96			Taitang	Taitang	112.4	50~60	55	72.14~74.84	73.49	0.748
97	Qingtang	Shika	Shika Plate Factory		98.5	36~46, 58~60	41	40.0~43.0	41.5	0.988
98	Gangbei	Gangcheng	Guigang Prison		115.4	60~70, 80~90	65	47.0~47.3	47.15	1.379

For the aforementioned 131 boreholes, 17 boreholes with water yield less than 1 m^3/h and 16 boreholes with doubtful aquifer depth are excluded. Results of the remaining 98 boreholes are analyzed, and the ratio of potential electrode spacing *MN* of FSM sounding anomaly to the first aquifer depth of drilling is calculated. When calculating the ratio, the spacing *MN* of sounding anomaly is the median value of the shallow anomaly, and the burial depth of aquifer is the median value of the first aquifer of drilling. The statistical results are shown in Table 1. The ratio has a minimum of 0.684 and a maximum of 1.429, the average value is 0.972, and the standard deviation is 0.177. It can be seen that the depth of groundwater and the potential electrode spacing *MN* of FSM sounding anomaly is in the ratio of almost 1:1.

4.2. Discussion

In the case of MT, it is generally known that the depth of investigation is dependent on frequency and resistivity of underground media, but not on the potential electrode spacing *MN*. Further to the above-noted applications of FSM sounding, the results show that the depth of investigation is closely related not only to frequency and resistivity, but also to the magnitude of the electrode spacing *MN*. This is in contradiction with the current theory of MT. The anomaly of FSM sounding curve is somewhat similar to that of conventional vertical electrical sounding, so it is necessary to carry out further theoretical discussion on this method. In addition, except for the increment of electrode spacing *MN* is 2 m when FSM sounding method is carried out in Shika Plate Factory of Qingtang District, the increment of *MN* is 10 m at other sounding points. If the increment of *MN* is smaller in practical application, the accuracy of the ratio of the potential electrode spacing *MN* of FSM sounding anomaly to the first aquifer depth may be higher (Table 1).

The author has made forward analysis on the causes of profile anomalies of FSM in recent years [27–29]. In this paper, the cause of anomaly in FSM sounding is also discussed with a simple sphere in homogeneous half space. For a conductive and magnetic spheres in the uniform half space medium as shown in Figure 5, the sphere is a, and its dielectric constant, electrical conductivity, and permeability are ε_1, σ_1, and μ_1, respectively. The corresponding parameters of surrounding rocks are ε_2, σ_2, and μ_2. According to the past research results, when FSM is used for exploration, only the natural primary electromagnetic field components, namely the horizontal alternating electric field E_x, E_y and the horizontal alternating magnetic field H_x, H_y act on the sphere, which come from the far field. To simplify the problem, it is assumed that $E_x = E_y = E_0 \times e^{i\omega t}$, $H_x = H_y = H_{p0} \times e^{i\omega t}$. Here ω is a circular frequency and $\omega = 2\pi f$, and f is the frequency. The coordinates of the observation points are $P(r, \theta, \varphi)$ in the spherical coordinate system, as shown in Figure 5.

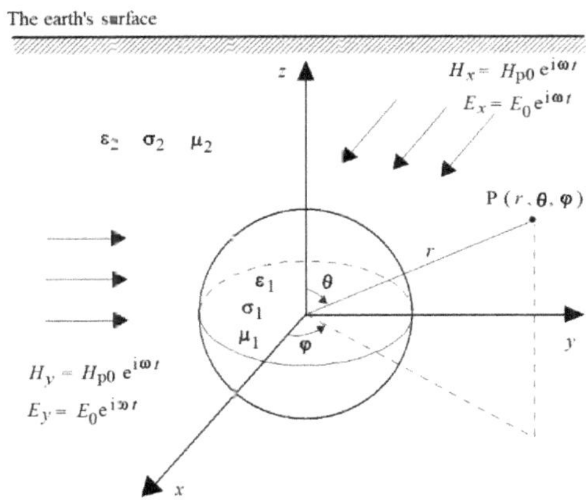

Figure 5. A sphere in natural alternating electromagnetic field in half space.

In the medium with electrical conductivity $\sigma \neq 0$, the free charge cannot be stacked in a certain place, so that the propagation of the magnetotelluric fields obeys the Maxwell's equations. Four-vector fields: the electric field ***E***, the magnetic field ***H***, the electric introduction ***D***, and the magnetic introduction ***B*** satisfy following equations:

$$\nabla \times \boldsymbol{E} = -\mu \cdot \frac{\partial \boldsymbol{H}}{\partial t} \tag{1}$$

$$\nabla \times \boldsymbol{H} = \sigma \cdot \boldsymbol{E} + \varepsilon \cdot \frac{\partial \boldsymbol{E}}{\partial t} \tag{2}$$

$$\nabla \cdot \boldsymbol{D} = 0 \tag{3}$$

$$\nabla \cdot \boldsymbol{B} = 0 \tag{4}$$

By using the separation variable method [27–29], the components of secondary electric field outside the sphere which produced by the primary magnetic field component Hx are given in Equations (5) and (6).

$$E_{2y} = -[C_1 J_{3/2}(k_2 r) + D_1 J_{-3/2}(k_2 r)]\frac{1}{\sqrt{k_2 r}}\frac{z}{\sqrt{x^2+y^2+z^2}} \tag{5}$$

$$E_{2z} = [C_1 J_{3/2}(k_2 r) + D_1 J_{-3/2}(k_2 r)]\frac{1}{\sqrt{k_2 r}}\frac{y}{\sqrt{x^2+y^2+z^2}} \tag{6}$$

Similarly, the components of secondary electric field outside the sphere generated by the primary magnetic field component H_y are given in Equations (7) and (8).

$$E_{2z} = -[C_1 J_{3/2}(k_2 r) + D_1 J_{-3/2}(k_2 r)]\frac{1}{\sqrt{k_2 r}}\frac{x}{\sqrt{x^2+y^2+z^2}} \tag{7}$$

$$E_{2x} = [C_1 J_{3/2}(k_2 r) + D_1 J_{-3/2}(k_2 r)]\frac{1}{\sqrt{k_2 r}}\frac{z}{\sqrt{x^2+y^2+z^2}} \tag{8}$$

where $J_{3/2}(kr)$ and $J_{-3/2}(kr)$ are the complex argument Bessel function, and k_2 is the wave number of surrounding rock, $k_2{}^2 = \omega^2\varepsilon_2\mu_2 - i\omega\mu_2\sigma_2$, i is imaginary unit, and $i^2 = -1$. The coefficients C_1, D_1 are written as:

$$C_1 = -\frac{i\omega 3\sqrt{\pi}\mu_1\mu_2 H_P k_1^2 a^3}{2\sqrt{2}\left\{(\mu_2-\mu_1)p_1\cos p_1 - \left[\mu_2\left(1-p_1^2\right)-\mu_1\right]\sin p_1\right\}}$$

$$D_1 = -\frac{a^3}{2\sqrt{2}} \frac{(2\mu_1 + \mu_2)p_1 \cos p_1 - \left[\mu_2\left(1 - p_1^2\right) + 2\mu_1\right] \sin p_1}{(\mu_2 - \mu_1)p_1 \cos p_1 - \left[\mu_2\left(1 - p_1^2\right) - \mu_1\right] \sin p_1} i\omega\sqrt{\pi}\mu_2 k_2^2 H_P$$

In the above two equations, $p_1 = k_1 a$, k_1 is the wave number of spherical media. If the displacement current is ignored, ${k_1}^2 = -i\omega\mu_1\sigma_1$.

For the effect of two horizontal alternating electric field $E_0 \times e^{i\omega\cdot t}$ in the horizontal direction in Figure 5, the solution can be approximately obtained by referring a conducting sphere in the uniform electrostatic field. If only the anomaly field is considered, the potential distribution of the anomalous field outside the sphere is written as:

$$U_p = 2\frac{\sigma_1 - \sigma_2}{\sigma_1 + 2\sigma_2}\left(\frac{a}{r}\right)^3 \frac{j_0}{\sigma_2} r \sin\theta e^{i\omega\cdot t} \tag{9}$$

According to Equation (9), the components of induced secondary electric field E_{2y}, E_{2x} on the main surface profile along the y axis ($x = 0$) in the rectangular coordinate system are calculated by Equations (10) and (11).

$$E_{2y} = 2\frac{\sigma_1 - \sigma_2}{\sigma_1 + 2\sigma_2} a^3 \frac{j_0}{\sigma_2} \frac{2y^2 - h_0^2}{\left(y^2 + h_0^2\right)^{5/2}} e^{i\omega\cdot t} \tag{10}$$

$$E_{2x} = 2\frac{\sigma_1 - \sigma_2}{\sigma_1 + 2\sigma_2} a^3 \frac{j_0}{\sigma_2} \frac{-1}{\left(y^2 + h_0^2\right)^{3/2}} e^{i\omega\cdot t} \tag{11}$$

where h_0 is the center point depth of the sphere, $j_0 = E_0/\rho_2$.

Assume that f = 170 Hz, h_0 = 30 m, a = 1 m, ρ_1 = 150 ohm-m, ρ_2 = 2500 ohm-m, $B_0 = 1 \times e^{i\omega\cdot t}$ T, and $E_0 = 25 \times 10^{-3} \times e^{i\omega\cdot t}$ V/m in the model shown in Figure 5. Here, ρ_1 and ρ_2 are resistivity of the sphere and the surrounding rock, respectively. It is assumed that the surrounding rock and the sphere are both nonmagnetic. Therefore, the secondary abnormal electric field potential difference ΔV of FSM sounding on the surface main profile along the y direction can be calculated by using Equations (5) and (10). Figure 6 reflects the variation of potential difference ΔV of FSM sounding with the magnitude of the electrode spacing MN and resistivity ρ_1 of the sphere, and the step length of MN in calculation process is 2 m. Because we are mainly interested in abnormal fields in electrical exploration, only the abnormal secondary electric field potential difference is plotted in Figure 6, and the primary electric field is not included. However, the measured results in Figure 4b contain the primary field. Since the primary field is considered to be a uniform field, its potential difference increases with the electrode spacing MN, and its curve is a slanted straight line.

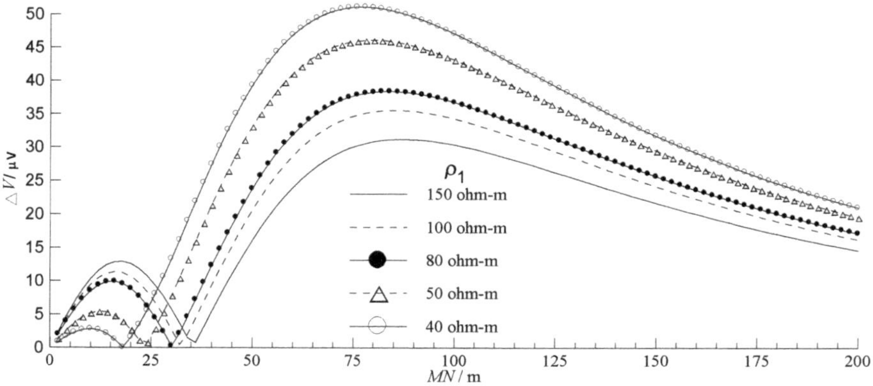

Figure 6. Forward calculating curves of FSM sounding for different resistivity ρ_1.

It can be seen that the anomalous electric field potential difference appears obvious relative low potential anomaly at MN = 36 m in Figure 6 when ρ_1 = 150 ohm-m, and the ratio of the anomalous polar distance (MN = 36 m) to the center point depth of the sphere (h_0 = 30 m) is 1.2. Assuming that only the resistivity ρ_1 of sphere is changed in the above model, as shown in Table 2, the shape of sounding curve of anomalous electric field remains unchanged (Figure 6), but the position of relative low potential anomaly will change (Figure 6, Table 2). The ratio MN/h_0 at abnormal points of sounding curve decreases with the decrease of resistivity ρ_1.

Table 2. Ratios of MN/h_0 at abnormal points of FSM sounding curve with different resistivity ρ_1 of sphere

ρ_1 (Ω·m)	150	100	80	50	40
MN of abnormal point position (m)	36	32	30	24	18
MN/h_0	1.20	1.07	1.00	0.80	0.60

Similarly, assuming that parameters of the model are f = 170 Hz, h_0 = 30 m, a = 1, ρ_1 = 80 ohm-m, $B_0 = 1 \times e^{i\omega \cdot t}$ T and $E_0 = 25 \times 10^{-3} \times e^{i\omega \cdot t}$ V/m, and only the resistivity ρ_2 of the surrounding rock is changed, as shown in Table 3 (forward calculating curves of FSM are shown in Figure 7), when ρ_2 is 4000 ohm-m or 1500 ohm-m, the relative low potential anomaly appears at MN = 30 m, and the ratio MN/h_0 at abnormal point of sounding curve is 1.0. In general, the ratio MN/h_0 decreases with the decrease of resistivity ρ_2, as shown in Table 3.

Table 3. Ratios of MN/h_0 at abnormal points of FSM sounding curve with different resistivity ρ_2 of surrounding rock.

ρ_2 (Ω·m)	4000	1500	1000	800	500
MN of abnormal point position (m)	30	30	28	28	24
MN/h_0	1.00	1.00	0.933	0.933	0.80

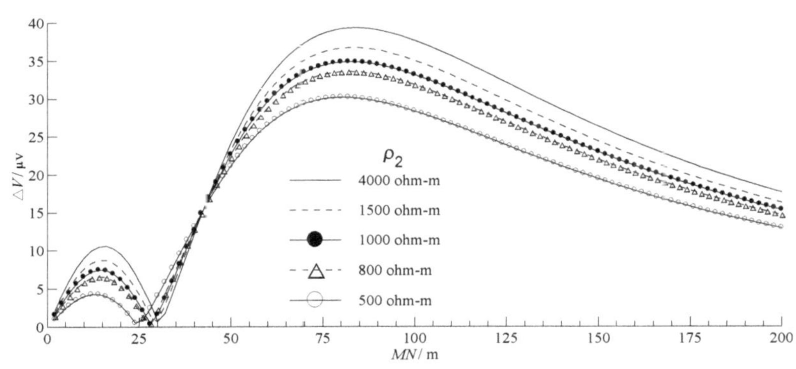

Figure 7. Forward calculating curves of FSM sounding for different resistivity ρ_2.

According to the calculating results in Tables 2 and 3, the ratio MN/h_0 at relative low potential abnormal point varies approximately around 1.0. This is in agreement with the practical application results in Table 1.

5. Conclusions

Statistical analysis of drilling results in 98 typical borehole wells shows that there is an approximate 1:1 relationship between the potential electrode spacing MN and the depth of aquifer at the relative low potential abnormal point of FSM sounding curves. This empirical relationship has a certain guiding role for future practice. Finally, the author carried out forward research on the FSM sounding method. A simplified geophysical model of a low resistance conductive sphere in a homogeneous half space was used in the study, and the secondary anomalous electric field on the surface main profile was calculated when the sphere was under the combined action of the horizontal alternating electromagnetic fields. Forward calculation results show that there will be a relatively low potential anomaly point in FSM sounding curve when the resistivity of the sphere is approximately equal to that of groundwater, and the ratio of electrode spacing (MN) to the center point depth (h_0) of the sphere at the relatively low potential anomaly point of sounding curves varies approximately around 1.0. The theoretical calculation results are in good agreement with the statistical results of practical application, which also explains the cause of anomaly of FSM sounding method.

The theoretical research indicates that the cause of FSM anomaly is the comprehensive effect of the natural 3D alternating electromagnetic signal underground. According to the application of FSM in the 12th Five-Year Plan of rural drinking water safety project in Guangxi Province, there are 114 boreholes with water yield more than 1 m^3/h, and the success rate is about 87.0%. This shows that FSM has obvious effectiveness in shallow (depth < 150 m) groundwater exploration, and it is an effective method to identify suitable borehole sites of groundwater in the future.

FSM belongs to the natural passive source method, and its field source is very complex. In this paper, the author used only a very simple geophysical model to discuss the anomaly of the FSM sounding curve, which is just a function of throwing out a brick to attract a jade. There are many related problems of FSM to be further studied in the future.

Author Contributions: Methodology, Y.T.; writing—original draft preparation, L.Y. (Lu Yulong); writing—review and editing, A.T.T.; funding acquisition, L.Y. (Liu Yang). All authors have read and agreed to the published version of the manuscript.

Funding: This research was funded by the Natural Science Foundation of Hunan Province (Grant No. 2022JJ30244) and the Research Project of Teaching Reform of Hunan Province (Grant No. HNJG-2022-0790).

Conflicts of Interest: The authors declare no conflict of interest.

References

1. Yang, T.C.; Xia, D.L.; Wang, Q.R.; Fu, G.H. *Theoretical Research and Application of Natural Electric Field Frequency Selection Method*; Central South University Press: Changsha, China, 2017.
2. Bao, G.S.; Li, D.Q.; Zhang, Y.S.; Wang, S.W. Research on the interfering electric field instrument. *Chin. J. Nonferrous Met.* **1994**, *4*, 9–13.
3. Zhang, J.; Li, K.; Huang, C.L.; Li, Z.Y. Detecting system of underground magneto fluid & application. *Comput. Technol. Autom.* **2010**, *29*, 119–122, 136.
4. Wang, R.G.; Liu, S.Y.; Shangguan, D.H.; Radic, V.; Zhang, Y. Spatial Heterogeneity in Glacier Mass-Balance Sensitivity across High Mounta in Asia. *Water* **2019**, *11*, 776. [CrossRef]
5. Deng, R.J.; Shao, R.; Ren, B.Z.; Hou, B.L.; Tang, Z.E.; Hursthouse, A. Adsorption of Antimony(III) onto Fe(III)-Treated Humus Sludge Adsorbent: Behavior and Mechanism Insights. *Pol. J. Environ. Stud.* **2019**, *28*, 577–586. [CrossRef] [PubMed]
6. Zhang, Y.; Ren, B.Z.; Hursthouse, A.; Deng, R.J.; Hou, B.L. Leaching and Releasing Characteristics and Regularities of Sb and As from Antimony Mining Waste Rocks. *Pol. J. Environ. Stud.* **2019**, *28*, 4017–4025. [CrossRef] [PubMed]
7. You, B.; Xu, J.X.; Shi, S.L.; Liu, H.Q.; Lu, Y.; Liang, X.Y. Treatment of coal mine sewage by catalytic supercritical water oxidation. *Fresenius Environ. Bull.* **2020**, *29*, 497–502.
8. Yu, S.Q.; Mo, Q.F.; Chen, Y.Q.; Li, Y.W.; Li, Y.X.; Zou, B.; Xia, H.P.; Jun, W.; Li, Z.A.; Wang, F.M. Effects of seasonal precipitation change on soil respiration processes in a seasonally dry tropical forest. *Ecol. Evol.* **2020**, *10*, 467–479. [CrossRef]
9. Wang, Z.H.; Huang, W.; Wang, C.; Li, X.S.; He, Z.H.; Hursthouse, A.; Zhu, G.C. Variation of Al species during water treatment: Correlation with treatment efficiency under varied hydraulic conditions. *AQUA Water Infrastruct. Ecosyst. Soc.* **2021**, *70*, 891–900. [CrossRef]
10. Zhang, J.; Huang, C.L.; Wang, J.; Zhou, H. Detecting system of underground magneto fluid and its application on mine area flood prevention. *Chin. J. Sci. Instrum.* **2009**, *30* (Suppl. S6), 807–810.
11. Cheng, H.; Bai, Y.C. Design and application of audio frequency natural electric field instrument. *Prog. Geophys.* **2014**, *29*, 2874–2879.
12. Zhang, K.Y.; Zhou, M.R.; Yan, P.C.; Wang, R. Research of electrical conductivity measurement system based on frequency selection method. *Instrum. Tech. Sens.* **2015**, *42*, 65–67.

13. Zhao, Y.L.; Luo, S.L.; Wang, Y.X.; Wang, W.J.; Zhang, L.Y.; Wan, W. Numerical analysis of karst water inrush and a criterion for establishing the width of water-resistant rock pillars. *Mine Water Environ.* **2017**, *36*, 508–519. [CrossRef]
14. Yang, T.C.; Zhang, Y.P.; Fu, G.H. Application research of comprehensive geophysical method to karst investigaion in a productive mine. *Prog. Geophys.* **2021**, *36*, 1145–1153.
15. Yang, Z.C. The least-square Fourier-series model-based evaluation and forecasting of monthly average water-levels. *Environ. Earth Sci.* **2018**, *77*, 328. [CrossRef]
16. Jin, C.S.; Liu, Y.X.; Li, Z.W.; Gong, R.Z.; Huang, M.; Wen, J.J. Ecological consequences of China's regional development strategy: Evidence from water ecological footprint in Yangtze River Economic Belt. *Environ. Earth Sci.* **2018**, *77*, 328. [CrossRef]
17. Zhou, J.Z.; Zhang, M.Q.; Ji, M.T.; Wang, Z.H.; Hou, H.; Zhang, J.; Huang, X.; Hursthouse, A.; Qian, G.R. Evaluation of heavy metals stability and phosphate mobility in the remediation of sediment by calcium nitrate. *Water Environ. Res.* **2020**, *92*, 1017–1026. [CrossRef] [PubMed]
18. Berdychevsky, M.H. *Telluric Current Prospecting*; China Industrial Press: Beijing, China, 1962.
19. Pham, V.N.; Boyer, D.; Xue, C.Y.; Liu, S.C. Application of telluric-telluric profiling combined with magnetotelluric and self-potential methods to geothermal exploration in the Fujian Province, China. *J. Volcanol. Geotherm. Res.* **1995**, *65*, 227–236. [CrossRef]
20. Mlynarski, M.; Zlotnicki, J. Fluid circulation in the active emerged Asal rift (east Africa, Djibouti) inferred from self-potential and Telluric-Telluric Prospecting. *Tectonophysics* **2001**, *339*, 455–472. [CrossRef]
21. Deng, R.J.; Jin, C.S.; Ren, B.Z.; Hou, B.L.; Hursthouse, A.S. The Potential for the Treatment of Antimony-Containing Wastewater by Iron-Based Adsorbents. *Wate* **2017**, *9*, 794. [CrossRef]
22. Wang, P.F.; Tan, X.H.; Zhang, L.Y.; Li, Y.J.; Liu, R.H. Influence of particle diameter on the wettability of coal dust and the dust suppression efficiency via spraying. *Process Saf. Environ. Prot.* **2019**, *132*, 189–199. [CrossRef]
23. Liu, T.; Chen, Z.S.; Li, Z.X.; Fu, H.; Chen, G.L.; Feng, T.; Chen, Z. Preparation of magnetic hydrochar derived from iron-rich Phytolacca acinosa Roxb. *for Cd removal Sci. Total Environ.* **2021**, *769*, 145–159. [CrossRef] [PubMed]
24. Yang, T.C.; Zhang, H. Study faults by natural electric field frequency selection method. *J. Hunan Univ. Sci. Technol. (Nat. Sci. Ed.)* **2013**, *28*, 32–37.
25. Yang, T.C.; Zhang, H. A study of the anomaly genesis for the frequency selection method in a natural electric field of a karst body. *Hydrogeol. Eng. Geol.* **2013**, *40*, 22–28.
26. Yang, T.C.; Shen, J.P.; Li, G.M.; Cao, Y.J.; Zhang, H. Application and analysis of natural electric field frequency selection method for water-filled karst investigation. *Coal Geol. Explor.* **2014**, *42*, 71–75.
27. Yang, T.C.; Zhang, Q.; Wang, Q.R.; Fu, G.H.; Liao, J.P. Study on the anomaly genesis of the frequency selection method for a sphere under natural electromagnetic field. *J. Hunan Univ. Sci. Technol. (Nat. Sci. Ed.)* **2016**, *31*, 58–65.
28. Chen, Z.C.; Deng, H.Y.; Yang, T.C. The observation and test of the daily variation based on the frequency selection method of natural electtric field. *Miner. Eng. Res.* **2020**, *35*, 61–66.
29. Yang, T.C.; Liang, J.; Cheng, H.; Cao, S.J.; Dong, S.Y.; Gong, Y.F. The effect and the anomaly analysis of shallow groundwater exploration based on the frequency selection method of natural electric field. *Geophys. Geochem. Explor.* **2018**, *42*, 1194–1200.
30. Zhou, Y.Y.; Ren, B.Z.; Hursthouse, A.S.; Zhou, S.J. Antimony Ore Tailings: Heavy Metals, Chemical Speciation, and Leaching Characteristics. *Pol. J. Environ. Stud.* **2019**, *28*, 485–495. [CrossRef]
31. Hou, B.L.; Liu, X.; Li, Z.; Ren, B.Z.; Kuang, Y. Heterogeneous fenton oxidation of butyl xanthate catalyzed by iron-loaded sewage sludge. *Fresenius Environ. Bull.* **2022**, *31*, 4125–4131.
32. Tang, J.T.; Zhou, C.; Xiao, X. Selection of minimum transmit-receive distance of CSAMT on complicated media. *Chin. J. Nonferrous Met.* **2013**, *23*, 1681–1693.
33. Shi, X.Y.; Ren, B.Z.; Hursthouse, A. Source identification and groundwater health risk assessment of PTEs in the stormwater runoff in an abandoned mining area. *Environ. Geochem. Health* **2022**, *44*, 3555–3570. [CrossRef] [PubMed]
34. Yuan, Q.; Zhu, G.C. A review on metal organic frameworks (MOFs) modified membrane for remediation of water pollution. *Environ. Eng. Res.* **2021**, *26*, 190435. [CrossRef]
35. Zhao, Y.L.; Zhang, C.S.; Wang, Y.X.; Lin, H. Shear-related roughness classification and strength model of natural rock joint based on fuzzy comprehensive evaluation. *Int. J. Rock Mech. Min. Sci.* **2021**, *137*, 104550. [CrossRef]
36. Liu, J.H.; Zhao, Y.L.; Tan, T.; Zhang, L.Y.; Zhu, S.T.; Xu, F.Y. Evolution and modeling of mine water inflow and hazard characteristics in southern coalfields of China: A case of Meitanba mine. *Int. J. Min. Sci. Technol.* 2022. [CrossRef]
37. Chongo, M.; Christiansen, A.V.; Tembo, A.; Banda, K.E.; Imasiku, A.N.; Larsen, F.; Peter, E.G. Airborne and ground-based transient electromagnetic mapping of groundwater salinity in the Machile-Zambezi Basin, southwestern Zambia. *Near Surf. Geophys.* **2015**, *13*, 383–395. [CrossRef]
38. Ikard, S.; Pease, E. Preferential groundwater seepage in karst terrane inferred from geoelectric measurements. *Near Surf Geophys.* **2019**, *17*, 43–53. [CrossRef]
39. Tian, Z.J.; Zhang, Z.Z.; Deng, M.; Yan, S.; Bai, J.J. Gob-Side Entry Retained with Soft Roof, Floor, and Seam in Thin Coal Seams: A Case Study. *Sustainability* **2020**, *12*, 1197. [CrossRef]
40. Zhang, Y.; Huang, F.Y. Indicative significance of the magnetic susceptibility of substrate sludge to heavy metal pollution of urban lakes. *ScienceAsia* **2021**, *47*, 374. [CrossRef]
41. Yang, J. Experimental results and theoretical study of the stray current method in karst area. *Geophys. Geochem. Explor.* **1982**, *6*, 41–54.

42. Li, Y.C.; Hu, X.X.; Ren, B.Z. Treatment of antimony mine drainage: Challenges and opportunities with special emphasis on mineral adsorption and sulfate reducing bacteria. *Water Sci. Technol.* **2016**, *73*, 2039–2051. [CrossRef]
43. Li, Y.C.; Xu, Z.; Ma, H.Q.; Hursthouse, A. Removal of Manganese(II) from Acid Mine Wastewater: A Review of the Challenges and Opportunities with Special Emphasis on Mn-Oxidizing Bacteria and Microalgae. *Wate* **2019**, *11*, 2493. [CrossRef]
44. Zheng, C.S.; Jiang, B.Y.; Xue, S.; Chen, Z.W.; Li, H. Coalbed methane emissions and drainage methods in underground mining for mining safety and environmental benefits: A review. *Process Saf. Environ. Prot.* **2019**, *127*, 103–124. [CrossRef]
45. Qin, B.T.; Li, L.; Ma, D.; Lu, Y.; Zhong, X.X.; Jia, Y.W. Control technology for the avoidance of the simultaneous occurrence of a methane explosion and spontaneous coal combustion in a coal mine: A case study. *Process Saf. Environ. Prot.* **2016**, *103*, 203–211. [CrossRef]
46. Liang, J.; Wei, Q.F.; Hong, J.; Zheng, S.Y.; Qin, Y.C.; Yan, F.S.; Feng, Y.X. Application of self-potential method to explore water in karst area. *Geotech. Investig. Surv.* **2016**, *44*, 68–78.
47. Zhao, Y.L.; Liu, Q.; Zhang, C.S.; Liao, J.; Lin, H.; Wang, Y.X. Coupled seepage-damage effect in fractured rock masses: Model development and a case study. *Int. J. Rock Mech. Min. Sci.* **2021**, *144*, 104822. [CrossRef]
48. Zhao, Y.L.; Zhang, L.Y.; Wang, W.J.; Tang, J.Z.; Lin, H.; Wan, W. Transient pulse test and morphological analysis of single rock fractures. *Int. J. Rock Mech. Min. Sci.* **2017**, *91*, 139–154. [CrossRef]

water

MDPI

Article

A Machine Learning Method for Engineering Risk Identification of Goaf

Haiping Yuan [1], Zhanhua Cao [1], Lijun Xiong [1], Hengzhe Li [1] and Yixian Wang [1,2,*]

[1] School of Civil Engineering, Hefei University of Technology, Hefei 230009, China
[2] State Key Laboratory of Explosion Science and Technology, Beijing Institute of Technology, Beijing 100081, China
* Correspondence: wangyixian2012@hfut.edu.cn

Abstract: The risk evaluation indexes of goaf are multi-source and have complex mutual internal correlations, and there are great differences in the risk identification of goaf from different mines among the various influencing factors. This paper mainly focuses on principal component analysis (PCA) and the differential evolution algorithm (DE), while a multi-classification support vector machine (SVM) is adopted to classify the risks of goaf. Then, the *K*-fold cross-validation method is used to prevent the overfitting of selection in the model. After the analysis, nine factors affecting the risk identification of goaf in a certain area of East China were determined as the primary influencing factors, and 120 measured goafs were taken as examples for classifying the risks. More specifically, the classification results show that: (1) SVM has the useful ability of generalization, especially when solving the problems of overfitting, and it is easy to fall into the local minima under the conditions of small samples; (2) PCA is employed to realize the intelligent dimensionality reduction and denoising of multi-source impact indicators for goaf risk identification, which immensely improves the prediction accuracy and classification efficiency of the model; (3) after using the DE, the optimal solutions of the problems to be optimized are automatically obtained through the global optimization search mechanism, namely, the kernel function parameter, 'γ', and the penalty factor, '*C*', of the SVM, which further verifies that the characteristics of clear logic, strong convergence, and good robustness can be found in the DE. As demonstrated, this method has the advantages of guiding significance and application value for goaf risk identification.

Keywords: goaf; risk assessment; support vector machine (SVM); principal component analysis (PCA); differential evolution algorithm (DE)

Citation: Yuan, H.; Cao, Z.; Xiong, L.; Li, H.; Wang, Y. A Machine Learning Method for Engineering Risk Identification of Goaf. *Water* **2022**, *14*, 4075. https://doi.org/10.3390/w14244075

Academic Editor: Dongmei Han

Received: 16 October 2022
Accepted: 8 December 2022
Published: 13 December 2022

1. Introduction

In recent decades, with the continuous exploitation of mines, the number and volume of goaf formations have been constantly rising, which has brought a tremendous increase in the potential risks regarding the safety of mines [1–5]. Therefore, research on the risk identification of goaf is of great significance to ensure the safe development of mines [6–9]. In general, there are many factors affecting the stability of goaf, such as the engineering and hydrogeological conditions, the exploitation depth, the ore block constituent elements, the goaf mining height, the pillar situation, the goaf formation time, the measured volume, the impact of blasting on the ore body, the goaf treatment rate, the distribution and scale of structural plane, the goaf treatment mode, the development of geological structure, the strength of the rock surrounding the goaf, the goaf's shape, the maximum exposure area, the maximum exposure height, the thickness of the roof and bottom protection, the mining method, etc., and each influencing factor will exert huge effects on the stability of mine goaf among the various areas [10–12]. Thus, the evaluation indexes of goaf in different mines should be explicitly analyzed according to the specific circumstances. In this paper, the risk types that are assessed mainly include the roof caving in, rockslides, falls due to collapses in the goaf, and other risks. Therefore, nine parameters that are closely related to these risk categories, such as the exploitation depth, the mining method, the goaf mining height, the maximum exposure area, the maximum exposure height, the maximum exposure span, the pillar locations, the measured volume, and the treatment rate are selected as the central influencing factors. However, when considering the numerous factors and when there is noise in the data, this will cause enormous inconvenience in terms of the analysis. Hence, the dimensionality reduction method can be used to process the data; as yet, the most widely used method in dimensionality reduction has been principal component analysis (PCA) [13]. Briefly, PCA is adopted to preprocess the input data, and the main information contained in the data is still retained in the principal component. Above all, not only can it reduce the dimension of the data, but also play a significant role in denoising, so as to make the prediction results more accurate.

In general, the analysis and research methods in the field of mine goafs regarding risk assessment have been applied by several scholars, the techniques used mainly include the fuzzy synthetic assessment method [14], the set pair analysis theory [15], the grey relation analysis method [16], the uncertainty measurement theory [17], etc. Nonetheless, all these non-machine learning algorithms have their own respective application scope; the fairly strong abilities of promotion and improvement are yet to be enhanced, and some of them need to further improve their generalization performance. Recently, data-driven machine learning models have been broadly utilized to work out the nonlinear problems, and some scholars have applied the techniques to the study of goaf risk. Hu et al. [18] proposed a Bayesian discrimination method (BDM) for the risk identification of goaf, and established the corresponding BDM for further research; Feng et al. [19] combined PCA and a neural network to evaluate the risk of goaf, so as to reduce the input variables, eliminate the correlation among variables, and improve the prediction accuracy; Wang et al. [20] constructed multi-classification support vector machine (SVM) models for goaf stability classification, according to the SVM theory and the 'one-against-one' method; Wang et al. [21] used the directed acyclic graph method to constitute the multi-classification SVM and acquired SVM models for goaf stability classification.

Currently, SVMs have been extensively utilized under nonlinear conditions and have achieved numerous satisfactory results. They apply the structural risk minimization principle to replace certain empirical risk minimization principles in traditional machine learning methods, which provides the useful ability of generalization. Compared with the neural network, there are evident advantages to the SVM in terms of solving the problems of overfitting and it is easy to fall into the local minima in the case of small samples, which is considered to be a superior theory for predictive learning [22]. Moreover, the SVM is a typical binary classification problem, and the multi-classification cases can be constructed using the 'one-against-all' method [23], the 'one-against-one' method [24], the directed acyclic graph method [25], the binary tree method [26], etc., while the cross-validation method can be used to choose the appropriate model. In this paper, the parameters of SVMs are optimized, based on the DE, which is an efficient global optimization algorithm on the basis of population and can realize a global search through competitive selection and differential mutation. Additionally, the technique of differential mutation can also avoid the problem of falling into local optimization due to a lack of mutation in the genetic algorithm (GA). On the whole, the DE is comprehensively applied in optimization problems owing to its clear structure, strong convergence, and good robustness.

In summary, the PCA adopted in this paper aims to preprocess the data in order to reduce the input variables and eliminate the correlation among them, so that the data-processing speed can be accelerated, and the prediction accuracy can be improved simultaneously. Subsequently, a multi-classification SVM is employed to train and predict the data, and the cross-validation principle is used to select the most preferable model. Furthermore, the DE is employed to optimize the parameters of the SVM. In the meantime, we have decided to take 120 measured goafs from a mine in a certain area of East China as an example to be classified and compare it with the phenomena after using the PCA of the Naive Bayes classification and the BP neural network.

Consequently, conclusions can be drawn that the classification results in this paper ideally reflect the actual classification of goaf stability, which has obvious beneficial guiding significance and application value for engineering.

2. Principal Component Analysis

2.1. Basic Principles

Principal component analysis (PCA) is a dimensionality reduction method that transforms the data represented by multiple related variables into a few unrelated ones through orthogonal mapping. After transformation, the variables are defined as 'principal components'. Among some multi-variable problems, the variables may be related to each other. Normally, a complex direct analysis may demonstrate inaccurate results. Therefore, the PCA method is adopted to replace the original multiple variables with a few principal components, while retaining most of the information involved in the data, so that it is convenient for further analysis.

2.2. Mathematical Model

Assuming that the m variables, cover x_1, x_2, ..., x_m, are included in the n samples, the original data matrix can be obtained, as follows:

$$x = \begin{bmatrix} x_{11} & x_{12} & \cdots & x_{1m} \\ x_{21} & x_{22} & \cdots & x_{2m} \\ \vdots & \vdots & \vdots & \vdots \\ x_{n1} & x_{n2} & \cdots & x_{nm} \end{bmatrix}. \tag{1}$$

Generally, there are different dimensions in the variables among the research questions, which may create some new problems. Thus, the original data should be standardized with Equation (2) before the PCA takes place:

$$x_{ij}^* = \frac{x_{ij} - \min(x_j)}{\max(x_j) - \min(x_j)} \tag{2}$$

where x_{ij} represents the data before standardization; x^* ij represents the data after standardization; $\max(x_j)$ and $\min(x_j)$ are the maximum and minimum values in the column j data, respectively.

After data standardization, the original data will become the values between 0 and 1 in subsequent calculations and analysis.

The matrix after data standardization is represented by y; namely, the m variables of $x = (x_1, x_2, \ldots, x_m)^T$ are denoted as the m new ones, and the new variables can be linearly expressed by the original ones as x_1, x_2,..., x_m, that is:

$$\begin{cases} y_{1m} = u_{11}x_1 + u_{12}x_2 + \cdots + u_{1m}x_m \\ y_{2m} = u_{21}x_1 + u_{22}x_2 + \cdots + u_{2m}x_m \\ \vdots \\ y_{nm} = u_{n1}x_1 + u_{n2}x_2 + \cdots + u_{nm}x_m \end{cases} \tag{3}$$

where y_{1m}, y_{2m},..., y_{nm} signifies the variables sourced through PCA, and u is the correlation coefficient matrix among the variables, which need to satisfy the following conditions:

(1) (1) $u_{k1}^2 + u_{k2}^2 + \cdots + u_{km}^2 = 1(k = 1, 2, \cdots, n)$;

(2) $\mathrm{cov}(y_i, y_j) = 0(i \neq j; i, j = 1, 2, \ldots, m)$, namely, the components of principal analysis are independent and there is no overlapping information;

(3) $\mathrm{var}(y_1) \geq \mathrm{var}(y_2) \geq \ldots \geq \mathrm{var}(y_m)$, namely, the principal components are sorted according to the standard deviation, where: y_1, y_2, ..., y_m, obtained through the

above process, can be determined as the principal components of $1, 2, \ldots, m$ of the original variables.

In general, the cumulative variance contribution rate represents the amount of original data information. However, one of the first of several principal component factors determines the number of principal components. In order to reduce the amount of calculation needed, the cumulative variance contribution rate can be about 80%.

2.3. Geometric Interpretation

In short, PCA is employed to acquire the linearly independent variables that are defined as the principal components from the correlated ones through orthogonal transformation. Specifically, by means of rotating and transforming the original coordinate system, the data is represented in the new system. As depicted in Figure 1, it shows the new coordinate system and the corresponding principal components obtained by rotating the data expressed by the two variables.

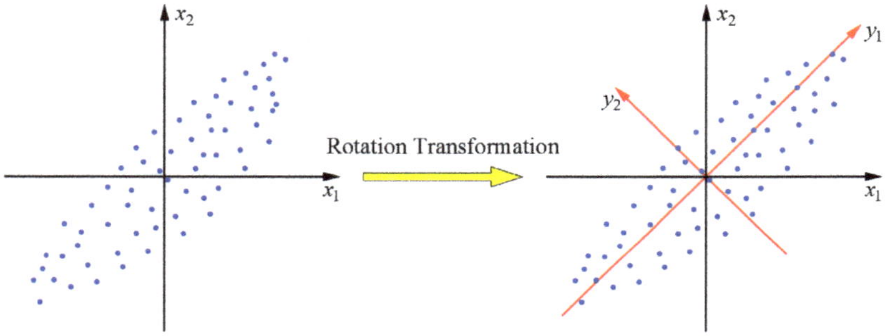

Figure 1. Geometric interpretation of PCA.

3. Multi-Classification Support Vector Machine (SVM)

3.1. Basic Principles of SVM

Notably, the learning strategy of the support vector machine (SVM) is to minimize the structural risk, based on a sound ability in terms of generalization. According to the machine learning theory [27], the basic model of the SVM is a binary classifier for linearly separable data sets, which can be applied to nonlinear problems by introducing the 'kernel' function. When linearly separating the data sets, the SVM is normally adopted to solve the optimal hyperplane that linearly separates the data sets according to interval maximization or the corresponding convex quadratic programming problems that need to be solved promptly.

Hence, the general optimal hyperplane can be determined:

$$wx + b = 0. \tag{4}$$

The corresponding classification decision function is:

$$f(x) = \text{sign}(wx + b). \tag{5}$$

To conclude, the basic model is denoted as the 'linear separable support vector machine', and interval maximization refers to 'hard interval maximization'.

Figure 2 demonstrates the classification problem within the two-dimensional feature space. In particular, the distance between the lines 'H_1' and 'H_2' is signified as 'interval', the size of which is related to the normal vector 'w' of line 'H_0', and where the value is equal to $2/||w||$.

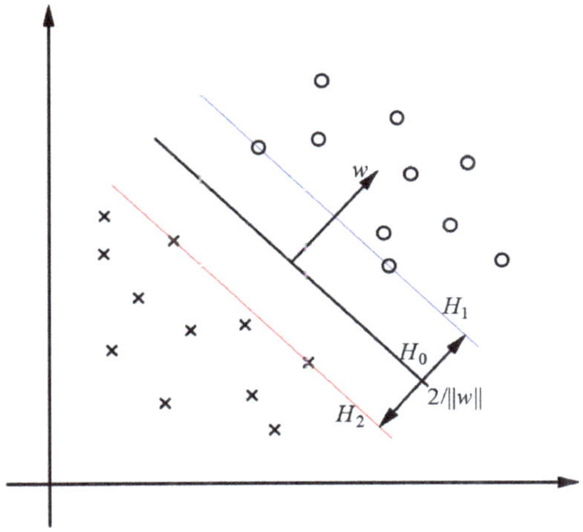

Figure 2. Schematic diagram of the linear separable support vector machine.

Considering the realistic conditions, the data sets cannot be completely linearly separated due to the existence of some special points among the data. As a consequence, the linear support vector machine can be structured by introducing the relaxation variables of '$\xi_i \geq 0$'. Thus, the interval maximization at this moment can be signified as 'soft interval maximization', the corresponding convex quadratic programming problem of which is:

$$\begin{aligned} \min_{w,b,\xi} \quad & \frac{1}{2}\|w\|^2 + C\sum_{i=1}^{N}\xi_i \\ \text{s.t.} \quad & y_i(wx_i + b) \geq 1 - \xi_i,\ i = 1, 2, \cdots, N \\ & \xi_i \geq 0, i = 1, 2, \cdots, N \end{aligned} \tag{6}$$

where C refers to the 'penalty factor'. The larger the value of C, the more prone it is to overfitting, and vice versa.

Regarding Equation (6) as the initial circumstance and using the 'Lagrange duality' to obtain the dual problem, the optimal solution of the original problem can be achieved by solving the dual problem. Indeed, the 'Lagrange' function of the original problem is determined as follows:

$$\begin{aligned} L(w, b, \xi, \alpha, \mu) \equiv \ & \frac{1}{2}\|w\|^2 + C\sum_{i=1}^{N}\xi_i - \\ & \sum_{i=1}^{N}\alpha_i(y_i(wx_i + b) - 1 + \xi_i) - \sum_{i=1}^{N}\mu_i\xi_i \end{aligned} \tag{7}$$

Among them, α_i refers to the 'Lagrange multiplier'; $\alpha_i \geq 0$ (i = 1, 2, .., N); $\mu_i \geq 0$. The dual problem in the original problem is:

$$\begin{aligned} \min_{\alpha} \quad & \frac{1}{2}\sum_{i=1}^{N}\sum_{j=1}^{N}\alpha_i\alpha_j y_i y_j (x_i x_j) - \sum_{=1}^{N}\alpha_i \\ \text{s.t.} \quad & \sum_{i=1}^{N}\alpha_i y_i = 0 \\ & 0 \leq \alpha_i \leq C, i = 1, 2, \cdots N \end{aligned} \tag{8}$$

Assuming $\alpha^* = (\alpha^*1, \alpha^*2, \ldots, \alpha^*N)^T$ is the solution of the dual problem, if there is a component $\alpha^* j$ of α^* that satisfies $0 < \alpha^* j < C$, then the optimal solution of the original problem can be determined as:

$$\begin{cases} w^* = \sum_{i=1}^{N} \alpha_i^* y_i x_i \\ b^* = y_j - \sum_{i=1}^{N} y_i \alpha_i^* (x_i x_j) \end{cases} . \tag{9}$$

After that, the classification decision function can be obtained as:

$$f(x) = \text{sign}(\sum_{i=1}^{N} \alpha_i^* y_i (x_i x) + b^*). \tag{10}$$

On account of the nonlinear real data, it is feasible to take advantage of the kernel technique to set up the nonlinear support vector machine, the classification decision function of which is:

$$f(x) = \text{sign}(\sum_{i=1}^{N} \alpha_i^* y_i K(x, x_i) + b^*) \tag{11}$$

where $K(x,z)$ represents the function of the positive definite kernel.

According to the SVM theory, there are four types of kernel function that are commonly used, namely, the 'linear' kernel function, the 'polynomial' kernel function, the 'Gaussian radial basis' function (the 'RBF' kernel function), and the 'Sigmoid' kernel function, respectively [28]. In particular, the kernel function adopted in this paper is the radial basis function:

$$K(x, z) = \exp(-\gamma \|x - z\|^2). \tag{12}$$

In addition, the corresponding classification decision function can be defined as:

$$f(x) = \text{sign}(\sum_{i=1}^{N} \alpha_i^* y_i \exp(-\gamma \|x - z\|^2) + b^*). \tag{13}$$

3.2. Constructing a Multi-Classification Support Vector Machine

As a matter of fact, the risk grade evaluation of goaf is a multi-classification problem rather than a simple binary classification problem. Generally, the most common construction methods of a multi-classification support vector machine include the 'one-against-one' method, the 'one-against-all' method, the directed acyclic graph method, the binary tree method, etc. In this paper, the 'one-against-one' method is implemented to construct the multiple classifiers. In this way, the basic principle of the 'one-against-one' method is to establish a binary classification support vector machine between any two types when there are n different classifications in the training set. Therefore, $n(n-1)/2$ binary classifiers will be accurately obtained. In addition, the test set is input into each classifier, and the category with the most votes will be selected as the final output category of the classifier. For instance, when the number of classifiers equals 4, the 6 binary classifiers will be established; the classification structure diagram is depicted in Figure 3.

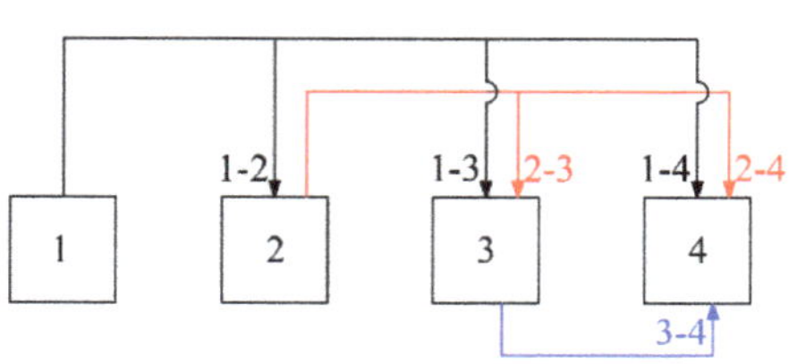

Figure 3. Schematic diagram of the classification structure.

3.3. Cross-Validation

In order to obtain a superior model under the conditions of small samples, a cross-validation technique can be implemented to make full use of the data for training and testing. As a rule, the most broadly used cross-validation method, namely, K-fold cross-validation, is adopted in this paper. To account for this, the specific principles can be summed up according to the following aspects: in the first place, it is viable to divide the data set into K subsets. Subsequently, each subset has the same size and does not intersect with any other subset. Later, one subset is taken as the test set and the remaining $K-1$ subsets are chosen as the training sets. In the end, after repeating the training for K times, the model with the smallest error among the K tests will be selected as the optimal choice.

3.4. Parameter Optimization of the Differential Evolution Algorithm

In fact, the differential evolution algorithm (DE) is an efficient global optimization algorithm, based on the population in question, which can realize a global search via competitive selection and differential mutations in the population. First and foremost, the DE is expected to be encoded and then randomly initializes the population of $M = (M_1, M_2, \ldots, M_N)$, where n denotes the size of the population. Note that the upper and lower bounds of the parameters should be set before initialization, then afterward the intermediate population can be accessibly obtained by mutating and crossing the parameters. Moreover, a greedy strategy is adopted to select the method of one-against-one between the two populations to obtain the new generation.

In this paper, the DE/rand/1/bin [29] is chosen as the form of DE, wherein the coding method adopts the real coding. One typical advantage of the real coding method that should be noted is that it does not need frequent coding and decoding, which can improve the accuracy and convergence speed when solving problems, and can effectively avoid some additional problems, such as 'Hamming cliffs', etc. Thus, the variation, crossover, and selection operations of this form can be concluded as follows.

Variation: An individual from the population will be randomly selected as the basis vector, and then it is necessary to take both other individuals as the difference vectors. After that, the variation operation can be performed:

$$n_i^{G+1} = m_{a1}^G + F(m_{a2}^G - m_{a3}^G) \tag{14}$$

where $a_1, a_2, a_3 \in \{1, 2, \ldots, N\}$, and $a_1 \neq a_2 \neq a_3$; N is the size of the population; F refers to the scaling factor, the value of which is a positive real number and also generally a random one between (0, 1), which can control the evolution rate of the population; G represents the current population, and $G + 1$ denotes the next generation.

Crossover: in order to increase the population diversity, the DE employs the binomial distribution crossover method to generate new individuals using predetermined parent individuals and mutated ones:

$$l_{i,j}^{G+1} = \begin{cases} n_{i,j}^{G+1}, rand \leq C \text{ } or \text{ } j = j_{rand} \\ m_{i,j}^G, otherwise \end{cases} \tag{15}$$

where, in this paper, $j \in \{1, 2\}$; $rand$ is a random number between [0, 1]; j_{rand} signifies an integer that is randomly generated in {1, 2} to ensure that at least one optimization parameter will mutate; C refers to the crossover factor, which is normally a random number between [0, 1].

Selection: A pair of survivors are competing for selection, and those with high fitness will be chosen for the next generation. In this case, the principle is that each individual in the population can only compare their fitness with those in the same position in another population. Hence, this can be expressed as:

$$m_i^{G+1} = \begin{cases} l_i^{G+1}, f(l_i^{G+1}) \leq f(m_i^G) \\ m_{i,j}^G, f(l_i^{G+1}) > f(m_i^G) \end{cases} \tag{16}$$

where $f(x)$ signifies the fitness function.

To sum up, through the operations of mutation, crossover, and selection, the population will evolve to form the next generation. In this way, it can reach the optimal after cycling and the optimal solution of the problem can be tackled. When the kernel function of SVM selects the radial basis function, the kernel parameter, 'γ', and the penalty factor, 'C', in the SVM model need to be determined. Therefore, the average score of cross-validation is taken as the goal to be optimized, and the parameters, such as 'C' and 'γ', are selected as the decision variables. Furthermore, the DE is adopted to find the optimal solution to the problem, and the optimal parameters can be obtained automatically, according to the optimization algorithm. Ultimately, the comprehensive workflow chart is displayed in Figure 4.

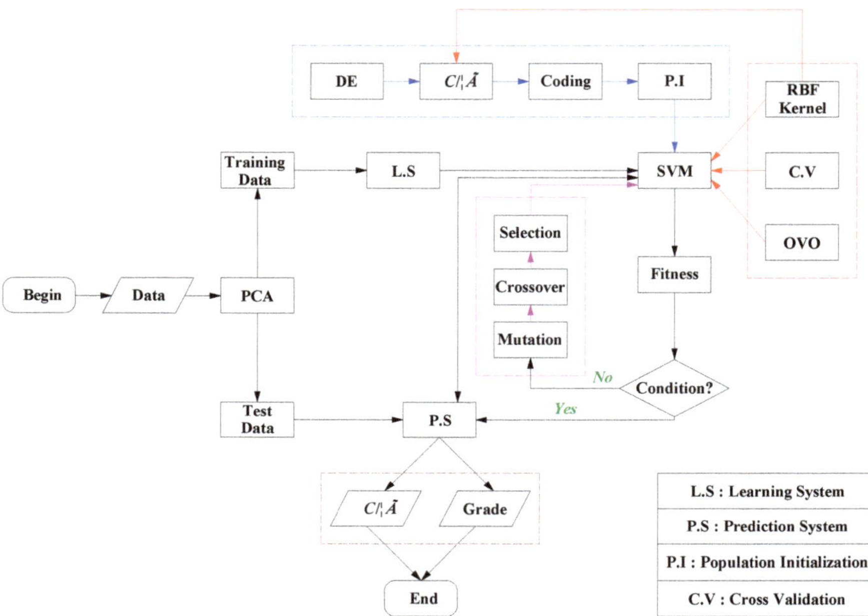

Figure 4. Integrated workflow chart.

4. Engineering Examples

After more than 20 years of mining, a certain amount of goaf from an iron mine in East China has been formed since the site was put into production. In this paper, 120 samples of sub-goaf are chosen as the data set and 9 parameters, such as the mining depth, the mining method, the goaf mining height, the maximum exposure area, the maximum exposure height, the maximum exposure span, the pillar situation, the measured volume, and the treatment rate are selected as the central influencing factors; sample data for goaf are listed in Table 1.

Table 1. Influencing factors of goaf stability and partial samples of the risk rank.

Sample Serial Number	Exploitation Depth X1/m	Mining Methods X2	Goaf Mining Height X3/m	Maximum Exposure Area X4/m^2	Maximum Exposure Height X5/m	Maximum Exposed Span X6/m	Pillar Situation X7	Measured Volume X8/m^3	Governance Rate X9	Risk Rank
1	130	1	35	3589	35	39	0	57,481.1	0.0	2
2	130	1	20	1208	0.99	24	1	12,141.3	94.4	1

Table 1. *Cont.*

Sample Serial Number	Exploitation Depth X1/m	Mining Methods X2	Goaf Mining Height X3/m	Maximum Exposure Area X4/m^2	Maximum Exposure Height X5/m	Maximum Exposed Span X6/m	Pillar Situation X7	Measured Volume X8/m^3	Governance Rate X9	Risk Rank
3	130	1	35	1735	5.97	28	0	31,595.7	96.3	1
4	130	1	35	1644	35	32	2	17,144.4	100.0	1
5	130	1	25	2439.5	25	39	2	19,377.7	100.0	1
					……					
119	220	1	15	349	15	17	0	3200	0.0	1
120	220	1	15	259	15	10	0	2867	0.0	1

When using the SVM to analyze data, the non-data factors can be transformed into data factors with the purpose of facilitating learning. Therefore, this paper deals with two factors: the mining method and the pillar situation. Based on this focus, the processing methods can be generalized into two aspects: (1) the mining method—the shallow mining method is recorded as +1, while the medium/deep-hole method is recorded as −1; (2) the pillar situation—a no-pillar situation is recorded as 0, while the boundary, intermediate, and mixed pillars are recorded as 1, 2, and 2, respectively.

According to the analysis and determinations of the field professionals, the risk level can be divided into three ranks, namely, 1, 2, and 3. In other words, the higher the rank, the more serious the risk that will occur.

As an illustration, the 9 factors affecting the stability of goaf are taken as the input factors, and the risk level is chosen to be the output factor. Afterward, the data in Table 1 were analyzed using the SPSS software. Meanwhile, the correlation coefficient adopts the Pearson correlation coefficient, while the two-tailed t-test is applied to the significance test.

The Pearson correlation coefficient matrix of each factor and the heat map of the correlation matrix are shown in Table 2 and Figure 5, respectively. It can be seen from Table 2 that several factors among the input factors have a strong correlation with each other. As a result, it is necessary to conduct the PCA using the input data.

Table 2. Pearson correlation coefficient matrix of each factor.

Index	X1	X2	X3	X4	X5	X6	X7	X8	X9
X1	1.000								
X2	−0.366	1.000							
X3	−0.389	0.273	1.000						
X4	0.001	−0.432	−0.084	1.000					
X5	−0.325	−0.045	0.512	0.089	1.000				
X6	−0.050	−0.465	0.097	0.695	0.227	1.000			
X7	−0.342	0.163	0.296	0.086	0.236	0.103	1.000		
X8	−0.046	−0.150	0.098	0.594	0.019	0.370	0.110	1.000	
X9	0.104	0.010	0.153	−0.039	−0.309	−0.093	−0.005	0.061	1.000

It is evident that there are different dimensions in the variables of the research issues, which may lead to some new problems. Therefore, the original data should be standardized using Equation (2) before the PCA takes place.

After using the SPSS software to conduct a PCA on the standardized data, the principal component gravel diagram (Figure 6) and the principal component list (Table 3) can be obtained promptly. According to Figure 6 and Table 3, the eigenvalues of the first five factors differ vastly from each other, and the cumulative contribution rate of the total variance is equal to 83.643%, which fulfills the requirement that the variance of the principal components accounts for 75–85% of the overall variance [30]; namely, the most information on the overall variance can be precisely summarized by the first five factors. Hence, it is practicable to select the first five components as the principal components to replace the original variables for analysis.

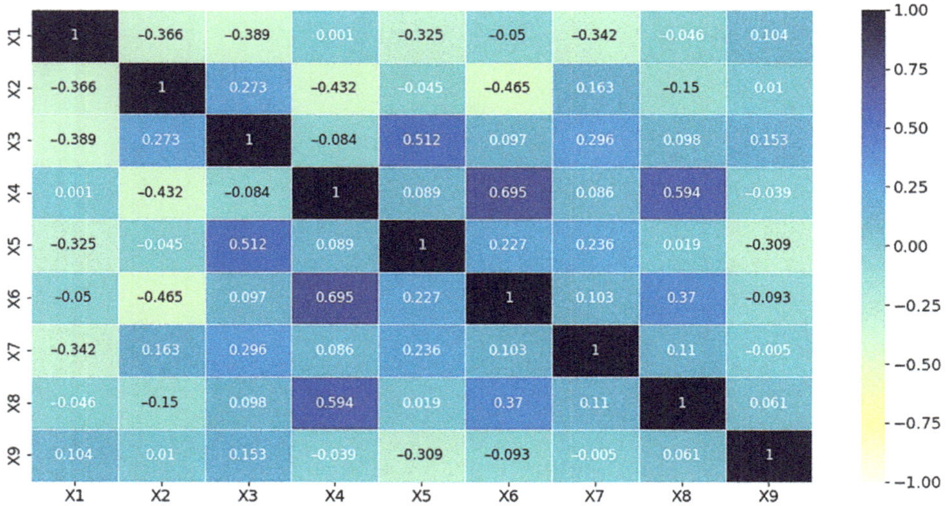

Figure 5. Correlation matrix heat map.

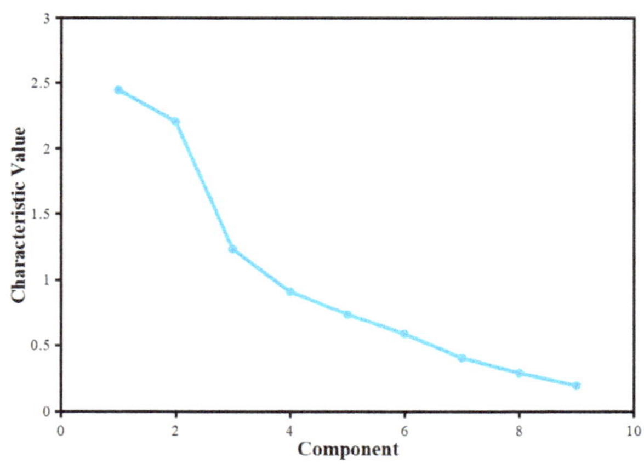

Figure 6. Gravel diagram of PCA.

Table 3. List of principal components.

Component	Initial Characteristic Value			Sum of Squares of Extracted Loads		
	Total	Percentage Variance	Accumulation/%	Total	Percentage Variance	Accumulation/%
1	2.444	27.150	27.150	2.444	27.150	27.150
2	2.208	24.528	51.678	2.208	24.528	51.678
3	1.233	13.698	65.377	1.233	13.698	65.377
4	0.911	10.127	75.504	0.911	10.127	75.504
5	0.733	8.139	83.643	0.733	8.139	83.643
6	0.584	6.493	90.136			
7	0.402	4.470	94.607			
8	0.290	3.217	97.824			
9	0.196	2.176	100.000			

As displayed in Table 4, the load matrix of the main component factors can be adopted to ascertain the relationship between the main component factors, Y1, Y2, Y3, Y4, and Y5, and the original variables. Hence, the expressions of the corresponding factors can be listed as follows:

$$Y1 = -0.059X1 - 0.565X2 + 0.100X3 + 0.884X4 + 0.321X5 + 0.858X6 + 0.188X7 + 0.664X8 - 0.118X9;$$
$$Y2 = -0.751X1 + 0.519X2 + 0.766X3 - 0.158X4 + 0.638X5 - 0.019X6 + 0.583X7 - 0.001X8 - 0.123X9;$$
$$Y3 = -0.031X1 + 0.219X2 + 0.211X3 + 0.110X4 - 0.454X5 - 0.065X6 + 0.191X7 + 0.405X8 + 0.846X9;$$
$$Y4 = 0.296X1 - 0.363X2 + 0.451X3 - 0.173X4 + 0.367X5 + 0.103X6 - 0.246X7 - 0.288X8 + 0.413X9;$$
$$Y5 = 0.117X1 - 0.251X2 - 0.182X3 - 0.043X4 - 0.067X5 + 0.060X6 + 0.697X7 - 0.343X8 + 0.096X9.$$

Table 4. Load matrix of the principal component factors.

Index	Principal Component				
	Y1	Y2	Y3	Y4	Y5
X1	−0.059	−0.751	−0.031	0.296	0.117
X2	−0.565	0.519	0.219	−0.363	−0.251
X3	0.100	0.766	0.211	0.451	−0.182
X4	0.884	−0.158	0.110	−0.173	−0.043
X5	0.321	0.638	−0.454	0.367	−0.067
X6	0.858	−0.019	−0.065	0.103	0.060
X7	0.188	0.583	0.191	−0.246	0.697
X8	0.664	−0.001	0.405	−0.288	−0.343
X9	−0.118	−0.123	0.846	0.413	0.096

According to the calculated factor expressions, it is necessary to conduct the PCA and calculation on the standardized data at a later time. This can be seen most obviously in Table 5 for the partial calculated data.

Table 5. Partial data after principal component calculation.

Sample Serial Number	Y1	Y2	Y3	Y4	Y5	Risk Rank
1	0.9692	1.8113	0.1422	0.3396	−0.6124	2
2	−0.1286	1.0236	1.2293	0.1273	0.0872	1
3	0.0849	1.2046	1.2507	0.5023	−0.4067	1
4	0.5255	2.3128	1.0500	0.6140	0.2771	1
5	0.6189	1.8470	1.1160	0.3447	0.3573	1
			......			
119	−0.2744	0.6973	0.0772	0.0678	−0.2842	1
120	−0.4014	0.7017	0.0838	0.0565	−0.2910	1

In view of the standardized data after PCA, the multi-classification SVM test will be carried out next. To begin with, the 'one-against-one' method was selected to construct the multi-classification classifiers. After adopting the cross-validation method of the *K*-fold, *K* = 5 is taken. Simultaneously, the principal components 'Y1', 'Y2', 'Y3', 'Y4', and 'Y5' are used as the input factors and the risk level is chosen as the output factors. Eventually, all 120 samples can be divided into the training sets and the test sets, in which the number of the former and the latter are equal to 80 and 40, respectively.

After optimizing the parameters of SVM by means of the DE, the real-number coding method has been employed. In this case, the number of individuals in the population and the maximum evolution algebra can be taken to be a value of 20 and 30, respectively. In the end, the final calculation result is compared with the classification results of SVM without PCA, and simultaneously, the results of the naive Bayes and the BP neural network after PCA are compared with each other.

As a matter of fact, the data prediction accuracy of BP neural network classification (Method 1) and Naive Bayes classification after PCA can both achieve 87.5% while the accuracy of the classification methods (Method 3) without PCA can reach 90%. Conversely, the data prediction accuracy of the method used in this paper (Method 4) can attain 92.5%.

More specifically, the relevant parameters after optimization by DE are C = 145.50 and γ = 0.26. As exhibited, the iterative effect of DE is depicted in Figure 7, and the confusion matrices for the test sets among each method are shown in Figure 8. In addition, the comparison of prediction accuracy is displayed in Figure 9.

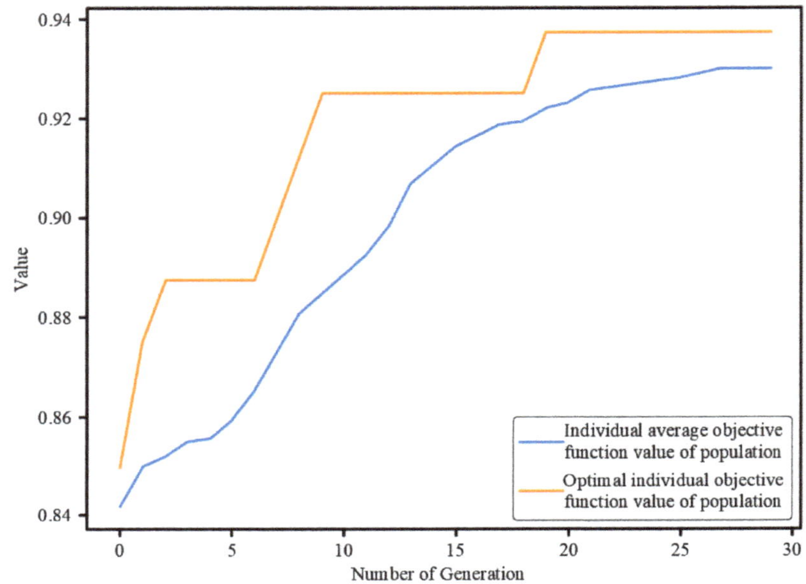

Figure 7. Iterative effect of the DE.

(a) Method 1

Actual Class \ Predicted Class	1	2	3
1	34	1	0
2	2	0	1
3	1	0	1

(b) Method 2

Actual Class \ Predicted Class	1	2	3
1	33	2	0
2	2	1	0
3	1	0	1

(c) Method 3

Actual Class \ Predicted Class	1	2	3
1	34	1	0
2	2	1	0
3	1	0	1

(d) Method 4

Actual Class \ Predicted Class	1	2	3
1	35	0	0
2	2	1	0
3	1	0	1

Figure 8. Confusion matrices for the test sets.

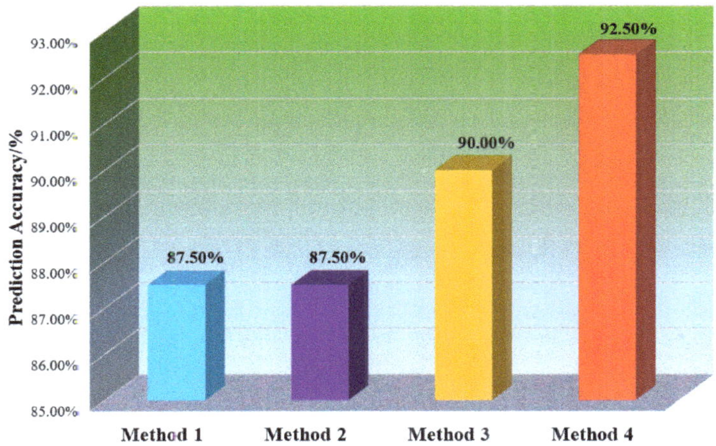

Figure 9. Comparison of the prediction accuracy.

As demonstrated, in the meantime, the PCA method can not only retain the most information involved in the original data but also play a vital role in denoising. Owing to the effects of the dimension reduction, the running time can be shortened, and the training efficiency can be enhanced. When there is noise in the large data sets, the PCA will significantly improve the tests' efficiency and accuracy. Although the neural network can deal with nonlinear issues due to the insufficient samples in practical application, the visible problems that will proceed, such as the local minimum, overfitting, etc., have reflected the limitation of weak robustness and low recognition accuracy.

5. Conclusions

In this paper, the support vector machine (SVM) based on the principal component analysis (PCA) and the differential evolution algorithm (DE) is adopted to identify the risk level of goaf, and the primary findings can be drawn as follows:

(1) The 'one-against-one' method is used to construct a multi-classification SVM. In order to prevent the overfitting of the model, the *K*-fold cross-validation method will be employed to select it. Above all, the research results reveal that the SVM has the desirable ability of generalization. Compared with the neural network, the apparent advantages lie in solving the problems of overfitting and it is easy to fall into the local minimum that can be detected in the SVM under the conditions of small samples.

(2) PCA is used to preprocess the original data of multi-source impact indicators for goaf risk identification, which can realize the dimensionality reduction and data denoising, and can simultaneously improve the prediction accuracy and classification efficiency while retaining the most information.

(3) Using the strategy of DE and a global optimization search mechanism, the optimal solution of the problems to be optimized will be automatically obtained, namely, the kernel function parameter of SVM, 'γ', and the penalty factor, 'C'. Moreover, the engineering calculation example further verifies that the DE has the characteristics of clear logic, strong convergence, and good robustness.

In simpler terms, the method embraced in this study is then discussed. Compared with some machine learning algorithms, the support vector machine (SVM) can be converted into high-dimensional feature space through nonlinear transformation, which can subtly avoid the curse of the dimensionality problem. In addition, since the algorithm can be transformed into a convex quadratic programming problem, it has an obvious ascendancy over the neural network in terms of eliminating the local extremum issues and small sample data sets. Therefore, the work efficiency of safety production in mines will be enormously improved by means of introducing this method into the domain of the risk identification of mine goafs, which has significant engineering instructive significance and widespread application values. Nevertheless, except for the aforementioned advantages, some limitations can be examined in SVM. For example, in contrast to other machine learning algorithms, when a huge level of capacity can be found in the sample data sets, SVM will perform inefficiently, and it is sensitive to missing the data among the data sets. Furthermore, standard algorithms cannot reasonably reflect the probability characteristics. Consequently, for further research routes, a number of rational improvements can be prompted for the SVM, for instance, introducing the probability characteristics can refine it into a probabilistic SVM, while replacing a single kernel function with a family of ones is enhanced as the multiple kernel SVM, etc.

Author Contributions: Conceptualization, H.Y. and Y.W.; methodology and software, Z.C.; validation, L.X.; investigation, H.L. All authors have read and agreed to the published version of the manuscript.

Funding: This research work was funded by the National Natural Science Foundation of China (51874112;42077249, 51774107), the University Synergy Innovation Program of Anhui Province (Grant. No.GXXT-2020-055), the Opening Project of the State Key Laboratory of Explosion Science and Technology, and the Beijing Institute of Technology (KFJJ21-03Z).

Data Availability Statement: The data supporting the findings of this study are not publicly available due to privacy.

Conflicts of Interest: The authors declare no conflict of interest.

References

1. Yi, H.; Zhang, X.; Yang, H.; Li, M.; Gao, Q.; Jinke. Goaf collapse vibration analysis and disposal based on a experiment of heavy ball touchdown. *Explos. Shock Waves* **2019**, *39*, 91–103.
2. Zhao, Y.; Tang, J.; Chen, Y.; Zhang, L.; Wang, W.; Wan, W.; Liao, J. Hydromechanical coupling tests for mechanical and permeability characteristics of fractured limestone in complete stress-strain process. *Environ. Earth Sci.* **2017**, *76*, 24. [CrossRef]
3. Zhao, Y.; Luo, S.; Wang, Y.; Wang, W.; Zhang, L.; Wan, W. Numerical Analysis of Karst Water Inrush and a Criterion for Establishing the Width of Water-resistant Rock Pillars. *Mine Water Environ.* **2017**, *36*, 508–519. [CrossRef]
4. Liao, Y.; Yu, G.; Liao, Y.; Jiang, L.; Liu, X. Environmental Conflict Risk Assessment Based on AHP-FCE: A Case of Jiuhua Waste Incineration Power Plant Project. *Sustainability* **2018**, *10*, 4095. [CrossRef]
5. Wu, H.; Jia, Q.; Wang, W.; Zhang, N.; Zhao, Y. Experimental Test on Nonuniform Deformation in the Tilted Strata of a Deep Coal Mine. *Sustainability* **2022**, *13*, 13280. [CrossRef]
6. Du, K.; Li, X.; Liu, K.; Zhao, X.; Zhou, Z.; Dong, L. Comprehensive evaluation of underground goaf risk and engineering application. *J. Cent. South Univ.* **2011**, *42*, 2802–2811.
7. Yuan, Z.; Zhai, J.; Li, S.; Jiang, Z.; Huang, F. A Unified Solution for Surrounding Rock of Roadway Considering Seepage, Dilatancy, Strain-Softening and Intermediate Principal Stress. *Sustainability* **2022**, *14*, 8099. [CrossRef]
8. Liu, Y.; Hao, Y.; Lu, Y. Improved Design of Risk Assessment Model for PPP Project under the Development of Marine Architecture. *J. Coast. Res.* **2021**, *9*, 74–80. [CrossRef]
9. Liu, S.; Nie, Y.; Hu, W.; Ashiru, M.; Li, Z.; Zou, J. The Influence of Mixing Degree between Coarse and Fine Particles on the Strength of Offshore and Coast Foundations. *Sustainability* **2022**, *14*, 9177. [CrossRef]
10. Chen, W.; Wan, W.; Zhao, Y.; Peng, W. Experimental Study of the Crack Predominance of Rock-Like Material Containing Parallel Double Fissures under Uniaxial Compression. *Sustainability* **2020**, *12*, 5188. [CrossRef]
11. Zhang, Y.; Chang, X.; Liang, J. Comparison of different algorithms for calculating the shading effects of topography on solar irradiance in a mountainous area. *Environ. Earth Sci.* **2017**, *76*, 295. [CrossRef]
12. Feng, T.; Chen, H.; Wang, K.; Nie, Y.; Zhang, X.; Mo, H. Assessment of underground soil loss via the tapering grikes on limestone hillslopes. *Agric. Ecosyst. Environ.* **2020**, *5*, 297. [CrossRef]
13. Chen, J.; Liu, L.; Zhou, Z.; Yong, X. Optimization of mining methods based on combination of principal component analysis and neural networks. *J. Cent. South Univ.* **2010**, *41*, 1967–1972.
14. Wang, X.; Duan, Y.; Peng, X. Fuzzy Synthetic Assessment of the Danger Degree of Mined-out Area Disaster. *Min. Res. Dev.* **2005**, *25*, 83–85.
15. Zhou, J.; Shi, X. Evaluation of the alternatives for mined out area disposal based on the identical degree of set pair analysis. *Met. Mine* **2009**, *396*, 10–13.
16. Wang, X.; Ding, D.; Duan, Y. Applications of the grey relation analysis in the evaluation of the risk degree of the underground mined-out stopes. *J. Saf. Sci. Technol.* **2006**, *2*, 35–39.

17. Gong, F.; Li, X.; Dong, L.; Liu, X. Underground goaf risk evaluation based on uncertainty measurement theory. *Chin. J. Rock Mech. Eng.* **2008**, *27*, 323–330.
18. Hu, Y.; Li, X. Bayes discriminant analysis method to identify risky of complicated goaf in mines and its application. *Trans. Nonferrous Met. Soc. China* **2012**, *22*, 425–431. [CrossRef]
19. Feng, Y.; Wang, X.; Cheng, A.; Zhang, Q.; Zhao, J. Method optimization of underground goaf risk evaluation. *J. Cent. South Univ.* **2013**, *44*, 2881–2888.
20. Wang, Z.; Guo, J.; Wang, L. Recognition of goaf risk based on support vector machines method. *J. Chongqing Univ.* **2015**, *38*, 85–90.
21. Wang, H.; Li, X.; Dong, L.; Liu, K.; Tong, H. Classification of goaf stability based on support vector machine. *J. Saf. Sci. Technol.* **2014**, *10*, 154–159.
22. Fang, X.; Ding, Z.; Shu, X. Hydrogen yield prediction model of hydrogen production from low rank coal based on support vector machine optimized by genetic algorithm. *J. China Coal Soc.* **2010**, *35*, 205–209.
23. Hsu, C.-W.; Lin, C.-J. A comparison of methods for multi-class support vector machines. *IEEE Trans. Neural Netw.* **2002**, *13*, 415–425. [PubMed]
24. Kebel, U. Pairwise Classification and Support Vector Machines. In *Advances in Kernel Methods-Support Vector Learning*; MIT Press: Cambridge, MA, USA, 1999; pp. 255–258.
25. Platt, J.C.; Cristianini, N.; Shawe-Taylor, J. Large Margin DAGs for Multi-class Classification. *Adv. Neural Inf. Process. Syst.* **2000**, *12*, 547–553.
26. Bennett, K.P.; Blue, J.A. A support vector machine approach to decision tree. *Rensselaer Polytech. Inst.* **1997**, *3*, 2396–2401.
27. Zhou, Z. *Machine Learning*; Tsinghua University Press: Beijing, China, 2017; pp. 121–139.
28. Liang, N.; Tuo, Y.; Deng, Y.; Jia, Y. Classification model of ice transport and accumulation in front of channel flat sluice based on PCA-SVM. *Chin. J. Theor. Appl. Mech.* **2021**, *53*, 703–713.
29. Chen, X.; Yang, G.; Huang, M. Real-coded Quantum Differential Evolution Algorithm. *J. Chin. Comput. Syst.* **2013**, *34*, 1141–1146.
30. Xu, Z.; Zhou, D.; Luo, Y. Fuzzy Neural Network Based on Principal Component. *Comput. Eng. Appl.* **2006**, *42*, 34–36.

water

MDPI

Article

Design and Field Monitoring of a Pile–Anchor–Brace Supporting System in a Soft Soil Area

Lin Sun [1], Ke Mao [2], Zhengzhen Wang [2,*], Shuaihua Ye [2], Tiantao Su [2], Guoliang Dai [3], Guangxiang Xu [1] and Jilong Sun [4]

[1] China State Construction Railway Investment & Engineering Group Co., Ltd., Beijing 100000, China
[2] School of Civil Engineering, Lanzhou University of Technology, Lanzhou 730050, China
[3] School of Civil Engineering, Southeast University, Nanjing 210096, China
[4] China Construction Third Engineering Bureau Group Co., Ltd., Wuhan 430000, China
* Correspondence: wangzz@lut.edu.cn

Abstract: With the continuous development of urbanization and the rapid development of science and technology, the requirements for foundation pit engineering are getting higher and higher. Foundation pit engineering is gradually developing in the direction of larger area and deeper excavation. In engineering examples, the combined supporting structure of a pile–brace and pile–anchor for foundation pits is widely used, while the engineering examples supported by a pile–anchor–brace supporting system are less frequently used. Based on a super-large deep foundation pit project in Yancheng City, Jiangsu Province, China, according to the surrounding environmental conditions, the foundation pit support scheme, and on-site construction situation, the design and on-site monitoring of the pile–anchor–brace supporting system were introduced and analyzed. The results show that: (1) the deformation of the pile–anchor–brace supporting system shows an obvious spatial effect, and the horizontal displacement of the pile and soil of the long side direction is greater than the short side direction; (2) in the initial state, the deep horizontal displacement of the soil is in the form of a 'cantilever', but in the later stage it changed to the form of a 'drum belly', and both the brace and anchor cable can limit the displacement of the soil effectively; (3) the axial force of the brace develops rapidly in the initial stage, but its development tends to be gentle after the completion of the first anchor cable construction. Through on-site monitoring, it was found that the axial force of the ring brace was larger than that of the corner brace, which was larger than the opposite brace; and (4) the development trend of the axial force for the two rows of anchor cables is quite different. The average axial force of the first row of anchor cables is greater than the second row of anchor cables, and the development trend of the first row of anchor cables is steep first and then gentle, while the change trend of the second row of anchor cables is just the opposite.

Keywords: foundation pit engineering; pile–anchor–brace supporting system; field monitoring; deformation; internal force

check for updates

Citation: Sun, L.; Mao, K.; Wang, Z.; Ye, S.; Su, T.; Dai, G.; Xu, G.; Sun, J. Design and Field Monitoring of a Pile–Anchor–Brace Supporting System in a Soft Soil Area. *Water* **2022**, *14*, 3949. https://doi.org/10.3390/w14233949

Academic Editor: Dan Ma

Received: 29 October 2022
Accepted: 1 December 2022
Published: 4 December 2022

1. Introduction

With the rapid development of the Chinese economy and the deepening of urbanization, the number and scale of foundation pit projects in coastal soft soil areas continues to increase. However, the high groundwater level, poor soil conditions, and soil disturbance sensitivity in soft soil areas bring great challenges to the design and construction of a deep foundation pit.

Some scholars have studied the force and deformation characteristics of foundation pits in soft soil areas. Clough et al. [1] summarized the displacement change law of diaphragm walls in soft soil areas by updating the existing database. By combining the monitoring data and finite element analysis results, Mana et al. [2] proposed a simplified method for predicting the displacement of the wall and supporting structures of the foundation pit in soft to medium clay strata. By selecting 10 foundation pit engineering cases with high construction quality and field monitoring data in a soft soil area, Ou et al. [3] generalized the characteristics of surface subsidence during excavation. Hashash et al. [4] used the nonlinear finite element method to analyze the deformation of internal brace diaphragm walls in deep clay layers. The above scholars started to study foundation pit engineering in soft soil areas in the early stages. Yang et al. [5] discussed the change in pre-stressed load of anchor cables in a practical project and analyzed the monitoring results to mainly study the lateral deformation characteristics of the foundation pit. Based on a deep foundation pit project in Nanjing in a soft soil area, Ju et al. [6] summarized the change law of displacement and the axial force through field monitoring. The results showed that the support system of the corner brace and the opposite brace can meet the requirements of the project. Based on the construction monitoring data of two adjacent deep foundation pits in soft soil areas, Xu et al. [7] analyzed the stress and deformation of the supporting structures during the simultaneous excavation of the two foundation pits. The analysis results showed that the two adjacent foundation pits have a great influence on the deformation and stress of each other. Taking a foundation pit project in Shanghai as an example, Li et al. [8] used the underground continuous wall and multi-channel internal brace to control the deformation of the foundation pit. Through the analysis of field monitoring data, it was found that the deformation control measures achieved the expected results. Zhang et al. [9] introduced an emergency reinforcement scheme dealing with the deformation of the foundation pit exceeding the warning value. The above scholars studied and summarized the application of some single supporting structures in foundation pit engineering in soft soil areas. Gong et al. [10] carried out field monitoring and analyzed the corresponding data of a foundation pit supported by a pile–anchor structure. According to the data, it was found that a pile–anchor support structure can control the deformation of the foundation pit in a complex environment well. Based on a deep foundation pit of a subway station in a soft soil area, Wei et al. [11] analyzed the measured the data of the deformation and stress of the supporting structure. The results showed that the axial force of the brace was not necessarily large in the upper row and was small in the lower row under the multiple braces condition. The spatial effect caused by the excavation of the foundation pit will lead to uneven settlement of the underground pipeline. Based on a foundation pit project in a soft soil area of Suzhou City, Li et al. [12] found that the foundation pit construction did not have a great impact on the surrounding environment, and each monitoring project did not exceed the alarm value. Li et al. [13] introduced the monitoring results of a deep foundation pit in a soft soil area during a construction period, and the results showed that the monitoring basically reflected the expected stress and deformation of the supporting structure and surrounding environment during the construction of the foundation pit, which can provide good guidance for construction. Using a foundation pit project in Nanning as the background, Jiang et al. [14] discussed the deformation of the foundation pit under a combined support structure and concluded that the deformation of the deep foundation pit was related to its surrounding environment. Based on a deep foundation pit project in a soft soil area, Long et al. [15] compared the deformation of pile–anchor support and soil nailing support and found that both of these support systems can control the deformation effectively. Through the monitoring of foundation pit engineering in a soft soil area, the above scholars studied the characteristics of some composite supporting structures.

It is easy to conclude through the introduction above that, at present, scholars have analyzed and studied the foundation pits in soft soil areas under various supporting forms (such as soil nailing support, anchor–pile support structures, and brace-pile support structures), but most of these foundation pits are a regular shape and relatively singularly supported. With the improvement in the scale and depth of the foundation pit, some more recent supporting structures (such as pile–anchor–brace supporting systems) should be extended. Therefore, based on a deep and large foundation pit in Yancheng City, Jiangsu Province, China, which is supported by a pile–anchor–brace supporting system (a relatively new supporting system), the design details and on-site monitoring results of this foundation pit were introduced, and the variation law of the deformation, stress, and other aspects of the support system were analyzed in detail.

2. Project Profile

2.1. Site Overview

This project, which has a two-story basement, is located in Yannan High-Tech Zone, Yancheng City, Jiangsu Province, China, and belongs to the second phase expansion project. It is adjacent to Daizhuang Road in the east, Renmin South Road in the west, Haiyang Road in the north, and Yuehai Road in the south. The total area of the project is about 113,600 m^2. The shape of the foundation pit is basically rectangular, 410 m long from east to west and 255 m wide from south to north. The area of the foundation pit is 102,900 m^2 and its perimeter is 1335 m.

2.2. Hydrogeological Conditions

The maximum excavation depth of the foundation pit is about 14 m. In the range of the exploration depth during the geological survey, the strata is divided into 15 layers, and the main affecting soil layers to the foundation pit are from Layer 1 to Layer 4B. The basic physical and mechanical parameters of each layer of soil are shown in Table 1 [16].

Table 1. Basic parameters of each rock and soil layer.

Soil Layer Sequence	Layer Name	Depth of Soil Layer/m	γ/(kN/m^3)	ω/(%)	c'/kPa	φ'/(°)
Layer 1	Plain fill (Q_4^{ml})	0.10~1.10	17.5	/	8.0	10.0
Layer 2	Silty clay (Q_4^{l})	1.30~2.20	18.9	29.9	20.7	8.1
Layer 3	Muddy silty clay (Q_4^{l})	12.20~19.20	17.3	43.7	12.5	7.1
Layer 4A	Clayey silt (Q_4^{m})	13.80~19.50	18.3	32.6	14.7	14.2
Layer 4B	Sandy silt (Q_4^{m})	15.00~22.80	19.0	28.1	8.3	27.4
Layer 5	Clayey silt (Q_4^{m})	18.20~24.10	18.3	33.3	14.8	16.8
Layer 6A	Clayey silt (Q_4^{m})	18.40~24.00	18.3	32.9	13.8	17.0
Layer 6B	Sandy silt (Q_4^{m})	21.70~25.70	18.9	28.9	8.2	27.0
Layer 6C	Clayey silt (Q_4^{m})	24.50~26.00	18.4	32.1	14.9	18.0

Note: (Definitions: γ is soil weight, ω is soil moisture content, c' is soil cohesion force, and φ' is the angle of internal friction).

3. Design and Construction

3.1. Selection of Supporting Structure

Considering the depth of the foundation pit, the surrounding environment, and hydrogeological conditions to ensure the stability and deformation requirements during the excavation of the foundation pit, the pile–anchor–brace supporting system (including an SMW pile, one row of concrete braces, and two rows of pre-stressed anchor cables) was selected to support this foundation pit (as shown in Figures 1 and 2).

3.2. SMW Pile

According to the characteristics of the foundation pit and the technical specification for a soil mixed wall [17], the diameter of the SMW pile was 850 mm and the size of the H-shaped steel was determined to be 700 mm × 300 mm. Finally, soil-cement pile mixed by five shafts and by three shafts inserted with H700 type steel were used in the foundation pit.

Figure 1. Aerial photograph of the foundation pit.

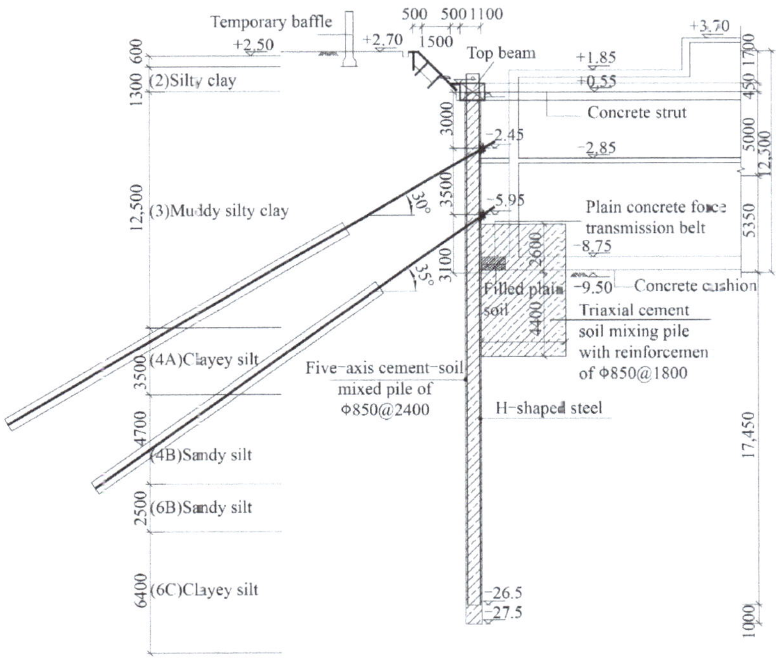

Figure 2. Typical support section (unit: mm).

3.3. Concrete Brace

In order to enable the supporting structure to better transfer its load and maintain the stability and safety of the foundation pit, and for the comprehensive consideration of the shape and size of the foundation pit [18], an internal brace system including an opposite brace, diagonal brace, corner brace, and ring brace was finally adopted. The parameters of each brace section are shown in Table 2. The concrete grade and reinforcement grade in the table are determined according to the Code for Design of Concrete Structures GB50204-2015 [19]. Figure 3 shows the steel bar binding for opposite brace types and Figure 4 shows the ring brace after construction.

Table 2. Sectional parameters of the braces.

Type of Brace	Concrete Grade	Sectional Height/mm	Section Width/mm	Reinforcement of Top of Brace	Reinforcement of Bottom of Brace	Longitudinal Stirrup	Lateral Stirrup
Opposite brace	C30	900	800	6C25	6C25	C8@200/400	C8@100/200
Top beam	C30	900	1300	6C20	8C20	4C8@300	4C8@150
Inner ring brace	C30	900	1400	6C20	8C20	4C8@300	4C8@150
Outer ring brace	C30	900	1300	6C20	8C20	4C8@300	4C8@150
Coupling beam	C30	900	700	5C25	5C25	C8@200/400	C8@100/200

Figure 3. Steel bar binding of the inner brace.

Figure 4. Ring brace.

3.4. Rotary Spray Anchor Cable

The project was located in a soft soil area. Considering the stress and load transfer of the anchor cable in soft soil and according to the requirements of the Technical Specification for Building Foundation Pit Support [20], a large-diameter rotary jet anchor cable was selected to support the foundation pit. Both of the two rows of anchor cables are composed of four bundles of steel strands with a nominal diameter of 15.2 mm and the diameter of each anchor cable is 500 mm. The other parameters of the anchor cables are shown in Table 3. The construction of drilling and grouting of the anchor cables can be seen in Figures 5 and 6.

Table 3. Parameters of the anchor cables.

Vertical Position	Dip Angle/(°)	Length of Free Section/m	Anchoring Length/m	Design Value of Axial Force for Anchor Cable/kN	Prestress of Anchor Cable/kN
First row	30	9	23	260	300
Second row	35	6	22	205	200

Figure 5. Drilling of the anchor cables.

Figure 6. Grouting of anchor cables.

3.5. Work Conditions

It took about 13 months from the excavation of the foundation pit to the completion of the underground structure. As the foundation pit safety can be ensured after the brace replacement conditions are formed for the construction of the underground second floor roof, this paper only introduces the monitoring results from the beginning of excavation to the completion of the construction of the underground second floor roof. All of the processes can be divided into five work conditions, as shown in Table 4.

Table 4. Work conditions.

Work Condition	Specific Construction Stage	Duration of Experience
I	Dig the first layer of soil to 3.5 m below the surface and construct the inner brace	90 d
II	Dig the second layer of soil to 6.5 m below the surface and construct the first row of anchor cables	85 d
III	Dig the third layer of soil to 9.5 m below the surface and construct the second row of anchor cables	75 d
IV	Dig the fourth layer of soil to 12 m below the surface and pour the cushion and floor	70 d
V	Construct underground structure to the negative two-layer roof	60 d

In the first work condition, the soil in the area of the inner brace was excavated at first and the soil was used as the framework for inner brace pouring. After the construction of the inner brace was completed and the design strength was reached, the excavation of other areas was carried out. In the second and third working conditions, the area around the anchor cable was excavated first, and then the anchor cable was constructed. Large-area excavation was carried out after the construction of the anchor cables was completed and the prestress was applied.

4. On-Site Monitoring and Data Analysis

The layout of the foundation pit monitoring points is shown in Figure 7. According to the foundation pit design drawings and the Technical Standards for Monitoring of Building Foundation Pit Engineering, there were 65 monitoring points for horizontal displacement of the pile top, which were numbered with QLi, and i was 1 to 65. They were arranged at the top of the coping beam with a spacing of 20 m. There were 26 monitoring points for deep horizontal displacement of the soil, which were numbered with CXi, and i was 1 to 26. They were buried in the soil around the foundation pit. There were 26 monitoring points for the groundwater level outside the foundation pit, which were numbered with SWi, and i was 1 to 26. They were buried in the soil around the foundation pit with a depth of 15 m. There were 62 monitoring points for road settlement, which were numbered with DLi, and i was 1 to 62. They were arranged at a distance of 20 m along the roadside adjacent to the foundation pit. There were 27 monitoring points for the axial force of brace, which were numbered with ZLi, and i was 1 to 27. They were laid on the braces. There were 28 monitoring points for the axial force of the anchor cables, which were numbered with MSi, and i is 1 to 28. They were divided into two rows and arranged near the anchor head.

The horizontal displacement monitoring points on the north side and on the west side were selected. Because the construction progress of both sides of the foundation pit was relatively fast, their results can be compared and studied. The monitoring points of the axial force of the brace were respectively selected on the opposite brace, angle brace, and circular brace, which made it convenient to study the various types of braces. The remaining monitoring points were arranged on four sides of the foundation pit, so as to facilitate mutual comparison and research. The monitoring results of the horizontal displacement of the pile top, deep horizontal displacement, underground water level outside the pit, road settlement, support axial force, and anchor cable axial force can be found in Figures 8–13.

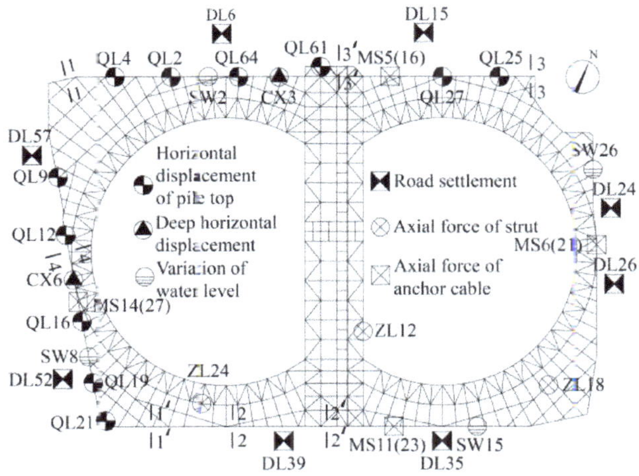

Figure 7. Layout of the monitoring points.

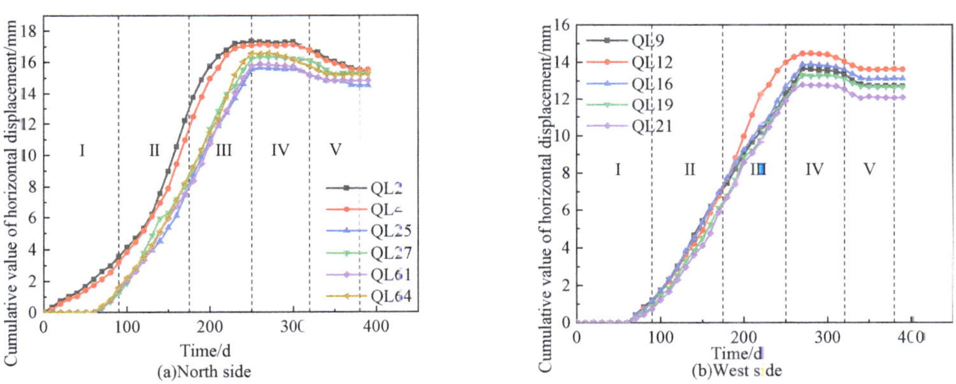

Figure 8. Horizontal displacement curve of the pile top.

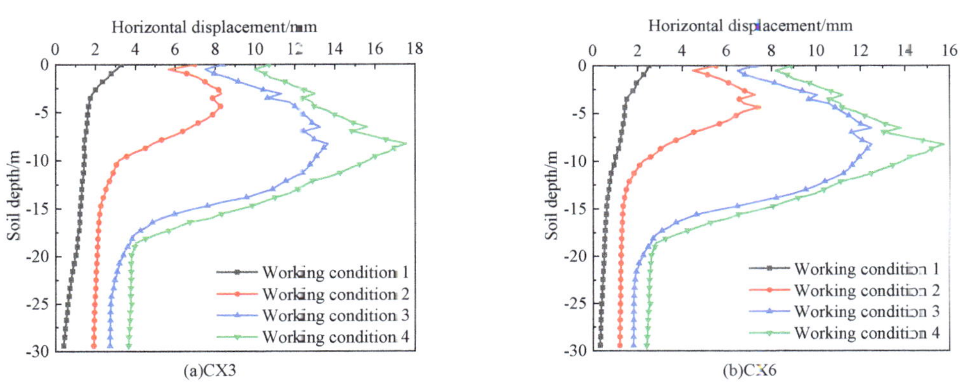

Figure 9. Deep horizontal displacement curve.

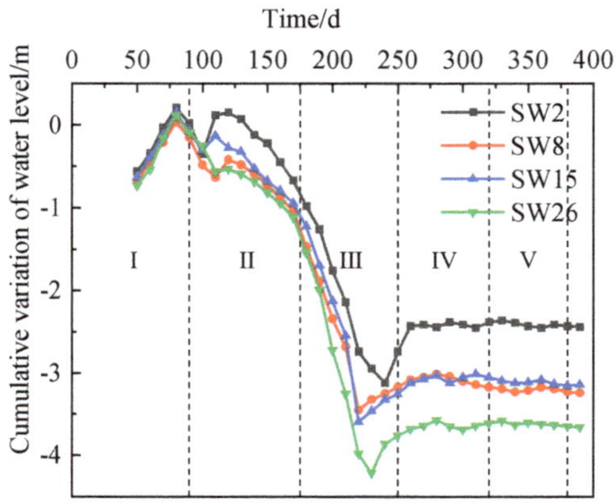

Figure 10. Change curve of the groundwater level outside the pit.

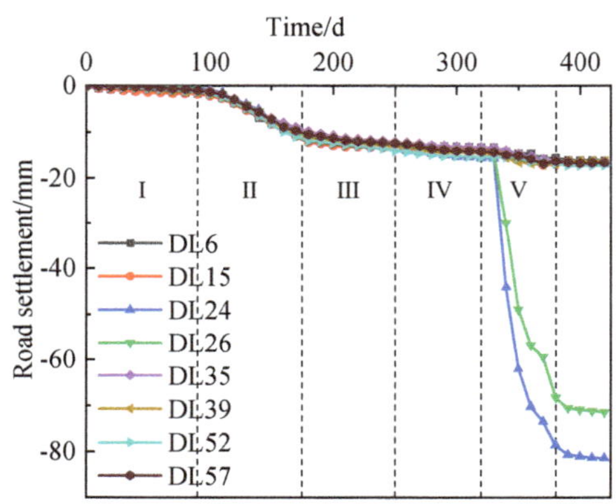

Figure 11. Road settlement curve.

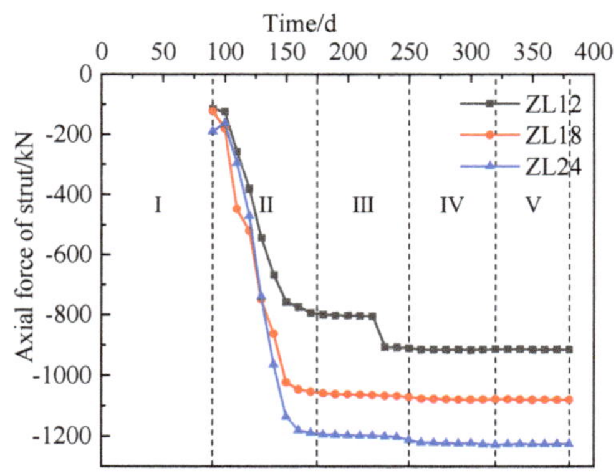

Figure 12. Variation curve of the axial force of the brace.

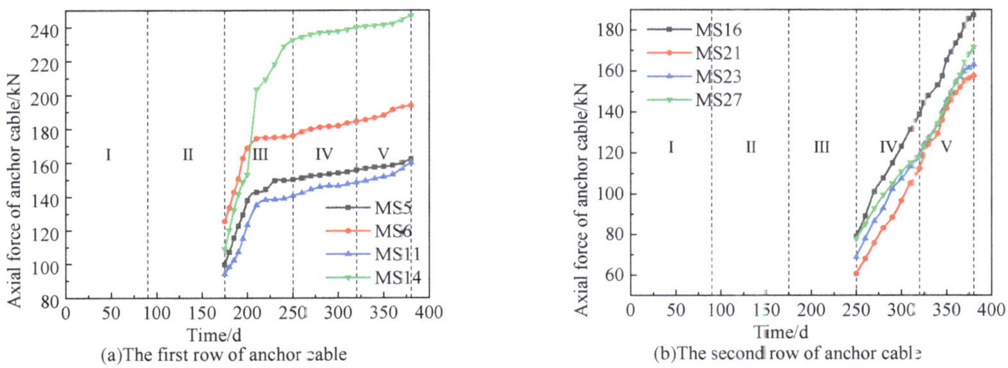

Figure 13. Variation curve of the axial force of the anchor cable.

4.1. *Horizontal Displacement of the Pile Top*

Figure 8 is the horizontal displacement curve of the pile top, where a positive value indicates that the displacement faces towards the inside of the foundation pit. The Roman numerals in the figures correspond to the working conditions of each stage and are reflected in the subsequent figures.

In the early stages of excavation, the horizontal displacement increment of each pile top was increased alternately without obvious differences, and the displacement variation was within 2 mm (before the 90th day). With the increase in the excavation depth, the horizontal displacement increased gradually, and the displacement change was between 2 mm and 17 mm (from the 90th day and the 250th day). The horizontal displacement tended to be stable after the completion of foundation slab construction (from the 250th day and the 320th day). During the construction of the underground structure inside the foundation pit, the horizontal displacement decreases (reverse displacement towards the outside of the pit appeared). This is because the underground structure construction imposes additional vertical stress on the soil inside the pit, which forced the pile to move towards the outside of the pit.

During the excavation of the foundation pit, the displacement of the pile top shows an obvious spatial effect [21]. In general, no matter on the long side or on the short side, the displacement of the middle part is large, and the pit angle is small. The closer the pit angle is, the smaller the displacement is. However, QL2 and QL4 in the northwest corner (as shown in Figure 7) did not follow the above rules. The reasons for these phenomena are as follows: (1) the excavation of the two areas was earlier than that of other areas, so the displacement of the pile top occurred earlier and faster than that of other areas, and (2) the pile top position of the two was constrained by the corner brace, which was slightly weaker than the ability of the opposite brace and the diagonal brace to constrain the deformation of the pile top. Among all monitoring points, the maximum horizontal displacement on the north side appeared in QL2, which was 17.35 mm, and 14.47 mm (in QL12) was the maximum horizontal displacement on the west side. There is an obvious difference between the two values, and the average horizontal displacement of the long side is greater than that of the short side.

4.2. *Deep Horizontal Displacement of the Soil*

Figure 9 shows the variation curve of the horizontal displacement of the soil along the depth direction. The monitoring points were arranged near the pile, and the positive value indicates that the deformation direction is toward the inside of the foundation pit.

It can be seen from Figure 9 that under work condition 1, the deep horizontal displacement of the two monitoring points mainly occurred above the inner brace. This is because the lower soil mass had not been excavated yet, and the supporting pile acted as the cantilever pile under the action of the brace. When it comes to work condition 2, the soil displacement mainly occurred in the soil layers 10 m below the surface, and the maximum displacement appeared at about 5 m below the surface. Under this work condition, the soil should have been excavated to 0.5 m below the first anchor cable, but due to the long construction period of the anchor cable, over-excavation was carried out, resulting in a further increase in the soil displacement and deepening of the overall soil displacement. For work condition 3, the soil displacement was further expanded in comparison to condition 2, and the overall displacement also developed deeper. Similar to condition 2, condition 3 also had a certain degree of over-excavation. Finally, for work condition 4, the overall horizontal displacement of the soil was bigger than condition 3, the maximum value of the horizontal displacement of the soil appears in the position of 8 m below the surface, and the trend of the curve was also close to condition 3. The reason for the increase in the overall horizontal displacement Is that the soil stress was continuously released from condition 3 to condition 4, which led to a further increase in the displacement of the soil. The reason why the two trends are close is that the foundation mat has been poured under condition 4, and the over-excavated surface of condition 3 is also close to the bottom of the foundation pit.

In addition, the development trend of the deep horizontal displacement of the soil at monitoring points CX3 and CX6 was basically the same. At the beginning, they all showed a 'cantilever' shape, and then they showed a 'drum belly' shape with the working conditions that were going on. The average value of the overall horizontal displacement of CX3 is larger than that of CX6. This is because monitoring point CX3 is on the long side and CX6 is on the short side. The spatial effect causes the difference between the two values.

4.3. Groundwater Level Outside the Foundation Pit

Figure 10 shows the curve of the underground water level outside the pit with time. In work condition 1 (before the 90th day), the depth of the lowered groundwater level in the pit is small, which has little impact on the groundwater level outside the pit. This condition happened to be in the rainy season, and there was more precipitation during the construction period, so the groundwater level rose. As the groundwater level in the pit continues to decrease, the groundwater level outside the pit also decreases.

It should be noted that the construction of two rows of anchor cables pierced the waterproof curtain of the foundation pit (from the 90th day and 250th day). Due to the high water head outside the pit, water leakage and sand leakage occurred at the site, of which the water leakage and sand leakage at the east side were serious. The range of the water level change was within 5 m.

With the treatment of water leakage and sand leakage, the groundwater level outside the pit gradually became stable (from the 250th day and the 380th day). Due to the differences in stratum distribution, the final groundwater level around the foundation pit was different. The groundwater level on the east side of the foundation pit was the shallowest (4.8 m below the surface), while the groundwater level on the north side of the foundation pit was the highest (3.5 m below the surface), and the west and south sides of the foundation pit were in the middle.

The warning value of the water level change is 1 m. Although the warning value had been greatly exceeded during the construction period, the project has not been greatly affected due to timely and appropriate treatment measures.

4.4. Road Settlement

Figure 11 is the road settlement curve. In the early stage of excavation, the settlement of the road around this foundation pit increased slowly. During the construction of the first row of anchor cables (in working condition 2), the largest road settlement reached to −15 mm, which was smaller than the alarm value (20 mm), and results of all monitoring points have the same change rule. However, the road settlements of different monitoring points showed obvious differences with the construction of the second row of anchor cables (working condition 3). At this stage, the settlement of the east road increased rapidly; the cumulative settlement of monitoring point DL24 on the east side of the foundation pit reached −80.64 mm from −17.35 mm, which accounted for 89.6% of the total settlement. while the roads settlement of the other three sides of this foundation pit showed a trend of steady growth.

The main reason for the difference in settlement is that relatively serious water leakage and sand leakage occurred during the construction of the second anchor cable on the east side. The flow of water and soil from outside into the pit provided space for road settlement outside the pit. In addition, the road on the east side is the main urban road with a large traffic flow and large additional load, which also aggravated the rapid increase in road settlement. With the effective disposal of the water leakage and sand leakage, the road settlement was stable after the completion of the second anchor cable construction, and the development of road settlement for the east side was not obvious under the fourth working condition. Although the total settlement value of the east side road exceeded the alarm value, this phenomenon did not have a significant impact on the entire project due to the implementation of road maintenance, the control of precipitation outside the foundation pit, and the setting of warning signs to prohibit the operation of large vehicles and heavy vehicles.

4.5. Axial Force of Brace

Figure 12 shows the change curve of brace axial force, in which a negative value means pressure. The monitoring of the axial force of the brace began in work condition 2, so all the curves started from the 90th day.

Axial force of all kinds of braces during work condition 2 (dig the second layer of soil to 6.5 m below the surface and construct the first row of anchor cable) grew rapidly However, for the work condition 3 (dig the third layer of soil to 9.5 m below the surface and construct the second row of anchor cable), the axial force remained stable and no longer increased. The reasons for this phenomenon may be as follows: at the beginning of excavation of the soil below the brace, the support structure of the foundation pit bore more unbalanced earth pressure. Since the anchor cable had not been constructed yet, the braces had to bear more force. With the completion of the construction of the first row of anchor cables, the unbalanced earth pressure caused by the excavation of the soil under the first row of anchor cables was borne by the anchor cable. At this time, the axial force would not change any more. So, the following conclusion can be drawn whereby the excavation of the foundation pit will only increase the axial force of the adjacent braces or anchor cable and will not affect the axial force of remote braces and anchor cables.

Among all kinds of braces, the axial force of the ring brace is the largest, reaching about 1250 kN, the opposite brace is the smallest, about 910 kN, and the angle brace is in the middle, about 1080 kN. Generally, the axial force of the opposite brace should be the maximum, but the excavation road is arranged on the opposite brace of this foundation pit, which was partially strengthened, so the maximum axial force appeared on the ring brace.

4.6. Axial Force of Anchor Cables

Figure 13 shows the change curve of axial force for the anchor cables. Since the monitoring of the axial force of the first row and the second row of anchor cables started in work condition 3 and work condition 4, the monitoring results of anchor cables were introduced from the 175th day and 250th day. It can be seen from Figure 13 that the axial force of the first row of anchor cables is generally greater than that of the second row of anchor cables. This is because during work condition 2, the water leakage and sand leakage reduced the water and soil pressure outside the foundation pit at the position of the second row of anchor cables.

With the excavation of the soil below the second row of anchor cables, the axial force of the first row of anchor cables tended to be stable, while the axial force of the second row of anchor cables increased, which is consistent with the conclusion in Section 3.5. After excavation to the bottom of the foundation pit, the axial force of the second row of anchor cables still increased, which was related to the time effect of the foundation pit. The soil outside the pit needed time to consolidate and deposit after construction disturbance.

According to the statistics of the results of all measuring points for anchor cables, it can be seen that the axial force of the first row of anchor cables ranged from 97.5 to 245.7 kN, and their average axial force was 171.6 kN, which is only 66% of the designed axial force (260 kN). The axial force of the second row of anchor cable varied from 62.4 to 187.5 kN, with an average of 124.9 kN, which is only 60.95% of its designed value (205 kN).

The maximum axial force of the two rows of anchor cables did not exceed the design value and there was a large margin, indicating that the anchor cables in some soil layers in soft soil areas can still play a very good supporting role and can be used as an alternative structure for foundation pit support in soft soil areas.

5. Conclusions

Based on the foundation pit engineering supported by the pile–anchor–brace supporting system in Yancheng City, Jiangsu Province, China, the following conclusions can be drawn through the analysis of the design and on-site monitoring of the pile–anchor–brace supporting system:

(1) The deformation of the foundation pit supported by the pile–anchor–brace supporting system also has an obvious spatial effect. The deformation of the supporting system in the middle of the foundation pit is greater than that in the corner and the deformation of long side of foundation pit is greater than of the short side. Therefore, the deformation monitoring of the supporting system in the middle part of the foundation pit should be strengthened in the process of construction.

(2) The deep horizontal displacement of the soil in the foundation pit supported by the pile–anchor–brace supporting system is initially in the shape of a 'cantilever'. With the advance of foundation pit excavation, it gradually presents a 'bulge' shape. At the position with the inner brace and anchor cable, the horizontal displacement has an obvious sudden change.

(3) The brace axial force would increase rapidly before the construction of anchor cable, but after the completion of the first anchor cable construction, the axial force would gradually become gentle. The axial force of the ring brace is larger than the corner brace, which is larger than the opposite brace.

(4) The change trend of the axial force of the two rows of anchor cables is different. The accumulated axial force of the first anchor cable is large and tends to be gentle after the foundation mat is poured, while the change in the axial force of the second row of anchor cables is relatively gentle. The axial force of the first row of anchor cables is generally greater than the second row. The axial force of the two anchor cables does not exceed their design value, so it can be used as a reference for other projects.

Author Contributions: Conceptualization, L.S. and G.X.; methodology, G.D.; validation, J.S., T.S. and S.Y.; formal analysis, K.M.; investigation, J.S.; resources, L.S.; data curation, G.X.; writing—original draft preparation, K.M.; writing—review and editing, Z.W.; supervision, G.D.; project administration, L.S.; funding acquisi-tion, Z.W. All authors have read and agreed to the published version of the manuscript.

Funding: This work was supported by the National Natural Science Foundation of China (Grant No. 52068048), Technology Project of Gansu Provincial Department of Housing and Urban-Rural Development Construction (Grant No. JK2022-03), Natural Science Foundation of Gansu Province (Grant No. 21JR11RM052), Youth Science and Technology Fund Program of Gansu Province (Grant No. 22JR5RA286), and the Hongliu Excellent Young Talents Program of Lanzhou University of Technology.

Data Availability Statement: The data of monitoring in the article is not freely available due to legal concerns and commercial confidentiality. Nevertheless, all the concepts and procedures are explained in the presented research and parts of the research may be available upon request.

Conflicts of Interest: The authors declare no conflict of interest.

References

1. Clough, G.W.; O'Rourke, T.D. Construction induced movements of in situ wall. In *Design and Performance of Earth Retaining Structures*; Geotechnical Special Publication; American Society of Civil Engineers (ASCE): Reston, VA, USA, 1990; pp. 439–470.
2. Mana, A.I.; Clough, G.W. Prediction of movement for braced cut in clay. *J. Geotech. Eng. Div.* **1981**, *107*, 759–777. [CrossRef]
3. Ou, C.Y.; Hsieh, P.G.; Chiou, D.C. Characteristics of ground surface settlement during excavation. *Can. Geotech. J.* **1993**, *30*, 758–767. [CrossRef]
4. Hashash, M.A.; Whittle, A.J. Ground movement prediction for deep excavations in soft clay. *J. Geotech. Eng.* **1996**, *122*, 474–486. [CrossRef]
5. Yang, Z.H.; Guo, Z.X. Monitoring analysis of pre-stressed load of anchor cables for deep excavations. *Chin. J. Geotech. Eng.* **2012**, *34*, 145–148.
6. Ju, C.; Li, J.C.; Zhou, X.J. Analysis of foundation pit supporting design and monitoring results in deep saturated soft soil area. *Build. Sci.* **2012**, *28*, 107–110.
7. Xu, W.; Xia, Q.W.; Xu, P.F.; Wang, J.F. Monitoring and analysis of synchronized excavation of extra large-scale adjacent riverside deep foundation pits in soft soil. *Chin. J. Rock Mech. Eng.* **2013**, *32*, 2676–2683.
8. Li, X.L.; Wei, X.; Liang, Z.R. Design practice and analysis of deformation control of deep excavations in soft soil areas. *Chin. J. Geotech. Eng.* **2014**, *36*, 160–164.
9. Zhang, Y.J.; Zeng, J.W. Dynamic monitoring and emergency treatment of deep foundation pit in ultra deep silt layer. *Chin. J. Geotech. Eng.* **2014**, *36*, 202–207.
10. Gong, H.; Xiong, Z.B.; Song, S.H.; Dai, W.K.; Wang, R.J. Structural design and monitoring analysis of deep foundation pit supporting under complex surroundings. *Chin. J. Undergr. Space Eng.* **2015**, *11*, 732–738.
11. Wei, G.; Hua, X.X.; Yu, X.F. Construction monitoring analysis of deep foundation pit excavation of a metro station in Hangzhou. *Eng. J. Wuhan Univ.* **2016**, *49*, 917–923.
12. Li, Y.B.; Dong, W.; Wang, B. Monitoring and analyzing deformation characteristics for deep foundation pit in soft soil area. *Bull. Surv Mapp.* **2017**, *2017*, 116–120.
13. Li, K.X.; Yang, Z.N.; Li, S.C. Monitoring analysis of deep foundation pit of shield launching section of highway tunnel in soft soil area. *Highway* **2017**, *62*, 282–288.
14. Jiang, J.; Xiao, M.; Liu, Z.Y.; Wang, Z.; Zhang, X.C.; Yang, D. Monitoring of deep foundation pit with a combination of various supporting systems under complicated environment. *J. Guangxi Univ.* **2018**, *43*, 269–278.
15. Long, L.; Li, Z.D. Monitoring and deformation law research of a deep foundation pit project in Changsha. *Build. Struct.* **2020**, *50*, 133–137.
16. Yancheng: Jiangsu Wanda Survey and Detection Design Co., Ltd. *Investigation Report of Yancheng First People's Hospital Phase II and Medical Complex Project*; Yancheng Publishing House: Yancheng, China, 2022.
17. *JGJ/T 199-2010*; Technical Specification for Soil Mixed Wall. Architecture Press: Beijing, China, 2010.
18. Liu, G.B.; Wang, W.D. *Manual for Foundation Pit Engineering*, 2nd ed.; Architecture Press: Beijing, China, 2009.
19. *GB50010-2010*; Code of Design of Concrete Structures. Architecture Press: Beijing, China, 2015.
20. China Academy of Building Sciences. *Technical Specification for Retaining and Protection of Building Foundation Excavations*; Architecture Press: Beijing, China, 2012.
21. Li, D.P.; Tang, D.G.; Yan, F.G.; Huang, M. Mechanics of deep excavation's spatial effect and pressure calculation method considering its influence. *J. Zhejiang Univ.* **2014**, *48*, 1632–1639.

water

MDPI

Article

Predicting Ground Surface Settlements Induced by Deep Excavation under Embankment Surcharge Load in Flood Detention Zone

Yixian Wang [1], Shi Chen [1], Jiye Ouyang [2], Jian Li [3], Yanlin Zhao [4], Hang Lin [5] and Panpan Guo [1,*]

1 College of Civil Engineering, Hefei University of Technology, Hefei 230009, China
2 Power China Zhongnan Engineering Corporation Limited, Changsha 410014, China
3 Anhui Rd & Bridge Grp Co., Ltd., Hefei 210029, China
4 School of Energy and Safety Engineering, Hunan University of Science and Technology, Xiangtan 411201, China
5 School of Resources and Safety Engineering, Central South University, Changsha 410083, China
* Correspondence: guopanpan@hfut.edu.cn

Abstract: In this paper, a simplified prediction formula of ground settlement induced by deep foundation pit excavation is proposed, especially suitable for ground overloading near a foundation pit, such as embankment surcharge load, which is carefully considered via the means of load equivalence. The ground settlement induced by foundation pit excavation and embankment surcharge load is determined by the modified skewness prediction formula and the simplified Boussinesq solution, respectively, and it is assumed that no coupling effect exists between the two settlement sources. In addition, this paper improves the determination of the maximum settlement location by combining calculus and curve fitting, replacing the existing prediction formula which relies heavily on engineering experience to determine the maximum settlement point. The predicted value obtained using this method comes close to the measured value, and the deviation of the maximum surface settlement value is controlled within about 5% in the three cases introduced, of which the accuracy is higher than the existing prediction formula.

Keywords: deep excavation; ground surface settlement; embankment; surcharge load; retaining wall

Citation: Wang, Y.; Chen, S.; Ouyang, J.; Li, J.; Zhao, Y.; Lin, H.; Guo, P. Predicting Ground Surface Settlements Induced by Deep Excavation under Embankment Surcharge Load in Flood Detention Zone. *Water* **2022**, *14*, 3868. https://doi.org/10.3390/w14233868

Academic Editor: Glen R. Walker

Received: 30 October 2022
Accepted: 24 November 2022
Published: 27 November 2022

1. Introduction

Deep foundation excavations are frequently constructed in urban areas, especially for areas such as subway stations, high-rise buildings and bridge pile foundations. The stability of foundation excavation itself and potential damage to surrounding structures are primary concerns in engineering construction [1,2]. As depicted in Figure 1, it has been indicated that excavation would result in adjacent ground differential settlement and even come into fissures, especially under the condition of existing surcharge load [3]. As for the ground surface settlement behind the wall caused by foundation pit excavation, many scholars [4–6] have conducted in-depth studies and proposed some empirical and semi-empirical methods based on field measurements, which are of great significance for the safe construction of foundation pit engineering. However, most of the previous studies on the ground settlement induced by excavation focus only on the impacts of the retaining wall horizontal displacement, whereas the effects of surcharge load have not been fully investigated. Therefore, further investigations with respect to this issue are necessary for safety assessment of the ground deformation.

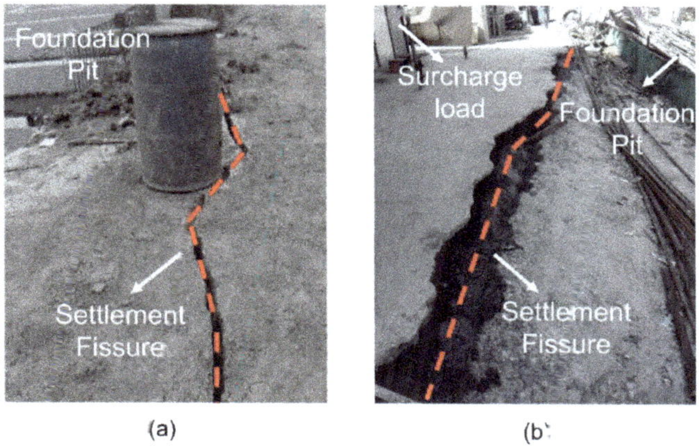

Figure 1. Settlement fissure induced by deep excavation: (**a**) condition without surcharge load; (**b**) condition under surcharge load.

For the construction of bridge pile foundations in river floodplain areas, the overloading of the embankment will aggravate the surface settlement around the foundation pit and the embankment pavement itself. The effects of surcharge load on the ground settlement caused by foundation excavation cannot be ignored, which pose an extra risk to the ground surface stability, in spite of relevant investigations into the surcharge load in regard to excavation being rarely covered. Previous research devoted to the effects of excavation on adjacent ground settlement is limited in the literature. The existing relevant investigations were performed adopting methods of theoretical analysis [7,8], field observation [9–11], centrifugal modeling experiments [12–14] and numerical simulation [15,16].

Usually, in previous studies, field observation is combined with theoretical analysis to determine the value range of some parameters in the derived formula. Peck [4] firstly proposed the principle of the stratum loss method in 1969, and systematically expounded the adjacent surface settlement caused by foundation pit excavation. This study assumed that the soil loss induced by the horizontal displacement of the foundation pit retaining wall is consistent with the total ground deformation outside the pit. Ou and Hsieh [17] found that the settlement influence zone is related to the depth and width of the foundation pit excavation, and proposed a simple prediction method for the ground surface settlement influence zone. Hu et al. [18] proposed a semi-theoretical and semi-empirical prediction method for the ground surface settlement of deep foundation pit excavation in the clay layer. In addition, other scholars have also carried out some theoretical investigations into the surface settlement caused by deep foundation excavation. These efforts provide new insights into the ground surface settlement prediction, which is beneficial to accurately evaluate the impact of foundation pit construction on the surrounding surface. However, the variation law of ground surface settlement with overloading on one side of foundation pit has not been systematically considered in previous studies, and some parameters in the existing prediction methods are mainly determined by engineering experience.

Numerical simulation provides a convenient method to perform comprehensive analysis of ground settlement rules related to geological and construction parameters. Based on field monitoring and finite difference analysis, Guo et al. [19] investigated the stress and deformation characteristics of a foundation pit under asymmetric surcharge loading, and concluded that asymmetric surcharge loading had adverse effects on the deformation of the supporting structure and the stability of the foundation pit. Rashidi and Shahir [20] carried out finite element analysis on the performance of an the anchor pile wall under additional load, indicating that the bending moment of pile body increased along with the increase in overload. Xu et al. [21] established a numerical model of a foundation pit under local load by PLAXIS, and studied the influence of the local load value and location on the deformation of the retaining structure. Moreover, some studies [22–24] focused on the dewatering effects during excavation construction, and proposed some rules of ground settlement with regard to dewatering settings. Nevertheless, difficulties in controlling the stress state and determining the required parameters may exist, hindering the application of numerical modeling in the prediction of the ground surface settlement under excavation.

In summary, the value determination of some parameters in the current theoretical empirical formula comes from past engineering experience, which is greatly affected by regional factors, and the accuracy cannot be guaranteed. It is necessary to revise some key parameters in the classical formula. As a matter of fact, the existence of an embankment would cause the foundation pit to bear asymmetric surcharge load. Therefore, under the disturbance of the embankment overload, significant differences exist in the settlement of the ground around the foundation pit, the lateral movement of the retaining structure, and the uplift on both sides of the foundation pit bottom, which would seriously threaten the stability of the foundation pit during operation. In view of this detrimental engineering condition, a simplified prediction formula for the ground surface settlement considering the effect of the nearby embankment surcharge load is proposed.

In this paper, three-dimensional foundation pit excavation deformation has been simplified to a two-dimensional plane strain problem. It is assumed that the ground loss induced by foundation pit excavation and the surface settlement caused by embankment surcharge load are two irrelevant factors, which can be considered separately. Accordingly, the influence of the retaining wall structure displacement and embankment load can be obtained by fitting the modified skewness distribution curve and Boussinesq solution, respectively. By superimposing the surface settlement curves under the above two influencing factors, the surrounding surface settlement caused by foundation pit excavation considering the effects of embankment load can be finally acquired. In addition, three typical case histories in Hangzhou [25], Shanghai [26] and Anqing, China are introduced to validate the rationality of the proposed method. According to the verification results, the proposed prediction formula for ground surface settlement under embankment surcharge load can provide guidance for the construction safety of riverside excavation.

In addition, the outline of this paper is as follows. Firstly, Section 1 (Introduction) generalizes the research progress about the topic, and poses what inadequacy exists in current investigations. Section 2 presents several typical surface settlement modes induced by excavation, which is the basis of the simplified approach in this paper. Then, Sections 3 and 4 describe the simplified method and verify the accuracy via three cases, respectively. Finally, Section 5 (Conclusions) summarizes the method proposed.

2. Typical Surface Settlement Modes Induced by Excavation

The excavation of foundation pit will induce ground settlement to varying degrees. As a matter of fact, the ground settlement mode is directly related to the lateral displacement pattern of the retaining wall. Qian and Gu [27,28] simplified the foundation pit deformation into a plane strain problem, and solved the equilibrium equation of ground settlement under foundation pit excavation by a variable separating solution.

As depicted in Figure 2, the deformation patterns of the retaining wall can be summarized into four typical modes: translation mode, rotation mode around the foot of the wall, rotation mode around the top of the wall, and flexible parabola deformation mode, resulting in different surface settlement patterns, respectively. It should be pointed out that the horizontal deformation of the foundation pit retaining wall is considered as a known displacement boundary, and there is no surcharge load on the surface behind the wall. In addition, it is assumed that the wall back is vertical, the ground behind the wall is horizontal, and the soil behind the wall is homogeneous and isotropic.

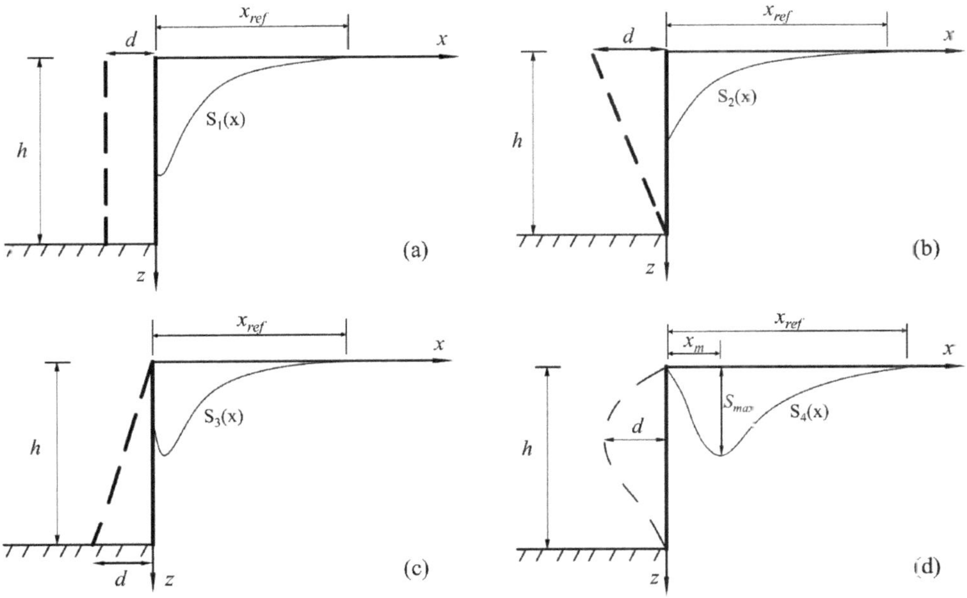

Figure 2. Typical surface settlement modes under different types of retaining-wall deformation: (**a**) translation mode; (**b**) rotation around the wall base mode; (**c**) rotation around the wall top mode; (**d**) flexible parabola deformation mode.

The surface settlement formulas under four typical foundation pit deformation conditions are presented below, corresponding to the four equations in Equation (1), respectively. In terms of the equation form involving complex integral transformation, the solution process of the explicit solution is quite complicated to directly obtain the distribution law of the surface settlement curve. In general, assuming that there is a linear relationship between the surface settlement range (i.e., x_{ref}) and the foundation pit depth (i.e., h), the surface settlement curves under the maximum horizontal displacement (i.e., d) of different retaining walls are drawn by numerical integration, and the transverse distribution value of the surface settlement would be ultimately obtained by normalization processing.

$$\left.\begin{array}{l}
S_1(x) = -\frac{2hd}{\pi^2}\int_{-\infty}^{+\infty}\frac{\xi^2-h^2}{(\xi^2+h^2)^2}\ln\left|\frac{\xi-x}{\xi-x_{ref}}\right|d\xi \\
S_2(x) = \frac{2d}{\pi^2}\int_{-\infty}^{+\infty}\left(\frac{h}{\xi^2+h^2}+\frac{1}{2h}\ln\frac{\xi^2}{\xi^2+h^2}\right)\ln\left|\frac{\xi-x}{\xi-x_{ref}}\right|d\xi \\
S_3(x) = -\frac{2d}{\pi^2 h}\int_{-\infty}^{+\infty}\left(\frac{1}{4}\ln\frac{\xi^4}{(\xi^2+h^2)^2}+\frac{2h^4}{(\xi^2+h^2)^2}\right)\ln\left|\frac{\xi-x}{\xi-x_{ref}}\right|d\xi \\
S_4(x) = -\frac{8d}{\pi^2 h^2}\int_{-\infty}^{+\infty}\left(\frac{h}{2}\ln\left(\frac{\xi^2}{\xi^2+h^2}\right)+3h+\frac{h\xi^2}{\xi^2+h^2}-4\xi\arctan\left(\frac{h}{\xi}\right)\right)\ln\left|\frac{\xi-x}{\xi-x_{ref}}\right|d\xi
\end{array}\right\} \quad (1)$$

In practical engineering, the deformation of a foundation pit retaining wall is much more complicated than that of the typical four modes. However, as depicted in Figure 3, the horizontal deformation of a general retaining wall can be generally viewed as a superimposition of the above four deformation modes. From the perspective of theoretical calculation, the curve of surface settlement behind the retaining wall can be obtained by fitting several basic settlement curves.

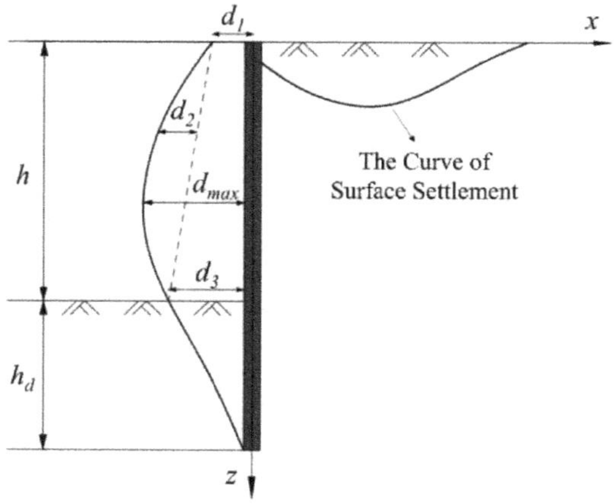

Figure 3. Ground settlement induced by superimposed displacement of retaining wall.

It is observed that in this solving method, there exists much inconvenience in calculation, such as difficulties in accurately analyzing the retaining-wall horizontal deformation curve, cumbersome calculation steps, wide margin of error, etc. However, it can be inspired to regard the retaining-wall lateral deformation as an arbitrary continuous curve, without considering the deformation mode of the foundation pit. The influence of retaining-wall deformation on the surface settlement would be explored from the perspective of differentiation, and specific solutions are elaborated in the next section.

3. Derivation of the Simplified Solution Method

3.1. Solution Thought

Under the condition of embankment load, the excavation of a foundation pit has more obvious influence on the surrounding surface settlement. In this paper, the influence of foundation pit excavation and bank overloading on the surface settlement is investigated, respectively. As shown in Figure 4, it is assumed that excavation and surface load do not interfere with each other, and there is no coupling effect during the surface settlement.

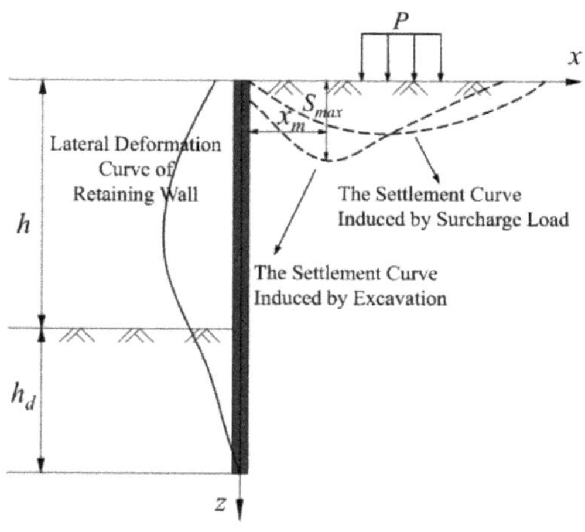

Figure 4. Surface settlement under the combined action of excavation and embankment load.

As a result, the ground surface settlement induced by a deep braced excavation adjacent to a river embankment consists of two parts (Figure 4). The first part is caused by the excavation of the foundation pit itself, and the second part is induced by the embankment surcharge load. In order to estimate and predict the surface settlement around the foundation pit near the embankment, this paper calculates the settlement of the two parts according to the ground loss method and the Boussinesq solution, respectively. The equation for ground settlement can be expressed as

$$S(x) = S_I(x) + S_{II}(x) \tag{2}$$

where $S(x)$ is the total formula of the ground settlement; $S_I(x)$ and $S_{II}(x)$ represent the settlement induced by adjacent excavation and embankment surcharge load, respectively.

According to Nie et al. [26], based on the concept of the formation loss method and a large number of field measurement data, the distribution rule of the surface displacement field around the soft soil deep foundation pit has been deeply investigated, and the surface settlement curve with skewed distribution is proposed as

$$S_I(x) = \frac{A_v}{\sqrt{2\pi}\varpi x}\exp\left(-\left(\ln\frac{x}{2x_m}\right)^2/\left(2\varpi^2\right)\right) \tag{3}$$

where x is expressed as the distance between the calculation point and the edge of the foundation pit; x_m represents the distance between maximum settlement point and the edge of foundation pit; and A_v is the envelope area of surface settlement curve behind the wall and ϖ is empirical coefficient (i.e., 0.6–0.7 for a soft soil foundation pit).

3.2. The Influence of Foundation Pit Excavation

In Equation (3), A_v and x_m are important parameters directly related to the surface settlement curve. However, in many studies and engineering applications in the past, the selection of these two parameters was mostly determined by engineering experience, which may bring out major errors, even resulting in wrong results. Therefore, the determination of A_v and x_m should be carefully considered, and a simplified method will be performed in the following sections.

3.2.1. Calculation of x_m

The ground settlement behind the retaining wall stems from the horizontal movement of the retaining wall in the foundation pit. If the retaining wall is assumed to have rigid translation, the basic analytical solution of the ground settlement induced by the retaining wall under the rigid translation displacement mode can be obtained. Although the possibility of rigid translation of the retaining wall in actual engineering is quite small, the surface subsidence degradation under the complex deformation mode can still be obtained based on the rigid translation mode of retaining wall [29]. Under the boundary condition of the retaining wall rigid translational displacement mode, the first formula in Equation (1) is solved by integration, and the explicit formula can be expressed as

$$w = \frac{2dh^2}{\pi}\left(\frac{1}{x^2+h^2} - \frac{1}{x_{ref}^2+h^2}\right) \tag{4}$$

where w represents the ground settlement; x is the distance from the foundation pit; and h is the excavation depth.

Through the application of the calculus method [30], the horizontal displacement curve of any retaining wall deformation form is divided into an infinite number of micro-segments, and the deformation of each micro-segment is regarded as the rigid translation mode of the retaining wall. According to the surface settlement solution under the retaining wall translational deformation mode, as expressed in Equation (4), the surface settlement caused by each micro-section can be, respectively, calculated. As presented in Figure 5, the surface settlement caused by horizontal deformation of the whole retaining wall would be finally obtained via summation calculation.

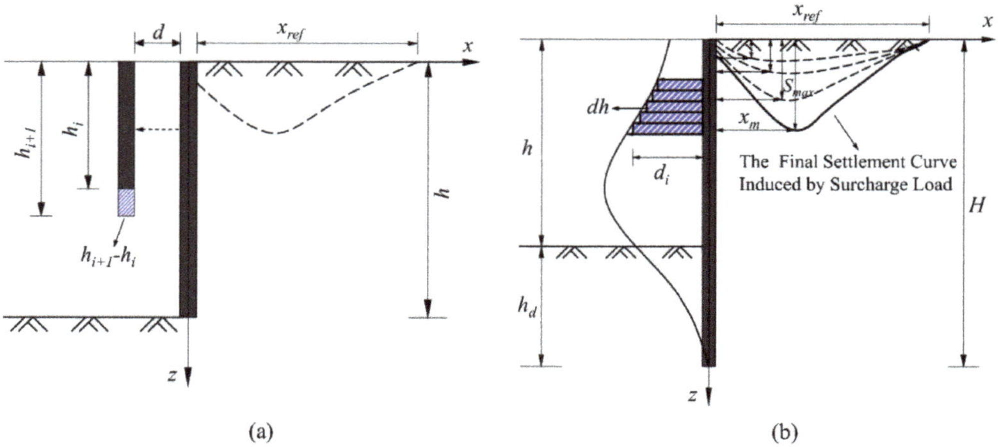

Figure 5. Infinitesimal calculus diagram of surface settlement induced by the retaining wall horizontal deformation: (**a**) equivalent calculation of surface settlement induced by arbitrary micro-section translation of retaining wall; (**b**) equivalent calculation of surface settlement induced by displacement of arbitrary retaining wall.

The retaining wall can be evenly divided into n segments along the depth direction, and it is assumed that the translation and displacement mode approximately occurred in each small section. For the segment i, the surface settlement value induced under the depth of h_i and h_{i+1} is calculated by Equations (5) and (6), respectively. The relevant equation could be expressed as

$$w_i = \frac{2d_i h_i^2}{\pi}\left(\frac{1}{x^2+h_i^2} - \frac{1}{x_{ref\_i}^2+h_i^2}\right) \tag{5}$$

$$w_{i+1} = \frac{2d_{i+1} h_{i+1}^2}{\pi}\left(\frac{1}{x^2+h_{i+1}^2} - \frac{1}{x_{ref\_i+1}^2+h_{i+1}^2}\right) \tag{6}$$

where w_i and w_{i+1} represent the ground settlement induced by segment i and segment i+1, respectively; d_i and d_{i+1} represent the retaining wall horizontal deformation of segment i and segment i+1, respectively; x_{ref_i} and x_{ref_i+1} represent the ground settlement range induced by segment i and segment i+1, respectively.

Based on the superposition principle, making a subtraction between Equations (5) and (6), and the surface settlement Δw_i caused by the translation of any retaining wall segment in the foundation pit is obtained, which can be simplified as

$$\Delta w_i = w_{i+1} - w_i = \frac{2}{\pi}\left[\frac{d_{i+1}h_{i+1}^2\left(x_{ref\_i+1}^2 - x^2\right)}{\left(x^2+h_{i+1}^2\right)\left(x_{ref\_i+1}^2+h_{i+1}^2\right)} - \frac{d_i h_i^2\left(x_{ref\_i}^2 - x^2\right)}{\left(x^2+h_i^2\right)\left(x_{ref\_i}^2+h_i^2\right)}\right] \tag{7}$$

Previous studies have indicated that there is a linear relationship between the influence range of surface settlement and the depth of foundation pit excavation. Therefore, for the retaining wall segment in different depths, the influence range of surface settlement caused by their translation can be set as m times of the corresponding depth (i.e., $x_{ref\_i} = mh_i$,

$x_{ref\_i+1} = mh_{i+1}$). Through the substitution of x_{ref_i} and x_{ref_i+1} into Equation (7), the final form of Δw_i can be simplified as

$$\Delta w_i = \frac{2}{\pi(m^2+1)}\left[\frac{d_{i+1}(m^2h_{i+1}^2 - x^2)}{x^2+h_{i+1}^2} - \frac{d_i(m^2h_i^2 - x^2)}{x^2+h_i^2}\right] \tag{8}$$

Ultimately, the surface settlement value Δw_i caused by the translation of each micro-section of the retaining wall can be superimposed and summed. As presented in Equation (9), the adjacent surface settlement formula under the horizontal deformation mode of any maintenance structure can be obtained.

$$w = \sum_{i=1}^{n}(w_{i+1} - w_i) = \sum_{i=1}^{n}\Delta w_i \tag{9}$$

Accordingly, x_m, the maximum position of surface settlement, can be obtained by w in Equation (9), which is easily realized by programming.

3.2.2. Calculation of A_h

According to a large number of experiments and engineering test data, there is a proportional relationship between the envelope area of the surface settlement curve behind the wall (i.e., A_v) and the envelope area of the displacement curve of excavation retaining wall (i.e., A_h) in the soft soil area, which can be expressed as

$$A_v = \beta A_h \tag{10}$$

where β represents the proportionality coefficient. According to the statistical analysis results and site engineering experience, when the wall penetration ratio h_d/h_1 (h_1 is the final excavation depth, and h_d represents the wall embedment depth below the excavation bottom) is less than 0.5, β can be taken as 1.0–1.2; otherwise, it may be within the range of 0.8 to 1.0.

In order to illustrate the expression of Equation (10) more vividly, Figure 6 describes the linear proportionality relation by different colors. The blue diagonal and the red diagonal represent A_v and A_h, respectively.

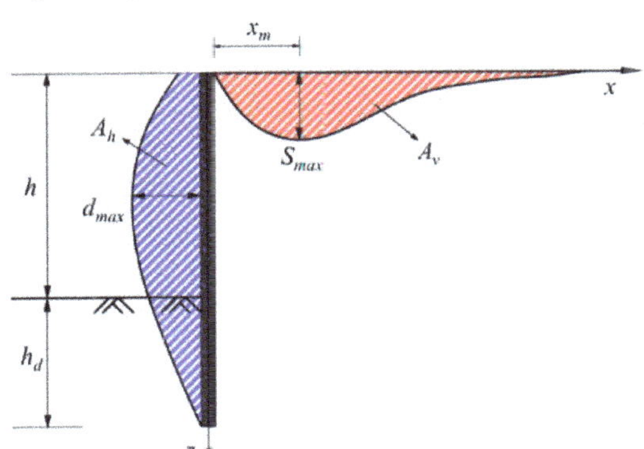

Figure 6. Schematic diagram of the proportional relationship between A_v and A_h.

Generally, the horizontal deformation of the supporting structure can be calculated by the elastic fulcrum method or obtained by actual measurement. Before calculating the envelope area of the displacement curve of the supporting structure (i.e., A_h), the displacement data of the calculated or measured points should be processed firstly to make it fit into a continuous smooth curve. In general, any function in a certain section can be approximated by polynomials. Therefore, assuming that the displacement point coordinates of the retaining wall calculated or measured along the depth direction of the foundation pit are (*h*, *d*), then the horizontal displacement fitting curve can be expressed as

$$d(h) = \sum_{i=0}^{n} a_i h^i \tag{11}$$

where n is the order of the polynomial and a_i represents the coefficient of the independent variable in the polynomial.

According to Nie and Hu's research [26,29], the lateral displacement value of supporting structure can be approximated by a quadratic parabolic function. Taking the first three terms of Equation (11), $d(h)$ can be expressed as

$$d(h) = a_0 + a_1 h + a_2 h^2 \tag{12}$$

Based on the least square method, the equation set can be obtained as

$$\left.\begin{aligned} a_0 n + a_1 \sum_{i=1}^{n} h_i + a_2 \sum_{i=1}^{n} h_i^2 &= \sum_{i=1}^{n} d_i \\ a_0 \sum_{i=1}^{n} h_i + a_1 \sum_{i=1}^{n} h_i^2 + a_2 \sum_{i=1}^{n} h_i^3 &= \sum_{i=1}^{n} h_i d_i \\ a_0 \sum_{i=1}^{n} h_i^2 + a_1 \sum_{i=1}^{n} h_i^3 + a_2 \sum_{i=1}^{n} h_i^4 &= \sum_{i=1}^{n} h_i^2 d_i \end{aligned}\right\} \tag{13}$$

where h_i and d_i represent the depth of the retaining wall and the corresponding horizontal displacement values acquired by calculating or measuring, respectively.

Substituting the known h_i and d_i into Equation (13), the values of the coefficients a_0, a_1 and a_2 would be obtained. According to the vertex coordinates (0, a_0) and extreme points (h_m, d_m) of the supporting structure, a_1 and a_2 can be finally derived as

$$\left.\begin{aligned} a_1 &= -\frac{2(a_0 - d_m)}{h_m} \\ a_2 &= \frac{a_0 - d_m}{h_m^2} \end{aligned}\right\} \tag{14}$$

By integral calculation of Equation (12), the area enclosed by the horizontal displacement curve of the supporting structure can be expressed as

$$A_h = \int_0^H \left(a_0 + a_1 h + a_2 h^2\right) dh = a_0 H + \frac{1}{2} a_1 H^2 + \frac{1}{3} a_2 H^3 \tag{15}$$

where H is the entire height of the retaining wall.

Substituting Equations (10) and (15) into Equation (3) yields

$$S_I(x) = \frac{\beta A_h}{\sqrt{2\pi}\varpi x} \exp\left(-\left(\ln \frac{x}{2x_m}\right)^2 / \left(2\varpi^2\right)\right) \tag{16}$$

3.3. *The Influence of Embankment Surcharge Load*

For the influence of the embankment load, the uniformly distributed load is simplified as a vertical concentrated force. Therefore, the embankment pavement settlement curve under the condition of foundation pit excavation can be obtained by the Boussinesq solution in semi-infinite space.

As depicted in Figure 7a, for an any point *M* (x, y, z) in the elastic half space, Boussinesq has derived the calculation formulas induced by the vertical concentrated load *P* acted on the surface. According to Boussinesq solution, vertical deformation (i.e., $u(z)$) can be expressed in the form of

$$u(z) = \frac{p(1+\mu)}{2\pi E}\left[\frac{z^2}{R^3} + 2(1-\mu)\frac{1}{R}\right] \tag{17}$$

where R is the distance from target calculation point to load point; μ is Poisson's ratio; and E is the elasticity modulus.

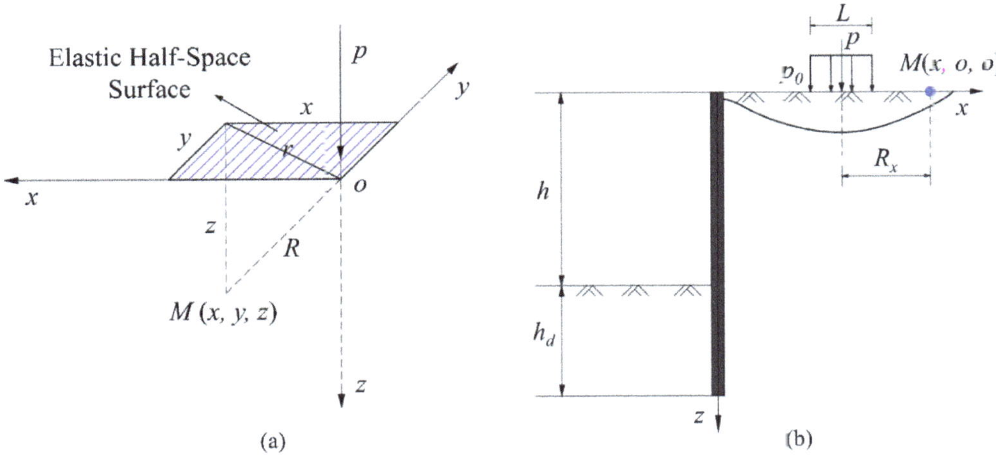

Figure 7. Diagram of surface settlement induced by the retaining wall horizontal deformation: (**a**) Boussinesq solution diagram; (**b**) simplification of embankment surcharge load.

When taking the embankment surcharge load into consideration, as depicted in Figure 7b, the embankment surcharge load is simplified as the vertical concentrated load perpendicular to the embankment surface, and the settlement at any point behind the wall can be figured out via Boussinesq solution.

For the condition of surface settlement, the value of z is 0. Similarly, R can be simplified as the horizontal distance between the target surface point and the equivalent concentrated force operation point, which is expressed with R_x. By assigning z to 0 in Equation (17), the expression for u (z) can be simplified to

$$u(z) = \frac{p(1-\mu^2)}{\pi E R_x} \tag{18}$$

In this paper, a three-dimensional foundation pit is analyzed as a plane strain problem. Accordingly, the embankment load is considered approximately as a uniform load with width L on the section. Therefore, the concentrated force P can be replaced by $p_0 L$ through integration, and $u(z)$ can be finally derived as

$$u(z) = \frac{p_0 L(1-\mu^2)}{\pi E R_x} \tag{19}$$

The settlement value of any point on the surface induced by the embankment load can be obtained by Equation (19), and the settlement value of other points can also be obtained by this method. Therefore, the settlement values at different locations on the surface are selected at equal intervals, and the complete settlement curve caused by the embankment surcharge load can be obtained through curve fitting (i.e., S_L (x)).

3.4. Optimized Solution Procedure

To clearly demonstrate the whole derivation process of the proposed analytical method, a flow chart is helpful, which is presented in Figure 8.

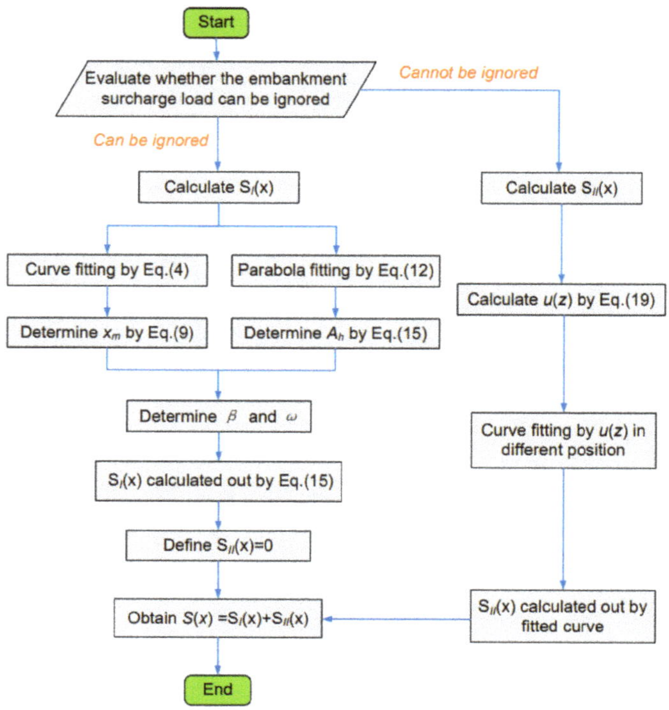

Figure 8. Flow chart of the proposed method.

4. Validation of the Simplified Analytical Method through Case Histories

In order to verify the applicability and accuracy of the simplified prediction formula in this paper, some specific engineering cases should be combined to meticulously analyze and illustrate. Case 1 and Case 2 are real projects involved in relevant sources from the literature [25,26]. The observation deformation data of the foundation pit supporting structure and the surface settlement caused by it were well recorded, which could well compare the advantages and disadvantages of the proposed method with the previous methods. Case 3 is an engineering project that the authors participated in personally. It mainly considers the influence of foundation pit excavation on surface deformation under the presence of embankment load, which is deserves special attention.

4.1. Case History 1

The underground part of a commercial and residential building in Hangzhou is a three-floor basement [25]. The project site is the alluvial plain landform of Qiantang River, with relatively flat terrain. The area affected by the excavation depth is dominated by saturated soft soil such as deep muddy clay, and the groundwater level is stable at about 1.5 m below the surface. The depth of this deep excavation pit varied between 14.85 and 17.35 m, and the bored pile walls with the size of 1.0 m × 38.0 m × 0.15 m (diameter × length × spacing) were installed to brace this deep excavation. Three levels of reinforced concrete struts were encountered for the most part of the excavation, while four levels of them were applied for local part where the excavation depth reached 17.35 m. In the foundation pit, triaxial cement mixing piles were used to strengthen the soil in the passive area. The width of the reinforcement strip varied from 7.2 m to 13.7 m according to the importance of the surrounding environment.

To validate the simplified analytical method proposed in this paper, the middle part of the west side of the foundation pit (i.e., CX2 section) was selected for research. For this part, the depth of the excavation was h = 17.35 m, and the reinforcement depth of mixing pile was 5 m. It should be noted that a six-story pile foundation building was located 20 m to the west of the retaining wall. As clearly presented in Figure 9a, the observed pile top displacement (i.e., d_1) at the CX2 section equals 25.9 mm, and the observed excavation basal displacement (i.e., d_3) equals 124.5 mm. In this project, the displacement at the bottom of pit is the maximum displacement of retaining pile. It is a common way to predict the ground settlement around the foundation pit based on the calculated or measured displacement data of the retaining wall. The observed surface settlement data, the prediction formula in the literature [26], and proposed method in this paper are compared, and the surface settlement curves are, respectively, depicted in Figure 9b. According to the correlation between the influence degree and the influence scope of the foundation pit excavation, five observation points were set, respectively, located at 0 m, 5 m, 10 m, 20 m and 40 m from the edge of the foundation pit.

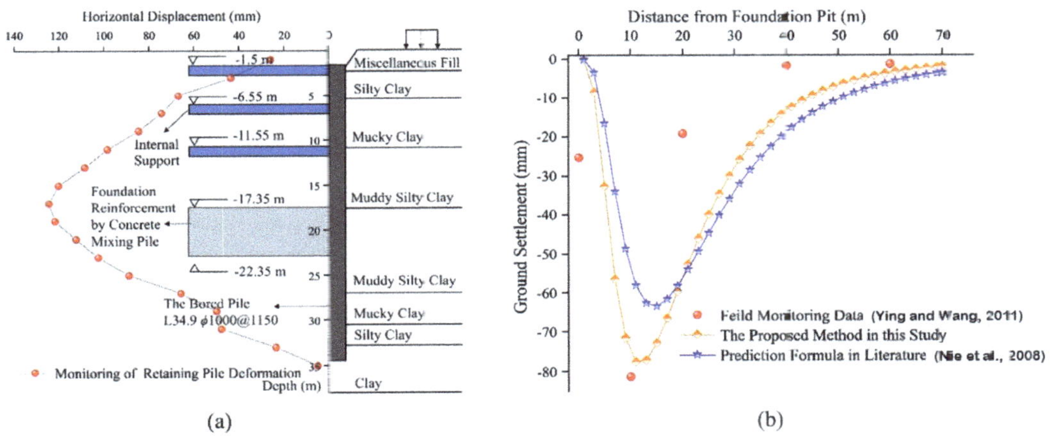

Figure 9. Validation based on the Case History 1 [25]: (**a**) field monitoring of the retaining pile deformation; (**b**) distribution of ground surface settlement under different methods [26]

Table 1 presents parameter values and corresponding ground settlement results for several methods. In the approach proposed in this paper, Equation (9) can be applied for fitting, which yields the maximum settlement location x_m = 11 m. By utilizing numerical integration method, the envelope area of the retaining structure lateral deformation curve can be obtained: A_h = 2314.5 m·mm, where β = 0.9, and ω = 0.65 according to experience. In the prediction formula [26], x_m is defined as a parameter linearly related to the foundation pit depth based on engineering experience and geology conditions, and x_m is set as 12.1 m in Case History 1, which equals to the excavation depth of 0.7 times. In addition, other parameters are consistent with the formula in this paper. The surface settlement curve can be obtained by substituting all parameters into Equations (15) and (16). As presented in Figure 9b, the method proposed in this paper performs well in approaching the observed results. Compared with the original settlement prediction formula, the simplified method used in this study is more accurate in fitting with the field monitoring values, especially the maximum settlement value, demonstrating the effectiveness of the proposed analytical method in this paper.

Table 1. Parameter values and results comparison of several methods in Case History 1 [25].

Methods	Key Parameter Values				Ground Settlement	
	A_h	β	A_v	ω	x_m	v_m
Field monitoring [25]	2295.8 m·mm	—	—	—	10.3 m	81.3 mm
Prediction formula in literature [26]	2256.9 m·mm	0.9	2031.2 m·mm	0.65	12.1 m	63.4 mm
Proposed method in this study	2314.5 m·mm	0.9	2083.0 m·mm	0.65	11.2 m	77.5 mm

A_h = envelope area of support structure displacement curve; A_v = envelope area of ground settlement curve; β = proportionality coefficient related to insertion ratio (i.e., A_v/A_h); ω = empirical coefficients related to soil quality; x_m = location of the maximum settlement point on the ground; v_m = maximum surface settlement.

4.2. Case History 2

This case history is the Shanghai Metro Line M8 Yanji Middle Road Station, located in the Yangpu District, Shanghai, China [26]. In this project, the diaphragm wall with 0.8 m in thickness, 27 m in length, and 0.76 m in penetration ratio, was constructed to serve as not only the bracing structure, but also the main structure side wall used during the in-service phase. The depth of the excavation in the standard segment adjacent to the monitoring point C28-1 was 15.3 m (i.e., h) and four levels of steel pipe struts were constructed along the depth direction. The observed horizontal wall displacement at the wall top was d_1 = 0.14 mm. The maximum horizontal wall displacement occurred approximately at the excavation bottom, with a magnitude of d_2 = 55.71 mm. As depicted in Figure 10a, the displacement at the top of the wall was quite small, and the maximum horizontal displacement was located at the bottom of the foundation pit.

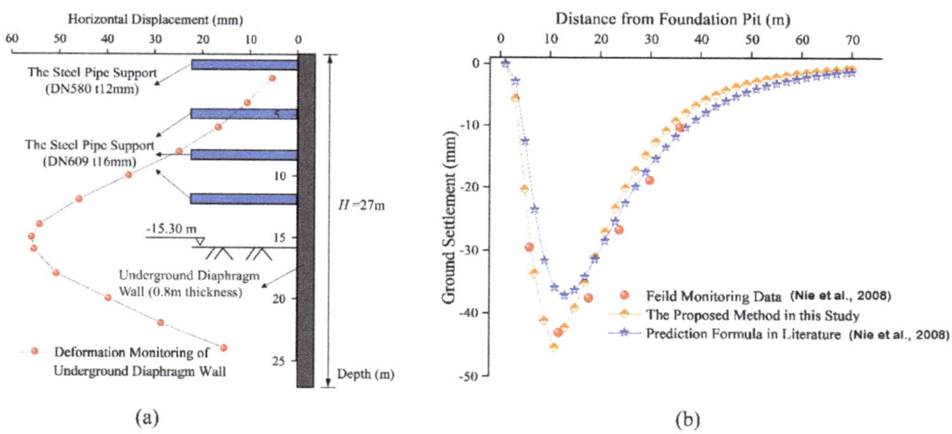

Figure 10. Validation based on the Case History 2 [26]: (**a**) field monitoring of the retaining wall deformation; (**b**) distribution of ground surface settlement under different methods.

Table 2 presents some key parameter settings and ground settlement results from field monitoring [26], prediction formula in the literature [26], and proposed method in this study. It is noted that the values of A_h and A_v are relatively close in the observed data, which is quite different from the conventional prediction formula. This may be caused by the vehicle dynamic load and material surcharge in the site. Due to the lack of available reference data, other parameters are obtained according to the literature [26]: A_h = 1083.8 m·mm, β = 0.9, and w = 0.65. Moreover, the prediction formula proposed in this paper also draws on some parameters in the literature above, such as β and w, which improve the accuracy of surface subsidence results.

Table 2. Parameter values and results comparison of several methods in Case History 2 [26].

Methods	Key Parameter Values				Ground Settlement	
	A_h	β	A_v	ω	x_m	v_m
Field monitoring [26]	892.1 m·mm	—	876.3 m·mm	—	11.7 m	43.1 mm
Prediction formula in literature [26]	1083.8 m mm	0.9	759.9 m·mm	0.65	13.2 m	37.2 mm
Proposed method in this study	925.0 m·mm	0.9	832.5 m·mm	0.65	10.9 m	45.5 mm

A_h = envelope area of support structure displacement curve; A_v = envelope area of ground settlement curve; β = proportionality coefficient related to insertion ratio (i.e., A_v/A_h); ω = empirical coefficients related to soil quality; x_m = location of the maximum settlement point on the ground; v_m = maximum surface settlement.

A multiple comparison of the ground surface settlements for this case history has been performed using two different analytical methods, and relevant data are depicted in Figure 10b. It can be indicated that, in Case History 2, the calculated results of the prediction formula proposed in this paper are in good agreement with the measured data, and the fitting result is more accurate than the method in the literature. Specifically, the observed A_v was 876.3 m·mm, while the method in the reference [26] predicted a value of 759.9 m·mm, with an error of 15.4%. This also causes the V_m of this method to differ significantly from the actual observed value, at 13.7%. In the method presented in this paper, A_v and V_m have relatively small deviations from the field-measured values, which are 5.0% and 5.6%, respectively. Further investigation indicates that x_m, a parameter closely related to surface settlement, was determined by engineering experience in the literature [26], whereas it was obtained by integral fitting in this paper. It is obvious that the prediction formula for the surface displacement field with skewed distribution behind the wall proposed in literature [26] is relatively simple. Up till now, most values of some parameters in the formula are obtained by engineering experience, which would increase the uncertainty of the prediction formula. In the proposed analytical method, the values of x_m and A_h are optimized to make the values of parameters more reasonable and the prediction results of surface settlement more accurate.

4.3. Case History 3

This case is based on a deep foundation pit project for the construction of bridge pile foundations, located in Anqing, Anhui Province, China. The length of the main structural part of Yangwanhe Bridge is 685 m, of which the 14 # main pier is located outside the Tongma Causeway of Yangwan River on the land. The cushion cap foundation pit is supported by cofferdam method, and the single row Larsen steel sheet pile (SP-IV. PU600 × 210) was selected as the maintenance structure. The steel sheet pile foundation pit is symmetrical dumbbell type, with length of 54.6 m, width of 25.2 m and height of 18 m. It is internally supported by two enclosing purlins, and the excavation depth is 8 m. Figure 11a describes the position relationship between the steel cofferdam and the embankment on the plane, and the horizontal distance between them is 6 m.

A typical steel cofferdam profile (A-A′) in Figure 11a is selected and depicted in Figure 11b. In view of the distance between the foundation pit and the bank embankment is only 6 m, the settlement deformation of the bank embankment pavement deserves attention. The profile of the bank embankment is a symmetrical trapezoid, with a pavement width of 7 m, a height of 5 m and a slope of 45°. Yangwan River is on the other side of the embankment, and the water level is relatively stable throughout the year. According to the engineering geological survey report, the stratum around the foundation pit is mainly cohesive soil.

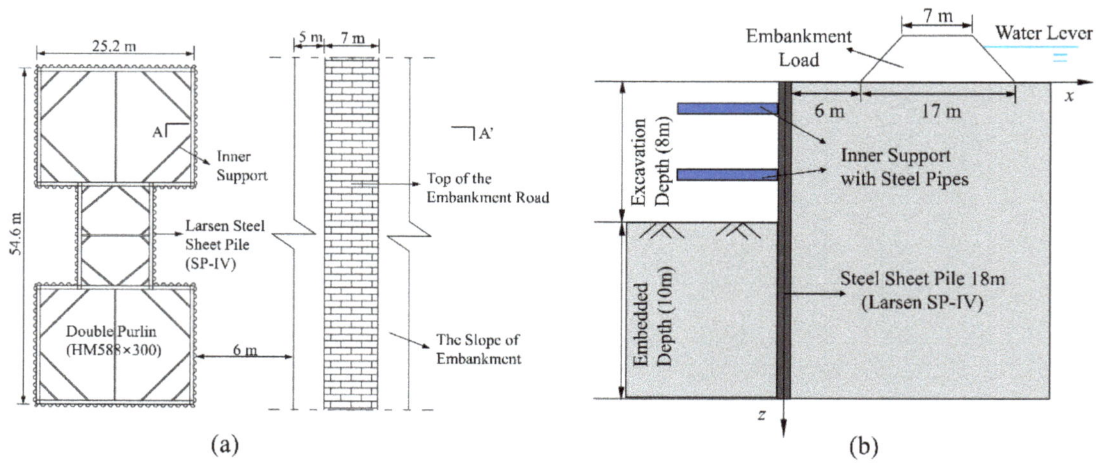

Figure 11. Relative spatial position of the foundation pit and the embankment in the case History 3: (**a**) relative plan layout of the steel cofferdam and the embankment; (**b**) A-A′ cross-section profile.

As depicted in Figure 12a, during the construction process, the maximum lateral deformation of steel sheet pile cofferdam, with the value of 25.8 mm, is at the top of pile, and the maximum deformation value at the bottom of foundation pit is 9.3 mm, both of which are within a reasonable deformation range. The area of the envelope formed by the sheet pile lateral movement (i.e., A_h) is 216.3 m·mm, and the empirical coefficient of the settlement envelope (i.e., β) is determined with 0.9 based on previous engineering experience. Therefore, the area of the surface settlement envelope curve (i.e., A_v) outside the foundation pit is 194.7 m·mm. The value of x_m can be obtained by superposition fitting of the micro-segment settlement curve calculated by Equation (8). In case history 3, the maximum settlement point calculated is 4.9 m away from the foundation pit. In addition, ω refers to the value of 0.65 in the reference [26]. By taking the above parameters into Equation (15) for calculation, the influence of foundation pit excavation on the surface settlement can be obtained. Furthermore, the influence of bank overloading can be easily obtained by the Boussinesq solution. Considering that the deformation of embankment pavement is the focus of attention, settlement observation points were set in the middle and both sides of the road.

Figure 12b presents the characteristic curve of surface settlement under the combined action of bank overload and foundation pit excavation. The predicted maximum settlement value of the embankment pavement coincides with the measured value very well, which strongly proves the effectiveness of this proposed method in predicting and evaluating the ground settlement induced by foundation pit excavation under the condition of existing embankment surcharge load.

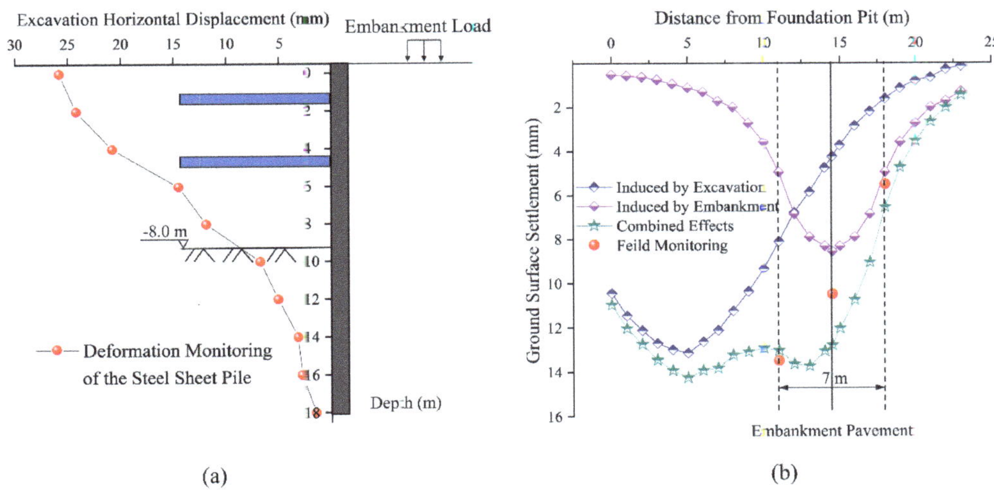

Figure 12. Validation based on the Case History 3: (**a**) field monitoring of the steel sheet pile deformation; (**b**) distribution of ground surface settlement under different methods.

5. Conclusions

In the method proposed in this paper, the ground settlement induced by foundation pit excavation and embankment surcharge load is determined using the modified skewness prediction formula and the simplified Boussinesq solution, respectively, and it is assumed that no coupling effect exists between the two settlement sources. Compared with the observed values, the error of the maximum settlement point and the corresponding maximum settlement value calculated by the method in this paper is almost controlled within 5%, of which the accuracy is much higher than the existing prediction formula.

- The ground overloading such as embankment overload has a certain influence on the surface settlement during the foundation pit excavation, which should be paid more attention to in the project. In this paper, the Boussinesq solution is applied to simplify the embankment load into the vertical concentrated force, and the ground settlement curve is easily calculated and fitted, which can well take into account the impact of ground overload.
- As for the surrounding surface settlement caused by foundation pit excavation, the value of x_m is generally determined by engineering experience in the previous partial settlement prediction formula, which is an important factor leading to a large deviation between the measured value and the theoretical prediction value. This paper improves the determination of x_m by the method of combining calculus with curve fitting, and the predicted value obtained by this method comes close to the measured one. Accordingly, in the three cases, the deviation between the calculated value of v_m and the actual observation value is 4.6%, 5.5% and 3.4%, respectively.
- Based on the classic prediction formula of surface subsidence skewness distribution, the simplified method in this paper draws on some empirical parameters, such as β and ω, which are derived from many engineering practices and have proved reliability among projects. However, the value range of these parameters is quite wide, and bring out many difficulties to determine the most reasonable parameter value in settlement prediction. As a result, it is necessary to make a more detailed division for the value range of empirical parameters in subsequent studies.

In conclusion, this simplified calculation method can significantly bring down the reliability and uncertainty of ground settlement prediction around the foundation pit. It is worth noting that although the accuracy of this method is demonstrated through the existing deformation data of the foundation pit retaining wall structure in the cases, the lateral displacement of the retaining wall can also be preliminarily determined via numerical simulation or theoretical formula calculation in engineering applications, and the ground settlement outside the foundation pit can be further derived.

Author Contributions: Conceptualization, Y.W. and S.C.; methodology J.O.; software, J.L.; validation, P.G.; investigation, Y.Z. and H.L. All authors have read and agreed to the published version of the manuscript.

Funding: This research work was funded by the National Natural Science Foundation of China (42077249, 51774107), the Opening Project of State Key Laboratory of Explosion Science and Technology, Beijing Institute of Technology (KFJJ21-03Z).

Institutional Review Board Statement: Not applicable.

Informed Consent Statement: Not applicable.

Data Availability Statement: The data support the findings of this study are not publicly available due to privacy.

Conflicts of Interest: The authors declare no conflict of interest.

References

1. Jiang, H.; Zhu, J.C.; Zhang, X.Y.; Zhang, J.X.; Li, H.L.; Meng, L.F. Wear Mechanism and Life Prediction of the Ripper in a 9-m-diameter Shield Machine Tunneling Project of the Beijing New Airport Line in a Sand-pebble Stratum. *Deep Undergr. Sci. Eng.* **2022**, *1*, 65–76. [CrossRef]
2. Guo, P.P.; Gong, X.N.; Wang, Y.X.; Lin, H.; Zhao, Y.L. Minimum Cover Depth Estimation for Underwater Shield Tunnels. *Tunn. Undergr. Space Technol.* **2021**, *115*, 104027. [CrossRef]
3. Ong, D.E.L.; Jong, S.C.; Cheng, W.C. Ground and Groundwater Responses Due to Shaft Excavation in Organic Soils. *J. Geotech. Geoenviron.* **2022**, *148*, 05022003. [CrossRef]
4. Peck, R.B. Deep Excavations and Tunneling in Soft Ground. In Proceedings of the 7th International Conference on Soil Mechanics and Foundation Engineering (Mexico), Mexico City, Mexico, 29 August 1969; pp. 225–290.
5. Hsieh, P.G.; Ou, C.Y. Shape of Ground Surface Settlement Profiles Caused by Excavation. *Can. Geotech. J.* **1998**, *35*, 1004–1017. [CrossRef]
6. Wang, J.H.; Xu, Z.H.; Wang, W.D. Wall and Ground Movements Due to Deep Excavations in Shanghai Soft Soils. *J. Geotech. Geoenviron. Eng.* **2010**, *136*, 985–994. [CrossRef]
7. Qian, J.G.; Tong, Y.M.; Mu, L.L.; Lu, Q.; Zhao, H.Q. A Displacement Controlled Method for Evaluating Ground Settlement Induced by Excavation in Clay. *Geomech. Eng.* **2020**, *20*, 275–285. [CrossRef]
8. Ou, C.Y.; Teng, F.C.; Li, C.W. A Simplified Estimation of Excavation-Induced Ground Movements for Adjacent Building Damage Potential Assessment. *Tunn. Undergr. Space Technol.* **2020**, *106*, 103561. [CrossRef]
9. Benson Hsiung, B.C. Observations of the Ground and Structural Behaviours Induced by a Deep Excavation in Loose Sands. *Acta Geotech.* **2020**, *15*, 1577–1593. [CrossRef]
10. Li, M.G.; Demeijer, O.; Chen, J.J. Effectiveness of Servo Struts in Controlling Excavation-Induced Wall Deflection and Ground Settlement. *Acta Geotech.* **2020**, *15*, 2575–2590. [CrossRef]
11. Ni, X.D.; Lu, J.F.; Wang, Y.; Shi, J.W.; Chen, W.C.; Tang, L.X. Field Investigation of the Influence of Basement Excavation and Dewatering on Ground and Structure Responses. *Tunn. Undergr. Space Technol.* **2021**, *117*, 104121. [CrossRef]
12. Mohammad Shoari Shoar, S.; Heshmati, A.A.; Salehzadeh, H. Prefailure Deformation of Nailed Deep Excavations under Surcharge by Centrifuge Model Test. *Adv. Civ. Eng.* **2021**, *2021*, 5569797. [CrossRef]
13. Zeng, C.F.; Song, W.W.; Xue, X.L.; Li, M.K.; Bai, N.; Mei, G.X. Construction Dewatering in a Metro Station Incorporating Buttress Retaining Wall to Limit Ground Settlement: Insights from Experimental Modelling. *Tunn. Undergr. Space Technol.* **2021**, *116*, 104124. [CrossRef]
14. Xiang, P.F.; Wei, G.; Zhang, S.M.; Cui, Y.L.; Guo, H.F. Model Test on the Influence of Surcharge, Unloading and Excavation of Soft Clay Soils on Shield Tunnels. *Symmetry* **2021**, *13*, 2020. [CrossRef]
15. Pal, A.; Roser, J.; Vulic, M. Surface Subsidence Prognosis above an Underground Longwall Excavation and Based on 3D Point Cloud Analysis. *Minerals* **2020**, *10*, 82. [CrossRef]
16. Ying, H.W.; Cheng, K.; Liu, S.J.; Xu, R.Q.; Lin, C.G.; Zhu, C.W.; Gan, X.L. An Efficient Method for Evaluating the Ground Surface Settlement of Hangzhou Metro Deep Basement Considering the Excavation Process. *Acta Geotech.* **2022**, *17*, 5759–5771. [CrossRef]
17. Ou, C.Y.; Hsieh, P.G. A Simplified Method for Predicting Ground Settlement Profiles Induced by Excavation in Soft Clay. *Comput. Geotech.* **2011**, *38*, 987–997. [CrossRef]
18. Hu, Z.F.; Chen, J.; Deng, Y.L.; Li, J.B. A Simplified Method for Predicting Ground Surface Settlement Induced by Deep Excavation of Clay Stratum. *J. Yangtze River Sci. Res. Instig.* **2019**, *36*, 60–67+72.

19. Guo, P.P.; Gong, X.N.; Wang, Y.X. Displacement and Force Analyses of Braced Structure of Deep Excavation Considering Unsymmetrical Surcharge Effect. *Comput. Geotech* **2019**, *113*, 103102. [CrossRef]
20. Rashidi, F.; Shahir, H. Numerical Investigation of Anchored Soldier Pile Wall Performance in the Presence of Surcharge. *Int. J. Geotech. Eng.* **2019**, *13*, 162–171. [CrossRef]
21. Xu, C.J.; Yin, M.; Lin, G. Characters Analysis of the Retaining Structure of the Foundation Pit under Local Load. *Appl. Mech. Mat.* **2014**, *477–478*, 448–452. [CrossRef]
22. Li, X.; Zhou, T.G.; Wang, Y.X.; Han, J.L.; Wang, Y.Q.; Tong, F.; Li, D.L.; Wen, J.M. Response Analysis of Deep Foundation Excavation and Dewatering on Surface Settlements. *Adv. Civ. Eng.* **2020**, *2020*, 8855839. [CrossRef]
23. Zhang, X.; Wang, L.H.; Wang, H.L.; Feng, C.L.; Shi, H.J.; Wu, S.Z. Investigating Impacts of Deep Foundation Pit Dewatering on Land Subsidence Based on CFD-DEM Method. *Eur. J. Environ. Civ. Eng.* **2022**, *26*, 6424–6443. [CrossRef]
24. Zheng, G.; Li, Q.H.; Cheng, X.S.; Ha, D.; Shi, J.C.; Shi, X.R.; Lei, Y.W. Diaphragm Wall Deformation and Ground Settlement Caused by Dewatering before Excavation in Strata with Leaky Aquifers. *Geotechnique* **2022**. [CrossRef]
25. Ying, H.W.; Yang, Y.W. Characteristics of A Large and Deep Soft Clay Excavation in Hangzhou. *Chin. J. Geotech. Eng.* **2011**, *33*, 1838–1846.
26. Nie, Z.Q.; Zhang, S.G.; Meng, S.P. Surface Settlement of Deep Foundation Pits by Excavation. *Chin. J. Geotech. Eng.* **2008**, *30*, 1218–1223.
27. Qian, J.G.; Wang, W.Q. Analytical Solutions to Ground Settlement Induced by Movement of Rigid Retaining Wall. *Chin. J. Rock Mech. Eng.* **2013**, *32*, 2698–2703.
28. Gu, J.B.; Qian, J.G. Analytical Theory of Ground Settlement Induced by Movement of Flexible Retaining Wall. *Chin. Rock Soil Mech.* **2015**, *36*, 465–470. [CrossRef]
29. Hu, Z.F.; Chen, J.; Deng, Y.F.; Li, J.B.; Zhou, X.T. Explicit Analytical Solution of Surface Settlement Induced by Horizontal Displacement of Retaining Wall. *Rock Soil Mech.* **2018**, *39*, 4165–4175. [CrossRef]
30. Shen, L.Y.; Qian, J.G.; Zhang, R.Z. Simplified Analytical Solution of Ground Settlement Induced by Horizontal Displacement of Retaining Wall. *Rock Soil Mech.* **2016**, *37*, 2293–2298. [CrossRef]

water

MDPI

Article

Stability of Braced Excavation Underneath Crossing Underground Large Pressurized Pipelines

Fangang Li [1], Panpan Guo [2,*], Ningning Geng [3], Lei Mao [4], Feng Lin [4], Yanlin Zhao [5], Hang Lin [6] and Yixian Wang [2,*]

1 Anhui Traffic Construction Co., Ltd., Hefei 230041, China
2 College of Civil Engineering, Hefei University of Technology, Hefei 230009, China
3 Hefei Rail Transit Group Co., Ltd., Hefei 230001, China
4 Hefei Urban Construction Investment Holding Co., Ltd., Hefei 230092, China
5 School of Energy and Safety Engineering, Hunan University of Science and Technology, Xiangtan 411201, China
6 School of Resources and Safety Engineering, Central South University, Changsha 410083, China
* Correspondence: guopanpan@hfut.edu.cn (P.G.); wangyixian2012@hfut.edu.cn (Y.W.); Tel.: +86-0551-6290-1660 (Y.W.)

Abstract: The practice of deep-braced excavation in congested urban environments involves frequently buried pipelines, which can exert a significant effect on the performance of the excavation. The objective of this paper is to investigate the performance of a 12.5-m-deep-braced excavation spanned by two shallowly buried large-diameter pressurized pipelines. A suspension structure is installed within the excavation to protect the in situ pipelines during the construction. The excavation performance is investigated by performing a three-dimensional finite element analysis. The finite element method is verified based on the observations at the site. The results indicate that, as expected, the excavation support structures displace together with varying degrees of deformation toward the excavated area. The strut shear forces are found to be distributed axially in linear manners, while the strut bending moments are in symmetric manners. The benefit of using the proposed pipeline suspension structure is demonstrated. By using this structure, pipeline deformation can be well controlled, and the structural integrity and safety of the pipelines can be ensured. This benefit depends on the convenient operation in that the elevation of the cork base of the pipeline suspension structure is stably lowered during the construction process.

Keywords: deep excavation; structural response; urban environment; buried pipeline; deformation analysis

check for updates

Citation: Li, F.; Guo, P.; Geng, N.; Mao, L.; Lin, F.; Zhao, Y.; Lin, H.; Wang, Y. Stability of Braced Excavation Underneath Crossing Underground Large Pressurized Pipelines. *Water* **2022**, *14*, 3867. https://doi.org/10.3390/w14233867

Academic Editor: Bruno Brunone

Received: 30 October 2022
Accepted: 23 November 2022
Published: 27 November 2022

1. Introduction

Deep-braced excavation is an important part of human civil engineering activities, playing an irreplaceable role in promoting the development and progress of human society. In recent years, with the continuous acceleration of urbanization and the rapid development of underground space development and utilization, deep-braced excavations for constructing high-rise buildings [1,2], metro stations [3–6], underground shopping centers [7], railway stations [8], three-dimensional underground garages [9], underground passages [10], bridge foundations [11], underground substations, and other infrastructures are booming. In order to meet the needs of large-scale engineering construction, the excavation depth, excavation scale (area, length), and other aspects of deep-braced excavation projects constantly refresh records.

At present, although the theory and technology in deep-braced excavations have made great progress [12,13], the collapse of deep-braced excavation and the instability and destruction of adjacent existing structures caused by various reasons are still common [14–16] and even serious casualties and economic property losses can happen. This is partially because the complicated performance of deep-braced excavation, which depends on countless factors and thus varies from case to case, has not been well understood by designers, engineers, and researchers [17–19]. In general, the performance of deep-braced excavation involves the behavior of the excavation support system, ground movements induced by excavation, and response of preexisting structures or facilities affected by the excavation.

The performance of deep-braced excavation has been extensively investigated by many investigators during the past several decades. The methods commonly used for investigating the performance of deep-braced excavation include mainly field monitoring data analysis [20–22], numerical simulations [23–25], an analytical method [26–28], and empirical or semiempirical methods [29–31]. It has been found based on these methods mentioned above that the performance of deep-braced excavation is affected by many factors, including excavation dimensions (e.g., depth, width, and shape), excavation support system, geological and hydrological conditions, and relative positions between excavation and preexisting structures or facilities (e.g., high-rise buildings, railways, metro tunnels, piles, and buried pipelines).

In the engineering practice of deep-braced excavations involving buried pipelines, the relative position between excavation and buried pipelines falls into two categories: one is the spanning case in which the excavation is spanned by the pipelines, and the other is the non-spanning case in which the pipelines are outside the excavation. Among them, the latter case is not rare in practical engineering, and relevant investigations into that case have been carried out. By performing finite element analysis, Hu et al. [32] investigated the influence of excavation-induced soil disturbance on pipeline displacement and proposed a practical measure for protecting the pipeline. Based on field monitoring data, Jiang et al. [33] numerically analyzed the response of buried gas pipelines to blasting excavation. Zhang et al. [34] proposed a stress-controlled method for estimating the deformation and bending moment of pipelines induced by deep excavation. In addition, measured results from typical case histories have also been reported to investigate the behavior of deep-braced excavation with buried pipelines located nearby and the response of the buried pipelines [35–38]. However, to the best of the authors' knowledge, the spanning case has not been reported in the literature. Moreover, the performance of deep-braced excavation spanned by buried pipelines still remains unclear due to a lack of relevant investigation.

This paper reports a deep-braced excavation case history in which the excavation is spanned by two shallowly buried large-diameter pressurized pipelines. The performance of this deep-braced excavation is investigated by adopting the finite element method verified based on the part of the observations at the site. Meanwhile, the effectiveness of the pipeline supporting structure for protecting the pipeline during excavation is discussed.

2. Project Overview

2.1. The Excavation Project

The Excavation Project is located at the intersection of YF Road and CS Road in Suzhou city, China. The Project serves as an independent deep-braced excavation engineering for the WY Tunnel, which is an open trench expressway tunnel with a length of 3450 m. This deep excavation is 38 m in length, 30.8 m in width, and 12.5 m in depth and is spanned by two shallowly buried large-diameter pressurized pipelines, i.e., DN1800 and DN1400. The angle of intersection between the pipelines and the excavation is approximately 80°. The DN1800 is the only sewage pipeline, while the DN1400 serves as one of the important clean water pipelines in the XC District. These two pipelines cannot be removed from the site due to their importance in ensuring the normal life of the residents as well as the limitation of the construction period for this project, and they need to be protected in situ during the construction of the YRNE Project. Table 1 summarizes some of the characteristics of the two pipelines. The positions of the two pipelines relative to the structure of the WY Tunnel are schematically presented in Figure 1. Figure 2 presents a photograph of the two pipelines in the YR Project.

Table 1. Summary of some of the typical characteristics of the pipelines in the YRNE Project.

Pipeline	Material	Cover Depth (m)	Diameter (m)	Wall Thickness (mm)	Internal Pressure (MPa)
DN1800	steel	1.45–1.55	1.8	16	0.3
DN1400	steel	1.60–1.74	1.4	14	0.2

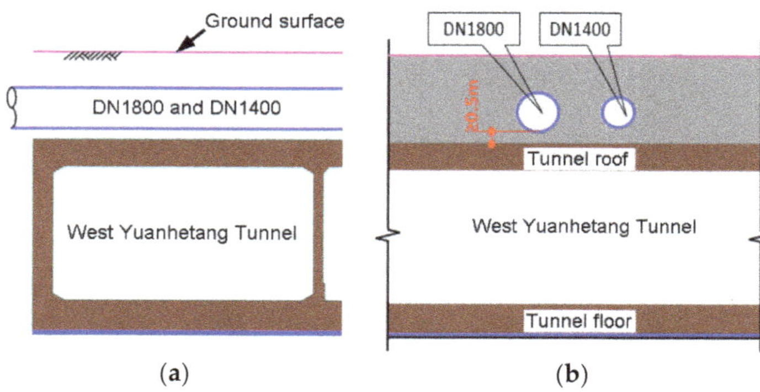

Figure 1. Schematic diagram of the relative position between pipelines and tunnel: (**a**) Transverse section; (**b**) Longitudinal section.

Figure 2. A photograph of the pipelines in the YRNE Project.

2.2. Excavation Support System

The deep excavation is supported by diaphragm walls of 0.6 m in thickness on its general segments of the perimeter, while on the specific segments of the perimeter with the interruption of the buried pipelines, the deep excavation is supported by the Metro Jet System (MJS) retaining wall. The MJS retaining wall extends to a distance of 8.5 m away from the excavation side. The net distance between the MJS retaining wall is no less than 0.5 m, and the net distance between the diaphragm wall and the pipelines is no less than 0.8 m. In addition, the excavation support system also includes concrete struts, steel struts, cast-in-situ bored (CISB) piles, lattice columns, and high-pressure jet grouting (HPJR) piles. A schematic diagram showing the excavation support system is presented in Figure 3.

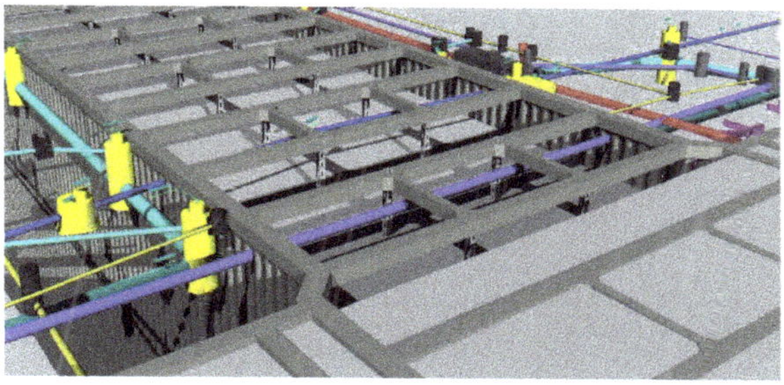

Figure 3. Schematic diagram of the excavation support system.

2.3. Pipeline Suspension Structure

During the construction of the deep excavation, the pipelines are protected in situ by using the suspension method. The pipeline suspension structure takes advantage of the excavation support system. Three of the concrete struts are used as the bearing system of the pipeline suspension structure, and the pipelines are suspended from the selected concrete struts by using the finely rolled rebars, as depicted in Figure 4. In order to achieve sufficient flexural rigidity of the three concrete struts being parts of the pipeline suspension structure, twelve lattice columns are installed below them. The longitudinal distance between two adjacent lattice columns is 6 m. The 32a U-bar and steel plate are combined to serve as the cork base of the pipeline suspension structure. The cork base is connected to the concrete struts through the fine-rolled rebars. A schematic diagram of the pipeline suspension structure is shown in Figure 4. Compared to the traditional ones, this pipeline suspension structure has the following advantages: (1) it reduces the construction cost by taking advantage of the excavation support system, and (2) the deformation of the protected pipeline can be dynamically adjusted during the excavation by controlling the length of the fine rolled rebar.

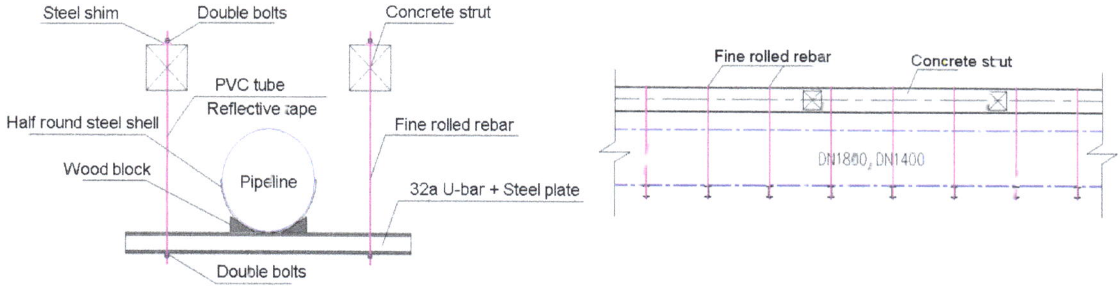

(a) (b)

Figure 4. Schematic diagram of the pipeline suspension structure: (**a**) Transverse section; (**b**) Longitudinal section.

2.4. Geological and Hydrologic Conditions

The soil layers involved in the YRNE Project, from the ground surface to the bottom of the MJS retaining wall, are, respectively, miscellaneous fill, plain fill, clay, silty clay, floury soil, silty clay, and clay, as schematically shown in Figure 5. The excavation bottom is mainly located within the floury soil layer. By performing in situ and laboratory tests, the basic properties of the soil are obtained. Table 2 summarizes the thickness and the basic properties of these soil layers. Note that the soil layer number has been depicted in Figure 5. The groundwater that has an effect on this project primarily includes phreatic aquifer, feeble confined water, and confined water. The groundwater table is about 5.0 m below the ground surface. The bottom of the deep excavation is mainly located within the feeble confined water layer.

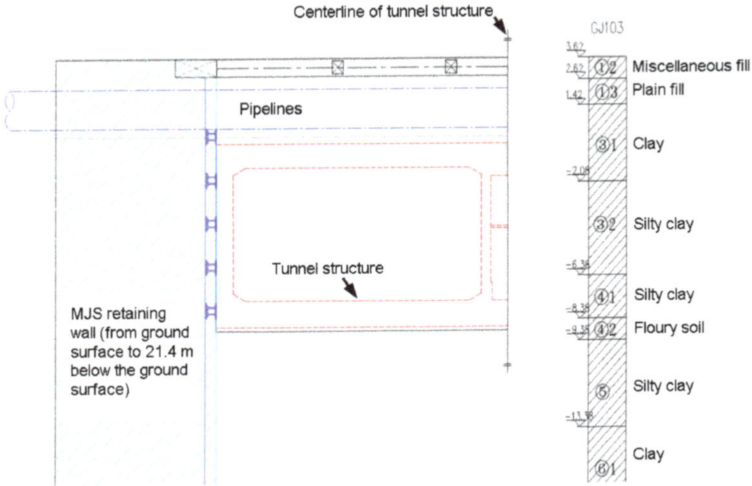

Figure 5. Profile of the soil layers at the site.

Table 2. Summary of the thickness and properties of the soil layers.

Soil Layer Number	Thickness (m)	Unit Weight (kN/m^3)	Elasticity Modulus (MPa)	Poisson's Ratio (-)	Friction Angle (°)	Cohesion (kPa)
①2	1	18.5	3.5	0.23	12	5
①3	1.2	19	4	0.2	10	12
③1	3.5	19.3	25	0.3	16.0	33.1
③2	4.3	18.6	20	0.33	17.0	19.8
④1	2	18.8	22	0.36	19.2	19.6
④2	1	18.4	30	0.25	17.7	17.2
⑤	4	18.5	20	0.33	15.9	39.4
⑥1	9.5	19.8	20	0.3	18.5	17.6

3. Numerical Modelling

3.1. Finite Element Analysis Software

The commercial software for finite element analysis (FEA), Midas GTS NX, is adopted in this study. It is characterized by real 3D geometry modeling, a powerful mesh generator, a fast analysis solver, and outstanding post-processing. This FEA software has been widely used for advanced geotechnical analysis in terms of geomaterial deformation and stability, groundwater flow, dynamic vibrations, and soil-structure interactions in 2D and 3D.

3.2. Meshing and Boundary Conditions

The meshing of the finite element analysis model established in this study is presented in Figure 6. Considering the three-dimensional size of the deep-braced excavation, the dimension of the overall model is selected as 200 m × 150 m × 40 m, which has eliminated the boundary effect. To simulate the deep-braced excavation in a reasonable way, the boundary conditions of the finite element analysis model are set as free at the top surface, rolled on the four vertical sides, and fixed at the bottom surface.

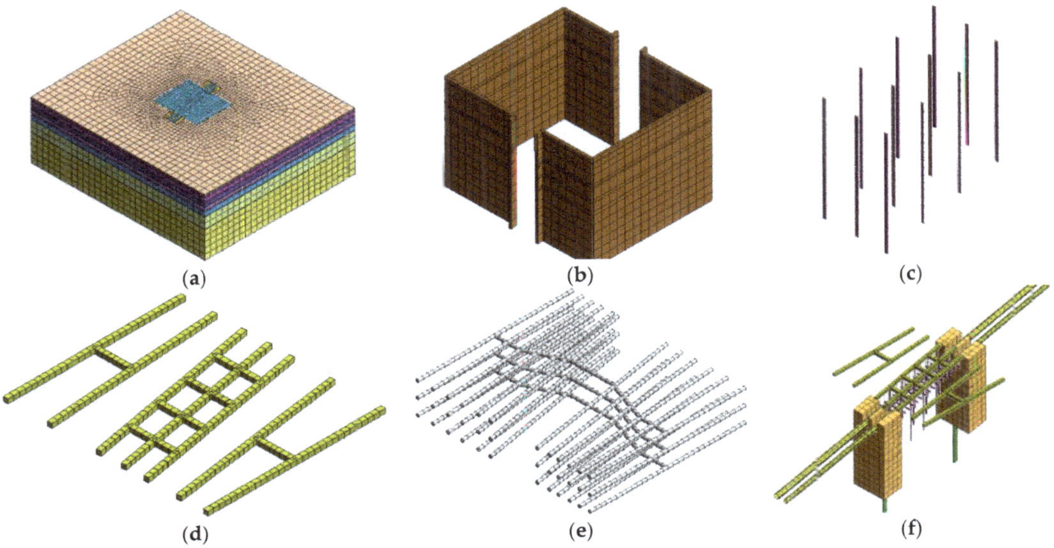

Figure 6. Meshing of the finite element analysis model: (**a**) Overall model; (**b**) Diaphragm walls; (**c**) Lattice columns; (**d**) Concrete struts; (**e**) Steel struts; (**f**) Pipeline suspension structure.

3.3. Parameters Used for Finite Element Analysis

The Mohr–Coulomb (MC) model is used for simulating the constitutive behavior of the soil in the YRNE Project. The MC model is the most widely used constitutive model in geotechnical simulations, thanks to the simplicity and the easy availability of the model parameter. The yield criterion of the MC model can be written as

$$\frac{\sigma_1 - \sigma_3}{2} = \frac{\sigma_1 + \sigma_3}{2} \sin\varphi + c\cos\varphi \tag{1}$$

where σ_1 and σ_3 are the major and the minor principal stresses, respectively; c is the cohesion; and φ is the angle of internal friction.

The soil parameters required for the MC model are summarized in Table 2. As for the excavation support system and the pipeline suspension structure, the linear elastic model is used to simulate their behavior, considering that the stress–strain relations of these structural components are always within the elastic domain during the construction of the deep-braced excavation Table 3 summarizes the parameters required for the simulations of the excavation support system and the pipeline suspension structure.

Table 3. Parameters required for the simulation of the excavation support system and the pipeline suspension structure.

Structure	Unit Weight (kN/m^3)	Elastic Modulus (MPa)	Poisson's Ratio (-)	Remark
Diaphragm wall	25	3×10^4	0.2	Thickness = 0.6 m; Depth = 25 m
MJS wall	20	1×10^4	0.2	From ground surface to −21.4 m
CISB pile	20	3×10^4	0.2	Diameter = 0.8 m; Length = 30 or 38 m
Ring beam	25	3×10^4	0.2	Sectional dimension = 1 m × 0.8 m
Concrete strut	25	3×10^4	0.2	Sectional dimension = 0.8 m × 0.8 m
Steel strut	78.5	2×10^5	0.25	Diameter = 609 mm; Wall thick. = 16 mm
Lattice column	78.5	2×10^5	0.25	Longitudinal interval = 6 m
Cork base	78.5	2×10^5	0.25	Using 32a U-bar
Fine rolled rebar	78.5	2×10^5	0.25	Diameter = 25 mm

3.4. Excavation Sequence Simulation

According to the actual excavation sequence, the construction process of the deep-braced excavation can be simulated in the following ten Excavation Steps:

Excavation Step 1: Reproduce the ground state of stress, and eliminate the ground displacement;

Excavation Step 2: Activate the excavation support structures, including the diaphragm wall, MJS retaining wall, and CISB piles;

Excavation Step 3: Excavate to the bottom of the concrete struts, and activate the ring beam and the concrete struts;

Excavation Step 4: Excavate to the elevation of 0.5 m below the pipelines, and activate the pipeline suspension structure;

Excavation Step 5: Excavate to the elevation of 0.5 m below the first-level steel struts, and activate the first-level steel struts;

Excavation Step 6: Excavate to the elevation of 0.5 m below the second-level steel struts, and activate the second-level steel struts;

Excavation Step 7: Excavate to the elevation of 0.5 m below the third-level steel struts, and activate the third-level steel struts;

Excavation Step 8: Excavate to the bottom of the excavation, and activate the cushion and the tunnel floor;

Excavation Step 9: Activate the side walls, middle walls, and roof of the tunnel structure; and

Excavation Step 10: Deactivate the pipeline suspension structure and the lattice columns, and backfill the soil.

4. Validation of Finite Element Method

The monitoring data for the performance of the deep-braced excavation observed at the site can be used to validate the finite element method adopted in this study. During the construction process, the monitored items include the ground surface settlements, pipeline displacements, displacements at the diaphragm wall top, lattice column displacements, groundwater level, and axial forces in concrete and steel struts. A plan view showing the arrangement of these monitored items is presented in Figure 7, where DB59-1–DB59-5 and DB60-1–DB60-5 denote ten monitoring points for ground surface settlements, JS1-1–JS1-15 denote fifteen monitoring points for DN1800 pipeline displacements, JS2-1–JS2-15 denote fifteen monitoring points for DN1400 pipeline displacements, P117 and P118 denote two monitoring points for displacements at diaphragm wall top, LZ58 denotes one monitoring points for lattice column displacements, W57 and W58 denote two monitoring points for groundwater level, ZL29 denotes one monitoring point for axial forces in concrete strut, and GL29 denotes one monitoring points for axial forces in steel strut. The devices used for monitoring these items and their versions are summarized in Table 4.

Table 4. Summary of the monitoring devices used in this deep-braced excavation and their versions.

Monitored Item	Device Used	Version
Horizontal disp. at diaphragm wall top	Total station instrument	TCRA1201
Vertical disp. at diaphragm wall top	Total station instrument	TCRA1201
Pipeline displacements	Total station instrument	TCRA1201
Ground surface settlements	Single-point settling meter	YH02-A20
Axial forces in struts	Vibrating string-type steel bar meter	GJJ10
Groundwater level	Pneumatic water level gauge	YH04-A06

In order to validate the finite element method adopted in this study, a comparison is made between the simulated results by the finite element method and the monitored results in the field in terms of displacements at diaphragm wall top, lattice column displacements, axial forces in struts, and ground surface settlements, as shown in Figure 8. The comparison indicates that the simulated results by the finite element method agree well with the monitored results, demonstrating the validity of the finite element method adopted in this study. Moreover, the slight difference between the simulated and the monitored results are caused by various factors, such as the simplification of the excavation support system, the inability of the MC model to describe the true behavior of the soil, and the complex geological conditions that cannot be reproduced in the finite element analysis.

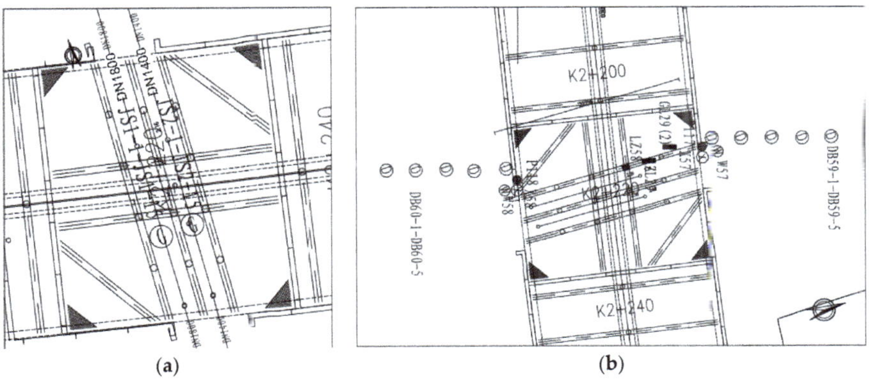

Figure 7. Plan view showing arrangement of the monitored items: (**a**) Pipeline displacements; (**b**) Ground surface settlements, displacements at diaphragm wall top, lattice column displacements, groundwater level, and axial forces in concrete and steel struts.

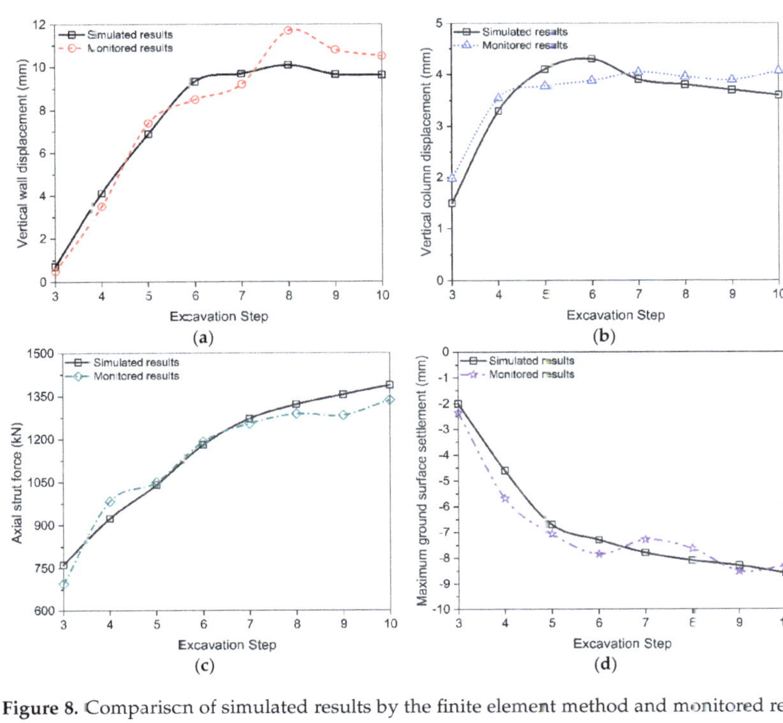

Figure 8. Comparison of simulated results by the finite element method and monitored results in terms of (**a**) Vertical wall displacement at P117; (**b**) Vertical column displacement at LZ58; (**c**) Axial strut force at ZL29; (**d**) Maximum ground surface settlement along DB59-1–DB59-5.

5. Results and Discussion

The simulated results describing the performance of the deep-braced excavation by the validated finite element method are presented in this section. In detail, this section analyzes the deformation characteristics of the diaphragm wall and the MJS retaining wall, the distribution of the internal forces in the concrete struts and steel struts, and the vertical displacements of the pipelines. In addition, a brief discussion of the supporting effect of the pipeline suspension structure is also presented.

5.1. Deformation Characteristics of Diaphragm Wall and MJS Retaining Wall

Figure 9 presents the displacement contours of the diaphragm wall and MJS retaining wall at four different Excavation Steps (i.e., Excavation Steps 3, 5, 6, and 8). A description of these Excavation Steps has been given in Section 3.4. It can be seen from Figure 8 that both the diaphragm wall and the MJS retaining wall tend to move towards the excavation with advancing excavation. The deformation of the diaphragm wall and the MJS retaining wall increases with increasing the excavation depth. When the final excavation depth (i.e., 12.5 m) is reached at Excavation Step 8, the maximum deformation of the diaphragm wall, 14 mm, is achieved.

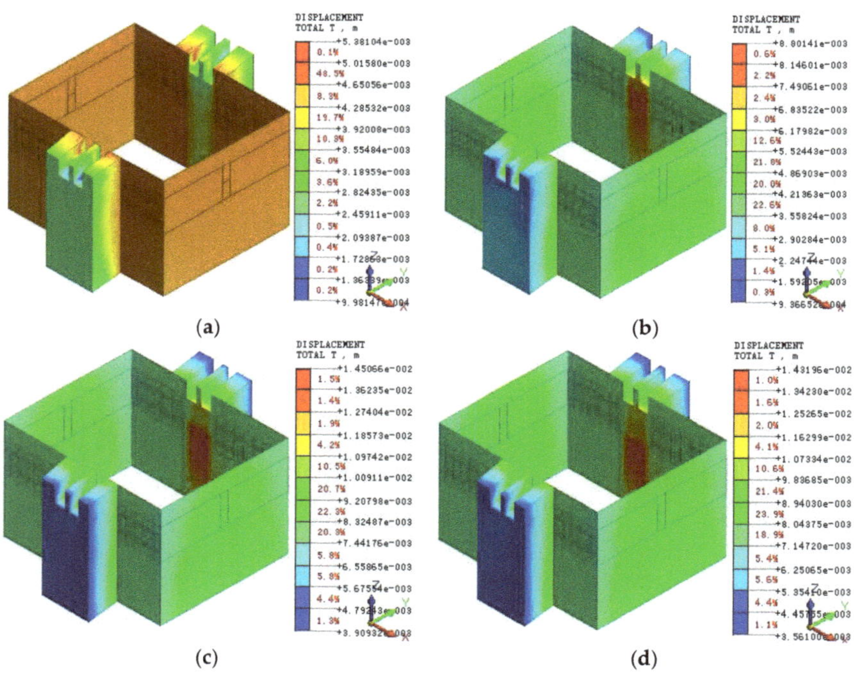

Figure 9. Displacement contours of diaphragm wall and MJS retaining wall at: (**a**) Excavation Step 3; (**b**) Excavation Step 5; (**c**) Excavation Step 6; (**d**) Excavation Step 8.

5.2. Distribution of Internal Forces in Concrete Strut

Figure 10 presents the contours of the internal forces in the concrete strut at Excavation Step 8. The internal forces include axial force (Figure 10a), shear force in the Y direction (Figure 10b), shear force in the Z direction (Figure 10c), the bending moment in the Y direction (Figure 10d), and bending moment in the Z direction (Figure 10e). The results indicate that the maximum magnitudes of the axial force, shear force in the Y direction, the bending moment in the Y direction, shear force in the Z direction, and bending moment in the Z direction in the concrete strut are, respectively, 269 kN, 397 kN, 79 kN·m, 24.5 kN, and

896 kN·m. Moreover, the distribution of the shear forces in the Y and Z directions along the axis of the concrete strut is in a linear manner, while the distribution of the bending moments in the Y and Z directions is in a symmetric manner.

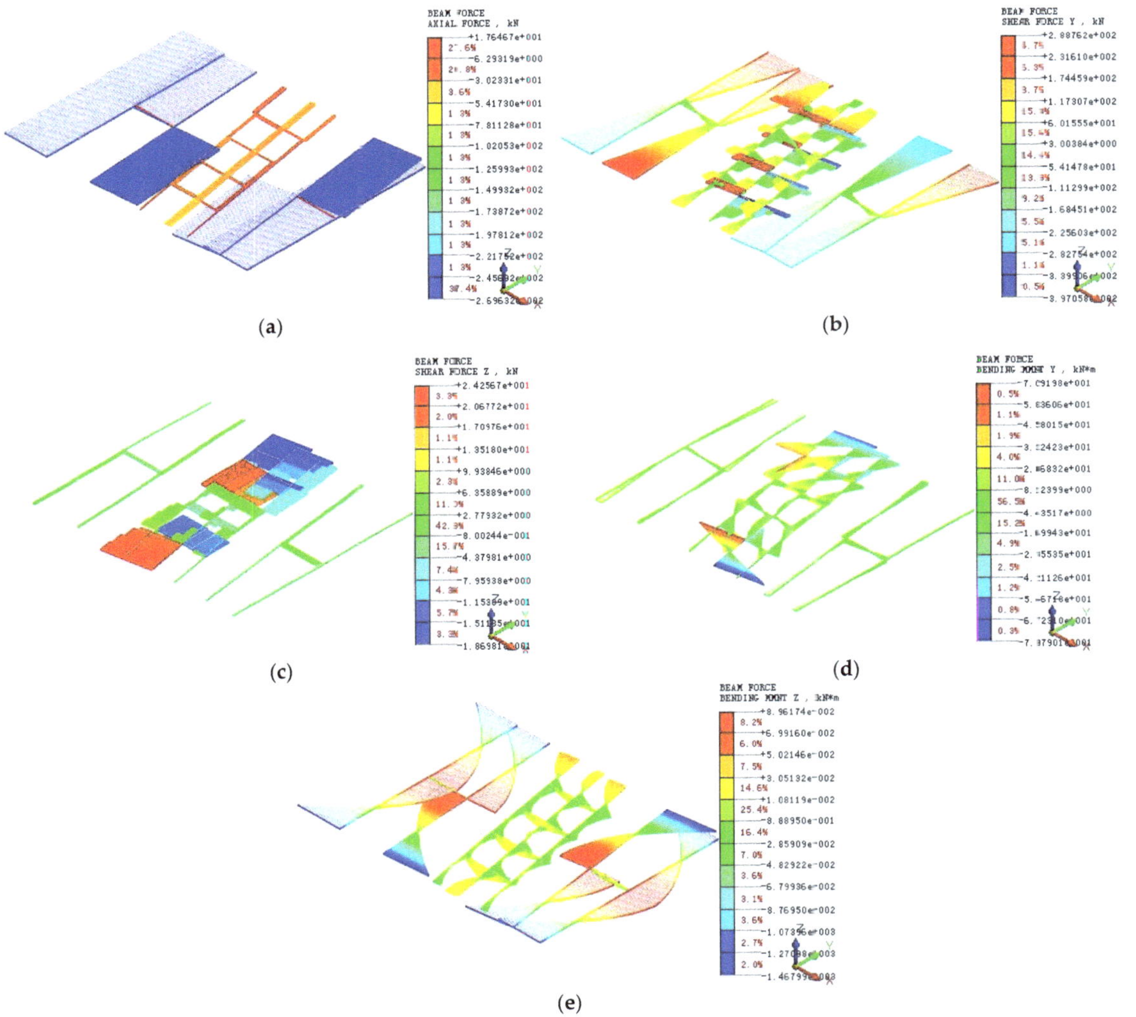

Figure 10. Contours of internal forces in concrete strut at Excavation Step 8: (**a**) Axial force; (**b**) Shear force in Y direction; (**c**) Shear force in Z direction; (**d**) Bending moment in Y direction; (**e**) Bending moment in Z direction.

5.3. Distribution of Internal Forces in Steel Strut

Figure 11 presents the contours of the axial force, shear force in the Y direction, shear force in the Z direction, the bending moment in the Y direction, and bending moment in the Z direction in the steel strut at Excavation Step 8. It can be indicated that the maximum axial force in the steel strut is 237 kN. The shear force in the Y direction is of linear distribution along the axis of the steel strut. The maximum shear force in the Y direction is located at the third steel strut, with its magnitude being 149 kN. However, compared to the magnitude of the maximum shear force in the Y direction, the maximum shear force in the Z direction is negligible, with its magnitude being merely 8.11 kN. The distribution of the bending moments in both Y and Z directions is almost symmetric with regard to the middle of the steel strut. The maximum magnitudes of the bending moments in Y and Z directions are, respectively, 23 and 315 kN·m, which are both reached at the third steel strut.

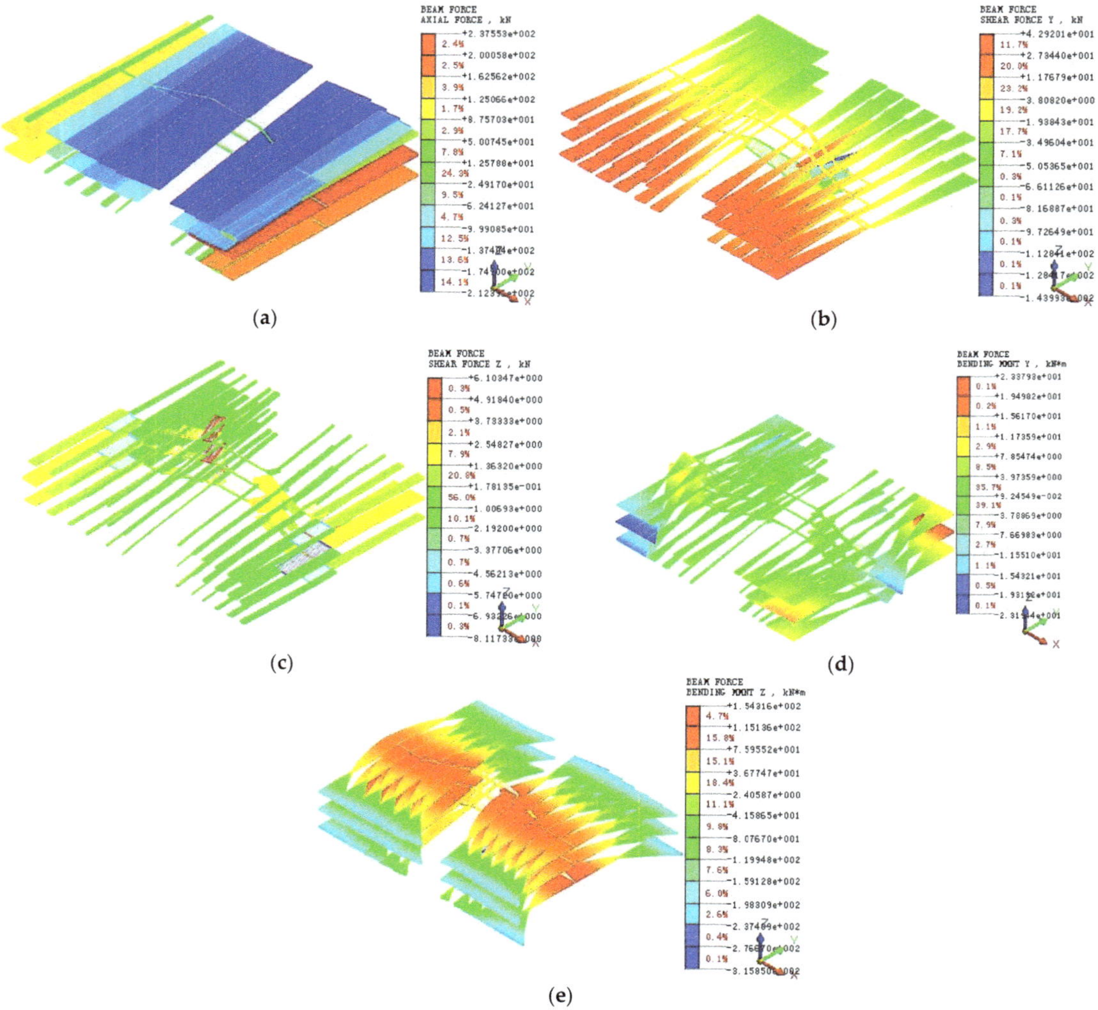

Figure 11. Contours of internal forces in steel strut at Excavation Step 8: (**a**) Axial force; (**b**) Shear force in Y direction; (**c**) Shear force in Z direction; (**d**) Bending moment in Y direction; (**e**) Bending moment in Z direction.

5.4. Vertical Displacements of Pipelines

Figure 12 presents the contours of the vertical displacements of pipelines at five different Excavation Steps (i.e., Excavation Steps 4, 5, 6, 7, and 8). It can be seen that at Excavation Steps 4 and 5, settlements of the pipelines occur for the entire length of the two pipelines, with the maximum settlement being 3 mm. This is attributed to the fact that the pipeline suspension structure has not taken much effect during the early phase after installation. At the following Excavation Steps, a heave of the pipelines occurs. The heave of the pipelines is primarily caused by the basal heave of the excavation after soil removal. Due to the occurrence of the basal heave, the lattice columns move upward, which raises the concrete struts. The rise of the concrete struts is eventually transferred to the pipelines by virtue of the pipeline suspension structure. With an increase in the excavation depth, the heave of the pipelines increases due to an increase in the basal heave of the excavation. The maximum heave of the pipelines, 13 mm, is achieved when the final excavation depth (i.e., 12.5 m) is reached. In addition, the maximum heave of pipeline DN1400 is 1 mm greater than that of pipeline DN1800, which may be attributed to the lighter weight of pipeline DN1400.

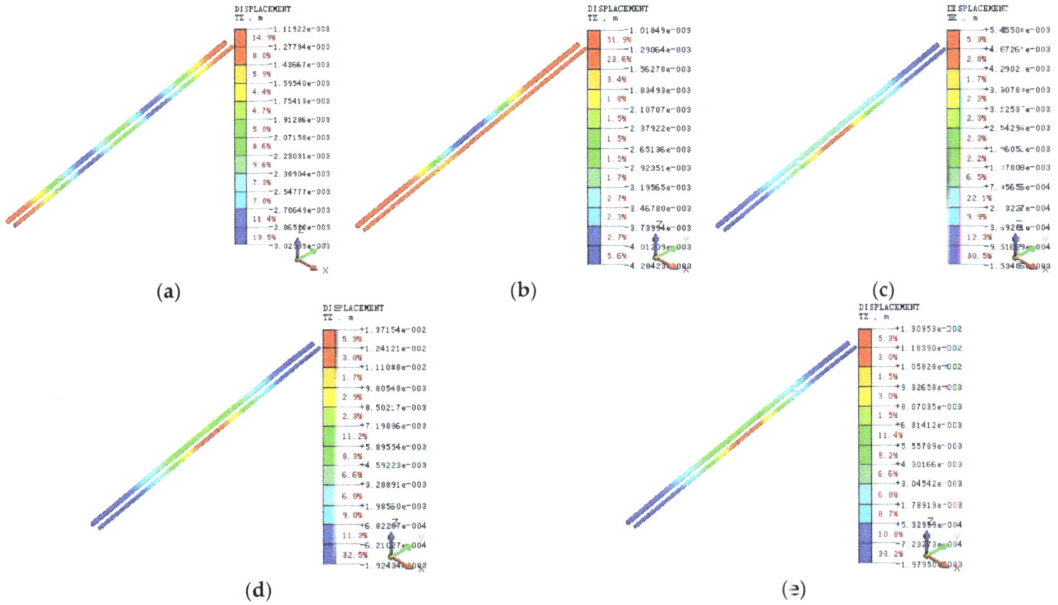

Figure 12. Contours of vertical displacements of pipelines at (**a**) Excavation Step 4; (**b**) Excavation Step 5; (**c**) Excavation Step 6; (**d**) Excavation Step 7; (**e**) Excavation Step 8.

5.5. Discussion of Supporting Effect of Pipeline Suspension Structure (PSS)

The effectiveness of the supporting system is discussed by comparing the results of two cases: one is the above-mentioned YRNE Project, where PSS is installed, and the other is the same project but without PSS. By performing finite element analysis using the material parameters, constitutive models, and excavation simulation sequence for the case with PSS, the results for the case without PSS are also obtained. Figure 13 compares the contours of the vertical displacements of pipelines at Excavation Step 8 between the cases with and without installing the pipeline suspension structure. The distributions of the vertical displacements of the pipelines along the distance from the excavation edge are compared in Figure 14 for the two cases described above.

From the results presented in Figures 13 and 14, it can be inferred that settlements of pipelines will occur for the case without installing PSS due to the action of gravity. The pipeline settlements peak in the middle of the excavation. Under the condition of no suspension support, both of the two pipelines exhibit a tendency for settlement. The maximum settlement of pipeline DN1400 is 9 mm, which is slightly greater than that of pipeline DN1800 (i.e., 6 mm). Under the condition where the PSS is installed, both of the two pipelines tend to be uplifted during the following Excavation Steps. The maximum heave of the pipeline occurs at pipeline DN1400 with a magnitude of 13 mm, which is slightly greater than that occurred at pipeline DN1800. During the construction process, the heave of the pipelines for the case with installing PSS can be eliminated by lowering the elevation of the cork base of the pipeline suspension structure. This operation should be slow and of uniform velocity, with the help of a chain hoist. Therefore, because of the existence of the pipeline suspension structure, the initial deformation pattern and stress state of the pipelines before the commencement of the construction can be maintained after the completion of the construction. However, in the case without PSS, the pipelines will always be in a curved state of deformation pattern during the operation life cycle of the WY Tunnel, which will jeopardize the structural integrity and the durability of the pipelines.

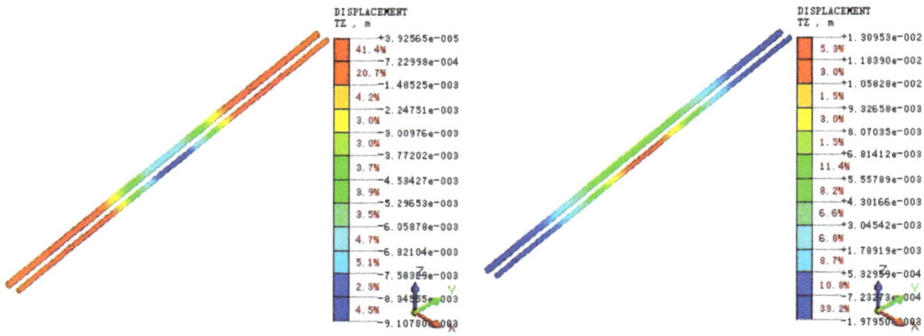

(**a**) (**b**)

Figure 13. Contours of vertical displacements of pipelines at Excavation Step 8: (**a**) without installing pipeline suspension structure; (**b**) with installing pipeline suspension structure.

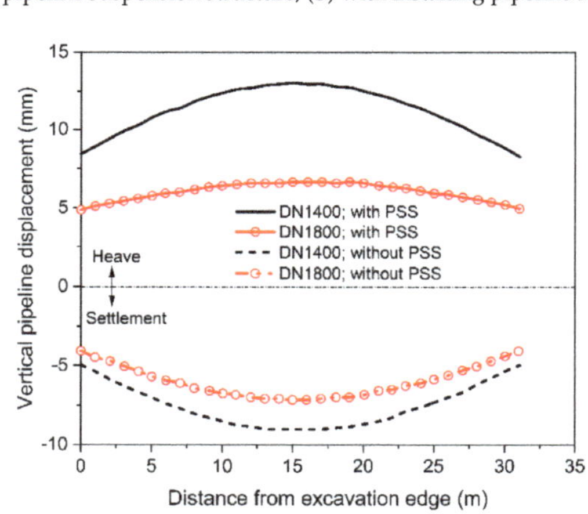

Figure 14. Comparison of vertical displacements of pipelines between cases with and without installing pipeline suspension structure (PSS).

6. Conclusions

An unusual deep-braced excavation case history in Suzhou, China, is reported. This excavation is spanned by two shallowly buried large-diameter pressurized pipelines. As the pipelines need to be protected in situ during the construction of the excavation, a pipeline suspension structure is installed in addition to the excavation support system. Three-dimensional finite element analysis is performed to capture the performance of this deep-braced excavation at different stages. Based on the part of the field monitoring data, the validity of the finite element method adopted has been demonstrated. The conclusions drawn from this study can be summarized as follows.

(1) Both the diaphragm wall and the MJS retaining wall have a tendency to deform towards the excavation with proceeding the Excavation Step. The deformation of these walls constantly increases with an increase in the excavation depth.
(2) When the final excavation depth is reached, the distribution of the shear forces in the Y and Z directions along the axis of the concrete strut is in a linear manner, while the distribution of the bending moments in the Y and Z directions is in a symmetric manner.
(3) At the final excavation depth, the shear force in the Y direction is of linear distribution along the axis of the steel strut, and the distribution of the bending moments in both Y and Z directions is almost symmetric with regard to the middle of the steel strut. The magnitude of the maximum shear force in the Z direction is negligible compared to the magnitude of the maximum shear force in the Y direction.
(4) With an increase in the excavation depth, the heave of the pipelines increases due to an increase in the basal heave of the excavation. The maximum heave of pipeline DN1400 is 1 mm greater than that of pipeline DN1800, which may be attributed to the lighter weight of pipeline DN1400.
(5) The installation of the pipeline suspension structure is beneficial for the structural integrity and the safety of the pipelines during both the construction phase and the operational phase of the tunnel. This benefit is obtained by implementing conveniently the operation that the elevation of the pipeline suspension structure cork base is stably lowered during the construction period.

Author Contributions: Conceptualization, P.G. and Y.W.; methodology, P.G.; software, P.G. and N.G.; validation, H.L., F.L (Fangang Li), and P.G.; investigation, F.L. (Fangang Li), F.L. (Feng Lin) and L.M.; supervision, Y.W.; project administration, Y.Z.; funding acquisition, Y.W. All authors have read and agreed to the published version of the manuscript.

Funding: This research work was funded by the National Natural Science Foundation of China (42077249, 51774107).

Data Availability Statement: The data presented in this study are available on request from the corresponding author P.G.

Acknowledgments: The authors thank the anonymous reviewers for their constructive comments.

Conflicts of Interest: The authors declare no conflict of interest.

References

1. Tan, Y.; Lu, Y.; Wang, D. Deep excavation of the Gate of the Orient in Suzhou stiff clay: Composite earth retaining systems and dewatering plans. *J. Geotech. Geoenviron. Eng.* **2018**, *144*, 05017009. [CrossRef]
2. Oztoprak, S.; Cinicioglu, S.F.; Oztorun, N.K.; Alhan, C. Impact of neighbouring deep excavation on high-rise sun plaza building and its surrounding. *Eng. Fail. Anal.* **2020**, *111*, 104495. [CrossRef]
3. Tan, Y.; Wei, B.; Zhou, X.; Diao, Y. Lessons learned from construction of Shanghai metro stations: Importance of quick excavation, prompt propping, timely casting, and segmented construction. *J. Perform. Constr. Facil.* **2015**, *29*, 04014096. [CrossRef]
4. Li, D.; Yan, C. Building deformation prediction based on ground surface settlements of metro-station deep excavation. *Adv. Civ. Eng.* **2018**, *2018*, 6050353. [CrossRef]
5. Liu, B.; Xu, W.; Zhang, D.; Zhang, Q. Deformation behaviors and control indexes of metro-station deep excavations based on case histories. *Tunn. Undergr. Space Technol.* **2022**, *122*, 104400. [CrossRef]
6. Shi, H.; Jia, Z.; Wang, T.; Cheng, Z.; Zhang, D.; Bai, M.; Yu, K. Deformation characteristics and optimization design for large-scale deep and circular foundation pit partitioned excavation in a complex environment. *Buildings* **2022**, *12*, 1292. [CrossRef]
7. Hou, Y.M.; Wang, J.H.; Zhang, L.L. Finite-element modeling of a complex deep excavation in Shanghai. *Acta Geotech.* **2009**, *4*, 7–16. [CrossRef]
8. Li, M.-G.; Chen, J.-J.; Xu, A.-J.; Xia, X.-H.; Wang, J.-H. Case study of innovative top-down construction method with channel-type excavation. *J. Constr. Eng. Manag.* **2014**, *140*, 05014003. [CrossRef]
9. Whittle, A.J.; Hashash, Y.M.A.; Whitman, R.V Analysis of deep excavation in Boston. *J. Geotech. Eng.* **1993**, *119*, 69–90. [CrossRef]
10. Huang, X.; Schweiger, H.F.; Huang, H. Influence of deep excavations on nearby existing tunnels. *Int. J. Geomech.* **2013**, *13*, 170–180. [CrossRef]
11. Wang, J.; Xiang, H.; Yan, J. Numerical simulation of steel sheet pile support structures in foundation pit excavation. *Int. J. Geomech.* **2019**, *19*, 05019002. [CrossRef]

12. Issa, U.; Saeed, F.; Miky, Y.; Alqurashi, M.; Osman, E. Hybrid AHP-fuzzy TOPSIS approach for selecting deep excavation support system. *Buildings* **2022**, *12*, 295. [CrossRef]
13. Hong, L.; Chen, L.; Wang, X. Reliability analysis of serviceability limit state for braced excavation considering multiple failure modes in spatially variable soil. *Buildings* **2022**, *12*, 722. [CrossRef]
14. Gong, X.-n.; Zhang, X.-c. Excavation collapse of Hangzhou subway station in soft clay and numerical investigation based on orthogonal experiment method. *J. Zhejiang Univ.-SCI A* **2012**, *13*, 760–767. [CrossRef]
15. He, L.; Liu, Y.; Bi, S.; Wang, L.; Broggi, M.; Beer, M. Estimation of failure probability in braced excavation using Bayesian networks with integrated model updating. *Undergr. Space* **2020**, *5*, 315–323. [CrossRef]
16. Do, T.-N.; Ou, C.-Y.; Chen, R.-P. A study of failure mechanisms of deep excavations in soft clay using the finite element method. *Comput. Geotech.* **2016**, *73*, 153–163. [CrossRef]
17. Chen, R.P.; Li, Z.C.; Chen, Y.M.; Ou, C.Y.; Hu, Q.; Rao, M. Failure investigation at a collapsed deep excavation in very sensitive organic soft clay. *J. Perform. Constr. Facil.* **2015**, *29*, 04014078. [CrossRef]
18. Tan, Y.; Lu, Y. Forensic diagnosis of a leaking accident during excavation. *J. Perform. Constr. Facil.* **2017**, *31*, 04017061. [CrossRef]
19. Hu, Q.; Xu, X.B.; Huang, T.M.; Lu, S.Q. Collapse of a deep excavation and its reconstruction in soft soil of Nanjing, China. *Proc. Inst. Civ. Eng.-Forensic Eng.* **2021**, *174*, 1–7. [CrossRef]
20. Liu, G.B.; Ng, C.W.W.; Wang, Z.W. Observed performance of a deep multistrutted excavation in Shanghai soft clays. *J. Geotech. Geoenviron. Eng.* **2005**, *131*, 1004–1013. [CrossRef]
21. Masini, L.; Gaudio, D.; Rampello, S.; Romani, E. Observed performance of a deep excavation in the historical center of Rome. *J. Geotech. Geoenviron. Eng.* **2021**, *147*, 05020015. [CrossRef]
22. Cheng, K.; Xu, R.; Ying, H.; Gan, X.; Zhang, L.; Liu, S. Observed performance of a 30.2 m deep-large basement excavation in Hangzhou soft clay. *Tunn. Undergr. Space Technol.* **2021**, *111*, 103872. [CrossRef]
23. Zhang, R.; Wu, C.; Goh, A.T.C.; Böhlke, T.; Zhang, W. Estimation of diaphragm wall deflections for deep braced excavation in anisotropic clays using ensemble learning. *Geosci. Front.* **2021**, *12*, 365–373. [CrossRef]
24. Guo, P.; Lei, G.; Luo, L.; Gong, X.; Wang, Y.; Li, B.; Hu, X.; Hu, H. Soil creep effect on time-dependent deformation of deep braced excavation. *Adv. Mater. Sci. Eng.* **2022**, *2022*, 5655592. [CrossRef]
25. Guo, P.; Gong, X.; Wang, Y. Displacement and force analyses of braced structure of deep excavation considering unsymmetrical surcharge effect. *Comput. Geotech.* **2019**, *113*, 103102. [CrossRef]
26. Chen, H.; Li, J.; Yang, C.; Feng, C. A theoretical study on ground surface settlement induced by a braced deep excavation. *Eur. J. Environ. Civ. Eng.* **2022**, *26*, 1897–1916. [CrossRef]
27. Liu, J.; Shi, C.; Lei, M.; Cao, C.; Lin, Y. Improved analytical method for evaluating the responses of a shield tunnel to adjacent excavations and its application. *Tunn. Undergr. Space Technol.* **2020**, *98*, 103339. [CrossRef]
28. Guo, P.; Liu, F.; Lei, G.; Li, X.; Zhu, C.; Wang, Y.; Lu, M.; Cheng, K.; Gong, X. Predicting response of constructed tunnel to adjacent excavation with dewatering. *Geofluids* **2021**, *2021*, 5548817. [CrossRef]
29. Zhang, D.-M.; Xie, X.-C.; Li, Z.-L.; Zhang, J. Simplified analysis method for predicting the influence of deep excavation on existing tunnels. *Comput. Geotech.* **2020**, *121*, 103477. [CrossRef]
30. Kung, G.T.C.; Juang, C.H.; Hsiao, E.C.L.; Hashash, Y.M.A. Simplified model for wall deflection and ground-surface settlement caused by braced excavation in clays. *J. Geotech. Geoenviron. Eng.* **2007**, *133*, 731–747. [CrossRef]
31. Ou, C.-Y.; Hsieh, P.-G. A simplified method for predicting ground settlement profiles induced by excavation in soft clay. *Comput. Geotech.* **2011**, *38*, 987–997. [CrossRef]
32. Hu, B.; Li, X.; Huang, D. Safety risk analysis and protective control of existing pipelines affected by deep pit excavation in metro construction. *Model. Simul. Eng.* **2019**, *2019*, 3643808. [CrossRef]
33. Jiang, N.; Zhu, B.; He, X.; Zhou, C.; Luo, X.; Wu, T. Safety assessment of buried pressurized gas pipelines subject to blasting vibrations induced by metro foundation pit excavation. *Tunn. Undergr. Space Technol.* **2020**, *102*, 103448. [CrossRef]
34. Zhang, Z.; Zhang, M.; Zhao, Q. A simplified analysis for deformation behavior of buried pipelines considering disturbance effects of underground excavation in soft clays. *Arab. J. Geosci.* **2015**, *8*, 7771–7785. [CrossRef]
35. Kog, Y.C. Buried pipeline response to braced excavation movements. *J. Perform. Constr. Facil.* **2010**, *24*, 235–241. [CrossRef]
36. Linehan, P.W.; Longinow, A.; Dowding, C.H. Pipe response to pile driving and adjacent excavation. *J. Geotech. Eng.* **1992**, *118*, 300–316. [CrossRef]
37. Tan, Y.; Lu, Y. Responses of shallowly buried pipelines to adjacent deep excavations in Shanghai soft ground. *J. Pipeline Syst. Eng. Pract.* **2018**, *9*, 05018002. [CrossRef]
38. Chen, H.; Li, J.; Li, L. Performance of a zoned excavation by bottom-up technique in Shanghai soft soils. *J. Geotech. Geoenviron. Eng.* **2018**, *144*, 05018003. [CrossRef]

water

MDPI

Article

Risk Assessment for Critical Flood Height of Pedestrian Escape in Subway Station

Yi Tang [1], Tianzhong Zhou [2], Youxin Zhong [2], Shengbin Hu [2], Jing Lin [2], Zhiyu Lin [1], Hongwei Liu [1], Baohua Liu [1], Yanlin Zhao [3], Yixian Wang [4] and Hang Lin [1,*]

1 School of Resources and Safety Engineering, Central South University, Changsha 410083, China
2 Nanning Rail Transit Co., Ltd., Nanning 530028, China
3 School of Energy and Safety Engineering, Hunan University of Science and Technology, Xiangtan 411201, China
4 School of Civil Engineering, Hefei University of Technology, Hefei 230009, China
* Correspondence: hanglin@csu.edu.cn

Abstract: The escape of pedestrians in the subway station is hampered by floods created by heavy rain. In order to explore the critical flood level in a subway station so that pedestrians can escape safely, the case study of the Mingxiu Road subway station in Nanning, China, was conducted using numerical simulation techniques. In total, 30 groups of sample pedestrians with different walking speeds and numbers were randomly generated by the Monte Carlo method, and 3D simulation software was used for escape simulation. The simulated escape data were put into the SVM model, and the maximum pedestrian capacity and minimum speed of pedestrians were solved successfully with different conditions of the Mingxiu Road subway station. Then, a 1:1 contour model of the pedestrian was constructed to simulate the flood resistance of the pedestrian escaping at the minimum speed. The flood resistance and the friction force between the pedestrian and the ground were compared to calculate the critical escape flood level height, and the critical escape flood level height of an adult, child, and elder was 87.4 cm,75.5 cm, and 83.0 cm, respectively.

Keywords: Monte Carlo method; 3D simulation; maximum pedestrian capacity; minimum speed of safe escape; critical escape flood level height

Citation: Tang, Y.; Zhou, T.; Zhong, Y.; Hu, S.; Lin, J.; Lin, Z.; Liu, H.; Liu, B.; Zhao, Y.; Wang, Y.; et al. Risk Assessment for Critical Flood Height of Pedestrian Escape in Subway Station. *Water* **2022**, *14*, 3409. https://doi.org/10.3390/w14213409

Academic Editor: Yurui Fan

Received: 23 August 2022
Accepted: 23 October 2022
Published: 27 October 2022

1. Introduction

In recent years, the rapid development of social economy and production technology has led to the rapid expansion of the urban construction scale [1–3], and the surface of the city is not sufficient for all kinds of infrastructure to be built [4–6]. Therefore, the development of underground space is an excellent direction for the sustainable development of the city. However, compared to the spacious surface, the underground space is a very closed place [7–9], with only a few auxiliary facilities (air shafts) and individual pedestrian entrances and exits, which are used by pedestrians connecting the underground space and the surface. In this case, once a disaster occurs, the space and time for pedestrians to escape will be very limited [10–12]. Compared with the surface, the characteristics of the underground space have three points [13,14]: (1) The underground space is very closed. It is only connected to the outside through a small number of water outlets, and drainage can only be carried out mechanically. (2) The interior structure of the underground space is complex, and the various buildings inside occupy a large area. (3) It is difficult to evacuate pedestrians safely [15]. Due to the complex internal structure of the underground space and the low terrain, the flood flow into the subway station cannot be eliminated in time [16–18], which poses a great threat to pedestrian safety [19,20]. In addition, compared with the movement on the surface, escaping from the underground space to the surface requires overcoming gravity to work, which costs more physical energy, and it is more difficult for pedestrians to evacuate safely [21].

Since the development of urban underground space, many accidents have occurred [22–24]. In January 2006, an underground garage in Brazil was flooded with a maximum flood depth of 2.2 m, killing six pedestrians [25]. Due to the closed structure of the underground space, pedestrians need to spend more time escaping when safety accidents such as floods occur in subway stations. Many scholars have carried out relevant studies on subway pedestrian evacuation. Yukawa et al. [26] proposed a method to evaluate evacuation plans for large urban floods based on equilibrium analysis. Zheng et al. [25] studied the dynamic evacuation of subway pedestrians with flood based on an improved floor-field model. Uno et al. [27] proposed a disaster evacuation simulation system based on a multi-agent model and applied the system to evacuation analysis of urban flood disasters. Simonovic et al. [28] constructed a computer simulation model based on system dynamics to identify human behavior in floods.

The above research, based on the methods of macroscopic simulation, dynamics, and evacuation model, has many valuable suggestions and experiences for pedestrian escape in subway stations with flood attacks. However, it must be noted that the rising flood in the subway station will cause great psychological pressure on the pedestrian, thus triggering a series of stampedes. In addition, the buoyancy and resistance generated by the flood will seriously hinder the normal escape of the pedestrian. When the flood height exceeds a critical level, the pedestrian will not be able to escape safely. Therefore, if the critical escape flood height of the pedestrian can be calculated, the pressure of evacuation will be relieved.

This paper proposed a method to calculate the critical flood height for the evacuation of pedestrians in subway stations. This method can calculate the maximum capacity and minimum escape speed of pedestrians in subway stations with different conditions and the critical flood height for pedestrians to achieve a safe escape. It is of practical significance.

2. Model and Method

2.1. Model of Subway Station

We used the Nanning Mingxiu Road Subway Station as an engineering model (Figures 1–7). Mingxiu Road station is a transfer station located in the bustling area of Nanning city. It consists of a surface plaza, 3 underground floors, and 12 exits leading to the city streets, forming a huge underground transportation system. The first basement floor is the one with the largest scale and the most commercial shops, named the property floor, and the second basement floor is the station hall floor, which is used for transfer and crowds entering the station. The third underground layer is the platform layer, which is used by pedestrians waiting for the arrival of the subway and entering the subway [29]. In addition, to ensure that the model can succinctly and intuitively reflect the pedestrian evacuation situation succinctly and intuitively, some contents need to be simplified. First of all, the components such as walls and columns that will hinder the observation of the evacuation process need to be simplified, they will be regarded as two-dimensional obstacles, and the height attribute will be removed. The elevator was deemed unusable. In addition, only public areas accessible to pedestrians are considered, and various facilities such as cable rooms and power distribution rooms in subway stations are not considered. The final model planes of each layer are shown in Figures 1–6. According to relevant regulations, the design of the station should ensure that the evacuation from the farthest location on the platform to a safe location can be completed in 6 min or less. Therefore, assuming that the pedestrian response time is 30 s, all pedestrians should be evacuated to the surface safety within $360 - 30 = 330$ (s) [30].

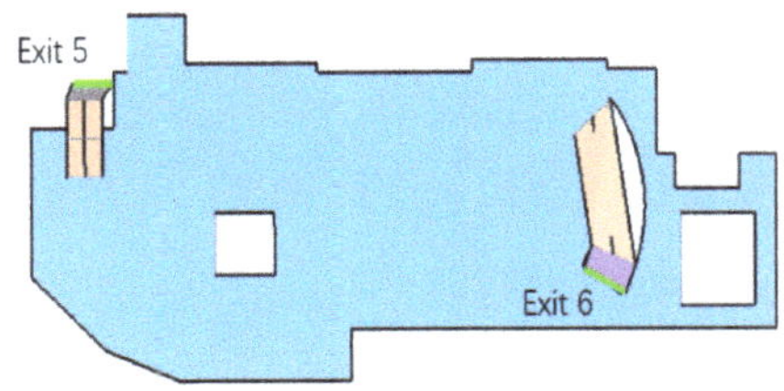

Figure 1. Surface model of the sunken square.

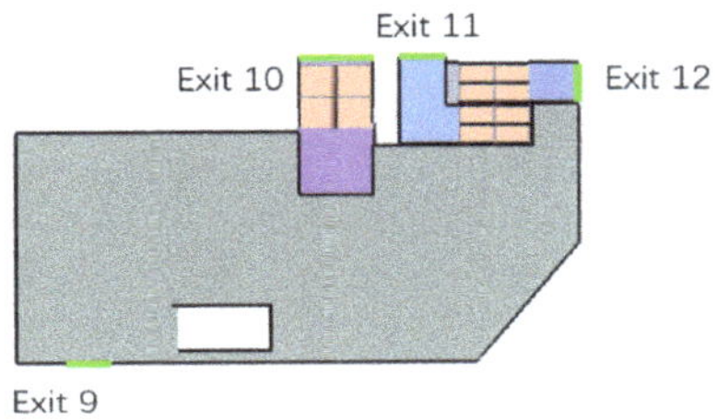

Figure 2. Surface model and export instructions.

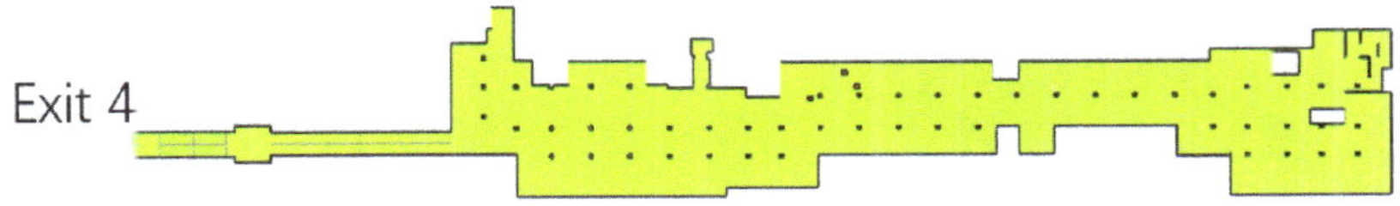

Figure 3. Property layer model diagram.

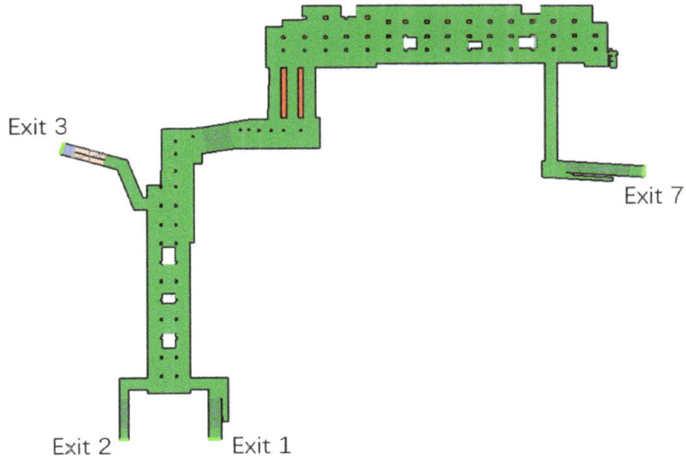

Figure 4. Station hall floor model plan.

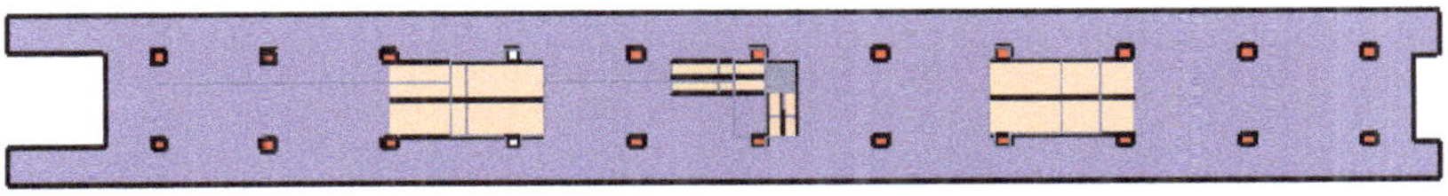

Figure 5. Model plan of platform 2.

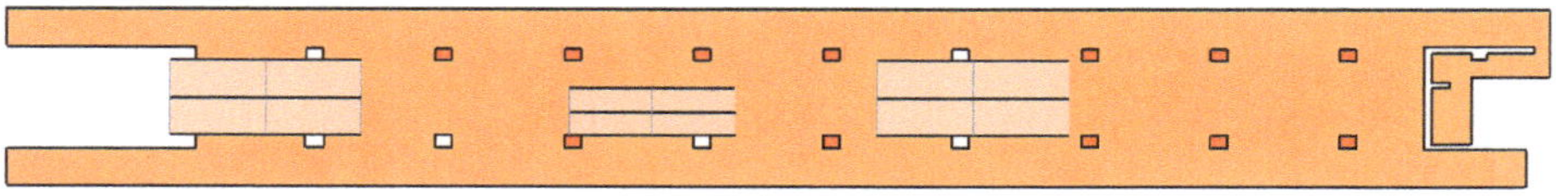

Figure 6. Model plan of platform 5.

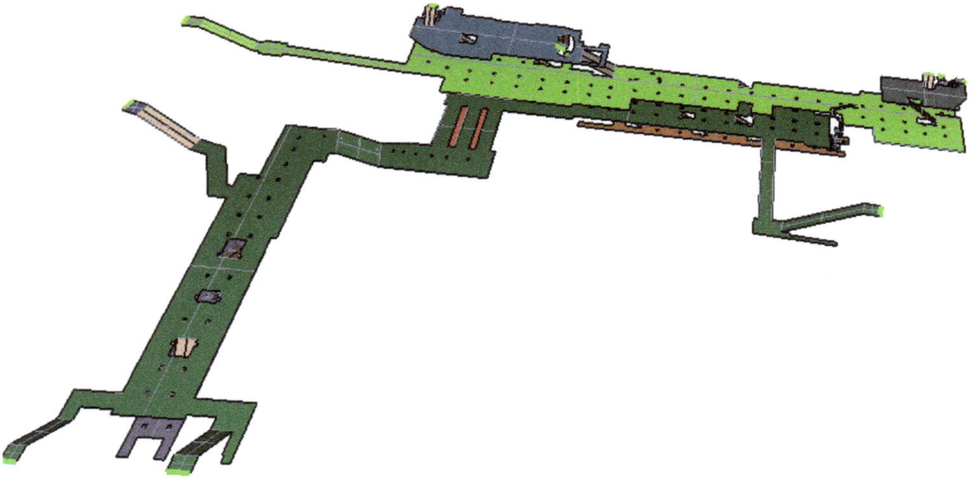

Figure 7. Schematic diagram of the overall 3D model.

2.2. Construction of Pedestrian Number and Speed Factor Based on Monte Carlo

The determination of constraint parameters mainly includes the determination of the number of evacuated pedestrians and indicators of evacuated pedestrians.

(1) According to the area of each region of the model, the pedestrian density range is set to 0.5~4.5 m^2/person. By setting a larger range of pedestrian density and expanding the range of evacuated pedestrians, the applicability and correctness of the model are improved. The specific parameters can be determined according to the specific conditions of the city and the station. Assuming that the pedestrian density is 1.5 m^2/person, the total number of pedestrians involved in the evacuation is 20,731.4/1.5 = 13,821.

(2) During the evacuation process, the number and walking speed of various groups of pedestrians will affect the evacuation time. From the count of the number of pedestrians of different ages taking the subway in literature [31], it can be seen that the subway pedestrian is mainly young and middle-aged, and about 86.6% are between 18 and 60 years old. Crowds aged 18 and below and over 60 accounted for a relatively small proportion of 6.6% and 6.8%, respectively. It is known that the normal walking speed of pedestrian walking is generally 5 km/h = 1.38 m/s [32], and this walking speed is changed according to the age of the pedestrian. For example, the walking speed of an adult is generally faster than that of a child and the elder, and the walking speed of a child is generally faster than that of the elder. Taking the reduction coefficients as 1, 0.8, and 0.75 [31], respectively, the basic walking speeds of the three groups of pedestrians are calculated to be 1.38 m/s, 1.10 m/s, and 1.035 m/s, respectively.

In addition, the normal distribution is used for the setting when establishing the simulated character model. For example, the adult height range is set to 1.55 m~1.85 m, the average is 1.73 m, and the standard deviation is 0.05 m. The specific parameter distribution of each population is shown in Tables 1 and 2.

Table 1. The height parameters of the simulated pedestrian.

Pedestrian Group	Distribution	Min (m)	Max (m)	Average (m)	Standard Deviation (m)
Adult	Normal distribution	1.55	1.85	1.73	0.05
Child	Normal distribution	1.0	1.6	1.4	0.05
Elder	Normal distribution	1.5	1.75	1.61	0.05

Table 2. The shoulder width parameters of the simulated pedestrian.

Pedestrian Group	Distribution	Min (m)	Max (m)	Average (m)	Standard Deviation (m)
Adult	Normal distribution	0.42	0.55	0.48	0.03
Child	Normal distribution	0.34	0.48	0.38	0.03
Elder	Normal distribution	0.39	0.51	0.44	0.03

The Monte Carlo method was used to randomly select 30 sample parameters in the density range of 0.5~4.5 (m^2/person). The training model is more feasible by setting a larger range of pedestrian's density to expand the range of evacuated pedestrians. For the walking speed, in the event of a disaster, it should be taken into account that some pedestrians are running at a faster pace or that some injured pedestrians are walking slower than normal. Therefore, the basic walking speed of each group is multiplied by the speed factor, and the coefficient range is between 0.5 and 3. Setting a larger speed factor range can meet the fast or slow walking speeds of pedestrians in different states. In total, 30 groups of data parameters were input and simulated, and the simulated parameters and evacuation time of each group are shown in Table 3.

Table 3. Sample data and escape time randomly selected based on the Monte Carlo method.

Case	Space Occupation (m^2/pers)	Density (pers/m^2)	Speed Factor	Number of Pedestrians (per)	Evacuation Time (s)
1	0.844	1.185	1.241	24,567	897.6
2	3.704	0.270	0.972	5597	410
3	4.216	0.237	0.959	4913	396.9
4	2.456	0.407	2.064	8438	279.9
5	1.448	0.691	0.703	14,325	1006.8
6	4.352	0.230	2.439	4768	152.3
7	1.694	0.509	2.323	12,240	286.4
8	3.760	0.266	1.266	5515	307.3
9	3.215	0.311	1.777	6447	274.6
10	1.963	0.508	2.487	10,532	280.2
11	0.652	1.534	1.447	31,802	870.1
12	4.152	0.241	1.832	4996	222.6
13	1.875	0.533	2.594	11,053	271.9
14	1.563	0.640	2.167	13,266	327.6
15	1.048	0.954	1.968	19,778	456.3
16	0.928	1.078	1.253	22,348	753.4
17	2.476	0.404	1.076	8375	491.5
18	3.360	0.298	0.987	6178	439.7
19	4.416	0.226	1.297	4685	295.9
20	3.296	0.303	1.589	6282	281.1
21	0.624	1.603	2.808	33,232	480
22	2.500	0.400	0.962	8293	560.8
23	4.120	0.243	2.949	5033	134
24	2.972	0.336	0.778	6966	623.7
25	3.720	0.269	1.522	5577	274
26	1.232	0.812	1.156	16,834	668.1
27	4.048	0.247	2.278	5121	201.3
28	3.028	0.330	0.794	6833	563.3
29	4.064	0.246	0.927	5100	400.6
30	0.740	1.351	1.770	28,008	648

Table 3 shows that among the 30 sets of randomly generated data, only 14 sets of samples meet the required safe evacuation time and can achieve safe escape for all pedestrians.

The rest of the sample data evacuation time is greater than 330 s, and the evacuation time of some samples is far more than 330 s, which means there is a huge security risk. The mapping relationship between the speed factor, the number of pedestrians evacuated, and the evacuation time is shown in Figure 8. It shows a relatively flat mapping curve for the speed factor, the number of pedestrians evacuated, and the evacuation time. When the number of pedestrians increases or the speed factor increases, the evacuation time will show a linear-plane growth or attenuation.

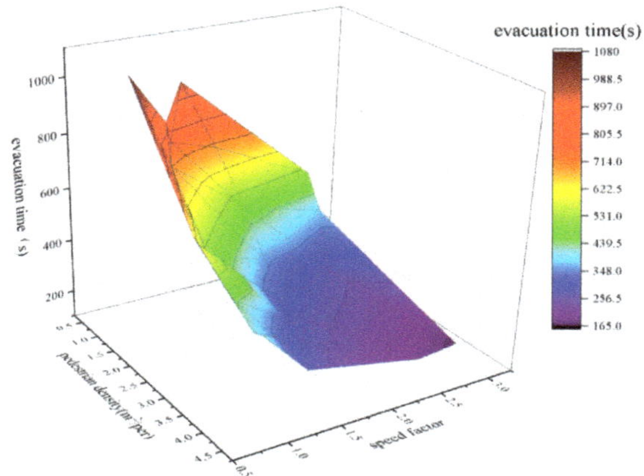

Figure 8. The relationship between speed factor, number of evacuees, and evacuation time.

2.3. Introduction to the Principle of Support Vector Machine

The support vector machine (SVM) is a mathematical method that can perform binary classification according to data characteristics and is a linear classifier with the largest interval defined in the feature space. The core idea is to divide the separation hyperplane with the largest geometric distance of different types of data according to the data samples, as shown in Figure 9.

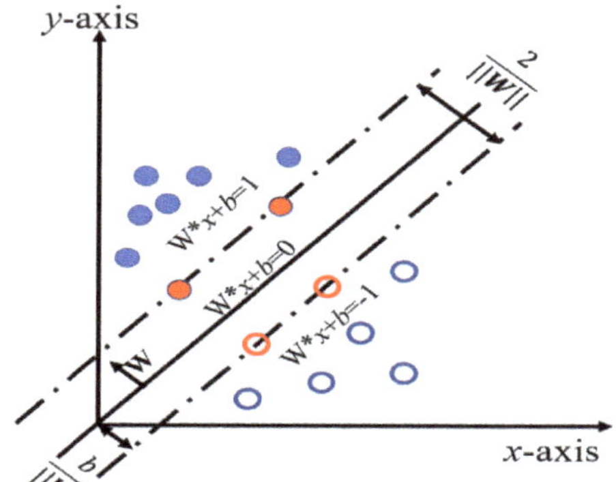

Figure 9. SVM schematic.

The blue circles in Figure 9 represent data points. Solid circles are classified as one type of data and hollow ones as another. The red circle is used as an important sample for classification.

3. Result Analysis

3.1. Analysis of Pedestrian Escape Time and Number in Each Sample

In order to further understand the changes in the number of pedestrians evacuated over time, according to the range of the number of random samples, the samples of 0–10,000, 10,000–20,000, and 20,000–30,000 pedestrians were divided into three groups, respectively. The specific grouping situation is shown in Table 4, and the relationship curve between the number of pedestrians who failed to escape and time is drawn, as shown in Figures 10–12.

Table 4. Sample groupings.

Number of Pedestrians	Group	Sample Serial Number
0~10,000	A	2, 3, 4, 6, 8, 9, 12, 17, 18, 19 20, 22, 23, 24, 25, 27, 28, 29
10,000~20,000	B	5, 7, 10, 13, 14, 15, 26
More than 20,000	C	1, 11, 16, 21, 30

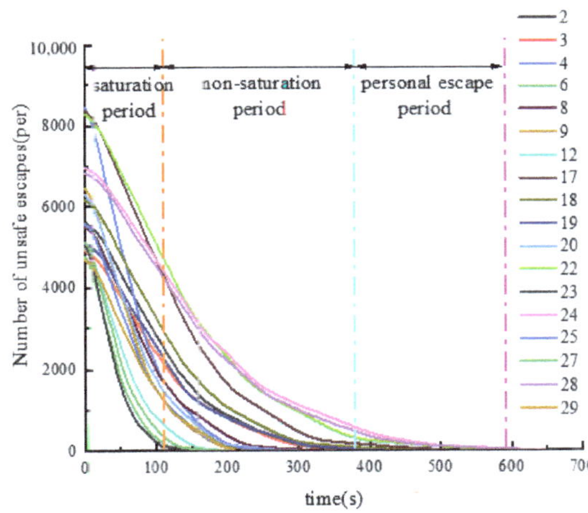

Figure 10. The curve of the number of pedestrians who failed to escape in group A with time.

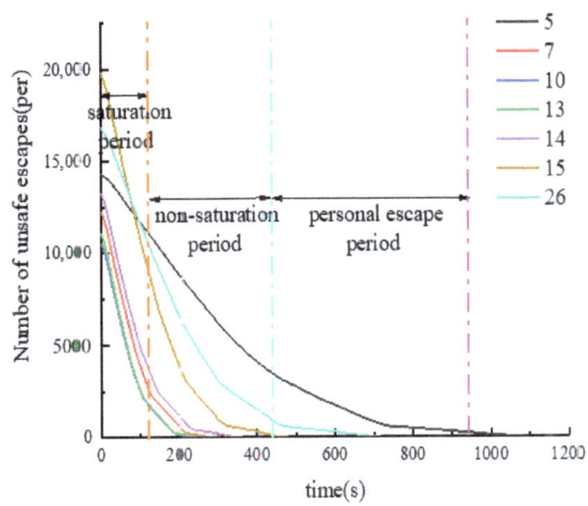

Figure 11. The curve of the number of pedestrians who failed to escape in group B with time.

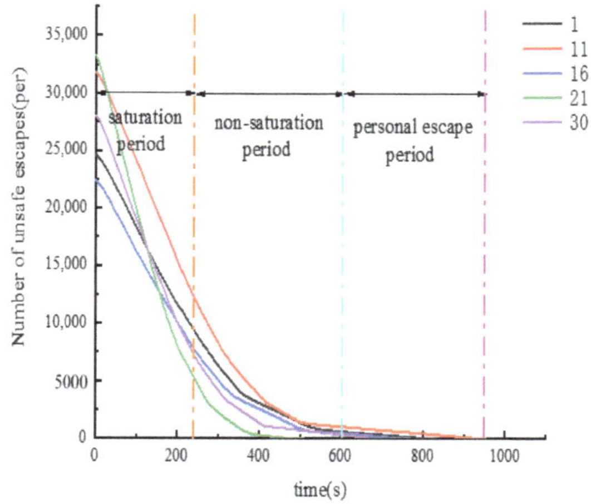

Figure 12. The curve of the number of pedestrians who failed to escape in group C with time.

Figures 10–12 show that all the curves basically show a concave change; that is, with the increase of time, the number of pedestrians fleeing the subway station per unit of time is gradually decreasing. A 3D visualization of the escape was performed (Figure 13). According to the changing form of the curve, pedestrian escape is divided into three periods: saturation period, non-saturation period, and personal escape. For example, within 0~110 s (saturation period) of the samples in group A, all the curves basically decline linearly, and the flow of the pedestrian at all exits in the subway reaches the maximum value at this time. The exit of the passage is in a saturated state, and a large number of pedestrians gather at the exit to escape (Figure 14a), resulting in congestion. Within 110~380 s (half-saturation period), the slope of the curve decreases. At this time, the pedestrian flow at each exit is relatively small (Figure 14b), and some exits have even been idle, and no one has passed. Within 380~590 s (personal escape period), the evacuation of pedestrians in the subway has basically been completed, and only a few pedestrians who are injured or have a slow walking speed are left to escape slowly.

(**a**) (**b**)

Figure 13. (**a**) Evacuation in platform 5; (**b**) Evacuation in Station hall floor.

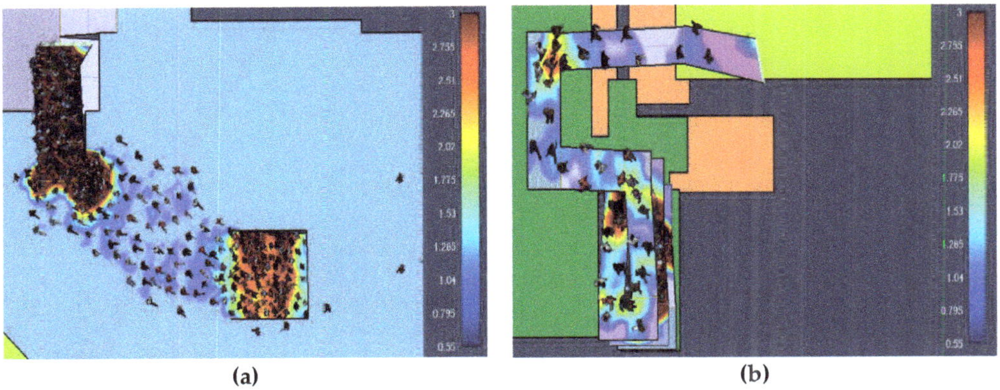

Figure 14. (**a**) Density of export pedestrians during saturation period; (**b**) Density of export pedestrians in unsaturated period.

3.2. Analysis of Pedestrian Passage Time and Number at Different Exits

Figures 10–12 show that the upper limit of escape time is determined by the walking speed of the pedestrian. For example, the total number of pedestrians in sample 5 in Figure 11 is 14,325, and the total escape time is as long as 1006.8 s, far exceeding the same group of samples. The single-person escape time accounts for more than half of the total time. The fundamental reason is that the walking speed is too slow. When pedestrians are distributed far from the safe passage, the time cost will increase significantly. In addition, due to the slower walking speed, the decreasing slope of the variation curve of sample 5 during the saturation period is also much lower than that of the same group. Although a slower walking speed can reduce stampede incidents caused by crowds, it will also greatly improve the safe evacuation time of all pedestrians. In order to further understand the pedestrian evacuation in the unsaturated period, samples 23 and 28 were randomly selected as examples to analyze the specific conditions of the 12 exits in the subway station during a safe evacuation, as shown in Figures 15 and 16.

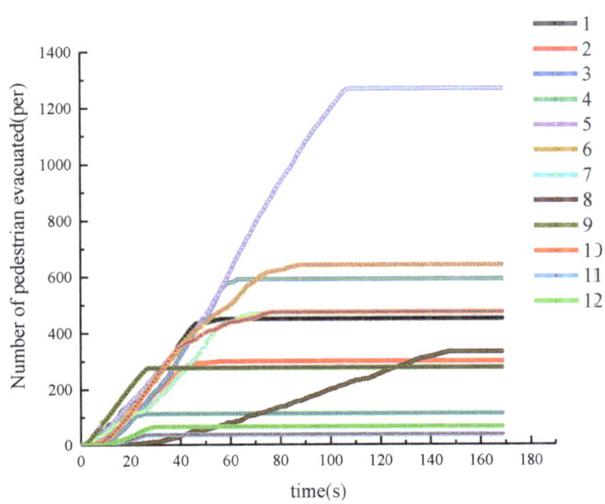

Figure 15. Number of pedestrians passing through each exit and travel time of sample 23.

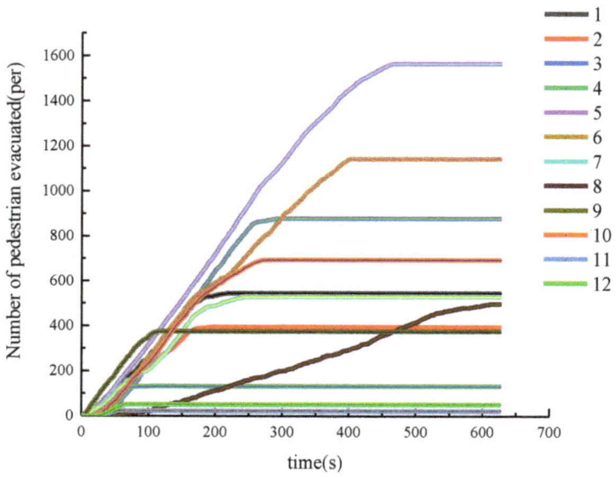

Figure 16. Number of pedestrians passing through each exit and travel time of sample 28.

Figure 15 shows the evacuation of pedestrians from all exits in sample 23. The total evacuation time for this simulation is 563 s. Exit 5 has the largest number of escapees, and the maximum flow of pedestrians exceeds seven pedestrians. A total of 1592 pedestrians pass through, accounting for 23.3% of the total number of pedestrians. It is the most important exit in this evacuation process. In addition, it can be seen that all exits in the early evacuation period (0~50 s) were fully utilized. With the advancement of time, Exits 4, 9, 11, and 12 have been unmanned within 50 to 110 s, and these exits will be idle in subsequent moments, and no one will pass. Exits 1, 2, 3, 7, and 10 end at 110~260 s, Exits 5 and 6 end at 400~450 s, and Exit 8 (530~540) is the last exit of this evacuation simulation work.

For comparison only from working hours, Exit 8 is about four times that of Exits 4, 9, 11, and 12, but the number of pedestrians is much lower than that of Exit 5. This result is because Exit 8 is an emergency escape exit, and the stairwell of exit 8 is narrower than other exits. During the evacuation process, the maximum flow of pedestrians at Exit 8 was only two pedestrian/s, while the peak flow of pedestrians at the other exits was greater than seven pedestrian/s. Pedestrian herd behavior leads to serious uneven distribution of evacuation resources, and the narrow stairwells and congestion greatly prolong the safe escape time of pedestrians.

The evacuation curves of sample 23 (Figure 15) and sample 28 (Figure 16) were compared, and their overall trend was almost the same. Among them, Exits 5 and 6 are the most important exits for evacuation, and they basically undertake 40% of the evacuation of pedestrians. They are very important exits, and the smooth flow of these two exits should be ensured at all times. In addition, the evacuation guidance of Exits 4, 9, 11, and 12 should be strengthened so as to prolong the passage time of the four exits. Reasonable allocation of evacuation resources to ensure the maximum efficiency of evacuation. The specific parameters of the number of pedestrians passing through each exit of samples 23 and 28 and the passage time are shown in Tables 5 and 6.

3.3. Calculation of Minimum Pedestrian Escape Speed Based on SVM

The above analysis shows that the number of pedestrians and walking speed are important factors affecting the overall evacuation time. In order to further study the internal criteria of whether pedestrians can escape safely during the evacuation, a support vector machine (SVM) [33] is introduced for data analysis.

The data in Tables 3 and 7 are brought into the SVM theory, the number of pedestrians evacuated and the walking speed factor are used as sample indicators, and the 330 s escape time in the safety specification is used as the critical time point. The escape time is greater than or equal to 330 s, which means that the escape failed. Otherwise, the escape is successful. See Table 7 for all the statistics.

Table 5. Number of pedestrians evacuated and evacuation time at each exit of sample 23.

Exit	Number of Pedestrians (per)	Proportion (%)	Evacuation Time (s)
1	448	8.90	51
2	309	6.14	50
3	586	11.64	59
4	112	2.23	26
5	1292	25.67	105
6	658	13.07	88
7	441	8.76	63
8	307	6.10	135
9	271	5.38	27
10	502	9.93	75
11	39	0.73	29
12	68	1.35	32

Table 6. Number of pedestrians evacuated and evacuation time at each exit of sample 28.

Exit	Number of Pedestrians (per)	Proportion (%)	Evacuation Time (s)
1	544	7.97	202
2	399	5.84	155
3	885	12.95	259
4	133	1.95	79
5	1592	23.30	450
6	1109	16.23	404
7	536	7.84	206
8	484	7.08	536
9	375	5.49	108
10	701	10.26	259
11	23	0.33	51
12	52	0.76	66

Table 7. Classification of escape results in each sample.

	Escape Result	Sample Serial Number
Failed to escape (time $\leq$ 330)	Success	4, 6,7, 8, 9, 10, 12, 13,14, 19 20, 23, 25, 27
Escape success (time > 330)	Fail	1, 2, 3, 5, 11, 15, 16, 17, 18, 21, 22, 24, 26, 28, 29, 30

The first 20 groups of samples are used as the training data, and the last 10 groups are used as the test data (only the index data is included in the test data, and the actual data is not included). The available classification result graph is shown in Figure 17, and the classification result data is shown in Table 6. The final classification hyperplane result equation is solved as follows:

$$y - 0.6762 \times 10^{-4}x - 0.8828 = 0 \ (x > 0) \quad (1)$$

Among them, y represents the speed factor, and x represents the number of pedestrians. For evacuation simulations with different numbers of pedestrians and walking speed coefficients, if the result is greater than or equal to 0 after it is brought into formula (1), it is determined that the escape is successful, and all pedestrians in the subway can escape safely. If the result is less than 0, it is determined that the escape failed, the subway pedestrians cannot achieve all safe escape.

Table 8 and Figure 17 show the classification results of this SVM, and the training samples simulated by this classification are all correctly divided. For the predicted samples, comparing the actual classification results in Table 8, it can be seen that the predicted results are completely correct, and the prediction accuracy is 100%. Therefore, it can be considered that the separation hyperplane equation of this SVM is accurate and effective. That is, by inputting the number of pedestrians in the subway and the walking speed of the pedestrians, it can be determined whether all the pedestrians in the subway can escape within the safety specification time by relying on the Formula (1).

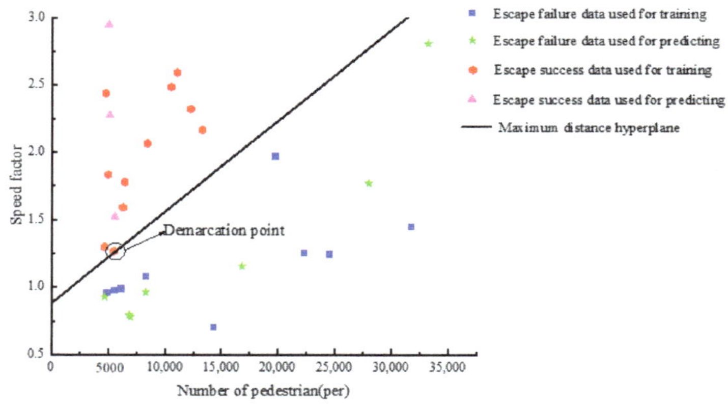

Figure 17. SVM classification results.

Table 8. Comparison of test classification results and SVM classification results.

Sample Serial Number	Simulation Results	SVM Classification Results
21	Fail	Fail
22	Fail	Fail
23	Success	Success
24	Fail	Fail
25	Success	Success
26	Fail	Fail
27	Success	Success
28	Fail	Fail
29	Fail	Fail
30	Fail	Fail

Observing Formula (1), we can see that when $x = 1$, y is almost equal to 0.8828. The physical meaning of this result is: When there is only one person in the subway station, and the position of the person satisfies the random distribution, then the minimum walking speed to ensure that he can escape successfully is 0.8828×1.035 m/s (the walking speed of the old man) = 0.9137 m/s.

In this paper, the basic pedestrian speed is set as 5 km/h. By setting the speed coefficient of 0.5~3, the actual range of pedestrian moving speed will be expanded to 0.52 m/s (1.035 m/s $\times$ 0.5)~4.14 m/s (1.38 ms $\times$ 3), and the minimum escape speed can be solved in this range. The basic speed of the pedestrian itself is not critical because the speed factor will make the speed range of the pedestrian include the minimum escape speed, and then the SVM can be solved. For example, set the pedestrian's base speed to be 1 m/s and the actual escape speed is 0.9 m/s, then the speed factor calculated by the SVM is 0.9. Set the pedestrian base speed to 0.5 m/s, and the actual escape speed is 0.9 m/s, then the velocity coefficient calculated by SVM is 1.8. Therefore, the basic speed of pedestrians is not critical. As long as the speed range includes the minimum escape speed, the speed factor will change accordingly, and the final calculated minimum escape speed will be basically the same.

Taking the escape situation of normal pedestrians as an example, assuming that the walking speed factor is set to 1.5. That is, the pedestrians in the subway station escape at a walking speed of 1.5 times the usual walking speed, then the Formula (1) shows that the maximum number of pedestrians carrying the Nanning Mingxiu Road Station is:

$$x = \frac{1.5 - 0.8828}{0.6762 \times 10^{-4}} = 9127\ (\text{per}) \quad (2)$$

In daily traffic, the number of pedestrians in the Mingxiu Road station should be less than 9127 to ensure the safe evacuation of all pedestrians in an emergency. In case of heavy rain or other extreme weather, the walking speed of pedestrians should be smaller, and the specific walking speed should be analyzed according to the actual situation.

3.4. Calculation of Critical Water Level Height Based on Minimum Escape Speed

In Section 3.3, the SVM model was used to solve the minimum escape speed of pedestrians as 0.914 m/s. However, there will be a flood in the subway station due to a rainstorm (Figure 18), and when the flood height exceeds the critical value, the resistance and buoyancy brought by the flood will make people unable to reach the minimum escape speed of 0.914 m/s, resulting in the failure of escape.

Figure 18. A sketch of a model of a man in a flood.

Therefore, when the flood height exceeds the critical value, the walking speed of pedestrians will not reach 0.914 m/s, so the escape will fail. In order to accurately solve the critical value of the height of the flood, a 1:1 contour height model was constructed for all groups (adult, child, elder). The model height was consistent with the simulated height in Section 3 (adult = 173 cm, child = 140 cm, and the elder = 161 cm). The flood height is set to 10 cm, 20 cm, 30 cm, 40 cm, 50 cm, 60 cm, 70 cm, 80 cm, 90 cm, and 100 cm. Since pedestrians keep relative movement with the flood when moving, the pedestrian velocity = 0.914 m/s and the flood velocity = 0 m/s can be converted to the flood velocity = 0.914 m/s and the pedestrian velocity = 0 m/s so as to simulate the impact resistance of flood flow to pedestrians at different flood heights. Dynamic flood pressure of all groups is shown in Figures 19–21, and dynamic flood resistance is shown in Table 9.

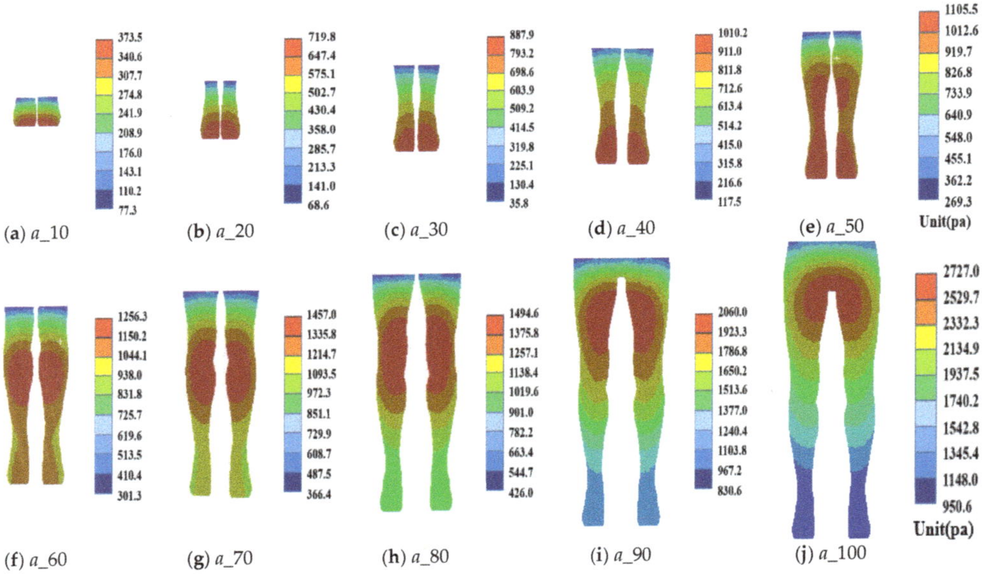

Figure 19. Hydrodynamic pressure of adult walking at a minimum safe speed at different flood heights.

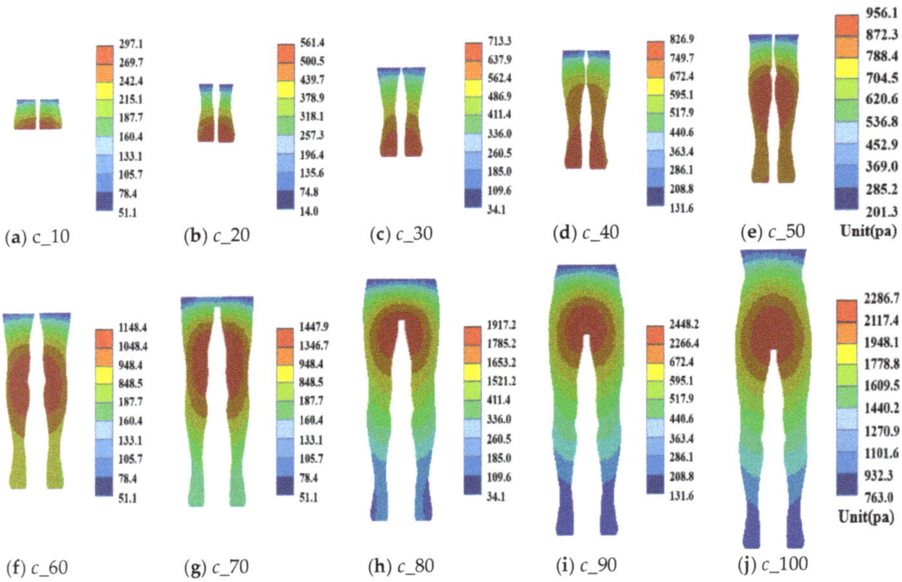

Figure 20. Hydrodynamic pressure of child walking at a minimum safe speed at different flood heights.

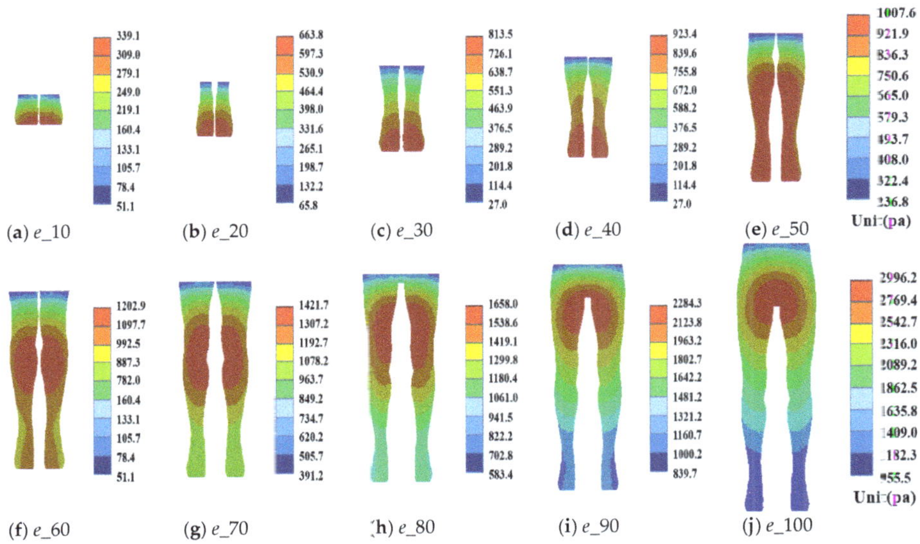

Figure 21. Hydrodynamic pressure of elder walking at a minimum safe speed at different flood heights.

Table 9. Flood resistance of all groups walking at a critical speed at different flood heights.

Flood Height (cm)	Adult	Child	Elder
10	3.6 N	2.0 N	2.9 N
20	6.4 N	6.7 N	9.7 N
30	21.6 N	15.3 N	18.8 N
40	40.9 N	30.8 N	36.6 N
50	67.9 N	48.4 N	59.8 N
60	94.9 N	73.9 N	86.3 N
70	116.5 N	116.4 N	124.2 N
80	162.3 N	175.8 N	172.7 N
90	261.3 N	252.0 N	269.0 N
100	372.2	275.6	386.2

The calculation formula of friction between pedestrians and the ground in flood is as follows:

$$F = f \times (Mg - W) \tag{3}$$

F represents the friction between the pedestrian and the ground, f represents the friction factor between the pedestrian and the ground, M represents the mass of the pedestrian, g represents the acceleration of gravity (9.8 m/s^2), W represents the buoyancy of the pedestrian in flood, and the specific expression of buoyancy W is:

$$W = \rho g V \tag{4}$$

ρ represents the density of the liquid, and V represents the volume of the flood displaced by the pedestrian standing in it (Table 10). As for M, the simplified BMI index of 21.5 [34] was used as the calculation. Based on the average height in Section 3, the standard weight of each group was calculated as follows: adult = 64.3 kg, child = 42.2 kg, and the elder = 55.7 kg.

$$\text{BMI} = \frac{M}{H^2} \tag{5}$$

Table 10. The drainage volume of all groups in flood at different flood heights.

Flood Height (cm)	Adult	Child	Elder
10	0.00136 m^3	0.00106 m^3	0.00124 m^3
20	0.0024 m^3	0.0019 m^3	0.0022 m^3
30	0.0036 m^3	0.0031 m^3	0.0034 m^3
40	0.0054 m^3	0.0047 m^3	0.0051 m^3
50	0.0073 m^3	0.0062 m^3	0.0069 m^3
60	0.0092 m^3	0.0080 m^3	0.0087 m^3
70	0.0113 m^3	0.0104 m^3	0.0109 m^3
80	0.014 m^3	0.0128 m^3	0.0136 m^3
90	0.017 m^3	0.0150 m^3	0.0164 m^3
100	0.02 m^3	0.0170 m^3	0.0190 m^3

BMI stands for pedestrian mass index, H is the height of the pedestrian, M is the mass of the pedestrian.

The drainage volume data in Table 10, adults = 64.3 kg, children = 42.2 kg, elderly = 55.7 kg, $g = 9.8\ \mathrm{m/s^2}$, $\rho = 1000$ kg/m3, $f = 0.5$ [35], are substituted into Equations (3) and (4). The friction force between the pedestrian and the ground at different flood heights is calculated and compared with the flood resistance of the pedestrian. Table 11 shows the comparison results, and Figures 22–24 show the comparison curve.

Table 11. Friction and flood resistance of all groups when walking at a critical speed at different flood heights.

Flood Height (cm)	Adult		Child		Elder	
	Friction	Flood Resistance	Friction	Flood Resistance	Friction	Flood Resistance
10	308.4	3.6	201.5	2.0	266.8	2.9
20	303.3	6.4	197.4	6.7	262.1	9.7
30	297.4	21.6	191.5	15.3	256.2	18.8
40	288.6	40.9	183.7	30.8	247.9	36.6
50	279.3	67.9	176.4	48.4	239.1	59.8
60	269.9	94.9	167.5	73.9	230.3	86.3
70	259.7	116.5	155.8	116.4	219.5	124.2
80	246.4	162.3	144.0	175.8	206.2	172.7
90	231.7	261.3	133.2	252.0	192.5	269.0
100	217.0	372.2	123.4	275.6	179.8	386.2

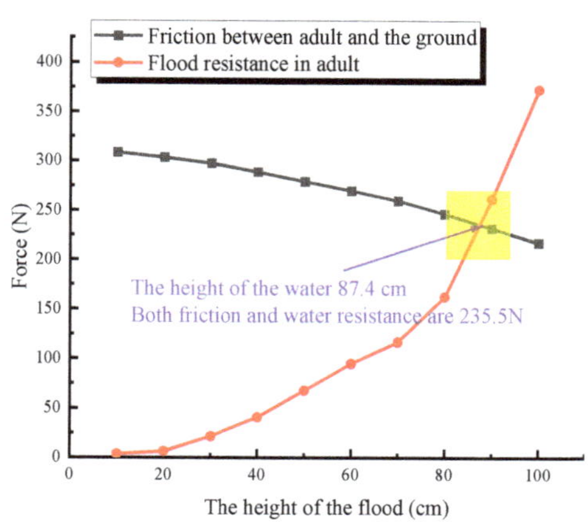

Figure 22. Curve comparison of the flood resistance and friction (adult).

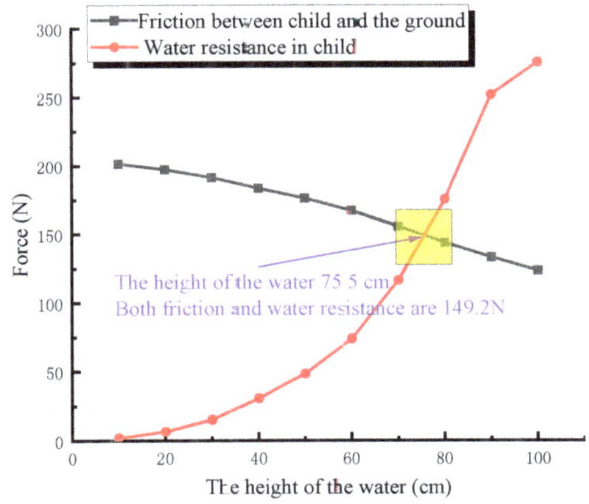

Figure 23. Curve comparison of the flood resistance and friction (child).

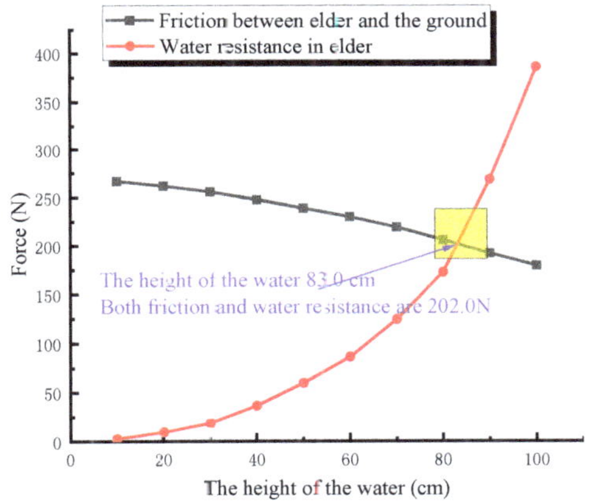

Figure 24. Curve comparison of the flood resistance and friction (elder).

The intersection of curves in Figures 22–24 was calculated, and the critical flood heights for adults, children, and the elder to escape were 87.4 cm,75.5 cm, and 83.0 cm, respectively. For an adult, when the height of the flood in the subway is greater than 87.4 cm, the walking speed of adults cannot reach 0.914 m/s, so the safety escape cannot be completed within the specified 330 s.

4. Conclusions

(1) Based on the model of the Mingxiu Road subway station in Nanning city, China, the maximum daily pedestrian capacity of the Mingxiu Road subway station is calculated to be 9127 using the Monte Carlo, 3D simulation, and SVM methods. The minimum speed of pedestrian safety escape is 0.914 m/s.

(2) The minimum escape speed was 0.914 m/s as the critical speed. Set different flood heights and compare the flood resistance and friction between pedestrians and the ground. Finally, the critical escape flood level heights of adults, children, and the elderly are 87.4 cm,75.5 cm, and 83.0 cm, respectively.

Author Contributions: Conceptualization, Y.T., T.Z. and Y.Z. (Youxin Zhong); methodology, Y.T. and S.H.; software, Y.T., J.L. and Z.L.; investigation, H.L. (Hang Lin), H.L. (Hongwei Liu) and B.L.;writing—original draft preparation, Y.T., H.L. (Hang Lin) and Y.Z. (Yanlin Zhao); writing—review and editing, Y.T. and H.L. (Hang Lin); funding acquisition, T.Z. and Y.Z. (Yanlin Zhao); project administration, S.H. and Y.W. All authors have read and agreed to the published version of the manuscript.

Funding: The Science and Technology Hunan Civil Air Defense Research Project (HNRFKJ-2021-07); Project (2021) of Study on Flood Disaster Prevention Model of Nanning Rail Transit.

Institutional Review Board Statement: Not applicable.

Informed Consent Statement: Not applicable.

Data Availability Statement: Some or all data, models, or code that support the findings of this study are available from the corresponding author upon reasonable request.

Acknowledgments: This paper obtained funding from the Science and Technology Hunan Civil Air Defense Research Project (HNRFKJ-2021-07); Project (2021) of Study on Flood Disaster Prevention Model of Nanning Rail Transit. The authors wish to acknowledge this support.

Conflicts of Interest: On behalf of all authors, the corresponding author states that there is no conflict of interest. This article does not contain any studies with human participants or animals performed by any of the authors.

References

1. Wang, H.-B.; Gong, X.-S.; Wang, B.-B.; Deng, C.; Cao, Q. Urban development analysis using built-up area maps based on multiple high-resolution satellite data. *Int. J. Appl. Earth Obs. Geoinf.* **2021**, *103*, 1102500. [CrossRef]
2. Li, X.-y.; Lu, Z.-h. Quantitative measurement on urbanization development level in urban Agglomerations: A case of JJJ urban agglomeration. *Ecol. Indic.* **2021**, *133*, 108375. [CrossRef]
3. Liu, B.; Lin, H.; Chen, Y.; Liu, J.; Guo, C. Deformation Stability Response of Adjacent Subway Tunnels considering Excavation and Support of Foundation Pit. *Lithosphere* **2022**, *2022*, 7227330. [CrossRef]
4. Löwe, R.; Mair, M.; Pedersen, A.N.; Kleidorfer, M.; Rauch, W.; Arnbjerg-Nielsen, K. Impacts of urban development on urban water management—Limits of predictability. *Comput. Environ. Urban Syst.* **2020**, *84*, 101546. [CrossRef]
5. Wang, H.; Qin, F.; Xu, C.; Li, B.; Guo, L.; Wang, Z. Evaluating the suitability of urban development land with a Geodetector. *Ecol. Indic.* **2021**, *123*, 1–13. [CrossRef]
6. Herrmann, C.R.; Maroko, A.R.; Taniguchi, T.A. Subway Station Closures and Robbery Hot Spots in New York City-Understanding Mobility Factors and Crime Reduction. *Eur. J. Crim. Policy Res.* **2021**, *27*, 415–432. [CrossRef]
7. Cui, J.; Broere, W.; Lin, D. Underground space utilisation for urban renewal. *Tunn. Undergr. Space Technol.* **2021**, *108*, 103726. [CrossRef]
8. Zhang, Y.-H.; Zhu, J.-B.; Liao, Z.-Y.; Guo, J.; Xie, H.-P.; Peng, Q. An intelligent planning model for the development and utilization of urban underground space with an application to the Luohu District in Shenzhen. *Tunn. Undergr. Space Technol.* **2021**, *112*, 103933. [CrossRef]
9. Xu, W.; Liu, B.; Liu, J.; Guo, C. Interactions of Foundation Pit on the Underlying Adjacent Existing Underground Structures. *Geofluids* **2022**, *2022*, 5675173. [CrossRef]
10. Liu, S.-C.; Peng, F.-L.; Qiao, Y.-K.; Zhang, J.-B. Evaluating disaster prevention benefits of underground space from the perspective of urban resilience. *Int. J. Disaster Risk Reduct.* **2021**, *58*, 102206. [CrossRef]
11. Han, G.; Leng, J.-W. Urban underground space planning based on FPGA and virtual reality system. *Microprocess. Microsyst.* **2021**, *80*, 103601. [CrossRef]
12. Gonzalez-Navarro, M.; Turner, M.A. Subways and urban growth: Evidence from earth. *J. Urban Econ.* **2018**, *108*, 85–106. [CrossRef]
13. Dou, F.-F.; Li, X.-H.; Xing, H.-X.; Yuan, F.; Ge, W.-Y. 3D geological suitability evaluation for urban underground space development—A case study of Qianjiang Newtown in Hangzhou, Eastern China. *Tunn. Undergr. Space Technol.* **2021**, *115*, 104052. [CrossRef]
14. Tan, F.; Wang, J.; Jiao, Y.-Y.; Ma, B.-C.; He, L.-L. Suitability evaluation of underground space based on finite interval cloud model and genetic algorithm combination weighting. *Tunn. Undergr. Space Technol.* **2021**, *108*, 103743. [CrossRef]
15. Yoon, Y.K. Usability Test for Mobile Subway Application. *J. Cult. Prod. Des.* **2018**, *53*, 89–97. [CrossRef]
16. van der Wal, C.N.; Robinson, M.A.; de Bruin, W.B.; Gwynne, S. Evacuation behaviors and emergency communications: An analysis of real-world incident videos. *Saf. Sci.* **2021**, *136*, 105121. [CrossRef]
17. Cheng, H.; Yang, X.-K. Emergency Evacuation Capacity of Subway Stations. *Procedia Soc. Behav. Sci.* **2012**, *43*, 339–348. [CrossRef]
18. Zhou, M.; Ge, S.-C.; Liu, J.-L.; Dong, H.-R.; Wang, F.-Y. Field observation and analysis of waiting passengers at subway platform—A case study of Beijing subway stations. *Phys. A Stat. Mech. Its Appl.* **2020**, *556*, 124779. [CrossRef]

19. Lyu, H.-M.; Shen, S.-L.; Zhou, A.-N.; Yang, J. Perspectives for flood risk assessment and management for mega-city metro system. *Tunn. Undergr. Space Technol.* **2019**, *84*, 31–44. [CrossRef]
20. Lyu, H.-M.; Shen, S.-L.; Zhou, A.-N.; Zhou, W.-H. Flood risk assessment of metro systems in a subsiding environment using the interval FAHP-FCA approach. *Sustain. Cities Soc.* **2019**, *50*, 101682. [CrossRef]
21. Qu, Y.-C.; Gao, Z.-Y.; Xiao, Y.; Li, X.-G. Modeling the pedestrian's movement and simulating evacuation dynamics on stairs. *Saf. Sci.* **2014**, *70*, 189–201. [CrossRef]
22. Wang, J.-H.; Yan, W.-Y.; Zhi, Y.-R.; Jiang, J.-C. Investigation of the Panic Psychology and Behaviors of Evacuation Crowds in Subway Emergencies. *Procedia Eng.* **2016**, *135*, 128–137. [CrossRef]
23. Yi, J.-X.; Pan, S.-L.; Chen, Q. Simulation of pedestrian evacuation in stampedes based on a cellular automaton model. *Simul. Model. Pract. Theory* **2020**, *104*, 102147. [CrossRef]
24. Feng, J.-R.; Gai, W.-M.; Yan, Y.-B. Emergency evacuation risk assessment and mitigation strategy for a toxic gas leak in an underground space: The case of a subway station in Guangzhou, China. *Saf. Sci.* **2021**, *134*, 105039. [CrossRef]
25. Ying, Z.; Xin-Gang, L.; Bin, J.; Rui, J. Simulation of pedestrians' evacuation dynamics with underground flood spreading based on cellular automaton. *Simul. Model. Pract. Theory* **2019**, *94*, 149–161.
26. Yukawa, S.; Hatayama, M.; Tatano, H. A study of evaluation method for large-scale flood evacuation planning using the network equilibrium analysis. *J. Jpn. Soc. Civ. Eng. Ser D3* **2011**, *67*, 177–188.
27. Uno, K.; Kashiyama, K. Development of Simulation System for the Disaster Evacuation Based on Multi-Agent Model Using GIS. *Tsinghua Sci. Technol.* **2008**, *13*, 348–353. [CrossRef]
28. Simonovic, S.P.; Ahmad, S. Computer-based Model for Flood Evacuation Emergency Planning. *Nat. Hazards* **2005**, *34*, 25–31. [CrossRef]
29. Lin, Z.; Hu, S.; Zhou, T.; Zhong, Y.; Zhu, Y.; Shi, L.; Lin, H. Numerical Simulation of Flood Intrusion Process under Malfunction of Flood Retaining Facilities in Complex Subway Stations. *Buildings* **2022**, *12*, 853. [CrossRef]
30. Qin, J.-W.; Liu, C.-C.; Huang, Q. Simulation on fire emergency evacuation in special subway station based on Pathfinder. *Case Stud. Therm. Eng.* **2020**, *21*, 100677. [CrossRef]
31. Shao, Z.-G.; Wang, X.-H.; Yu, D.-H. Study on the simulation of subway station safety evacuation based on Revit Pathfinder. *J. Qingdao Univ. Technol.* **2021**, *42*, 143–148
32. Zhao, J.-P.; Wang, L. Practical Research on College Students' Increasing Load Running Platform. *Contemp. Sport. Sci. Technol.* **2020**, *10*, 242–244.
33. Kikuchi, T.; Abe, S. Comparison between error correcting output codes and fuzzy support vector machines. *Pattern Recognit. Lett.* **2005**, *26*, 1937–1945. [CrossRef]
34. He, Y.-J.; Tao, Q.-S.; Li, X.-T.; Sun, F.; Zhang, S.-Y. Influence of sex and age on the diagnostic criteria of overweight and obesity in adults. *China Public Health* **2009**, *25*, 441–443.
35. Yang, Y.-Q.; Sun, J.-L.; Jia, L.-Q. Preparation of national Engineering industry standard of Technical Specification for Anti-skid of Building Ground. In *Proceedings of China Building Materials Science and Technology Special Issue of 2013 Floor*; Standards Press of China: Beijing, China, 2000; pp. 66–69.

MDPI
St. Alban-Anlage 66
4052 Basel
Switzerland
www.mdpi.com

Water Editorial Office
E-mail: water@mdpi.com
www.mdpi.com/journal/water

www.ingramcontent.com/pod-product-compliance
Lightning Source LLC
Chambersburg PA
CBHW061608220926
51850CB00005B/457